PETROLEUM RESERVOIR ENGINEERING

Physical Properties

PETROLEUM RESERVOIR ENGINEERING

Physical Properties

JAMES W. AMYX
DANIEL M. BASS, JR.
ROBERT L. WHITING
The Agricultural and Mechanical College of Texas

McGRAW-HILL BOOK COMPANY
New York   Toronto   London
1960

PETROLEUM RESERVOIR ENGINEERING

9 10 11 12 – MAMM – 7 5

ISBN 07-001600-3

PREFACE

This book, the first of two volumes on petroleum-reservoir engineering, presents concepts and applications of rock and fluid properties which are fundamental to engineering analyses of petroleum reservoirs. In addition the organization and evaluation of laboratory and field data for reservoir analyses are presented together with applications of such ordered data to the determination of the volume of hydrocarbons "in place."

The book is arranged so that it can be used as a text or as a reference work. If it is used as a text, the organization of material permits flexibility in designing course offerings. Although planned for use in a course or courses comprising a total of four semester hours of work and presuming a prerequisite course in oil-field development, the book can be used in a first course in petroleum engineering if desired.

Chapter 1 is comprised of an introductory discussion of hydrocarbon accumulations, oil-field development, and production methods. Chapters 2 and 3 present rock properties in a complete and cohesive independent unit. Chapters 4 to 6 present a study of fluid properties also as a complete, cohesive independent unit. Chapter 7 deals with evaluation of rock and fluid properties both from laboratory and field data. Chapter 8 presents the derivation of the material balance and the applications of such balances to the determination of volume of hydrocarbon in place.

This volume is a direct outgrowth of a multilithed book used in a series of summer schools in reservoir engineering offered to industry personnel during 1956 and 1957.

Much of the material on which the book is based was drawn from the literature published by the Society of Petroleum Engineers of the American Institute of Mining, Metallurgical and Petroleum Engineers, the Division of Production of the American Petroleum Institute, the Natural Gasoline Association of America, and the American Association of Petroleum Geologists. In addition, the authors are indebted to the host of authors who have contributed to the petroleum literature in various other publications. Core Laboratories, Incorporated, and Shell Oil Company provided additional data not generally available in the literature.

The authors are indebted to the following companies who supported the industry summer courses: Argo Oil Corporation, The California Company,

The California Standard Company, Canadian Seaboard Company, Champlin Oil and Refining Company, Core Laboratories, Inc., Delhi-Taylor Oil Corporation, Honolulu Oil Corporation, Humble Oil and Refining Company, Kewanee Oil Company, Lion Oil Company, Magnolia Petroleum Company, Mound Company, Murphy Corporation, McAlester Fuel Company, Plymouth Oil Company, Pure Oil Company, Railroad Commission of Texas, Republic National Bank of Dallas, Richmond Exploration Company, Rotary Engineers Laboratories, Shell Oil Company, Southern Minerals, Inc., Sun Oil Company, Sunray–Mid-Continent Oil Company, Standard Oil Company of California, Standard Oil Company of Texas, Western Leaseholds, Ltd., and Western Operations.

Mr. Donald A. Flanagan, Mr. Robert L. Ridings, and Dr. Denton R. Wieland worked out many of the numerical examples. Mrs. Wilmoth Boring, Mrs. Lora Watson, Mrs. Joan Hodges, Mrs. Gloria Conrad, and Mrs. Betty Short typed and retyped the many drafts of the manuscript.

We are especially indebted to our wives, Mrs. Louise Amyx, Mrs. Dorothy Bass, and Mrs. Sharon Whiting, for their patience and encouragement during the preparation of this volume.

James W. Amyx
Daniel M. Bass, Jr.
Robert L. Whiting

CONTENTS

Hydrocarbon Volume 536
Evaluation of Porosity and Permeability—Water Saturation—Calculation of Hydrocarbon Volume.

8. The Material Balance 561

Introduction 561
Derivation of Material-balance Equation 562
Solution-gas Drive—Solution-gas–Gas-cap Drive—Simple Solution-gas–Gas-cap–Water-drive Reservoirs—Solution-gas–Gas-cap–Water-drive with Fluid Injection—Slightly Compressible Hydrocarbon Reservoirs—Gas Reservoir—Comparison of Drives.
Data for Material Balance 575
Fluid-production Data—Reservoir Temperatures—Reservoir Pressures—Fluid Analysis—Core Analysis and Laboratory Rock Data.
Calculation of Oil in Place Using the Material-balance Equation 589
Estimation of Oil in Place for a Solution-gas-drive Reservoir—Estimation of Oil in Place for Slightly Compressible Fluids—Estimates of Gas in Place from Material Balance.

Name Index 599

Subject Index 603

KEY TO SYMBOLS

ENGLISH LETTER SYMBOLS

a	constant in equation of state
A	area
A	Avogadro's number
A_T	adhesion tension
A_a	actual area
b	constant in equation of state
B	formation voulme factor
B_g	gas formation volume factor
B_{gc}	gas-cap formation volume factor
B_g'	injected gas formation volume factor
B_o	oil formation volume factor
B_t	total formation (two-phase) volume factor
B_w	water formation volume factor
c	compressibility
C	constant
C	flow coefficient
c_b	bulk compressibility
c_f	formation (rock) compressibility
c_g	gas compressibility
c_o	oil compressibility
c_p	pore-volume compressibility
c_r	compressibility of rock
c_r	pseudo-reduced compressibility
c_w	water compressibility
C'	function of tortuosity
d	diameter
D	depth
D	diffusion coefficient
e	influx (encroachment) rate
e_g	gas-influx (encroachment) rate
e_o	oil-influx (encroachment) rate
e_w	water-influx (encroachment) rate
f	friction factor

f	fraction (such as the fraction of a flow stream consisting of a particular phase)
F	formation factor
F	frequency
F	force
f'	fugacity
g	acceleration of gravity
G	total initial gas in place in reservoir
G	gas gravity (air = 1)
g_t	geothermal gradient
G_{dh}	gradient of heavier fluid
G_{dl}	gradient of lighter fluid
G_e	cumulative gas influx (encroachment)
ΔG_e	gas influx (encroachment) during a time period
G_i	cumulative gas injected
ΔG_i	gas injected during a time period
G_p	cumulative gas produced
ΔG_p	gas produced during a time period
G_{pc}	cumulative gas-cap gas produced
G_{ps}	cumulative solution-gas produced
G_{re}	oil gradient
h	height
h	net pay thickness
H	gross pay thickness
h_c	closure of structure
H_c	depth to interface
H_p	depth of perforations
H_t	depth to top of interval
i	injection rate
I	amperes (electric current)
I	injectivity index
i_g	gas-injection rate
i_w	water-injection rate
I_s	specific-injectivity index
J	productivity index
J_s	specific-productivity index
k	absolute permeability
K	equilibrium ratio (y/x)
K	constant of proportionality
k_g	effective permeability to gas
k_l	permeability to a single liquid phase
k_o	effective permeability to oil
k_{rg}	relative permeability to gas

k_{ro}	relative permeability to oil
k_{rw}	relative permeability to water
k_w	effective permeability to water
k_z	Kozeny constant
L	moles of liquid phase
L	length
L_a	actual flow path
m	mean hydraulic radius
m	mass
m	mass rate of flow
m	ratio initial-reservoir free gas volume to initial-reservoir oil volume
m	exponent
m	total number of moles of a mixture existing in two phases
M	molecular weight
n	exponent
n	total moles of a mixture in the gas state
n	number of tubes
N	initial oil in place in reservoir
N_e	cumulative oil influx (encroachment)
ΔN_e	oil influx (encroachment) during a time period
N_p	cumulative oil produced
ΔN_p	oil produced during a time period
P	pressure
P^*	external pressure
$\overline{P}$	average pressure
P_a	atmospheric pressure
P_{ar}	areal weighted pressure
P_b	bubble-point (saturation) pressure
P_b	pressure at bottom of interval
P_c	critical pressure
P_c	capillary pressure
P_{cf}	casing pressure, flowing
P_{ch}	parachor
P_{cs}	casing pressure, static
P_d	dew-point pressure
P_D	dimensionless pressure
P_e	pressure at external boundary
P_f	pressure at the front of interface
P_i	initial pressure
P_r	reduced pressure
P_{sc}	pressure at standard conditions
P_{sp}	separator pressure
P_t	pressure at top of interval

P_{tf}	tubing pressure, flowing
P_{ts}	tubing pressure, static
$\overline{P}_{VL}$	volumetrically weighted pressure
P_w	bottom-hole pressure, general
P_{wf}	bottom-hole pressure, flowing
P_{ws}	bottom-hole pressure, static
Q	volumetric flow rate
q_D	dimensionless production rate
q_g	gas-production rate
q_o	oil-production rate
q_w	water-production rate
r	radial distance
r	resistance (electrical logging symbol)
R	producing gas-oil ratio
R	radius of curvature
R	universal gas constant (per mole)
R	resistivity (electrical logging symbol)
r_D	dimensionless radial distance
r_e	external-boundary radius
r_o	oil-field radius
r_w	well radius or internal boundary of a cylindrical flow system
R_c	resistivity of clay
R_L	liberated gas-oil ratio
R_o	resistivity of rock saturated with water (electrical logging symbol)
R_{oa}	resistivity of shaley sand saturated with water
R_p	cumulative gas-oil ratio
R_s	solution-gas–oil ratio (gas solubility in oil)
R_{sw}	gas solubility in water
R_T	total gas-oil ratio
R_w	resistivity of water (electrical logging symbol)
$°R$	degrees Rankine
s	direction
S	saturation
S	standard deviation
S_g	gas saturation
S_{gc}	critical gas saturation
S_{gr}	residual gas saturation
S_o	oil saturation
S_{or}	residual oil saturation
S_p	internal surface per unit pore volume
S_p	shrinkage due to change in pressure
S_t	shrinkage due to change in temperature
S_w	water saturation

S_{wc}	critical water saturation
S_{wr}	residual water saturation
t	temperature, °Fahrenheit
t	time
T	temperature, °Rankine
t_D	dimensionless time
T_c	critical temperature
T_r	reduced temperature
T_{sc}	temperature, standard conditions
u	volumetric velocity (flow rate per unit area)
v	velocity
v	specific volume
V	moles of vapor phase
V	volume
v_s	velocity along direction s
v_x	velocity along direction x
V_B	bulk volume
V_m	volume per mole
V_p	pore volume
V_s	solid volume
w	width
w	weight
W	initial water in place in reservoir
W_e	cumulative water influx (encroachment)
ΔW_e	water influx (encroachment) during a time period
W_i	cumulative water injected
ΔW_i	water injected during a time period
W_p	cumulative water produced
ΔW_p	water produced during a time period
x	mole fraction of a component in liquid phase
X	salinity correction factor
y	mole fraction of a component in vapor phase
Y	salinity of water
z	mole fraction of a component in mixture
Z	gas deviation factor (compressibility factor, $Z = PV/nRT$)

GREEK LETTER SYMBOLS

α	(alpha) mobility ratio[1]
β	(beta) thermal expansion coefficient
γ	(gamma) specific gravity

[1] When the mobilities involved are on opposite sides of an interface, the mobility ratio is defined as the ratio of the displacing phase mobility to the displaced phase mobility or as the ratio of the upstream mobility to the downstream mobility.

γ_o (gamma) oil specific gravity (water = 1)

γ_g (gamma) gas specific gravity (air = 1)

Δ (delta) difference ($\Delta x = x_2 - x_1$ or $x_1 - x_2$)

θ (theta) angle

η (eta) hydraulic diffusivity ($k/\phi c\mu$)

λ (lambda) mobility (k/μ)

λ (lambda) lithology factor

λ_g (lambda) gas mobility

λ_o (lambda) oil mobility

λ_w (lambda) water mobility

μ (mu) viscosity

μ_g (mu) gas viscosity

μ_o (mu) oil viscosity

μ_w (mu) water viscosity

ν (nu) kinematic viscosity

ρ (rho) resistivity (electrical logging symbol)

ρ (rho) density

ρ_g (rho) gas density

ρ_o (rho) oil density

ρ_w (rho) water density

σ (sigma) surface tension (interfacial tension)

σ (sigma) conductivity

τ (tau) tortuosity

ϕ (phi) porosity

Φ (phi, capital) potential

Ψ (psi, capital) stream function

SUBSCRIPT LETTER SYMBOLS

av   average

a atmospheric

a air

a actual

b bubble point, or saturation

b base conditions

B bulk (used with volume only)

c capillary (used in P_c only)

c cumulative

c critical

cf casing, flowing (used with pressure only)

cp critical point

cs casing, static (used with pressure only)

d datum

d dew point

d	differential separation
D	dimensionless quantity
e	cumulative influx or encroachment
e	external boundary conditions
f	flash separation
f	front, or interface
f	formation or rock
f	flowing
g	gas
hc	hydrocarbon
i	initial value, or conditions
i	ith component, etc.
i	cumulative injected
L	laboratory
L	liquid
m	mean
m	mercury
m	mixture
max	maximum
min	minimum
M	molal
nwt	nonwetting
o	oil
p	pseudo (preceding)
p	cumulative produced
p	pure
p	perforations
p	pore (used with volume only)
pv	pore volume
r	reduced
r	relative
r	reservoir
r	residual
R	residual
R	reservoir
s	gas-oil solution (used in R_s only)
s	shut-in
s	solid
sc	standard conditions
sp	separator conditions
st	stock tank
sw	gas-water solution (used in R_{sw} only)
t, T	total

T threshold

T tension (used with adhesion tension)

tf tubing, flowing (used with pressure only)

ts tubing, static (used with pressure only)

v vapor

VL volumetric

w water

wt wetting

w well conditions

wf bottom hole, flowing (used with pressure only)

ws bottom hole, static (used with pressure only)

ABBREVIATIONS

av   average

°API   degrees on the American Petroleum Institute modified Baumé
          scale for liquids

bbl   barrel (oil field, 42 U.S. gallons per barrel)

cc   cubic centimeter

cm   centimeter

cp   centipoise

cu   cubic

ft   feet

FVF   oil formation volume factor

gm   gram

GOC   gas-oil contact

GOR   gas-oil ratio, standard cubic feet per stock-tank barrel

gpM   gallons per thousand standard cubic feet

lb   pound

M   thousands

MM   millions

Mscf   thousands of standard cubic feet

PI   productivity index

psi   pounds per square inch

res   reservoir

scf   standard cubic feet

sec   second

sep   separator

SG   specific gravity

SPI   specific productivity index

sq   square

STO   stock-tank oil

WOC   water-oil contact

MATHEMATICAL NOTATIONS

dx	notation of derivative of x
$f(\ )$	function of
$g(\ )$	function of
ln	logarithm to the base e (natural logarithm)
log	logarithm to the base 10 (common logarithm) or with appropriate subscript to base indicated by subscript.
s	distance along direction of flow
x, y, z	notation of coordinate axes in three-dimensional space
∂x	notation of partial derivative of x
Δ	finite increment
Σ	(sigma, capital) summation

MODIFYING SIGN

$\bar{x}$	average, or mean, value of a quantity x

CHAPTER 1

INTRODUCTION

Beginning with the Industrial Revolution of the early nineteenth century, man has turned more and more to the use of mineral fuels to supply the energy to operate his machines. The first commercial well drilled solely for oil was completed in the United States in 1859. The drilling was supervised by Col. Edwin L. Drake; thus the well came to be known as the Drake Well. Following the success of the Drake Well, petroleum production and processing rapidly grew into a major industry in the United States. In the early history of the petroleum industry, petroleum products were largely used for lubricants and for illuminating fuel.

With the development of internal-combustion engines and other devices, the use of petroleum for fuel became increasingly important. In 1900, the total mineral energy production in the United States was 7,643 trillion British thermal units (Btu); of this, 92 per cent came from coal, about 5 per cent from oil, and 3 per cent from natural gas. By 1925, mineral energy production in the United States reached 21,000 trillion Btu, of which 73 per cent was from coal, 21 per cent from oil, and 6 per cent from natural gas. In 1950, the demand for energy reached 33,000 trillion Btu in the United States; of this, 45 per cent was supplied by coal, 35 per cent from oil, and 20 per cent from natural gas.[1]*

Through 1956, the cumulative crude-oil production for the world was 95 billion barrels, of which about 55 billion barrels was produced in the United States. Today, petroleum is used not only as a fuel and a source of lubricants but as a raw material for many modern industrial materials, such as paints, plastics, rubber, and so forth.

General Composition of Petroleum

What is petroleum? Petroleum is a mixture of naturally occurring hydrocarbons which may exist in the solid, liquid, or gaseous states, depending upon the conditions of pressure and temperature to which it is subjected. Virtually all petroleum is produced from the earth in either liquid or gaseous form, and commonly, these materials are referred to as either crude oil or natural gas, depending upon the state of the hydrocarbon mixture.

* Superscript numbers refer to references at end of chapter.

1

Crude oil is the material most sought after of these naturally occurring hydrocarbons, but natural gas is commonly produced along with the crude oil. In the early years of the petroleum industry, natural gas was considered to be a nuisance and was burned at the well site. In recent years with the advent of transcontinental transmission lines and petrochemical industries, the demand for natural gas as a fuel and a raw product has increased the value of natural gas to the point where it is no longer a nuisance but a valuable raw material.

Petroleum consists chemically of approximately 11 to 13 wt % hydrogen and 84 to 87 wt % carbon. Traces of oxygen, sulfur, nitrogen, and helium may be found as impurities in crude petroleum. Although all petroleum is constituted primarily of carbon and hydrogen, the molecular constitution of crude oils differs widely. About 18 series of hydrocarbons[2] have been recognized in crude petroleum. In Table 1-1 are listed the group formulas

TABLE 1-1. HYDROCARBON SERIES FOUND IN PETROLEUM[2]

No. of carbon atoms	Pennsylvania	Mid-Continent	California and Gulf Coast
5	C_nH_{2n+2}	C_nH_{2n+2}	C_nH_{2n} and C_nH_{2n+2}
10	C_nH_{2n+2}	C_nH_{2n+2} and C_nH_{2n}	C_nH_{2n} and C_nH_{2n-6}
15	C_nH_{2n+2}	C_nH_{2n-2}	C_nH_{2n-2}
20	C_nH_{2n}	C_nH_{2n-4}	C_nH_{2n-4}
25	C_nH_{2n} and C_nH_{2n-2}	C_nH_{2n-4}	C_nH_{2n-4}
30	C_nH_{2n} and C_nH_{2n-4}	C_nH_{2n-8}	C_nH_{2n-8}
35	C_nH_{2n-4} and C_nH_{2n-8}	C_nH_{2n-8} and C_nH_{2n-12}	C_nH_{2n-12}
40	C_nH_{2n-4} and C_nH_{2n-8}	C_nH_{2n-8} and C_nH_{2n-12}	C_nH_{2n-12} and C_nH_{2n-16}
50	C_nH_{2n-8}	C_nH_{2n-8} and C_nH_{2n-12}	C_nH_{2n-16}
80	C_nH_{2n-8}	C_nH_{2n-16}	C_nH_{2n-20}

of series identified in petroleum. Of these series, the most commonly encountered are the paraffins, the olefines, the polymethylenes, the acetylenes, turpenes, and benzenes. Natural gas is composed predominantly of the lower-molecular weight hydrocarbons of the paraffin series.

Hydrocarbons can be classified into essentially four categories depending on the structural formula. Two of the categories refer to the structural arrangement of the carbon atoms in the molecule. These are (1) open chain and (2) ring or cyclic compounds. The remaining two categories refer to the bonds between the carbon atoms. These are (1) saturated or single bond and (2) unsaturated or multiple-bond compounds.

The names of the various individual hydrocarbon molecules are derived in a systematic fashion from rules established by the International Union of Chemistry. The established names of the individual hydrocarbons of the paraffin series are utilized for compounds having the same number of carbon atoms but with word endings and prefixes designating the group to which the compound belongs. The word ending "ane" designates saturated hydrocarbons while "ene" designates unsaturated hydrocarbons that have double bonds between carbon atoms. If more than one double bond ex-

TABLE 1-2. CLASSIFICATION OF HYDROCARBONS[2]

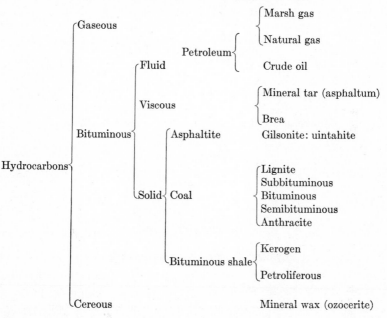

ists in unsaturated hydrocarbons, the ending is modified to indicate the number of double bonds; thus, two double bonds are designated by "diene," three double bonds by "triene," etc.

Ring or cyclic compounds are designated by adding the prefix "cyclo" to the name of the compound as derived from the above rules. However, the cyclic aromatic hydrocarbons, benzenes, retain the customary names except that the ending "ene" is used rather than the older forms, benzol, etc. The structural formulas of various hydrocarbons that have six carbon atoms are shown below. The group name and group formula of each series are designated

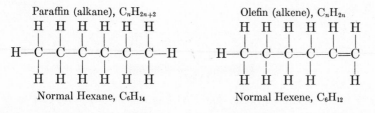

Paraffin (alkane), C_nH_{2n+2}

Normal Hexane, C_6H_{14}

Olefin (alkene), C_nH_{2n}

Normal Hexene, C_6H_{12}

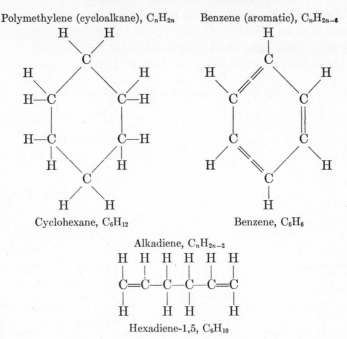

Polymethylene (cycloalkane), C_nH_{2n} Benzene (aromatic), C_nH_{2n-6}

Cyclohexane, C_6H_{12} Benzene, C_6H_6

Alkadiene, C_nH_{2n-2}

Hexadiene-1,5, C_6H_{10}

Physical Properties of Hydrocarbons

The detailed analysis of a crude oil is virtually impossible to obtain. Therefore, crude oils are classified according to their physical properties. Among the physical properties commonly considered in various classifications are color, refractive index, odor, density, boiling point, freezing point, flash point, and viscosity. Of these, the most important physical properties from a classification standpoint are the density (specific gravity) and the viscosity of the liquid petroleum. The specific gravity of liquids is defined as the ratio of the density of the liquid to the density of water, both at specified conditions of pressure and temperature.

The specific gravity of crude oils ranges from about 0.75 to 1.01. Since crude oils are generally lighter than water, a Baumé-type scale is used in the petroleum industry. This scale is referred to as the API or (American Petroleum Institute) scale for crude petroleum and relates the specific gravity through a modulus to an expression of density called API gravity. Expressed mathematically

$$\gamma = \frac{141.5}{131.5 + °API}$$

or
$$°API = \frac{141.5}{\gamma} - 131.5$$

where γ is the specific gravity and °API is the API gravity. It may be

TABLE 1-3. RELATION OF API, SPECIFIC GRAVITY, AND WEIGHT
PER GALLON OF CRUDE OIL[2]

Degrees API	Degrees of specific gravity	Weight of gallon, lb	Degrees API	Degrees of specific gravity	Weight of gallon, lb	Degrees API	Degrees of specific gravity	Weight of gallon, lb
0	1.076	8.962	36	0.8448	7.034	71	0.6988	5.817
1	1.068	8.895	37	0.8398	6.993	72	0.6953	5.788
2	1.060	8.828	38	0.8348	6.951	73	0.6919	5.759
3	1.052	8.762	39	0.8299	6.910	74	0.6886	5.731
4	1.044	8.698	40	0.8251	6.870	75	0.6852	5.703
5	1.037	8.634						
6	1.029	8.571	41	0.8203	6.830	76	0.6819	5.676
7	1.022	8.509	42	0.8155	6.790	77	0.6787	5.649
8	1.014	8.448	43	0.8109	6.752	78	0.6754	5.622
9	1.007	8.388	44	0.8063	6.713	79	0.6722	5.595
10	1.0000	8.328	45	0.8017	6.675	80	0.6690	5.568
11	0.9930	8.270	46	0.7972	6.637	81	0.6659	5.542
12	0.9861	8.212	47	0.7927	6.600	82	0.6628	5.516
13	0.9792	8.155	48	0.7883	6.563	83	0.6597	5.491
14	0.9725	8.099	49	0.7839	6.526	84	0.6566	5.465
15	0.9659	8.044	50	0.7796	6.490	85	0.6536	5.440
16	0.9593	7.989	51	0.7753	6.455	86	0.6506	5.415
17	0.9529	7.935	52	0.7711	6.420	87	0.6476	5.390
18	0.9465	7.882	53	0.7669	6.385	88	0.6446	5.365
19	0.9402	7.830	54	0.7628	6.350	89	0.6417	5.341
20	0.9340	7.778	55	0.7587	6.316	90	0.6388	5.316
21	0.9279	7.727	56	0.7547	6.283	91	0.6360	5.293
22	0.9218	7.676	57	0.7507	6.249	92	0.6331	5.269
23	0.9159	7.627	58	0.7467	6.216	93	0.6303	5.246
24	0.9100	7.578	59	0.7428	6.184	94	0.6275	5.222
25	0.9042	7.529	60	0.7389	6.151	95	0.6247	5.199
26	0.8984	7.481	61	0.7351	6.119	96	0.6220	5.176
27	0.8927	7.434	62	0.7313	6.087	97	0.6193	5.154
28	0.8871	7.387	63	0.7275	6.056	98	0.6166	5.131
29	0.8816	7.341	64	0.7238	6.025	99	0.6139	5.109
30	0.8762	7.296	65	0.7201	5.994	100	0.6112	5.086
31	0.8708	7.251	66	0.7165	5.964			
32	0.8654	7.206	67	0.7128	5.934			
33	0.8602	7.163	68	0.7093	5.904			
34	0.8550	7.119	69	0.7057	5.874			
35	0.8498	7.076	70	0.7022	5.845			

Kinematic viscosity, cs	Equivalent Saybolt Universal viscosity, sec			Kinematic viscosity, cs	Equivalent Saybolt Universal viscosity, sec		
	At 100°F (basic values, see Note)	At 130°F	At 210°F		At 100°F (basic values, see Note)	At 130°F	At 210°F
2.0	32.6	32.7	32.8	31	145.3	145.6	146.3
2.5	34.4	34.5	34.6	32	149.7	150.0	150.7
3.0	36.0	36.1	36.3	33	154.2	154.5	155.3
3.5	37.6	37.7	37.9	34	158.7	159.0	159.8
4.0	39.1	39.2	39.4	35	163.2	163.5	164.3
4.5	40.7	40.8	41.0				
5.0	42.3	42.4	42.6				
6.0	45.5	45.6	45.8	36	167.7	168.0	168.9
7.0	48.7	48.8	49.0	37	172.2	172.5	173.4
8.0	52.0	52.1	52.4	38	176.7	177.0	177.9
9.0	55.4	55.5	55.8	39	181.2	181.5	182.5
10.0	58.8	58.9	59.2	40	185.7	186.1	187.0
11.0	62.3	62.4	62.7	41	190.2	190.6	191.5
12.0	65.9	66.0	66.4	42	194.7	195.1	196.1
13.0	69.6	69.7	70.1	43	199.2	199.6	200.6
14.0	73.4	73.5	73.9	44	203.8	204.2	205.2
15.0	77.2	77.3	77.7	45	208.4	208.8	209.9
16.0	81.1	81.3	81.7	46	213.0	213.4	214.5
17.0	85.1	85.3	85.7	47	217.6	218.0	219.1
18.0	89.2	89.4	89.8	48	222.2	222.6	223.8
19.0	93.3	93.5	94.0	49	226.8	227.2	228.4
20.0	97.5	97.7	98.2	50	231.4	231.8	233.0
21.0	101.7	101.9	102.4	55	254.4	254.9	256.2
22.0	106.0	106.2	106.7	60	277.4	277.9	279.3
23.0	110.3	110.5	111.1	65	300.4	301.0	302.5
24.0	114.6	114.8	115.4	70	323.4	324.0	325.7
25.0	118.9	119.1	119.7				
26.0	123.3	123.5	124.2	Over 70	Saybolt	Saybolt	Saybolt
27.0	127.7	127.9	128.6		sec = cs	sec = cs	sec = cs
28.0	132.1	132.4	133.0		× 4.620	× 4.629	× 4.652
29.0	136.5	136.8	137.5				
30.0	140.9	141.2	141.9				

NOTE: To obtain the Saybolt Universal viscosity equivalent to a kinematic viscosity determined at t°F, multiply the equivalent Saybolt Universal viscosity at 100°F by $1 + (t - 100)0.000064$; for example, 10 cs at 210°F is equivalent to 58.8×1.0070 or 59.2 Saybolt Universal seconds at 210°F.

noted that the API gravity yields numbers greater than 10 for all materials having specific gravities less than 1. Since the density of a liquid is a function of temperature and pressure, it is necessary to designate standard conditions for reporting specific gravity and API gravity. The petroleum industry has adopted as standards a temperature of 60°F and atmospheric pressure. Table 1-3 lists the relationship between API gravity and other commonly used expressions of the density of petroleum liquids.

The viscosity of crude oil ranges from about 0.3 centipoise for a gas-saturated oil at reservoir conditions to about 1,000 centipoises for a gas-free crude oil at atmospheric pressure and 100°F. Viscosities of crude-oil and liquid-petroleum products are frequently reported in terms of the time of efflux, in seconds, of a known volume of liquid through a standardized orifice. The times reported depend on the instrument employed such as Saybolt Universal, Saybolt Furol, Engler, or other similar device. The time of efflux from such instruments has a complex functional relationship to the kinematic viscosity, which is usually expressed in centistokes. The absolute viscosity in centipoises is obtained by multiplying the kinematic viscosity in centistokes by the density of the fluid in grams per cubic centimeter. Table 1-4 gives the relationship between the Saybolt Universal viscosity and centistokes. Viscosity is dependent on temperature. Therefore, standard tests with the Saybolt viscosimeter are conducted at 100°F.

Other physical properties of liquid petroleum are frequently correlated with API gravity and viscosity. In general, such correlations have rather limited application.

Crude oils are frequently classified by "base." The earliest such classification system provided three classifications:

1. Paraffin-base, or oils containing predominantly paraffin series hydrocarbons
2. Asphalt-base, or oils containing predominantly polymethylene or olefin series hydrocarbons
3. Mixed-base, or oils containing large quantities of both paraffin and polymethylene series hydrocarbons

The U.S. Bureau of Mines[3] introduced a somewhat more elaborate system of classification which provides for nine possible classifications. This system is based on a modified Hempel distillation of the crude oil and upon the API gravity of certain fractions obtained upon distillation.

The distillation is conducted in two phases: one at atmospheric pressure and one at an absolute pressure of 40 mm of mercury. The fraction boiling between 482 and 527°F at atmospheric pressure is key fraction 1. The fraction boiling between 527 and 572°F at 40 mm absolute is key fraction 2. The nine possible classifications of a crude oil are summarized in Table 1-5.

The U.S. Bureau of Mines reported the average results of distillations of

TABLE 1-5. U.S. BUREAU OF MINES CLASSIFICATION OF CRUDE OILS[3]

Oil	Key fraction 1, °F	Key fraction 2, °F
Paraffin	40 or lighter	30 or lighter
Paraffin—intermediate	40 or lighter	20–30
Intermediate—paraffin	33–40	30 or lighter
Intermediate	33–40	20–30
Intermediate—naphthene	33–40	20 or heavier
Naphthene—intermediate	33 or heavier	20–30
Naphthene	33 or heavier	20 or heavier
Paraffin—naphthene	40 or lighter	20 or heavier
Naphthene—paraffin	33 or heavier	30 or lighter

303 crude-oil samples from throughout the world. These results appear in Table 1-6. Analyses of this type are useful in evaluating crude oils for refining purposes. Note that of the 303 samples analyzed, 109 samples are classified as intermediate and 83 samples are naphthene base.

Natural gas is composed largely of hydrocarbons of the paraffin series. Methane and ethane frequently comprise 80 to 90 per cent by volume of a natural gas. Other hydrocarbons, ranging in molecular weight from 44 (propane) to in excess of 142 (decane), together with impurities compose the remaining percentage. Carbon dioxide, nitrogen, and hydrogen sulfide are the more common impurities found in natural gas. Helium and other inert rare gases occasionally occur in small concentrations in natural gases.

Gas gravity is widely used to characterize natural gases. Gas gravity is the ratio of the density of a gas at atmospheric pressure and temperature to the density of air at the same condition of pressure and temperature. Since at atmospheric pressure and temperature the densities of gases are directly proportional to the molecular weight, the gravity is the ratio of the molecular weight of the gas to the molecular weight of air. The molecular weight of methane is 16. Therefore, the gravity of pure methane is 0.55 or $16 \div 29$. Gas gravities for natural gases range from 0.6 to 1.1, depending on the relative concentration of the heavier hydrocarbons present in the gas.

Compositional analyses of natural gases are readily obtained by low-temperature distillation, chromatography, or mass spectrometry. Volume or mole percentages of the individual components present are ordinarily reported through heptanes plus. The heptanes-plus fraction includes heptane and all heavier hydrocarbons.

Natural gases are also described as dry or wet gases depending on the amount of condensable hydrocarbons present in the mixture. Pentane and heavier components are considered to be condensable hydrocarbons, as at atmospheric pressure and temperature pure pentane exists as a liquid.

TABLE 1-6. AVERAGE DISTILLATION FACTORS OF THE VARIOUS CLASSES OF CRUDE PETROLEUM[3]

Class	Sulfur, %	Gravity of key fraction 1, °API	Gravity of key fraction 2, °API	Gasoline, light ends, %	Total gasoline and naphtha, %	Total gasoline and naphtha gravity, °API	Kerosene distillate, %	Kerosene distillate gravity, °API	Gas oil, %	Gas oil gravity, °API	Lubricating distillate, %
Paraffin, wax-bearing	0.22	43.0	32.7	6.1	22.6	60.1	17.5	45.1	11.0	39.0	18.3
Paraffin—intermediate, wax-bearing	0.50	40.6	28.0	5.8	28.9	58.4	18.5	43.1	10.1	35.6	17.4
Intermediate—paraffin, wax-bearing	0.19	37.8	31.0	5.6	21.9	52.0	8.8	42.4	24.9	36.3	23.3
Intermediate, wax-bearing	1.00	37.0	25.2	5.4	25.0	55.3	7.5	42.4	18.2	34.9	20.4
Intermediate, wax-free	0.36	34.9	23.4	4.5	25.1	52.5	5.4	41.6	19.4	34.2	19.4
Intermediate—naphthene, wax-bearing		37.1	19.0	9.3	38.1	56.0	12.0	42.3	12.0	33.7	14.7
Intermediate—naphthene, wax-free		34.3	17.3		11.2	50.9			26.9	34.9	20.3
Naphthene—intermediate, wax-bearing	0.35	31.5	23.4	3.3	24.7	49.4			29.8	32.2	23.0
Naphthene—intermediate, wax-free	0.80	31.1	22.6		6.8	48.6			22.3	31.7	29.5
Naphthene, wax-bearing	1.50	29.4	16.7	4.7	21.5	51.0			27.7	30.6	23.0
Naphthene, wax-free	1.07	28.8	16.0	2.9	15.1	50.3			24.3	30.3	26.6

The lighter hydrocarbons—methane, ethane, propane, and butane—exist in the gaseous state at atmospheric conditions.

Origin of Petroleum

Petroleum is a mineral substance and is produced from the earth. How, then, does petroleum exist in the earth? The liquid or gas, as the case might be, is contained in the pore space or interstices of rock materials. These rock materials are referred to as reservoir rocks. The rocks in which petroleum is found are sedimentary materials, generally sandstones or limestones. Crude petroleum has been found at various depths throughout the great sedimentary basins of the earth at depths as shallow as 40 and as deep as 21,000 ft. Crude oil and natural gas are produced from oil and/or gas reservoirs by a number of well bores drilled to the reservoir rock. It is well known that oil and gas do not underlie all the surface of the earth but rather are found in localized accumulations under certain conditions of geologic environment.

Two questions, in general, arise in considering the distribution of petroleum accumulations. One, what is the origin of petroleum; the other, what geologic conditions are required in order that the oil or gas can accumulate in a localized area? Many theories of the origin of petroleum have been advanced. As yet, a definite solution of the problem has not been obtained. The theories of the origin of petroleum may be classified as either inorganic or as organic. The inorganic theories attempt to explain the formation of petroleum by assuming chemical reactions among water, carbon dioxide, and various inorganic substances, such as carbides and carbonates, in the earth. The organic theories assume that petroleum evolved from decomposition of vegetable and animal organisms that lived during previous geologic ages. The various theories of petroleum formation are summarized in Table 1-7.

Although some of the inorganic theories appear to be plausible and, in fact, petroleum can be produced in the laboratory by reactions of inorganic materials, geologic evidence indicates that these materials are not present in the earth in sufficient quantities to produce petroleum accumulations. In general, scientists have abandoned the inorganic theory as untenable. The organic theory, conversely, is supported by much geologic evidence. Oil and gas are commonly found in sedimentary basins, and furthermore, these accumulations are found in the vicinity of beds which contain large amounts of organic matter. Beds which are rich in organic material are called *source beds*.

To have a petroleum accumulation it is necessary to have source beds and a reservoir or storage bed. A suitable reservoir rock is porous and permeable. That is, the pores interconnect so that fluids can migrate through the rock. The occurrence of petroleum further indicates that the

TABLE 1-7. THEORIES ADVANCED IN EXPLANATION OF THE ORIGIN OF PETROLEUM[2]

Name of theory or its originator	Salient features	Evidence
Inorganic theories		
Berthelot's alkaline carbide theory	Deep-seated deposits of alkaline metals in the free state react with CO_2 at high temperatures, forming alkaline carbides. These, on contact with water, liberate acetylene, which, through subsequent processes of polymerization and condensation, forms petroleum	Evidence lacking. Neither free alkaline metals nor carbides found in nature
Mendeleev's carbide theory	Iron carbides within the earth on contact with percolating waters form acetylene, which escapes through fissures to overlying porous rocks and there condenses	See above. Magnetic iron oxides would also be formed as a product of these reactions. Magnetic irregularities have been noted in the vicinity of some oil fields
Moissan's volcanic theory	Moissan suggests that volcanic explosions may be caused by the action of water on subterranean carbides	Small quantities of petroleum noted in volcanic lavas near Etna and in Japan. Petroleum also associated with volcanic rocks in Mexico and Java
Sokolov's cosmic theory	Petroleum considered to be an original product resulting from the combination of carbon and hydrogen in the cosmic mass during the consolidation of the earth	Small quantities of hydrocarbons occasionally found in meteorites
Limestone, gypsum, and hot-water theory	Reactions between carbonate and sulfate of lime in the presence of water at temperatures sufficient to dissociate the water theoretically may form hydrocarbons	Practically, it has been found impossible to demonstrate this reaction in the laboratory
Organic theories		
Engler's animal-origin theory	Petroleum formed by a process of putrefaction of animal remains. Nitrogen thus eliminated and residual fats converted by earth's heat and pressure into petroleum. Activity of anaerobic bacteria thought to play a part in the reactions	Oils resembling petroleum can be distilled from sediments containing fish remains. Many petroleum deposits associated with marine sediments contain an abundance of foraminifera
Hofer's vegetable-origin theory	Petroleum formed by decay of accumulated vegetable refuse under conditions which prevent oxidation and evaporation of the liquid products formed	Deposits of petroleum found in close association with sedimentary deposits containing diatoms, seaweed, peat, lignite, coal, and oil shale of known vegetable origin. Oils closely resembling petroleum can be distilled from these substances
Hydrogenation of coal or other carbonaceous materials	Solid organic materials converted into liquid hydrocarbons by combination with free hydrogen at high pressures and temperatures in the presence of a suitable catalyzer, such as nickel	Hydrogenation of coal in the laboratory and in commercial plants. The ash of some petroleums is chiefly nickel. However, the existence of free hydrogen in nature is yet to be demonstrated

11

petroleum must migrate from the source beds to the reservoir rock in order that sufficient quantities accumulate to form the commercial deposits that the petroleum industry exploits. This evidence of migration indicates a third requirement: a carrier bed. The carrier bed may be a part of the reservoir rock in which the accumulation occurs, or it may be an adjacent reservoir rock having interconnected pores.

Traps

The primary forces causing the migration of petroleum are bouyancy and capillarity. As oil and gas are lighter than the ground water which permeates the porous rocks below the water table, it is evident that the upward movement of petroleum must be restricted in order that accumulations exist at depth. A natural barrier, or trap, must exist for a petroleum accumulation to form. Traps associated with oil fields are, in general, complex.

Wilhelm[4] proposed a classification system for traps which differentiates between factors indicating the structural environment of a reservoir in an area and the actual attitude or situation of the reservoir bed. The classification system is expressed by means of a group of structural environment indicators and by a group of trap indicators.

Trap indicators are grouped as follows:

A. Convex trap reservoirs, which are completely surrounded by edgewater, as the porosity extends in all directions beyond the reservoir areas. The reservoir peripheries are therefore defined by uninterrupted edgewater limits. The trap is due to convexity alone.

B. Permeability trap reservoirs, with a periphery partly defined by edgewater and partly by the barrier resulting from the loss of permeability in the reservoir layer. In the extreme case, the reservoir may be entirely surrounded by such a permeability barrier.

C. Pinchout* trap reservoirs, with the periphery partly defined by edgewater and partly by the margin due to the pinchout of the reservoir bed.

F. Fault trap reservoirs, with the periphery partly defined by edgewater and partly by a fault boundary.

G. Piercement trap reservoirs, with the periphery partly defined by edgewater and partly by piercement contact.

Figure 1-1 shows elementary reservoir traps in sectional view.

The structural environment indicators are as follows:

1. Dome and anticline, representing the most important types of uplifts in reservoir structures
2. Structural salient, nose, arch, or promontory
3. Structural terrace or platform

* Pinchout refers to the wedging out of the formation against another.

4. Monocline-homocline flexure
5. Plunging syncline
6. Absence of controlling structural conditions

Figure 1-2 presents contours of structural environments.

A complete classification, then, of a petroleum trap is a combination of

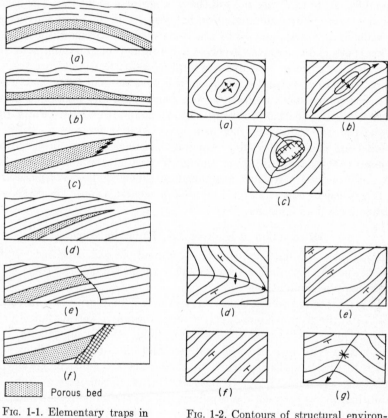

Fig. 1-1. Elementary traps in sectional view. (a) Simple convex trap (by folding); (b) simple convex trap (by differential thickness); (c) permeability trap; (d) pinchout trap; (e) fault trap; (f) piercement trap. (*From Wilhelm.*[4])

Fig. 1-2. Contours of structural environments. (a) Dome; (b) anticline; (c) piercement salt dome; (d) structural salient; (e) structural terrace; (f) monocline; (g) plunging syncline. (*From Wilhelm.*[4])

one of the elementary reservoir traps and one of the structural environment indicators. That is, the structure in the area in which the trap exists defines the structural environment; the actual attitude or configuration of a petroleum reservoir determines the trap classification. Some reservoirs, of course, are complex and result from a combination of two or more of the elementary trap features.

For an accumulation of oil or gas to exist, there must be closure* in the trap. The classification system discussed recognizes this requirement for closure and further recognizes that the reservoir rock must be overlain by impermeable beds so that oil or gas accumulation cannot seep from the trap and migrate to higher beds in the stratigraphic sequence. The classification system allows differentiation between an oil field and a reservoir. An oil field is defined simply by its areal extent on the surface of the ground, while a petroleum reservoir involves a continuity of reservoir rock and fluid. An oil field may encompass several oil reservoirs.

In general, in petroleum exploration, it is possible to evaluate and define structural environment indicators from geophysical and regional geologic information. Furthermore, it is possible to define from such information some of the trap indicators, such as a convex trap, a fault trap, or a piercement trap. The permeability trap and the pinchout trap, however, are disclosed only by the drill. Even though traps exist, there is no assurance that a commercial petroleum accumulation exists. A trap is a necessary but not sufficient condition for a petroleum accumulation. Therefore, not all traps form petroleum reservoirs.

Distribution of Hydrocarbon Fluids in Traps

Porous rocks are fluid-permeated, containing oil, gas, or water. Gravitational and capillary forces largely control the distribution of these fluids in petroleum accumulations.

The gravitational forces cause the less dense fluids to seek the higher positions in the trap. Capillary forces tend to cause a wetting fluid to rise into pore space containing a nonwetting fluid. Water, in general, is a wetting fluid with respect to oil and gas, and oil is a wetting fluid with respect to gas. Capillarity tends to counteract the force of gravity in segregating the fluids. Prior to the disturbance of the accumulation, an equilibrium exists between the capillary and gravitational forces.

Typical fluid distributions resulting from the equilibrium of these forces are shown schematically by sectional views of a domal trap in Fig. 1-3. Also illustrated in the figure are the possible modes of occurrence of gas: (1) solution gas, (2) associated free gas, (3) nonassociated free gas.

The accumulation of crude oil is shown in Fig. 1-3a. At the conditions of pressure and temperature existing in the trap, only oil and water are present. The oil is accumulated in the top of the trap and is underlain by water. Between the oil zone and the water zone, an oil-water transitional zone exists. The pore space of the rock in the oil zone contains a small amount of water (commonly called connate water). The fraction of the

---

* Closure is the height between the lowermost closed contour and the top of the trap.

pore space occupied by water increases with depth in the transitional zone so that the base of the transitional zone is delineated by completely water-saturated pore space. Natural gas initially occurs in such a reservoir only as solution gas.

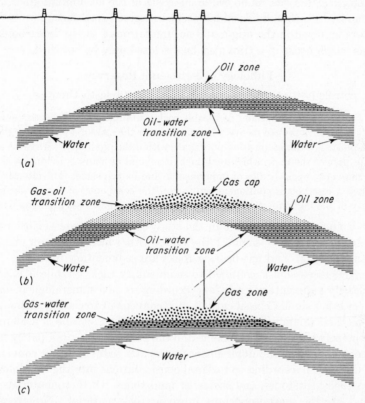

FIG. 1-3. Sketch showing typical fluid distributions in hydrocarbon reservoirs. (a) Oil reservoir; (b) associated oil-gas reservoir; (c) nonassociated gas reservoir.

An accumulation in which both crude oil (liquid phase) and natural gas (free gas phase) occur at the conditions of pressure and temperature existing in the trap is shown in Fig. 1-3b. The natural gas occupies the highest position of the trap and forms a "gas cap." The crude oil occupies an intermediate position and is, in turn, underlain by water.

Transitional zones are shown both between the gas and the oil and between the oil and water. These are zones of variable saturation in the wetting and nonwetting fluids. Connate water exists in the gas cap as well as in the oil zone. The natural gas occurring in such an accumulation is comprised of the associated free gas in the gas cap and the solution gas dissolved in the crude oil.

An accumulation of only natural gas is shown in Fig. 1-3c. The gas zone is underlain by a gas-water transitional zone and water. The gas zone contains connate water, which increases with depth in the transitional zone to complete water saturation in the water zone. The gas in this accumulation is nonassociated gas, as no crude oil exists in the accumulation.

The hydrocarbon fluids which occur in a given trap depend on unknown factors involved in the origin and accumulation of the hydrocarbons. The fluids which occur in a trap may be disclosed only by the drill.

Lithology of Petroleum Reservoirs

A petroleum reservoir may be defined according to Uren[5] as:

. . . a body of porous and permeable rock containing oil and gas through which fluids may move toward recovery openings under the pressures existing or that may be applied. All communicating pore space within the productive formation is properly a part of the rock, which may include several or many individual rock strata and may encompass bodies of impermeable and barren shale. The lateral expanse of such a reservoir is contingent only upon the continuity of pore space and the ability of the fluids to move through the rock pores under the pressures available.

It is appropriate therefore in the study of petroleum-reservoir engineering to review briefly some of the rocks with which engineers must deal. Virtually all oil- and gas-bearing rocks are sedimentary in origin. Therefore, this discussion is confined to sedimentary rocks. While it is possible to classify sedimentary materials according to their mineralogical composition or other similar classifications, it is convenient to classify them according to their origin. Two major subdivisions, then, can be considered: the dominantly fragmental sediments and the partly fragmental, partly precipitated sediments. The dominantly fragmental sediments can conveniently be subdivided according to textural considerations into the conglomerates, sandstones, siltstones, and shales, or mudstones. Of these materials, sandstones are the most important reservoir rock material. Limestones and dolomites are important petroleum reservoir rocks which are partly fragmental, partly precipitated in origin. Limestones and dolomites may be of mechanical or chemical origin or may be developed as a result of both processes of deposition.

The three most common sedimentary rocks associated with petroleum reservoirs are sandstone, shale, and limestone. In fact, these sedimentary rocks are so common in the subsurface with respect to petroleum reservoirs that it is convenient to think of all the sedimentary rocks as being composed of these materials. Figure 1-4 shows lithologic relationship of these common rock materials. The nomenclature used is common to the oil field and indicates the gradation from one type of rock to another.

Sandstones can further be divided into three classifications with respect to origin, as proposed by Krynine[7] and described by Pirson.[8] Kyrnine has classified sandstones into orthoquartzite, graywacke, and arkose. An orthoquartzite is a sedimentary quartzite developed as a result of excessive silicification without the impress of metamorphism and is comprised primarily of quartz and other stable minerals. The cementing material is

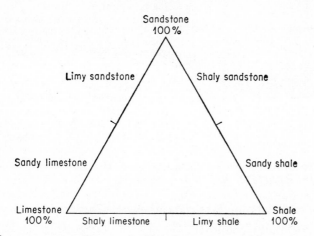

FIG. 1-4. Three-component diagram of sedimentary rock constituents. (*From Vance.*[6])

primarily carbonate or silica, and the orthoquartzites are relatively clean sediments, that is, free from shales and clay. Figure 1-5 shows the composition of quartzite sediments and the minerals present in such a rock material. Such sediments, according to Krynine, are derived from relatively

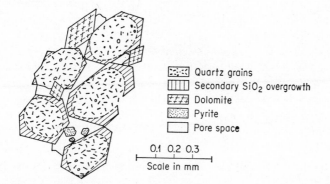

FIG. 1-5. Dolomite quartzite of Wilcox, Okla., oil-sand type. (*From Pirson.*[7])

low-lying coastal plains bordered by shallow seas in periods of quiescence. In general, quartzites are extensive in area and, owing to the relative quiescence of the depositional environment, are usually quite uniform. Local variation in properties may be due to secondary cementation with gypsum, glauconite, or other such material. The idealized conditions giving rise to orthoquartzite sediments are illustrated in Fig. 1-6.

The second subdivision of sandstones is graywacke, which is composed of large angular grains, mainly quartz, feldspar, and rock fragments. The

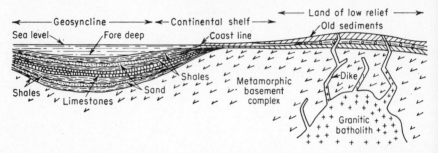

Fig. 1-6. Idealized land and sea conditions which give rise to quartzose-type sediments. (*From Pirson.*[8])

cementing materials are clays and carbonate. The land and sea conditions which give rise to graywacke-type sediments are depicted in Fig. 1-7. Note that the land from which the sediments are derived is of moderate relief. Because of the more rapid erosion and transport of the rock fragments from the land area to the site of deposition, a greater variety of rock

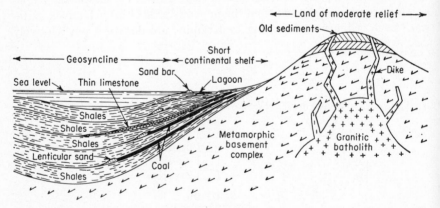

Fig. 1-7. Idealized land and sea conditions which give rise to graywacke-type sediments. (*From Pirson.*[8])

fragments remains unaltered than in an orthoquartzite. Figure 1-8 illustrates the rocks and rock materials comprising a typical graywacke, such as encountered in the Gulf Coast area. Note in particular the occurrence of clay and other micaceous material. Illite is believed to be the principal clay mineral occurring in graywacke. Graywackes are frequently lenticular and occur as numerous thin sand bodies in a thick sequence of sediments. The Frio formation of the Gulf Coast of Texas and Louisiana is a typical graywacke.

An arkose or arkosic sandstone contains 25 per cent or more feldspar

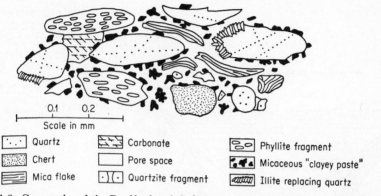

Fɪɢ. 1-8. Graywacke of the Bradford and Gulf Eocene oil-sand type. (*From Pirson.*[8])

derived from acid igneous rock. Arkose is usually coarse grained and is derived from lands of steep relief during periods of intense diastrophism. Since transport of the materials from the uplands to the site of deposition is relatively rapid, many unstable minerals do not decompose. The cementing material is chiefly clay containing a large percentage of kaolinite but also high proportions of reactive clays, such as montmorillonite. The sediments are characterized by thick sections of poorly sorted material. Because of the poor sorting and the variety of minerals composing arkose, the physical properties of the rock are quite variable. Figure 1-9 illustrates the conditions giving rise to arkosic sediments, while Fig. 1-10 illustrates the minerals and rock materials comprising a typical arkose. Note the poor sorting and the relative angularity of the materials comprising an arkose.

Limestones, dolomites, and other carbonate reservoir rock materials are frequently derived by precipitation. Limestones are typically extensive and massive. A pure limestone or dolomite rarely occurs owing to the presence of varying amounts of detrital material. Carbonate reservoir rocks can be divided into the following lithologic types: oolitic limestone, limestone, chalk, dolomitic limestones, dolomites, and cherty limestones and dolomites. Of these materials the physical properties of only the oolitic limestone are largely determined by the depositional environment. The remaining carbonate rocks are largely finely crystalline, and their physical properties depend greatly on such processes as deformation and solution after deposition. A limestone has little resistance to tension, and when it is subjected to tension forces, fractures develop, thus allowing subsurface waters to percolate through those fractures, subjecting the carbonate material to processes of secondary solution and deposition.

Shales are of little importance as reservoir rocks but comprise a large

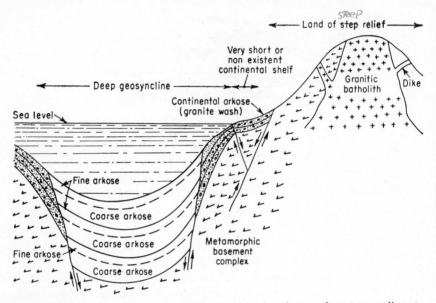

FIG. 1-9. Idealized land and sea conditions which give rise to arkose-type sediments. (*From Pirson.*[8])

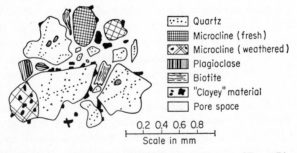

FIG. 1-10. Arkose of the Stevens, Calif., oil-sand type. (*From Pirson.*[8])

proportion of the caprock or impermeable seals which are necessary for any petroleum reservoir. Shales are quite fine-grained and offer high resistance to migration of fluids. Since shales behave as plastics under loading, fracturing occurs infrequently.

Wilhelm[4] presented the following list of reservoir rocks:

LIST OF RESERVOIR ROCKS[4]

1. Sand, conglomeratic sand, and gravel in varying state of consolidation, porosity due to fragmental textures, common
   a. Clean sands, etc., pore space between sand grains uncontaminated
   b. Argillaceous sands, etc., pore space partly filled with argillaceous matter
   c. Silty sands, etc., pore space partly filled with silt
   d. Lignitic sands, etc., pore space partly filled with lignitic matter
   e. Bentonitic sands, pore space partly filled with volcanic ash
2. Porous calcareous sandstone and siliceous sandstone, porosity due to incomplete cementation, frequent
3. Fractured sandstone and fractured conglomerate, porosity due to fracturing in tight sandstones or hard conglomerates caused by faulting or sharp folding, infrequent
4. Arkosic (feldspathic) sand, arkose, arkosic conglomerate (granite wash), porosity due to fragmental texture, infrequent
5. Detrital limestone (calcitic and dolomitic), porosity due to fragmental texture and frequently increased by solution, common
6. Porous crystalline limestone (calcitic and dolomitic), porosity due mainly to solution, common
7. Cavernous crystalline limestone (calcitic and dolomitic), porosity due to strong solution effects, common. Note: 5, 6, and 7 are not sharply separable
8. Fractured limestone (calcitic, dolomitic, and siliceous), porosity due to open fissures along fracture patterns, frequent
9. Sugary dolomite, "saccharoidal" porosity possibly due to volume shrinkage in the process of formation of dolomite from calcitic sediment, common
10. Oolitic limestone, porosity due to oolitic texture with uncemented or partially cemented interstices, frequent
11. Coquina and shell breccia, porosity due to fragmental texture, infrequent
12. Crinoidal limestone, a variety of coquina, porosity due to fragmental texture, infrequent
13. Porous cap rock on shallow salt plugs, porosity due to solution, infrequent
14. Honeycombed anhydrite, porosity due to leaching, rare
15. Fractured shale, porosity due to fracturing of brittle siliceous shale under sharp folding, rare
16. Fractured chert, porosity due to fracturing under sharp folding, rare
17. Porous tectonic breccia, formed along fault and thrust zones, porosity mainly due to incomplete cementation or subsequent solution, rare

18. Contact-metamorphic shales, porosity due to volume shrinkage after "baking," rare

19. Porous igneous rock, porosity primary as in tuffs or due to fracturing as in basalt or due to decomposition, rare

The references—common, frequent, infrequent, rare—following each major rock type indicate the relative frequency of occurrence.

Drilling

Oil and gas are produced from the earth by means of wells drilled to the reservoir rock. Any drilling method must meet two requirements: (1) a means of breaking or abrading the formations to be penetrated and (2) a means of removing the cuttings or the rock fragments which are produced in the drilling operation. Although many methods may be conceived which can accomplish these two purposes, oil-well drilling has been restricted largely to two methods. These methods may be identified as (1) churn drilling and (2) rotary drilling. A third category is sometimes included, a combination of the two foregoing drilling methods.

Cable-tool Drilling

While there are many variations of the churn-drilling method, that commonly used in the United States is known as cable-tool drilling. Approximately 15 per cent of all the holes drilled in the United States each year are drilled by the cable-tool method. Cable-tool drilling is used to a great extent in the Appalachian area. The cable-tool rigs used are spudders, drilling machines, or American Standard cable-tool rigs. The principal components of the cable-tool drilling rig together with a sectional view of the hole with the drilling tools are shown in Fig. 1-11. The drilling tools are comprised of a bit, a drill stem, jars, and rope socket suspended on a wire rope. To drill, the tools are lowered to the bottom of the hole and the drilling line attached to the walking beam which imparts a reciprocating motion to the tools. This reciprocating motion causes the tools to strike repeated blows on the formations at the bottom of the hole, thus causing breaking or abrading of the formation by a simple pounding or chipping action. After a certain amount of material has been broken from the formation, the drilling motion is interrupted and the drilling tools are removed from the hole.

After the tools have been retrieved to the surface and set back, a bailer is lowered into the hole to remove the broken formation material or cuttings. Water is added to the hole periodically as drilling progresses. Thus the cuttings in cable-tool drilling are suspended in a thin mud slurry. In cable-tool drilling, the fluid to suspend the cuttings is maintained at a low level, only partially filling the hole. More effective blows are struck by the tools if a low head of fluid is maintained during the drilling operation.

It is apparent that as the formations are penetrated, the fluids contained within those formations may readily enter the well bore, since the pressure in the bore hole is only that of a low head of fluid. It is necessary in many

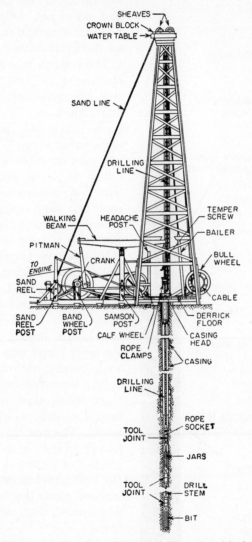

FIG. 1-11. Principal components of a cable-tool drilling rig with drilling tools in the hole. (*From Uren.*⁹)

areas to insert casing in the hole as the drilling operation proceeds to exclude water from the well bore in order that the hole be maintained relatively free of fluid and the drilling operation be unimpeded. In cable-tool drilling, commercial oil- or gas-bearing sands are indicated by the entry of oil and gas into the well bore. For many engineering and geological purposes, however, the mere knowledge of the presence of oil and gas is not sufficient. Certain physical measurements are required on the reservoir rock material.

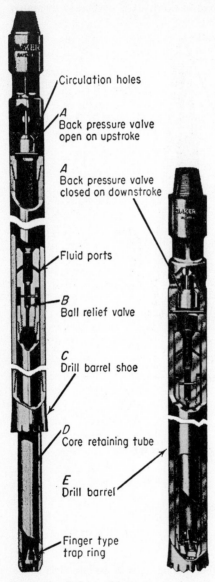

The cuttings from normal drilling operations are quite small. In addition, materials such as shales and clays, which may be included in the rocks penetrated, become readily dispersed in water. Two techniques are used in cable-tool drilling to obtain larger formation samples. The first is cable-tool coring; the second is chip coring. Cable-tool coring is conducted by attaching to the string of tools, not a bit, but a cable-tool core barrel, such as the Baker cable-tool core barrel illustrated in Fig. 1-12. The Baker cable-tool core barrel is composed of an inner barrel and an outer barrel. The drilling motion drives the inner barrel into the formation while causing the outer barrel to excave formation from around the inner barrel, allowing a cylindrical section of rock to enter the inner barrel. The core barrel is approximately 10 ft in length. After that amount of formation has been cut, the core barrel is retrieved to the surface, where the cylindrical core is extruded from the inner core barrel.

Circulation holes

A
Back pressure valve open on upstroke

A
Back pressure valve closed on downstroke

Fluid ports

B
Ball relief valve

C
Drill barrel shoe

D
Core retaining tube

E
Drill barrel

Finger type trap ring

FIG. 1-12. Baker cable-tool core barrel. (*Courtesy of the Baker Oil Tool Company.*)

Chip coring utilizes a bit specially sharpened to wedge point so that, in the drilling action, the fragments of formations obtained will be somewhat larger than those obtained during the regular drilling motion. In addition, the hole is bailed more frequently with a device known as a sand pump. The sand pump is so designed that a suction is created to aid in picking up rock fragments contained in the well bore.

While core samples represent satisfactorily the physical properties of the formations penetrated, the fluid contents of the core are not those of the undisturbed rock. The core has been subjected to two processes which disturb the fluid contents of the rock. The processes are (1) pressure reduction, allowing the fluids contained within the formation to expand and be expelled from the core, and (2) flushing by the drilling fluid as the contents of the rock tend to come to pressure equilibrium with the well bore fluid. If the rock contains gas and oil and the well fluid is water, the water will tend to enter the rock and occupy space voided by oil or gas. Thus, the core sample obtained does not contain the original reservoir fluid.

Rotary Drilling

The rotary drilling method has, in the last fifty years, largely supplanted the cable-tool drilling method in the United States. About 85 per cent of the wells drilled in the United States are drilled by the rotary method. As its name implies, the rotary drilling method utilizes the rotational motion of a bit operating in the hole to break or abrade the formations. This bit is attached by means of one or more drill collars to a string of drill pipe which extends to the surface. At the surface, a rotary motion is imparted to the drill pipe by means of a rotary table and a special joint of pipe known as the Kelly joint. The cuttings are removed from the hole by means of a circulating fluid, commonly a water-base fluid or drilling mud. In normal circulation, the drilling fluid is pumped down through the Kelly joint, drill pipe, and the bit, returning to the surface in the annular space between the drill pipe and the wall of the hole. The cuttings are transported to the surface by the circulating fluid in the annular space. A typical rotary rig is shown in Fig. 1-13.

In contrast to cable-tool drilling, the hole in rotary drilling is filled with a fluid. This fluid exerts a hydrostatic pressure on the formations penetrated which is much greater than the hydrostatic pressure exerted by the relatively low head of fluid used in cable-tool drilling. The formation, as it is drilled, is broken into small fragments and can be recovered at the surface from the drilling fluid. It may be noted that these cuttings are subjected to flushing throughout their transport from the bottom of the hole to the surface, and in fact, owing to the pressure exerted by the column of drilling fluid, there is flushing ahead of the bit. Also, the cuttings are subjected to pressure reduction as they rise in the drilling fluid and are brought to the surface. Therefore, cuttings obtained from rotary drilling are flushed in the same fashion essentially as are the cuttings from cable-tool drilling. In the normal course of rotary drilling, the formation fluids cannot enter the well bore, as the hydrostatic pressure of the mud column is greater than formational fluid pressures. Therefore, it is possible to drill through oil- and gas-bearing formations without detecting them in the course of drilling.

Rotary drilling fluids can be grouped into three broad categories, de-

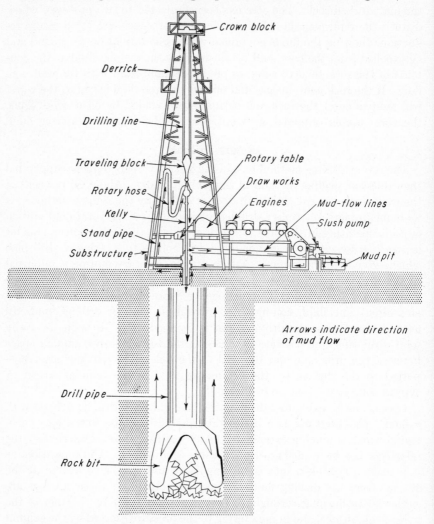

FIG. 1-13. Rotary drilling rig. (*Adapted from "Primer of Oil Well Drilling."* [10])

pending upon the base of the drilling fluid. These categories are (1) water-base fluids, (2) oil-base fluids, and (3) air or gas in the order of frequency of use. Both water-base and oil-base muds consist of a continuous liquid phase, the base, and dispersed solids. The dispersed solids increase the density of the fluid and impart desirable colloidal properties to the mud fluid. When a porous formation is penetrated, the dispersed solids form a filter cake which restricts the entry of the drilling fluid into the formation. A portion of the liquid phase, however, is filtered out in forming the cake. This liquid phase is called filtrate and is water and oil for water-base and oil-base muds, respectively. The formation cuttings are flushed primarily with filtrate.

In order to obtain samples of sufficient size for the measurement of physical properties of the formation rock, it is necessary in rotary drilling to core the formations. There are essentially two types of rotary coring devices: (1) the bottom-hole type device and (2) the side-wall-type device. The bottom-hole coring device, as the name implies, is used to core formations as the hole is drilled deeper. The side-wall coring device is used to obtain samples of formations that have previously been penetrated by the drill. Figure 1-14 shows a conventional rotary core barrel used for bottom-hole coring. This type of device obtains a sample approximately 3 in. in diameter and up to 70 ft in length. Side-wall coring devices, however, obtain smaller samples, ranging from about ¾ in. in diameter and 2 in. long to about 1 in. in diameter and about 6 in. long. It is apparent that cores cut either with bottom-hole coring devices or with side-wall coring devices are subjected to the same processes that the cuttings are subjected to, that is, flushing and expulsion of fluids on pressure reduction. In the early 1940s, a pressure core barrel was developed in order to investigate the original fluid contents of formations cut with rotary core barrels. Numerous field tests with the pressure core barrel proved that flushing occurred ahead of the bit.

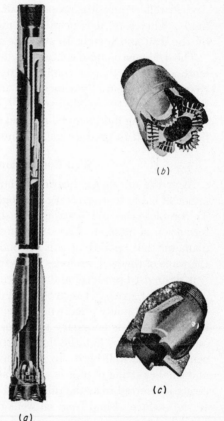
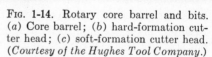

FIG. 1-14. Rotary core barrel and bits. (a) Core barrel; (b) hard-formation cutter head; (c) soft-formation cutter head. (*Courtesy of the Hughes Tool Company.*)

A more detailed discussion of drilling methods and drilling fluids is included in other texts such as Uren[9] and Brantly.[11]

In rotary-drilled wells, examination of the cuttings and core samples is not sufficient to evaluate fully the formations penetrated. Various "logging" methods, such as electrical logging and radioactive logging, yield additional information useful to the engineer and geologist. These devices reflect physical properties of the formations and of the formation fluids. Therefore, it is possible to identify the lithologic units penetrated by the drill and further to evaluate the fluid content of these formations.

The evidence on fluid contents as determined from cores and from logging methods can further be confirmed by use of the drill-stem test. The drill-stem test is essentially a means of making a temporary completion using the drill pipe. By making such a temporary completion, the hydrostatic head of drilling fluid may be relieved from the formation and the formation fluids allowed to produce under control into the drill pipe.

Well Completion and Production

After the oil- or gas-bearing formation has been identified, in the case either of cable-tool or rotary drilling, it is necessary to complete the well. To complete the well, a string of casing is ordinarily run to or through the formation of interest. The string of casing is simply steel pipe of sufficient diameter that operations can be conducted within it. The casing supports the walls of the well, excludes fluids from intervals other than that in which it is desired to produce, and confines the produced fluids to the well bore. The annular space between the wall of the well bore and the outside of the casing is commonly filled with cement. If the casing has been set through a formation, it is necessary to perforate the casing and the cement in the annular space so that the fluid contained within the objective formation can enter the well bore. In addition to the string of casing, an auxiliary and smaller string of pipe is usually suspended in the string of casing. This string is referred to as the tubing and is used to conduct the produced fluids to the surface. Fluid from oil wells may be expelled to the surface by the available energy of the reservoir fluids or may be artificially lifted. The reservoir pressure and gas in solution determine the available energy. Artificial lift is accomplished either by pumping or by introducing extraneous gas into the well bore to gas-lift the fluid.

Since the fluid produced from an oil well is comprised of both crude oil and natural gas, provisions must be made at the surface for separating the fluids when they are obtained. The fluid normally flows from the well head to an oil and gas separator, where the gas is separated from the oil. The oil is then conducted to stock tanks for lease storage. The gas is normally gathered and sent to a gasoline plant, where it is processed further into liquid components and into residue gas, which is either returned to the formation or sold for fuel. Quantities of water are also normally produced during the life of a well. Therefore, provisions must be made for removing the water from the well stream. Water is removed by gunbarrel or other types of water knockouts and then siphoned off from the liquid petroleum. The generalized production system for an oil well is shown in Fig. 1-15.

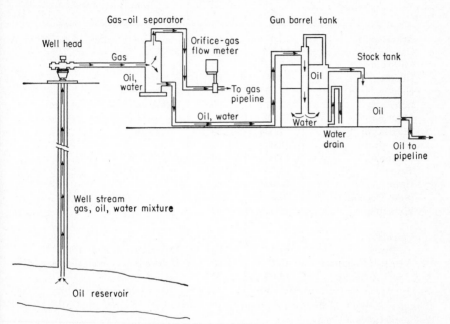

FIG. 1-15. Generalized production system. (*Adapted from "Primer of Oil and Gas Production."* [12])

Crude oil is normally gauged in the stock tanks in which it is gathered and stored after being separated from the gas and water with which it may have been produced. Crude-oil production in the United States is reported in barrels of stock-tank oil. An oil field barrel is defined as 42 U.S. gal. The standard conditions for reporting oil volumes are 60°F and atmospheric pressure.

The gas off the separator is ordinarily metered by means of an orifice meter and is reported in cubic feet at standard conditions of pressure and temperature. Standard conditions for gas measurement are defined by statute in the various states but approximates atmospheric pressure and temperature. In Table 1-8, statutory standards are shown for several states. In most fields, the stock-tank vapors, that is, gas released from the oil in the stock tank, is not measured, so that the measured gas volumes reflect the volume of gas produced from the separator. Water, of course, is an extraneous fluid of no value and is rarely measured with precision. Water volumes are reported in barrels. Accurate records of the produced

TABLE 1-8. STANDARD PRESSURES FOR GAS MEASUREMENT
IN VARIOUS STATES

State or province	Base pressure, psia	Base temp., °F	Correction for deviation from Boyle's law
Arkansas	14.65	60	
California	14.73	60	
Colorado	15.025	60	
Illinois	14.65	60	
Kansas	14.65	60	Above 100 psig
Louisiana	15.025	60	Above 200 psig
Michigan	14.73	60	
Mississippi	15.025	60	
New Mexico	15.025	60	
Oklahoma	14.65	60	Above 200 psig
Texas	14.65	60	Above 100 psig
Utah	15.025	60	
West Virginia	14.85	60	
Wyoming	15.025	60	
Canada:			
Alberta	14.4	60	If deviation is more than 1%
British Columbia	14.4	60	If deviation is more than 1%
Saskatchewan	14.65	60	

SOURCE: Compiled from data supplied by state agencies, U.S. Bureau of Mines, and Phillips Petroleum Company.

fluid volumes and of reservoir pressure are necessary for engineering analysis of well and reservoir problems. Typical field data are shown graphically in Fig. 1-16.

Reservoir Performance

The reservoir engineer is, of course, concerned with the production of oil and gas from the reservoir and primarily with the methods of stimulating or increasing the recovery from the reservoir as a whole. A basic understanding of drilling and production operations is required in reservoir engineering, as the hydrocarbon fluids are withdrawn from the earth through the well bore. In addition, virtually all the information upon which a reservoir engineer can base his studies must be obtained from these same well bores in terms of well logs, formation samples, samples of oil and gas, oil- and gas-production statistics, and reservoir pressures.

Efficient drilling and completion operations depend upon the physical properties of the rocks which are penetrated and in particular upon the properties of the producing formation. In addition, efficient production operations depend on a knowledge of formation characteristics, distribution of fluids within the formation, and the data requirements of the reservoir engineer. While this text is essentially a text on reservoir engineering, its purpose is also to report and discuss those subjects in which reservoir engineering, drilling engineering, and production engineering have common interests.

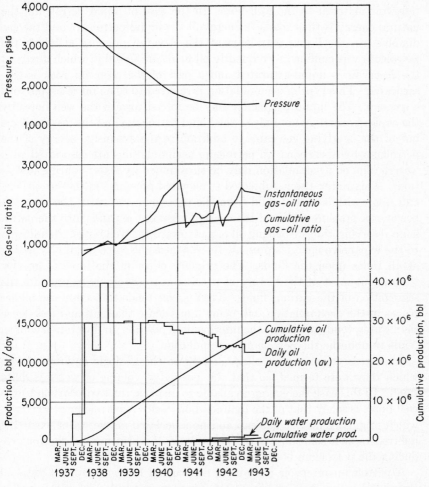

FIG. 1-16. Typical field production history.

An oil field is comprised of an aggregate of well bores penetrating one or more petroleum reservoirs in the subsurface. Modern development methods involve drilling of wells on a spacing of one well to each 20 to 40 acres. The question then arises, What forces the petroleum hydrocarbons to the well bore so that they can be produced to the surface? Several sources of energy exist in the formation. One of these sources is the expansive energy of the hydrocarbon fluid. In the case of gas the material is confined under pressure, and when the formation is opened to a well bore existing at a lower pressure, the fluid will tend to expand and flow toward the pressure sink. This, too, is true of crude oil or liquid petroleum. If the well constitutes a pressure sink, then fluid will migrate through the porous reservoir material to the well bore. In the case of liquid petroleum, the natural energy is the expansive energy of the liquid petroleum and the gas dissolved in the liquid petroleum at the elevated pressure at which the petroleum was confined. Frequently oil fields are found in which a part of the reservoir is liquid-saturated and a part of the reservoir rock is gas-saturated. This type of accumulation is referred to as an oil reservoir with a gas cap. The liquid petroleum may be forced toward the well bores by the expansive energy not only of the liquid petroleum and the dissolved gas but of the overlying gas cap. In addition to the expansive energy of the petroleum hydrocarbons, all petroleum accumulations are associated with water. The oil accumulation may be surrounded by water-bearing formations. This water also is subjected to elevated pressures in the subsurface, and upon withdrawal of fluid from the petroleum reservoir, the reservoir becomes a pressure sink and the contiguous water expands into the petroleum reservoir, thus displacing oil or gas toward the well bores. In addition to the expansive energies present, there is also the force of gravity acting at all times upon the fluids. The primary effect of the force of gravity throughout most of the history of petroleum reservoirs is to promote the segregation of the various fluids. That is, gas tends to occupy the higher places in the accumulation; oil, being more dense than gas and less dense than water, tends to occupy the intermediate position; and water, of course, tends to underlie the petroleum accumulation.

Some reservoirs may be closed, owing to the geologic environment in which they were formed, so that the associated volume of water is quite small. In this case the energy available to displace the hydrocarbon to the well bores is solely that of the hydrocarbon itself. A petroleum reservoir in which originally no free gas cap and no associated active water existed is referred to as a solution-gas–drive reservoir; the principal energy for producing the petroleum is that of the gas in solution in the oil.

A petroleum reservoir containing an original free gas cap but with no associated active water produces by a process or drive which is referred to as solution-gas–gas-cap drive.

A petroleum reservoir which is associated with water-bearing formations that are so active that little or no pressure drop occurs in the petroleum reservoir on the withdrawal of hydrocarbon fluids is referred to as a water-drive reservoir. That is, water from the surrounding aquifer enters the

reservoir almost as rapidly as the hydrocarbon fluid is withdrawn, therefore preventing any substantial decline in pressure.

The force of gravity does not become important as a driving mechanism
until the reservoir becomes substantially depleted. However, as previously
mentioned, gravitational forces are present in all the three preceding mechanisms and play a substantial role in the distribution of the fluids in that
gravity tends to promote segregation of the fluids contained within the
reservoir.

The various drives are characterized by pressure-production history.
Typical pressure-production histories of the three major drives are compared in Fig. 1-17. The solution-gas drive is characterized by a rapid pres-

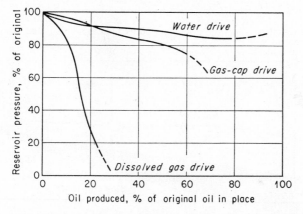

Fig. 1-17. Typical pressure-production histories of the three major drives. (*From Clark.*[13])

sure decline and a low recovery efficiency. In the gas-cap–drive reservoir
pressure is maintained at higher levels than in the solution-gas drive, and
recovery efficiency is thus improved. The degree of improvement depends
on the size of the gas cap relative to the oil zone and on the production
procedure used. Water drive is the most efficient in maintaining reservoir
pressure and usually yields the highest recovery efficiency. However, a
gas-cap drive, managed to secure the greatest aid from gravitational forces,
may yield a greater recovery efficiency than a water drive.

Most petroleum reservoirs are subjected to one or more drives either
simultaneously or at various times throughout the life of the reservoir.
For instance, a large reservoir after initial discovery may behave in its
early life as if it were solution-gas drive. Then after a short period of
production, the associated gas cap becomes effective and contributes substantially to the energy of the reservoir. Furthermore, after substantial
withdrawals have been made, enough pressure drop may have been estab-

lished in the adjacent aquifer so that water drive may become an important part of the reservoir mechanism. Reservoirs having more than one type of drive present are referred to as combination-drive reservoirs. Petroleum reservoirs containing only material in the gaseous phase at reservoir conditions are generally referred to as gas or condensate reservoirs.

The reservoir engineer must identify the drive mechanisms of a reservoir and develop production procedures to secure the maximum economic recovery efficiency. The production procedures recommended may include supplementation of natural energy by fluid injection. The fluid injection may involve the return of gas, water, or gas and water to the reservoir. One of the many possible injection procedures is shown schematically in Fig. 1-18.

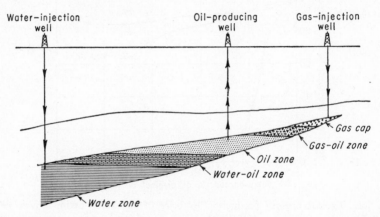

Fig. 1-18. Diagram showing the supplementing of natural reservoir energy by water injection into the water zone and gas injection into the gas cap.

If fluid injection is undertaken prior to the substantial depletion of the natural reservoir energy, the process is usually referred to as pressure maintenance. Fluid injection into a depleted reservoir is usually termed secondary recovery.

This text and its companion volume will develop systematically the fundamental concepts from which reservoir analyses can be made. The measurement, collection, and reduction of data will be discussed. Methods of evaluating well and reservoir performance will be developed, and applications presented. These methods will be extended to the prediction of reservoir performance under various modes of operation.

REFERENCES

1. Petroleum Productive Capacity: A Report of the National Petroleum Council, 1952.

2. Hager, Dorsey: "Practical Oil Geology," 6th ed., McGraw-Hill Book Company. Inc., New York, 1951.

3. Kraemer, A. J., and E. C. Lane: Properties of Typical Crude Oil from Fields of the Eastern Hemisphere, *U.S. Bur. Mines Bull.* 401, 1937.

4. Wilhelm, O.: Classification of Petroleum Reservoirs, *Bull. Am. Assoc. Petro. Geologists*, vol. 29, 1945.

5. Uren, L. C.: "Petroleum Production Engineering: Oil Field Exploitation," 3d ed., McGraw-Hill Book Company, Inc., New York, 1953.

6. Vance, Harold: "Elements of Petroleum Subsurface Engineering," Educational Publishers, Inc., Saint Louis, Mo., 1950.

7. Krynine, P. D.: Sediments and the Search for Oil, *Mineral Ind.*, vol. 13, no. 3, December, 1943.

8. Pirson, S. J.: "Elements of Oil Reservoir Engineering," 1st ed., McGraw-Hill Book Company, Inc., New York, 1950.

9. Uren, L. C.: "Petroleum Production Engineering: Oil Field Development," 4th ed., McGraw-Hill Book Company, Inc., New York, 1956.

10. "Primer of Oil Well Drilling," Industrial and Business Training Bureau, and the A.A.O.D.C., Texas Education Agency, July, 1951.

11. Brantly, J. E.: "Rotary Drilling Hand Book," 4th ed., Palmer Publications, Los Angeles, 1948.

12. "Primer of Oil and Gas Production," American Petroleum Institute, New York, 1954.

13. Clark, N. J.: Review of Reservoir Engineering, *World Oil*, May, 1951.

CHAPTER 2

FUNDAMENTAL PROPERTIES OF
FLUID-PERMEATED ROCKS

INTRODUCTION

Naturally occurring rocks are in general permeated with fluid, water, oil, or gas or combinations of these fluids. The reservoir engineer is concerned with the quantities of fluids contained within the rocks, the transmissivity of fluids through the rocks, and other related properties.

These properties depend on the rock and frequently upon the distribution or character of the fluid occurring within the rock. In this and the following chapter, properties of rocks containing fluids will be discussed.

This chapter deals with the properties which are considered fundamental and from which other properties and concepts can be developed. The properties discussed are the porosity—a measure of the void space in a rock; the permeability—a measure of the fluid transmissivity of a rock; the fluid saturation—a measure of the gross fluid distribution within a rock; and the electrical conductivity of fluid-saturated rocks—a measure of the conductivity of the rock and its contained fluids to electrical current. These properties constitute a set of fundamental parameters by which the rock can be quantitatively described.

POROSITY

From the reservoir-engineering standpoint, one of the most important rock properties is porosity, a measure of the space available for storage of petroleum hydrocarbon. Porosity is defined as the ratio of the void space in a rock to the bulk volume of that rock multiplied by 100 to express in per cent. Porosity may be classified according to the mode of origin as (1) original and (2) induced. Original porosity is that developed in the deposition of the material, while induced porosity is that developed by some geologic process subsequent to deposition of the rock. Original porosity is typified by the intergranular porosity of sandstones and the intercrystalline and oolitic porosity of some limestones. Induced porosity is typified by fracture development as found in some shales and limestones and by the vugs or solution cavities commonly found in limestones. Rocks having original porosity are more uniform in their characteristics than those rocks in which a large part of the porosity is induced. For direct quantitative measurement of porosity, reliance must be placed on formation samples obtained by coring.

Early investigations of porosity were conducted to a large extent by investigators in the fields of ground-water geology, chemical engineering, and ceramics. Therefore, much of the interest was centered on the investigation of the porosity of unconsolidated materials. In an effort to determine approximate limits of porosity values, Slichter[1] and, later, Graton and Fraser[2] computed the porosity of various packing arrangements of uniform spheres. Unit cells of two of the packings studied are shown in Fig. 2-1. The porosity for cubical packing (the least compact arrangement) is 47.6 per cent, and that for rhombohedral (the most compact arrangement) is 25.96 per cent. Considering cubic packing, the porosity can be calculated as follows:

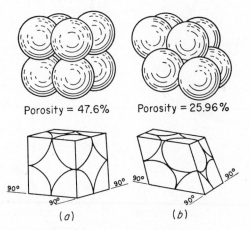

Porosity = 47.6%   Porosity = 25.96%

(a) (b)

Fig. 2-1. Unit cells and groups of uniform spheres for cubic and rhombohedral packing. (a) Cubic, or wide-packed; (b) rhombohedral, or close-packed. (*After Graton and Fraser.*[2])

The unit cell is a cube with sides equal to $2r$ where r is the radius of the sphere. Therefore,

$$\text{Bulk volume} = (2r)^3 = 8r^3$$

Since there are 8 ($\frac{1}{8}$) spheres in the unit cell,

$$\text{Sand-grain volume} = \frac{4\pi r^3}{3}$$

the porosity, therefore, is

$$\frac{\text{Pore volume}}{\text{Bulk volume}} \times 100 = \frac{\text{bulk volume} - \text{grain volume}}{\text{bulk volume}} \times 100$$

$$= \frac{8r^3 - \frac{4}{3}\pi r^3}{8r^3} \times 100 = 1 - \frac{\pi}{2(3)} \times 100 = 47.6 \text{ per cent}$$

Of particular interest is the fact that the radii cancel and the porosity of packings of uniform spheres is a function of packing only.

The investigators recognized that naturally occurring materials were composed of a variety of particle sizes and that not only the arrangement but the angularity and distribution of particle size would affect porosity. The angularity of particles comprising a sandstone is shown in the thin section of Fig. 2-2. The configuration of the pore space is obviously dif-

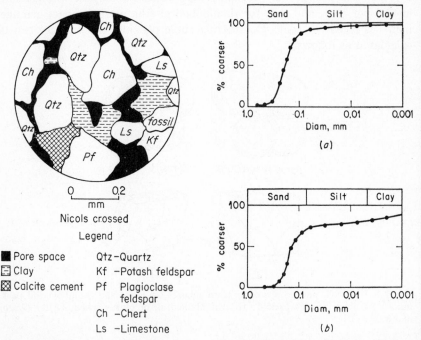

Nicols crossed

Legend

■ Pore space	Qtz – Quartz
▤ Clay	Kf – Potash feldspar
▨ Calcite cement	Pf Plagioclase feldspar
	Ch – Chert
	Ls – Limestone

FIG. 2-2. Thin section of sandstone illustrating grain and pore configuration. (*From Nanz.*[3])

FIG. 2-3. Cumulative grain-size distributions of graywacke. (*a*) Sand; (*b*) shaly sand. (*From Nanz.*[3])

ferent from that which would be obtained from the packing of uniform spheres. Furthermore, a portion of the space is filled with clay and cementing material. The diversity of particle size characteristic of a graywacke sandstone is illustrated by the particle-size-distribution curves of Fig. 2-3. The data were obtained by a standard sieve analysis.

The particle-size-distribution curves of Fig. 2-3 were obtained by Nanz[3] in the study of the origin and genesis of a Gulf Coast graywacke. The nomenclature sand and shaly sand is used in the common oil-field sense and implies firm sandstone and shaly sandstone, respectively. The shaly sand may be represented, as in Fig. 2-4, as being composed of a framework

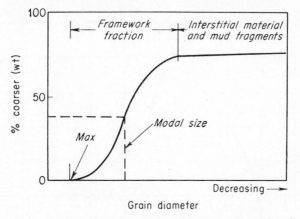

FIG. 2-4. Generalized texture of sand deposits. (*From Nanz.*[3])

fraction and interstitial material. Other physical measurements on the rocks indicate that increasing interstitial material reduces the pore space of the rock.

Grain-size distributions may be characterized in part by skewness of the distribution. Skewness is a statistical measure of the uniformity of the distribution of a group of measures. Tickell[4] has presented experimental data indicating that for packings of Ottawa sand, porosity was a function of skewness of the grain-size distribution (see Fig. 2-5). Other investigators have measured the effects of distribution, grain size, and grain shape. In general, smaller grain size and greater angularity tend to increase the porosity while an increase in range of particle size tends to decrease porosity.

In dealing with reservoir rocks (usually consolidated sediments) it is necessary, because the cementing materials may seal off a part of the pore volume, to define (1) total porosity and (2) effective porosity. *Total porosity is the ratio of the total void space in the rock to the bulk volume of the rock; effective porosity is the ratio of the interconnected void space in the rock to the bulk volume of the rock, each expressed in per cent.* From the reservoir-engineering standpoint, effective porosity is the quantitative value desired, as this represents the space which is occupied by mobile fluids. For intergranular materials, poorly to moderately well cemented, the total porosity is approximately equal to the effective porosity. For more highly cemented materials and limestones, significant difference in total porosity and effective porosity values may occur.

In Fig. 2-6 are presented photographs of impregnated rocks having essentially intergranular porosity.[5] The pore configuration is complex, but the pores are relatively uniformly distributed. Complex pore configurations arise from the interaction of many factors in the geologic environment of the deposit. These factors include the packing and particle-size distribution

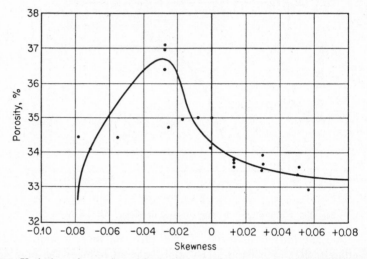

FIG. 2-5. Variation of porosity with skewness of Ottawa sand. (*From Tickell et al.*[4])

of the framework fraction, the type of interstitial material, and the type and degree of cementation. The influence of these various factors can be evaluated as statistical trends. However, a quantitative evaluation of the porosity resulting from the interaction of the various factors is possible only by laboratory measurements.

Materials having induced porosity such as the carbonate rocks shown in Fig. 2-7 have even more complex pore configurations. In fact, two or more systems of pore openings may occur in such rocks. The basic rock material is usually finely crystalline and is referred to as the matrix. The matrix contains uniformly small pore openings which comprise one system of pores. This system is the result of the crystalline structure of the rock. One or more systems of larger openings usually occur in carbonate rocks as a result of leaching or fracturing of the primary rock material. Vugular pore openings are frequently as large as an ordinary lead pencil and are usually attributed to leaching of the rock subsequent to deposition. Fractures also may be quite large and contribute substantially to the volume of pore openings in the rock. Both fractures and vugs may be closed or partially closed by precipitated calcite or other similar material. Vugs and fractures are highly variable in size and in distribution. Therefore, even more than

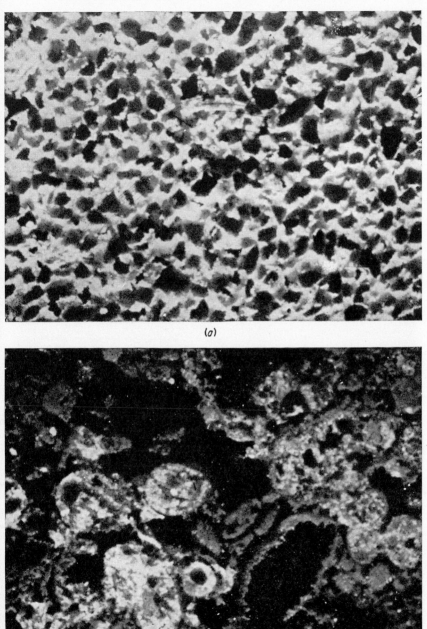

(a)

(b)

Fig. 2-6. Cast of pore space of typical reservoir rocks. (a) Fine intergranular sandstone; (b) coarse intergranular sandstone. (*From Nuss and Whiting.*[5])

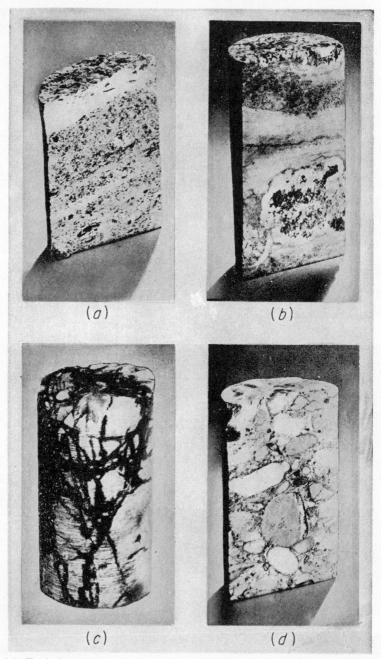

FIG. 2-7. Typical carbonate reservoir rocks. (*a*) Vugular porosity; (*b*) vugular with pin-point porosity; (*c*) fractured dense dolomite; (*d*) conglomerate. (*From Core Laboratories, Inc.*⁶)

for intergranular materials, laboratory measurements are required for quantitative evaluation of porosity.

Laboratory Measurement of Porosity

Numerous methods have been developed for the determination of the porosity of consolidated rocks having intergranular porosity. Most of the methods developed have been designed for small samples, roughly the size of a walnut. As the pores of intergranular material are quite small, a determination of the porosity of such a sample involves measuring the volume of literally thousands of pores. The porosity of larger portions of the rock is represented statistically from the results obtained on numerous small samples.

In the laboratory measurement of porosity it is necessary to determine only two of the three basic parameters (bulk volume, pore volume, and grain volume). All methods of determination of bulk volume are, in general, applicable to determining both total and effective porosity.

Bulk Volume. Although the bulk volume may be computed from measurements of the dimensions of a uniformly shaped sample, the usual procedure utilizes the observation of the volume of fluid displaced by the sample. This procedure is particularly desirable, as the bulk volume of irregular-shaped samples can be determined as rapidly as that of shaped samples.

The fluid displaced by a sample can be observed either volumetrically or gravimetrically. In either procedure it is necessary to prevent fluid penetration into the pore space of the rock. This can be accomplished (1) by coating the rock with paraffin or a similar substance, (2) by saturating the rock with the fluid into which it is to be immersed, or (3) by using mercury, which by virtue of its surface tension and wetting characteristics does not tend to enter the small pore spaces of most intergranular materials.

Gravimetric determinations of bulk volume can be accomplished by observing the loss in weight of the sample when immersed in a fluid or by observing the change in weight of a pycnometer when filled with mercury and when filled with mercury and the core sample. The details of gravimetric determinations of bulk volume are best summarized by example calculations.

———•◆•———

Example 2-1. Coated Sample Immersed in Water.

A = weight dry sample in air = 20.0 gm

B = weight dry sample coated with paraffin = 20.9 gm (density of paraffin = 0.9 gm/cc)

C = weight coated sample immersed in water at 40°F = 10.0 gm (density of water = 1.00 gm/cc)

Weight of paraffin = $B - A$ = 20.9 − 20.0 = 0.9 gm

Volume of paraffin = 0.9/0.9 = 1 cc
Weight of water displaced = $B - C$ = 20.9 − 10.0 = 10.9 gm
Volume of water displaced = 10.9/1.0 = 10.9 cc
Volume of water displaced − volume of paraffin = 10.9 − 1.0 = 9.9 cc
Bulk volume of rock = 9.9 cc

Example 2-2. Water-saturated Sample Immersed in Water.
A = weight dry sample in air = 20 gm
D = weight saturated sample in air = 22.5 gm
E = weight saturated sample in water at 40°F = 12.6 gm
Weight of water displaced = $D - E$ = 22.5 − 12.6 = 9.9 gm
Volume of water displaced = 9.9/1.0 = 9.9 cc
Bulk volume of rock = 9.9 cc

Example 2-3. Dry Sample Immersed in Mercury Pycnometer.
A = weight dry sample in air = 20.0 gm
F = weight of pycnometer filled with mercury at 20°C = 350.0 gm
G = weight pycnometer filled with mercury and sample at 20°C =
235.9 gm (density of mercury = 13.546 gm/cc)

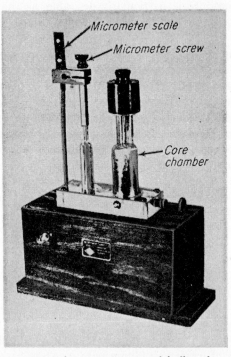

FIG. 2-8. Electric pycnometer for measurement of bulk volume of small samples.
(*Courtesy of Refinery Supply Co.*)

Weight of sample + weight of pycnometer filled with mercury = $A + F = 20 + 350 = 370$ gm

Weight of mercury displaced $= A + F - G = 370 - 235.9 = 134.1$ gm

Volume of mercury displaced $= 134.1/13.546 = 9.9$ cc

Bulk volume of rock $= 9.9$ cc

———·•••·———

Determination of bulk volume volumetrically utilizes a variety of specially constructed pycnometers or volumeters. An electric pycnometer from which the bulk volume can be read directly is shown in Fig. 2-8. The sample is immersed in the core chamber, which causes a rise in the level of the connecting U tube. The change in level is sensed by the micrometer screw. The resulting change in level is read directly in volume from the micrometer scale. Either dry or saturated samples may be used in the device.

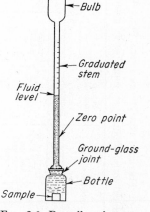

Fig. 2-9. Russell volumeter for determining grain and bulk volumes of rock samples. (*From Russell.*[9])

The Russell volumeter shown in Fig. 2-9 also provides for direct reading of the bulk volume. A saturated sample is placed in the sample bottle after a zero reading is established with fluid in the volumeter. The resulting increase in volume is the bulk volume. Only saturated or coated samples may be used in the device.

An evaluation of the foregoing and other methods of determining bulk volume is presented in Table 2-1. Careful procedure with all the methods listed yields sufficiently reliable results for engineering purposes.

Sand-grain Volume. The various porosity methods are usually distinguished by the means used to determine the grain or pore volume. Several of the oldest methods of porosity determination are based on determination of grain volume.

The grain volume can be determined from the dry weight of the sample and the sand-grain density. For many purposes, results of sufficient accuracy can be obtained by using the density of quartz (2.65 gm/cc) as the sand-grain density.

For more rigorous determination either the Melcher-Nutting[8] or Russell[9] methods can be employed. In each, the bulk volume of a sample is determined, then either that sample or an adjacent sample is reduced to grain size, and the grain volume determined. In the Melcher-Nutting technique, all the measurements are determined gravimetrically, utilizing the principle of bouyancy. The Russell method utilizes an especially designed volumeter (Fig. 2-9), and the bulk volume and grain volume are determined

TABLE 2-1. A COMPARISON OF METHODS FOR DETERMINING BULK VOLUME[7]

(Bulk volume based on cubical cores, about 10 cc; round cores, about 7 cc)

Method	Accuracy, cc (estimated)	Precision on 90% of determinations, cc	Time required per core, min	Effects on physical properties of core			Average deviation from saturation or mercury pycnometer method	Possible errors in bulk volume, cc			
				Preparation	Effects	Reclamation		Shape		High permeability	Miscellaneous
								Cylindrical	Cubical		
Saturation with tetrachloroethane	±0.03	±0.02	5	Saturate with tetrachloroethane	Possible oil extraction	Core must be dried 2 hr at 100° and evacuated 30 min		Drainage ±0.015	Drainage ±0.03	Excessive drainage −0.05	1. Cores with rough or irregular surfaces are more difficult to drain than those with smooth surfaces 2. This method overcomes error of possible incomplete saturation 3. Temperature-density errors of saturant, ±0.02 cc
Mercury pycnometer	±0.02	±0.01	10	Cores with high permeability saturate with tetrachloroethane to prevent mercury penetration	Nil—very slight mercury contamination, saturation effects	None—except for saturated cores		Trapped air, 0.01	Trapped air, 0.02	Possible mercury penetration −0.05 Incomplete drainage of saturated cores, −0.03	Care must be taken not to trap an air bubble in the pycnometer 0.03 cc Temperature-density errors of mercury ±0.02 cc
Russell volumeter using tetrachloroethane	±0.06	±0.06	5	Saturate with tetrachloroethane	Possible oil extraction	Core must be dried 2 hr at 100° and evacuated 30 min	±0.02 cc (round cores) 0.04 cc (square cores) of mercury pycnometer value	Drainage ±0.15	Drainage ±0.03	Possible excessive drainage −0.05	1. Tedious procedure 2. Volumeter drainage, and air trapped by glass joint ±0.03 cc
Russell volumeter using mercury	±0.05 to 0.20 (with correction factor of ±0.16 cc) 0.05	±0.03	0.5	Cores with high permeability saturate with tetrachloroethane to prevent mercury penetration	Nil—very slight mercury contamination, saturation effects	None—except for saturated cores	0.16 cc higher than mercury pycnometer method (with correction factor 0.04 cc)	Trapped air, 0.20	Trapped air, 0.20	Possible mercury penetration, −0.05 incomplete drainage saturated cores, −0.03 cc.	Cores trap varying amounts of air up to 0.20 cc

46

Method										Remarks
Mechanical measurement	±0.15	±0.08	5	Grind surface to 90° angles	None	None	None	Measures maximum dimensions (high spots) 0.15	Method not applicable to square cores	None	This method may be used only on round cores with smooth, flat surfaces. Surfaces must be at 90° angles
Ruska instrument	−0.30 to +0.20	±0.03	4	None	Nil—slight mercury contamination	None	0.10 cc higher than mercury pycnometer method	Trapped air, 0.04	Trapped air, 0.20	Mercury penetration −0.30	Filling pycnometer with mercury ±0.01 cc
Coating with collodion: volume by K and F instrument[a]	Insufficient experimental work has been conducted to perfect this method. Accuracy would depend upon K and F instrument and errors in coating procedure 0.10	0.08		Coating-dipping and drying	None	Extraction of coating	0.09 cc higher than saturation method (tight cores) for cores with high permeability (−0.01 to −0.26 cc)			Method not applicable	This method is not recommended for accurate work in porosity determinations. Combined errors of two readings and errors in coatings and errors in coating procedure would reduce accuracy to approximately 2%
Loss of weight in water by paraffin-coated core	Insufficient experimental data	Est. 0.04		Coating	Contamination with paraffin	Extraction of paraffin			Not applicable paraffin penetration	Not applicable paraffin penetration	Imperfect seal—water penetration while weighting suspended in water
Displacement of water	−0.10 to +0.30	±0.03	5	Saturation with water	Possible leaching or disintegration	Intermittent drying and evacuation 2 hr	Readings average 0.20 cc higher than Ruska instrument and Russell volumeter	Drainage ±0.03	Drainage ±0.05	Excessive drainage −0.05	1. Errors in reading graduate ±0.05 cc 2. This method gives high values

[a] Instrument devised by Kaye and Freeman. Not available commercially.

volumetrically. The porosity determined is total porosity. In sands of relatively uniform characteristics, the grain density can be determined from the above measurements, and that density, together with an observed dry weight and bulk volume, can be used to calculate the porosity of adjacent samples of the same lithology. The results are highly reproducible. The Melcher-Nutting method is illustrated by Example 2-4.

Example 2-4. Sand-grain Volume by Melcher-Nutting Method.

A = weight dry crushed sample in air = 16.0 gm
A' = weight crushed sample plus absorbed water = 16.1 gm
B = weight pycnometer filled with water at 40°F = 65.0 gm
C = weight pycnometer filled with sample and water at 40°F = 75 gm
Weight of pycnometer filled with water plus weight of crushed sample = $B + A$ = 65.0 + 16.0 = 81.0 gm
Weight of water displaced = $B + A - C$ = 81.0 − 75 gm = 6.0 gm
Volume of water displaced = 6.0/1.0 = 6.0 cc
Grain volume of sample = 6.0 cc $= V_m$

The porosity is then computed by combining results of the grain-volume determination (Example 2-4) and the bulk-volume determination (Example 2-1). Such a solution for porosity is given in Example 2-5.

Example 2-5. Determination of Total Porosity.
From Example 2-4,
  Sand-grain density 16/6.0 = 2.67 gm/cc
From Example 2-1,
  Weight of dry sample in air = 20 gm
  Bulk volume of sample = 9.9 cc

$$\text{Grain volume of sample} = \frac{\text{wt of dry sample in air}}{\text{sand-grain density}} = \frac{20}{2.67} = 7.5 \text{ cc}$$

$$\text{Total porosity} = \phi_t = \frac{\text{bulk volume} - \text{grain volume}}{\text{bulk volume}} \times 100$$

$$= \frac{9.9 - 7.5}{9.9} \times 100 = 24.2 \text{ per cent}$$

The methods of determining grain volume described above when combined with an observation of bulk volume yield total porosity values. The Stevens porosimeter is a means of measuring the "effective" grain volume. The porosimeter, shown in Fig. 2-10, consists of a core chamber which can be sealed from atmospheric pressure and closed from the remaining parts of the porosimeter by a needle valve. The volume of the core chamber is known accurately. In operation a core is placed in the core chamber; a vacuum is established in the system by manipulating the mercury reservoir; the air in the core and chamber is expanded into the evacuated system and then measured at atmospheric pressure in the graduated tube. The difference in volume of the core chamber and of the air extracted is the

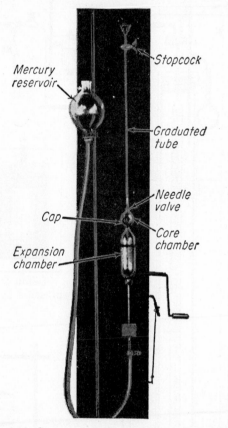

FIG. 2-10. Stevens porosimeter. (*From Stevens.*[10])

"effective" grain volume, the volume of the grains plus any sealed pore space. Thus if the effective grain volume is subtracted from the bulk volume, the volume of interconnected or effective pore space is obtained. The Stevens method is an adaptation of the Washburn-Bunting procedure, which will be described in the section on measurement of pore volume.

Example 2-6. Determination of Grain Volume by Gas Expansion. Stevens porosimeter:

A = volume of core chamber = 15 cc

Volume of air (1st reading) = 6.970

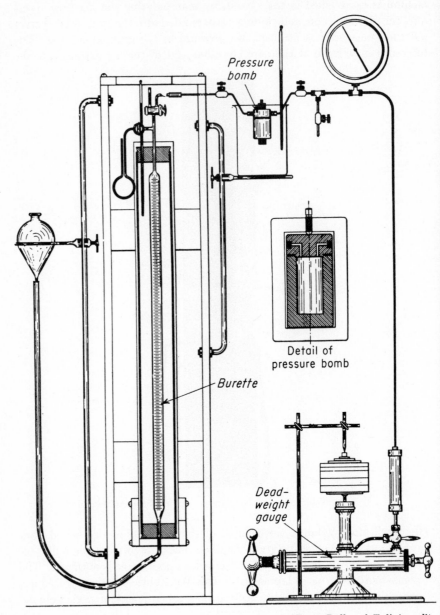

FIG. 2-11. Bureau of Mines gas-expansion porosimeter. (*From Rall and Taliaferro.*[11])

Volume of air (2d reading) = 0.03
Volume of air (3d reading) = 0
B = total of readings = 7.00 cc
Effective grain volume = $A - B$ = 8 cc
Bulk volume of sample (from pycnometer) = 10 cc
Effective porosity = ϕ = $[(10 - 8)/10] \times 100$ = 20 per cent

The Bureau of Mines gas expansion porosimeter (see Fig. 2-11) also measures the effective grain volume and thus yields effective porosities.

Pore Volume.   All the methods of measuring pore volume yield effective

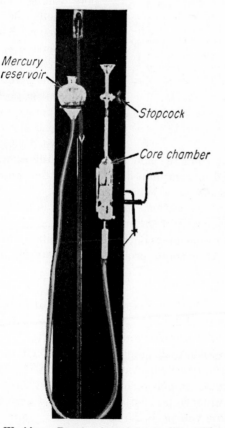

Fig. 2-12. Washburn–Bunting porosimeter. (*From Stevens.*[10])

porosity. The methods are based on either the extraction of a fluid from the rock or the introduction of a fluid into the pore space of the rock.

The Washburn-Bunting porosimeter, shown in Fig. 2-12, measures the volume of air extracted from the pore space by creating a partial vacuum in the porosimeter by the manipulation of the attached mercury reservoir. The core is exposed to contamination by mercury and is therefore not suitable for additional testing. The Stevens method previously illustrated is a modification of the Washburn-Bunting procedure especially designed to prevent contamination of the core.

A number of other devices have been designed for measuring the pore volume, including the Kobe porosimeter and the mercury pump porosimeter. The mercury pump porosimeter is so designed that the bulk volume may be obtained as well as the pore volume.

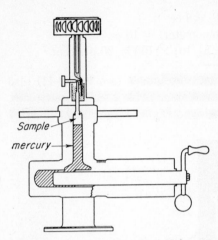

Fig. 2-13. Schematic sketch of Kobe porosimeter. (*From Beeson.*[26])

The saturation method of determining porosity consists of saturating a clean dry sample with a fluid of known density and determining the pore volume from the gain in weight of the sample. The sample is usually evacuated in a vacuum flask to which the saturation fluid may be admitted by means of a separatory funnel. If care is exercised to achieve complete saturation, this procedure is believed to be the best available technique for intergranular materials.

The Kobe porosimeter and the mercury pump are illustrated in Fig. 2-13 and 2-14. An example problem will illustrate the saturation technique.

Example 2-7. Effective Porosity by the Saturation Method. From the data of Example 2-2:

A = weight dry sample in air = 20 gm
D = weight saturated sample in air = 22.5 gm
Density saturating fluid (water) = 1.00 gm/cc
Weight of water in pore space = $D - A$ = 22.5 − 20 = 2.5 gm
Volume of water in pore space = 2.5 gm/(1 gm/cc) = 2.5 cc
Effective pore volume = 2.5 cc
Bulk volume (Example 2-2) = 9.9 cc
Effective porosity = (2.5/9.9) × 100 = 25.3 per cent

A comparison of several methods of determining effective porosity is summarized in Table 2-2.

Fig. 2-14. Mercury porosimeter and capillary-pressure apparatus. (*Courtesy of Core Laboratories, Inc.*)

Precision of Porosity Measurements

To investigate the precision with which porosities were determined, five major company laboratories participated in a porosity check program.[12] Ten selected samples were circulated among the laboratories, and the porosity of each sample determined by routine methods normally used in each laboratory. The methods used were either gas-expansion or saturation techniques. Figure 2-15 and Table 2-3 summarize the results of the check. Note that the gas-expansion method is consistently higher than the saturation method. This is undoubtedly due to the fact that the errors inherent in each tend to be in opposite directions. In the case of gas expansion, errors due to gas adsorption would cause high values to be obtained, while incomplete saturation of the sample would result in low values in the case of the saturation methods. The difference in the average values obtained by the two methods is about 0.8 per cent porosity. The spread between the high and low values ranges from 0.07 to almost 2 per cent porosity. While the differences in the average values are not disturbing, the spread in high and low is of sufficient magnitude to contribute appreciable errors if single observations and a small number of samples are used

TABLE 2-2. METHODS OF DETERMINING POROSITY

	Effective porosity	Effective porosity	Effective porosity	Effective porosity	Effective porosity (best method)	Effective porosity	Effective porosity	Total porosity
Method	Washburn-Bunting	Stevens	Kobe porosimeter	Boyle's law porosimeter	Saturation	Core laboratories Wet sample	Core laboratories Dry sample	Sand density
Type sampling	One to several pieces per increment (usually one)	One to several pieces per increment (usually one)	One to several pieces per increment (usually one)	One to several pieces per increment (usually one)	One to several pieces per increment (usually one)	Several pieces for retort, one for mercury pump	One to several pieces per increment (usually one)	Several pieces per increment Extraction, then in 2d step, crush sample to grian size
Preparation	Solvent extraction and oven drying. Occasionally use retort samples	Solvent extraction and oven drying. Occasionally use retort samples	Solvent extraction and oven drying. Occasionally use retort samples	Solvent extraction and oven drying. Occasionally use retort samples	Solvent extraction and oven drying. Occasionally use retort samples	None	Solvent extraction and oven drying. Occasionally use retort samples	
Functions measured	Pore volume and bulk volume	Sand-grain volume and unconnected pore volume and bulk volume	Sand-grain volume and unconnected pore volume and bulk volume	Sand-grain volume and unconnected pore volume and bulk volume	Pore volume and bulk volume	Volumes of gas space, oil and water, and bulk volume	Sand-grain volume and unconnected pore volume and bulk volume	Bulk volume of sample and volume of sand grains
Manner of measurement	Reduction of pressure on a confined sample and measurement of air evolved. Bulk volume from mercury pycnometer	Difference in volume of air evolved from a constant-volume chamber when empty and when occupied by sample. Bulk volume by Russell tube	Difference in volume of air evolved from a constant-volume chamber when empty and when occupied by sample. Bulk volume by Russell tube	Difference in volume of air evolved from a constant-volume chamber when empty and when occupied by sample. Bulk volume by Russell tube	Weight of dry sample, weight of saturated sample in air, weight of saturated sample immersed in saturant	Weight of retort sample, volume of oil and water from retort sample, gas volume and bulk volume of M.P.S.	Difference in volume of air evolved from a constant-volume chamber when empty and when occupied by sample	Weight of dry sample, weight of saturated sample immersed weight, and volume of sand grains
Errors	Air from dirty mercury, possible leaks in system, incomplete evacuation due to rapid operation or tight sample	Mercury does not become dirty. Possible leaks in system, incomplete evacuation due to rapid operation or tight sample	Mercury does not become dirty. Possible leaks in system, incomplete evacuation due to rapid operation or tight sample	Mercury does not become dirty. Possible leaks in system, incomplete evacuation due to rapid operation or tight sample	Possible incomplete saturation	Obtain excess water from shales. Loss of vapors through condensers	Possible leaks in system, incomplete evacuation due to rapid operation or tight sample	Possible loss of sand grains in crushing. Can be reproduced most accurately

54

TABLE 2-3. CHARACTERISTICS OF SAMPLES USED IN POROSITY-MEASUREMENT COMPARISONS[12]

Type of material	Sample No.	Approximate gas permeability, millidarcys	Porosity, %				
			Average	Average from gas methods	Average from saturation methods	Value from high observation	Value from low observation
Limestone	1	1	17.47	17.81	16.95	18.50	16.72
Fritted glass	2	2	28.40	28.68	27.97	29.30	27.56
Sandstone	3	20	14.00	14.21	13.70	15.15	13.50
Sandstone	4	1,000	30.29	31.06	29.13	31.8	26.8
Semiquartzitic sandstone	BZE	0.2	3.95	4.15	3.66	4.60	3.50
Semiquartzitic sandstone	BZG	0.8	3.94	4.10	3.71	4.55	3.48
Alundum	61-A	1,000	28.47	28.78	28.00	29.4	27.8
Alundum	722	3	16.47	16.73	16.08	17.80	16.00
Chalk	1123	1.6	32.67	33.10	32.03	33.8	31.7
Sandstone	1141-A	45	19.46	19.68	19.12	20.2	18.8

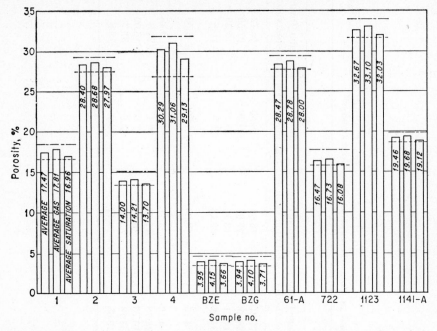

Fɪɢ. 2-15. Results of porosity check. Dot-dash lines represent maximum values; dash lines represent minimum values. Compare with Table 2-3 for more information on the samples. (*After Dotson et al.*[12])

to characterize a reservoir rock. However, it is felt that all the methods commonly used to determine effective porosity yield results with desired degree of accuracy if carefully performed.

Carbonate Rocks

The foregoing discussion is applicable to materials of intergranular-type porosity. Carbonate rocks are more heterogeneous, as was shown in Fig. 2-7. Small samples, such as used in the routine techniques, yield values of porosity which do not include the effect of vugs, solution cavities, etc. The saturation methods of determining pore volume and bulk volume are unsatisfactory, as drainage will occur from the larger pore spaces. The various other techniques also have inherent errors when applied to vugular materials. It is necessary, therefore, to use larger core samples and to determine the bulk volume by measurement of the core dimensions. The effective grain volume is obtained by using a large gas-expansion porosimeter similar to the Bureau of Mines type. Kelton[13] reported results of whole core analysis, a method utilizing large sections of the full diameter core. Figure 2-16 and Table 2-4 summarize a part of Kelton's work.

Matrix porosity is that determined from small samples; total porosity is that determined from the larger whole core. Whole-core analysis satisfactorily evaluates most carbonate rocks. However, no satisfactory technique is available for the analysis of extensively fractured materials.

In coring materials which *in situ* are extensively fractured, the core frequently breaks along the natural fracture planes. Therefore, it is difficult to determine the fraction of pore space contributed to the reservoir by such a fracture system. Recent developments in formation evaluation by production tests indicate that laboratory determinations on highly fractured pays give minimum values of porosity.

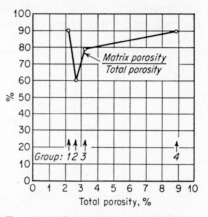

Fig. 2-16. Comparison of large core analysis with conventional analysis. (*After Kelton.*[13])

TABLE 2-4. MATRIX AND WHOLE-CORE DATA FOR ELLENBURGER, FULLERTON FIELD[13]

Group	1	2	3	4
Matrix porosity, % bulk	1.98	1.58	2.56	7.92
Total porosity, % bulk	2.21	2.62	3.17	8.40

Compressibility of Porous Rocks

The porosity of sedimentary rocks has been shown by Krumbein and Sloss[14] to be a function of the degree of compaction of the rock. The compacting forces are a function of the maximum depth of burial of the rock. The effect of natural compaction on porosity is shown in Fig. 2-17. The porosity of shales are greatly reduced by compaction largely because "bridging" is eliminated by the greater forces. The effect illustrated in Fig. 2-17 is principally due to the resulting packing arrangement after compaction. Thus sediments which have been deeply buried, even if subsequently uplifted, exhibit lower porosity values than sediments which have not been buried a great depth.

Apart from the effect of compaction on grain arrangement, rocks are also compressible. Geerstma[15] states that three kinds of compressibility must be distinguished in rocks: (1) rock matrix compressibility, (2) rock bulk compressibility, (3) pore compressibility.

Rock matrix compressibility is the fractional change in volume of the solid rock material (grains) with a unit change in pressure. Rock bulk compressibility is the fractional change in volume of the bulk volume of the rock with a unit change in pressure. Pore compressibility is the frac-

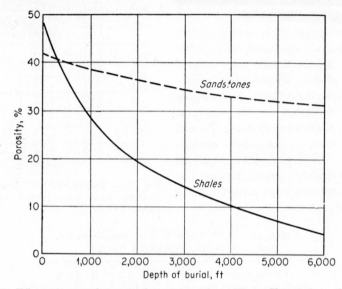

FIG. 2-17. Effect of natural compaction on porosity. (*From Krumbein and Sloss.*[14])

tional change in pore volume of the rock with a unit change in pressure.

Rocks buried at depth are subjected to internal stress exerted by fluids contained in the pores and to external stress which is in part exerted by the overlying rocks. The internal stress is hydrostatic, while the external stresses may have different values in different directions.

The depletion of fluids from the pore space of a reservoir rock results in a change in the internal stress in the rock, thus causing the rock to be subjected to a different resultant stress. This change in stress results in changes in the grain, pore, and bulk volume of the rock. Of principal interest to the reservoir engineer is the change in the pore volume of the rock. The change in bulk volume may be of importance in areas where surface subsidence could cause appreciable property damage.

Geertsma[15] and others have developed a theory of rock compressibility which provides an insight into the mechanics of rock deformation under oil-field conditions.

The theory can best be explained in terms of experimental techniques. First consider the material forming the grains or solid portion of the rock. The solids when subjected to a hydrostatic stress will deform uniformly. The bulk deformation of this material can be expressed as a compressibility

$c_r = (1/V_r)(dV_r/dP)$, where V_r is the volume of solids and P is the hydrostatic pressure. The value of c_r for a particular rock can be determined simply by saturating the rock with a fluid, immersing the saturated rock in a pressure vessel containing the saturating fluid, then imposing a hydrostatic pressure on the fluid and observing the change in volume V_r of the rock sample. The compressibility of the solids is considered for most rocks to be independent of the imposed pressure.

Reservoir rocks are subjected to other conditions of loading than described above. Therefore, it is necessary to introduce other compressibility concepts. A rock buried at depth is subjected to an overburden load due to the overlying sediments. This overburden load may be considered to exert an external hydrostatic stress, which is in general greater than the internal hydrostatic stress of the formation fluids.

In the laboratory it is possible to design an experiment utilizing equipment such as illustrated in Fig. 2-18.[16] A core sample is enclosed in a copper jacket which is then placed in a pressure vessel and connected to a Jerguson sight gauge. The hydraulic-pressure system is arranged so that a saturated core can be subjected to variable internal pressures and overburden or external pressures. The resulting internal volume changes are indicated

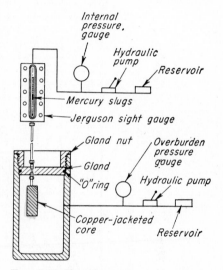

Fig. 2-18. Experimental equipment for measuring pore volume compaction and compressibility. (*From Fatt.*[16])

by the position of the mercury slug in the sight gauge. Carpenter and Spencer[17] used similar equipment; however, in their equipment the internal pressure was maintained at atmospheric pressure.

It is possible with the equipment illustrated to vary either the external or internal pressure or both.

Carpenter and Spencer in testing Woodbine cores with similar equipment varied only the external pressure. Typical curves obtained are shown in Fig. 2-19. The ordinate is the reduction in pore space resulting from a change in overburden load. The change in pore space V_p was determined by measuring the volume of water expelled from the jacketed core on increasing the overburden pressure. V_B is the bulk volume, and ϕ the porosity fraction. Therefore, $V_B\phi = V_p$, the pore volume. The slope of the curves shown is a compressibility of the form $(1/V_p)(\partial V_p/\partial P^*)_P$ where P^*

is the external pressure and P is the internal pressure. It may be noted that the slope of the curves can be considered constant over most of the pressure range above 1,000 psi.

Hall[18] performed tests similar to those of Carpenter and Spencer. The compressibility term $(1/V_p)(\partial V_p/\partial P^*)_P$, he designated as the formation

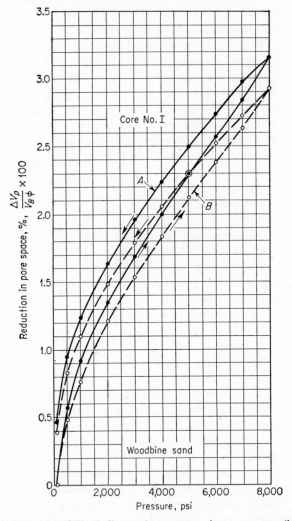

FIG. 2-19. Rock compressibility. Curve A: average of two compressibility tests at 91°F; curve B: compressibility test No. 3 at 146°F. Core data: Magnolia Petroleum Co., John Radford well No. 5, Margaret Tennison Survey, Gregg County, Tex.; top of producing stratum, 3,708 ft; total depth of well, 3,715 ft; depth at which core was cut, 3,711± ft; initial daily production, 12,000 bbl. (*From Carpenter and Spencer.*[17])

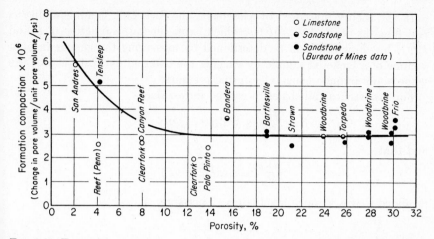

FIG. 2-20. Formation compaction component of total rock compressibility. (*From Hall.*[18])

compaction component of the total rock compressibility. In Fig. 2-20 is presented a correlation of this function with porosity. The correlation includes the data of Carpenter.

In addition, Hall investigated the compressibility $(1/V_p)(\partial V_p/\partial P)_{P*}$ at constant overburden pressure. This he designated as effective rock compressibility and correlated with porosity. The correlation is presented in Fig. 2-21. In both Figs. 2-20 and 2-21, it may be noted that the compressibility decreases as the porosity increases.

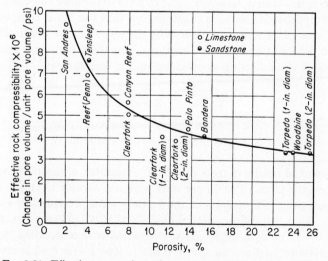

FIG. 2-21. Effective reservoir rock compressibilities. (*From Hall.*[18])

In the theory presented by Geertsma, the rock bulk compressibility c_b is defined as $(1/V_B)(\partial V_B/\partial P^*)_P$, which is, in general, a function of P and P^*. The value of c_b can be determined by measuring the change in bulk volume of a jacketed sample by varying the external hydrostatic pressure while maintaining a constant internal pressure.

For sandstones and shales, it can be shown that

$$\frac{dV_p}{V_p} \approx \frac{c_b}{\phi}(dP^* - dP)$$

and that

$$\frac{dV_B}{V_B} \simeq c_b(dP^* - dP)$$

provided that c_r is much less than c_b. Therefore,

$$\frac{dV_p}{V_p} \approx \frac{1}{\phi}\frac{dV_B}{V_B}$$

since

$$V_p = \phi V_B \qquad dV_p \approx dV_B$$

This states simply that the total change in volume is equal to the change in pore volume. Carpenter made use of this principle in his experimental technique.

Furthermore, if the external hydrostatic pressure P^* on a rock is constant in all directions as in a laboratory test of a jacketed core, $dP^* = 0$, and

$$\frac{dV_p}{V_p} = -\frac{c_b}{\phi}dP \qquad \text{or} \qquad -\frac{1}{V_p}\frac{dV_p}{dP} = \frac{c_b}{\phi}$$

such that c_b/ϕ may be defined as the pore volume compressibility c_p.

Geertsma has stated, however, that in a reservoir only the vertical component of hydrostatic stress is constant and that the stress components in the horizontal plane are characterized by the boundary condition that there is no bulk deformation in those directions. For these boundary conditions, he developed the following approximation for sandstones:

$$-\frac{1}{V_p}\frac{dV_p}{dP} \approx \frac{1}{2}\frac{c_b}{\phi} = \frac{1}{2}c_p$$

Thus, the effective pore compressibility for reservoir rocks on the depletion of internal pressure is only one-half of that determined by present methods in the laboratory.

Fatt reported results of tests on a limited number of samples having porosities ranging from 10 to 15 per cent with one sample having a porosity near 20 per cent. The results of these studies are presented in Figs. 2-22 and 2-23. Fatt,[16] in effect, determined $(1/V_p)(\partial V_p/\partial P)_{P^*}$ for a range of values of P^*. The data reported are correlated with a so-called "net overburden pressure" defined as $(P^* - 0.85P)$. The factor 0.85 is introduced to take into account that the internal pressure does not wholly react against

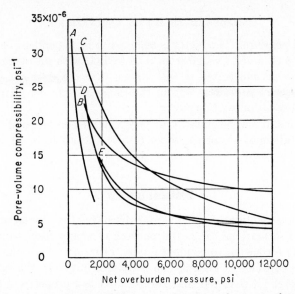

FIG. 2-22. Pore volume compressibility as a function of net overburden pressure ($P^* - 0.85P$). (A) Unconsolidated sand, 28- to 35-mesh flint shot, porosity 36 per cent; (B) basal Tuscaloosa sandstone, Mississippi, porosity 13 per cent; (C) sandstone from wildcat, Santa Rosa County, Fla., porosity 15 per cent; (D) sandstone from Ventura Basin Field, Calif., porosity 10 per cent; (E) sandstone from West Montalvo Area Field, Calif., porosity 12 per cent. (*From Fatt.*[16])

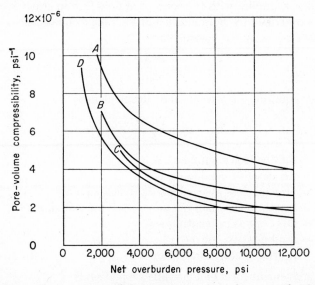

FIG. 2-23. Pore volume compressibility as a function of net overburden pressure ($P^* - 0.85P$). (A) Weber sandstone, Rangely Field, Colo., porosity 12 per cent; (B) sandstone from Nevada wildcat, porosity 13 per cent; (C) Strawn sandstone, Sherman Field, Tex., porosity 13 per cent; (D) Bradford sandstone, Pennsylvania, porosity 15 per cent. (*From Fatt.*[16])

the external pressure. The factor is believed to be dependent on the structure of the rock and to range from 0.75 to 1.00 with an average of 0.85.

Fatt found, as illustrated in the curves, that the pore compressibility was a function of pressure. Within the range of data considered, he did not find a correlation with porosity.

In summary, it can be stated that pore volume compressibilities of consolidated sandstones are of the order of 5×10^{-6} to 10×10^{-6} reciprocal psi.

PERMEABILITY

The previous section of this chapter discussed the storage capacity of underground formations. It is the purpose of this section to discuss the ability of the formation to conduct fluids. From usage the name for fluid conductance capacity of a formation is permeability. In the introduction to API Code 27[19] it is stated that permeability is a property of the porous medium and is a measure of the capacity of the medium to transmit fluids. The measurement of permeability, then, is a measure of the fluid conductivity of the particular material. By analogy with electrical conductors, the permeability represents the reciprocal of the resistance which the porous medium offers to fluid flow.

The reader is familiar with the concept of fluid flow in circular tubes and conduits as described by Poiseuille's and Fanning's equations.

Poiseuille's equation for viscous flow:

$$v = \frac{d^2 \, \Delta P}{32 \mu L} \tag{2-1}$$

Fanning's equation for viscous and turbulent flow:

$$v^2 = \frac{2d \, \Delta P}{f \rho L} \tag{2-2}$$

where v = fluid velocity, cm/sec
  d = diameter of conductor, cm
  ΔP = pressure loss over length L, dynes/sq cm
  L = length over which pressure loss is measured, cm
  μ = fluid viscosity, centipoises
  ρ = fluid density, gm/cc
  f = friction factor, dimensionless

A more convenient form of Poiseuille's equation is

$$Q = \frac{\pi r^4 \, \Delta P}{8 \mu L} \tag{2-3}$$

where r is the radius of the conduit in centimeters, Q is the volume rate of flow in cubic centimeters per second, and the other terms are as previously defined.

If the reservoir rock system is considered to be a bundle of tubes such that the flow could be represented by a summation of the flow from all the tubes, then the total flow would be

$$Q_t = n \frac{\pi r^4 \, \Delta P}{8 \mu L}$$

where n is the number of tubes of radius r. If the rock consists of a group of tubes of different radii, then

$$Q_t = \sum_{j=1}^{k} n_j \frac{\pi r_j^4 \, \Delta P}{8 \mu L}$$

where n_j = number of tubes of radius r_j
k = number of groups of tubes of different radii

The previous equation reduces to

$$Q_t = \frac{\pi \, \Delta P}{8 \mu L} \sum_{j=1}^{k} n_j r_j^4 \tag{2-4}$$

If $\dfrac{\pi}{8} \sum_{j=1}^{k} n_j r_j^4$ is treated as a flow coefficient for the particular grouping of tubes, the equation reduces to

$$Q_t = C \frac{\Delta P}{\mu L} \tag{2-5}$$

where

$$C = \frac{\pi}{8} \sum_{j=1}^{k} n_j r_j^4 \tag{2-6}$$

If the fluid-conducting channels in a porous medium could be defined as to the dimension of the radii and the number of each radii, it might be possible to use Poiseuille's flow equation for porous media. As there are numerous tubes and radii involved in each segment of porous rock, it is an impossible task to measure these quantities on each and every porous-rock sample.

In the attempt to use Poiseuille's flow equation to define flow in a porous rock, it was assumed that a series of tubes of length L comprised the flow network. If these tubes are interconnected and are not individual tubes over the length L, then the derivation would have to account for the interconnection of the flow channels.

A cast of the flow channels in a rock formation is shown in Fig. 2-24. It is seen that the flow channels are of varying sizes and shapes and are randomly connected. It is impossible to define the exact dimension of all the flow channels and their flow relationship to one another. It becomes apparent that some means other than Poiseuille's law had to be found to define the flow coefficient of a rock.

In the preceding section on porosity, it was shown that the porosity was independent of sand-grain size but dependent on the mode of packing. In the same section, it was shown that the size of the openings between the sand grains decreased as the sand-grain size decreased. Therefore, perhaps

Fig. 2-24. Metallic cast of pore spaces in a consolidated sand. (*Courtesy of Humble Oil & Refining Co.*)

it is possible to derive a flow equation in terms of the mean diameter of the sand grain. Fancher, Lewis, and Barnes[20] conducted experimental work on porous systems to determine the relationship between sand-grain size and fluid conductance of porous media. The resulting data were correlated using Fanning's flow equation so as to account for both turbulent and viscous flow. The results of their study are shown in Fig. 2-25.

For unconsolidated sands it was found that an expression of the friction factor f could be obtained in terms of Reynolds number. But for consolidated sandstones it was found that a different relationship existed between the friction factor and Reynolds number for each sample investigated. If a single relationship could have been obtained for consolidated sandstones as was obtained for unconsolidated sandstones, then it would have been necessary to classify rocks only as to average grain diameter and whether consolidated or unconsolidated. As this is not possible, it again becomes evident that another method of expressing fluid conductance of rocks must be used.

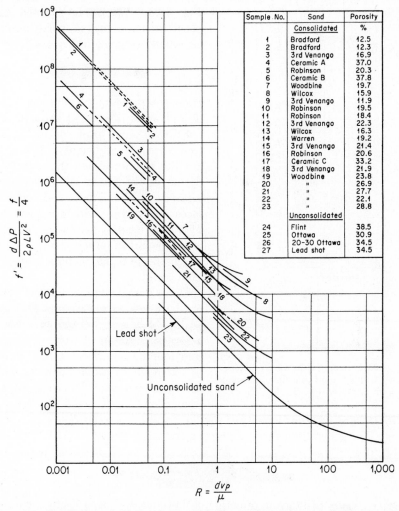

Sample No.	Sand	Porosity
	Consolidated	%
1	Bradford	12.5
2	Bradford	12.3
3	3rd Venango	16.9
4	Ceramic A	37.0
5	Robinson	20.3
6	Ceramic B	37.8
7	Woodbine	19.7
8	Wilcox	15.9
9	3rd Venango	11.9
10	Robinson	19.5
11	Robinson	18.4
12	3rd Venango	22.3
13	Wilcox	16.3
14	Warren	19.2
15	3rd Venango	21.4
16	Robinson	20.6
17	Ceramic C	33.2
18	3rd Venango	21.9
19	Woodbine	23.8
20	"	26.9
21	"	27.7
22	"	22.1
23	"	28.8
	Unconsolidated	
24	Flint	38.5
25	Ottawa	30.9
26	20-30 Ottawa	34.5
27	Lead shot	34.5

$$f' = \frac{d\,\Delta P}{2\rho L V^2} = \frac{f}{4}$$

$$R = \frac{d v \rho}{\mu}$$

FIG. 2-25. Correlation of friction factor with Reynolds number for flow of homogeneous fluids through porous media, where d is defined as the diameter of the average grain and V is the apparent velocity, i.e., *volume rate of flow/total cross-sectional area*. (*After Fancher, Lewis, and Barnes.*[20])

The preceding attempts to determine a means of calculating the conductance of a rock were made to augment or supplant the empirical relationship of permeability as developed by Darcy.[21] The pore structure of rocks does not permit simple classification, and therefore empirical data are required in most cases.

In 1856, Darcy[21] investigated the flow of water through sand filters for water purification. His experimental apparatus is shown schematically in Fig. 2-26.[22] Darcy interpreted his observations so as to yield results essentially as given in Eq. (2-7).

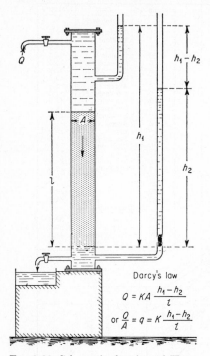

$$Q = KA \frac{h_1 - h_2}{L} \qquad (2\text{-}7)$$

Here, Q represents the volume rate of flow of water downward through the cylindrical sand pack of cross-sectional area A and height l. h_1 and h_2 are the heights above the standard datum of the water in manometers located at the input and output faces respectively and represent the hydraulic head at points 1 and 2. K is a constant of proportionality and was found to be characteristic of the sand pack.

Darcy's investigations were confined to flow of water through sand packs which were 100 per cent saturated with water. Later investigators found that Darcy's law could be extended to other fluids as well as water and that the constant of proportionality K could be written as k/μ where μ is the viscosity of the fluid and k is a property of the rock alone. The generalized form of Darcy's law as presented in API Code 27 is Eq. (2-8).

Darcy's law

$$Q = KA \frac{h_1 - h_2}{l}$$

$$\text{or } \frac{Q}{A} = q = K \frac{h_1 - h_2}{l}$$

FIG. 2-26. Schematic drawing of Henry Darcy's experiment on flow of water through sand. (*From Hubbert.*[22])

$$v_s = -\frac{k}{\mu}\left(\frac{dP}{ds} \underset{\text{minus}}{-} \frac{\rho g}{1.0133}\frac{dz}{ds} \times 10^{-6}\right) \qquad (2\text{-}8)$$

Here, s = distance in direction of flow and is always positive, cm
   v_s = volume flux across a unit area of the porous medium in unit time along flow path s, cm/sec
   z = vertical coordinate, considered positive downward, cm
   ρ = density of the fluid, gm/cc
   g = acceleration of gravity, 980.665 cm/sec²
   dP/ds = pressure gradient along s at the point to which v_s refers, atm/cm

μ = viscosity of the fluid, centipoises

k = permeability of the medium, darcys

$$1.0133 \times 10^6 = \text{dynes/(sq cm)(atm)}$$

dz/ds can be expressed as $\sin \theta$ where θ is the angle between s and the horizontal. v_s can further be defined as Q/A where Q is the volume rate of flow and A is the average cross-sectional area perpendicular to the lines of flow. The coordinate system applicable to Eq. (2-8) is shown in Fig. 2-27. The convention of sign is that v_s should be positive when the fluid is flowing toward increasing values of the coordinate s. The quantity of Eq. (2-8) in parentheses can be interpreted as the total pressure gradient minus the gradient due to a head of fluid. Thus if the system is in hydrostatic equilibrium, there is no flow and the quantity inside the parentheses will

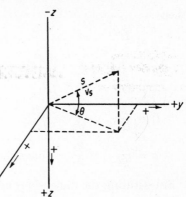

FIG. 2-27. Coordinate system to which generalized Darcy's law is referred.

be zero. Equation (2-8) can be written as follows:

$$v_s = \frac{k}{\mu} \frac{d}{ds} \left(\frac{\rho g z \times 10^{-6}}{1.0133} - P \right) \tag{2-9}$$

The quantity $(d/ds)[(\rho g z \times 10^{-6}/1{,}0133) - P]$ can be considered to be the negative gradient of a function ϕ, where

$$\Phi = P - \frac{\rho g z \times 10^{-6}}{1.0133} \tag{2-10}$$

Φ is a potential function such that flow will occur from higher values of Φ toward lower values of Φ. M. King Hubbert[22] defines a potential function

$$\Phi^1 = g z + \frac{P}{\rho} \tag{2-11}$$

which is equivalent to the above except the positive direction of z is taken upward. Muskat[23] defines a velocity potential function

$$\Phi^{11} = \frac{k}{\mu} (P \pm \rho g z) \tag{2-12}$$

where the plus sign corresponds to the upward direction of the positive z coordinate and the minus sign to the downward direction of the positive z coordinate. The concept of a flow potential is shown to be useful in later applications.

The dimensions of permeability can be established by substituting the units of the other terms into Eq. (2-8).

Let L = length
M = mass
T = time

then

$$v_s = \frac{L}{T}$$

$$\mu = \frac{M}{LT} \qquad \rho = \frac{M}{L^3}$$

$$P = \frac{M}{LT^2} \qquad \frac{dP}{ds} = \frac{M}{L^2T^2}$$

$$g = \frac{L}{T^2} \qquad \frac{dz}{ds} = \text{dimensionless}$$

Substituting the dimensions in Eq. (2-8)

$$\frac{L}{T} = \frac{k}{M/LT} \left(\frac{M}{L^2T^2} - \frac{M}{L^3} \frac{L}{T^2} \right)$$

$$= \frac{kLT}{M} \left(\frac{M}{L^2T^2} - \frac{M}{L^2T^2} \right)$$

$$= \frac{k}{LT}$$

$$k = L^2$$

A rational unit of permeability in the English system of units would be the foot squared and in the cgs system, the centimeter squared. Both were found to be too large a measure to use with porous media. Therefore, the petroleum industry adopted as the unit of permeability, the darcy, which is defined as follows:

A porous medium has a permeability of one darcy when a single-phase fluid of one centipoise viscosity that completely fills the voids of the medium will flow through it under conditions of viscous flow at a rate of one cubic centimeter per second per square centimeter cross-sectional area under a pressure or equivalent hydraulic gradient of one atmosphere per centimeter.

Conditions of viscous flow mean that the rate of flow will be sufficiently low to be directly proportional to the pressure or hydraulic gradient.

Darcy's law holds only for conditions of viscous flow as defined above. Furthermore, for the permeability k to be a property of the porous medium alone, the medium must be 100 per cent saturated with the flowing fluid when the determination of permeability is made. In addition, the fluid and the porous medium must not react; that is, if a reactive fluid flows through a porous medium, it alters the porous medium, therefore changing the permeability of the medium as flow continues.

Equation (2-8) is a useful generalization of Darcy's law. However, several simple flow systems are so frequently encountered in the measurement and application of permeability that it is appropriate to obtain the integrated form for these systems.

Horizontal Flow

Horizontal rectilinear steady-state flow is common to virtually all measurements of permeability. Consider a block of a porous medium as in Fig. 2-28. Here Q, the volume rate of flow, is uniformly distributed over the inflow face of area A. If the block is 100 per

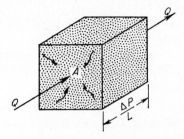

FIG. 2-28. Sand model for rectilinear flow of fluids.

cent saturated with an incompressible fluid and is horizontal, then $dz/ds = 0$, $dP/ds = dP/dx$, and Eq. (2-8) reduces to

$$v_x = -\frac{k}{\mu}\frac{dP}{dx}$$

$$= \frac{Q}{A} = -\frac{k}{\mu}\frac{dP}{dx} \tag{2-13}$$

separating variables,

$$\frac{Q}{A}\,dx = -\frac{k}{\mu}\,dP$$

integrating between the limits 0 and L in x and P_1 and P_2, where P_1 is the pressure at the inflow face and P_2 the pressure at the outflow face,

$$\frac{Q}{A}\int_0^L dx = -\frac{k}{\mu}\int_{P_1}^{P_2} dP$$

$$\frac{Q}{A}(L-0) = -\frac{k}{\mu}(P_2 - P_1) = \frac{k}{\mu}(P_1 - P_2)$$

or

$$Q = \frac{kA(P_1 - P_2)}{\mu L} \tag{2-14}$$

If kA is permitted to equal the flow coefficient C defined with Poiseuille's equation (2-5), it is seen that the two expressions are identical such that

$$kA = \frac{\pi}{8}\sum_{j=1}^{k} n_j r_j^4$$

If a compressible fluid flows through a porous medium, Darcy's law, as expressed in Eq. (2-8), is still valid. However, for steady flow, the mass rate of flow is constant through the system rather than the volume rate of flow. Therefore, the integrated form of the equations differs. Considering rectilinear flow and steady flow of compressible fluids, Eq. (2-8) becomes

$$\rho v_x = -\frac{k\rho}{\mu}\frac{dP}{dx} \tag{2-15}$$

where both the right-hand and left-hand members of Eq. (2-8) are multiplied by the density.

For steady flow ρv_x is a constant.

For a slightly compressible liquid the equation of state can be expressed as

$$\rho = \rho_0 e^{cP} \tag{2-16}$$

differentiating with respect to x

$$\frac{d\rho}{dx} = c\rho_0 e^{cP}\frac{dP}{dx} \qquad \text{since } \rho = \rho_0 e^{cP}$$

$$= c\rho\frac{dP}{dx} \qquad \frac{1}{c\rho}\frac{d\rho}{dx} = \frac{dP}{dx}$$

by definition

$$v_x = \frac{Q}{A}$$

substituting the above quantities in Eq. (2-15),

$$\rho\frac{Q}{A} = -\frac{k\rho}{\mu}\left(\frac{1}{c\rho}\frac{d\rho}{dx}\right)$$

by definition $\rho Q = m = $ mass rate of flow; separating variables,

$$\frac{m}{A}dx = -\frac{k}{c\mu}d\rho \tag{2-17}$$

integrating,

$$\frac{m}{A}\int_0^L dx = -\frac{k}{c\mu}\int_{\rho_1}^{\rho_2}d\rho$$

$$\frac{m}{A}(L-0) = -\frac{k}{c\mu}(\rho_2 - \rho_1)$$

$$m = \frac{kA}{c\mu}\frac{\rho_1 - \rho_2}{L}$$

if

$$\rho_1 = \rho_0 + \rho_0 cP_1$$

$$\rho_2 = \rho_0 + \rho_0 cP_2$$

then

$$\rho_1 - \rho_2 = \rho_0 c(P_1 - P_2)$$

then

$$\rho_0 Q_0 = m = \frac{kA}{c\mu}\frac{\rho_0 c(P_1 - P_2)}{L}$$

therefore
$$Q_0 = \frac{kA}{\mu}\frac{P_1 - P_2}{L} \tag{2-18}$$

For isothermal flow of ideal gases Eq. (2-15) again applies.

$$\rho v_x = -\frac{k}{\mu}\rho\frac{dP}{dx} \tag{2-19}$$

since $v_x = Q/A$,

$$\rho\frac{Q}{A} = -\frac{k}{\mu}\rho\frac{dP}{dx}$$

but $\rho Q = \rho_b Q_b$ = constant where Q and Q_b are defined at flowing temperature, and $\rho = \rho_b(P/P_b)$.

Therefore
$$\rho_b\frac{Q_b}{A} = -\frac{k}{\mu}\rho_b\frac{P}{P_b}\frac{dP}{dx}$$

$$\frac{P_bQ_b}{A} = -\frac{k}{\mu}P\frac{dP}{dx}$$

separating variables and integrating,

$$\frac{P_bQ_b}{A}L = \frac{k}{\mu}\frac{(P_1^2 - P_2^2)}{2}$$

$$Q_b = \frac{kA}{2\mu L}\frac{(P_1^2 - P_2^2)}{P_b} \tag{2-20}$$

Define $\overline{P}$ as $(P_1 + P_2)/2$ and $\overline{Q}$ as the volume rate of flow at $\overline{P}$. Then $\overline{P}\overline{Q} = P_bQ_b$. Substituting in (2-20) above

$$P_bQ_b = \overline{P}\overline{Q} = \frac{kA}{2\mu L}(P_1^2 - P_2^2)$$

or
$$\frac{P_1 + P_2}{2}\overline{Q} = \frac{kA}{\mu L}(P_1 - P_2)\frac{(P_1 + P_2)}{2}$$

$$\overline{Q} = kA\frac{(P_1 - P_2)}{\mu L} \tag{2-21}$$

which is the same form as (2-14). Therefore flow rates of ideal gases can be computed from the equations for incompressible liquids as long as the volume rate of flow is defined at the algebraic mean pressure.

Vertical Flow

Figures 2-29 to 2-31 illustrate three vertical flow systems frequently encountered in practice. Each system is of uniform cross-sectional area A. (In the developments which follow the fluids are considered incompressible.) First consider the case when the pressures at the inlet and outlet are equal (free flow) such that only the gravitational forces are driving the fluids (Fig. 2-29).

$$s = z \quad \text{and} \quad \frac{dz}{ds} = 1$$

For these conditions

$$\frac{dP}{ds} = 0 \text{ by definition of flowing conditions}$$

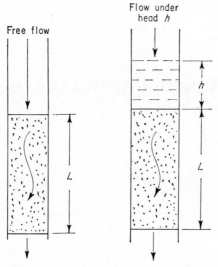

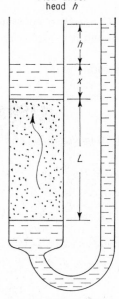

FIG. 2-29. Vertical free flow downward.

FIG. 2-30. Vertical flow downward with head.

FIG. 2-31. Vertical flow upward with head.

Therefore Eq. (2-8) reduces to

$$v_s = \frac{k}{\mu} \rho g = \frac{Q}{A}$$

$$Q = \frac{kA}{\mu} \rho g \tag{2-22}$$

Next consider the case of downward flow when the driving head (difference in hydraulic head of inlet and outlet) is h (Fig. 2-30). Then

$$\frac{dz}{ds} = 1 \quad \frac{dP}{ds} = \frac{dp}{dz} - \frac{\rho g h}{L}$$

Therefore from Eq. (2-8)

$$v = \frac{k}{\mu}\left(\frac{hg}{L} + \rho g\right) = \frac{Q}{A}$$

$$Q = \frac{kA}{\mu}\rho g\left(\frac{h}{L} + 1\right) \tag{2-23}$$

When the flow is upward and the driving head is h (Fig. 2-31) (z defined as positive downward),

$$\frac{dz}{ds} = -1 \qquad \frac{dP}{ds} = -\frac{dP}{dz} = -\frac{\rho gh}{L} - \rho g$$

$$v = +\frac{k}{\mu}\left(+\frac{\rho gh}{L} + \rho g - \rho g\right)$$

$$= \frac{k\rho gh}{\mu L}$$

$$\frac{Q}{A} = v_s$$

Therefore
$$Q = \frac{kA\rho gh}{\mu L} \tag{2-24}$$

Example 2-8. Linear Vertical Flow. In a city water-filtration plant, it was desired to filter 5,000 gal of water per hour through a sand filter bed to remove all the suspended matter and solids from the water. A vertical cross-sectional view of the filtration unit is shown in Fig. 2-32.

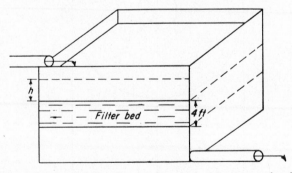

FIG. 2-32. Vertical flow through filter bed with constant head.

Data:

Quantity of water to be filtered = 5,000 gal/hr
Thickness of sand filter bed = 4 ft
Permeability of sand filter bed = 1,200 millidarcys
Cross-sectional area of pit (also sand filter bed) = 1,800 sq ft
Viscosity of water at 20°C = 1.0 centipoise
Barometric pressure = 14.7 psi, 760 mm Hg, 29.92 in. Hg

What level of water must be kept in the pit above the sand filter bed in order to filter 5,000 gal of water per hour? Assume that the solids removed from the water do not alter the permeability of the bed.

$$Q = \frac{kA\rho g}{1.0133 \times 10^6 \mu}\left(\frac{h}{L} + 1\right)$$

$$\frac{5,000\ (3,785)}{3,600} = \frac{1.2\ (1,800)(30.48)^2(1)(980)}{1\ (1.0133 \times 10^6)}\left(\frac{h}{L} + 1\right)$$

$$\frac{h}{L} + 1 = 2.72$$

$$h = 1.72\ (4) = 6.88$$

Radial Flow

A radial-flow system, analogous to flow into a well bore from a cylindrical drainage region, is idealized in Fig. 2-33.

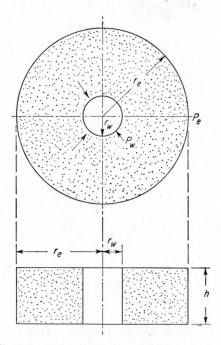

Fig. 2-33. Sand model for radial flow of fluids to central well bore.

Considering the cylinder horizontal and radial flow inward as equal to r, it is possible to integrate Eq. (2-8) and obtain an equation for steady-state radial flow of an incompressible fluid.

By definition

$$ds = -dr \qquad \frac{dz}{ds} = 0$$

Therefore, from Eq. (2-8)

$$v_s = +\frac{k}{\mu}\frac{dP}{dr} \tag{2-25}$$

$$\frac{Q}{A} = \frac{Q}{2\pi rh} = +v_s$$

Therefore

$$\frac{Q}{2\pi rh} = \frac{k}{\mu}\frac{dP}{dr}$$

Separating variables and integrating,

$$\frac{Q}{2\pi h}\int_{r_w}^{r_e}\frac{dr}{r} = \frac{k}{\mu}\int_{P_w}^{P_e}dP$$

$$Q = \frac{2\pi kh(P_e - P_w)}{\mu \ln (r_e/r_w)} \tag{2-26}$$

where Q = volume rate of flow, cc/sec
k = permeability, darcys
h = thickness, cm
μ = viscosity, centipoises
P_e = pressure at external boundary, atm
P_w = pressure at internal boundary, atm
r_e = radius to external boundary, cm
r_w = radius to internal boundary, cm
$\ln$ = natural logarithm, base e

Equation (2-26) can be appropriately modified for the flow of compressible fluids. The details of modifying the foregoing equation are omitted, as they are essentially the same as the ones used in the horizontal, rectilinear-flow systems.

After modification for variations in flowing volumes with changing pressures Eq. (2-26) becomes

1. Slightly compressible fluids:

$$m = \frac{2\pi kh(\rho_e - \rho_w)}{c\mu \ln (r_e/r_w)} \tag{2-27}$$

or

$$Q_\theta = \frac{2\pi kh(P_e - P_w)}{\mu \ln (r_e/r_w)}$$

where Q_θ is defined at the pressure P_θ where the density is ρ_θ.

2. Ideal gases:

$$Q_b = \frac{\pi kh(P_e{}^2 - P_w{}^2)}{\mu P_b \ln (r_e/r_w)} \tag{2-28}$$

or
$$\overline{Q} = \frac{2\pi kh(P_e - P_w)}{\mu \ln (r_e/r_w)} \tag{2-29}$$

where $\overline{Q}$ is the volume rate of volume at the algebraic mean pressure

$$\frac{P_e + P_w}{2}$$

Conversion of Units in Darcy's Law. In Darcy's law and the special flow equations developed therefrom, the units were presumed to be either a consistent set or those obtained from the definition of the darcy, the accepted unit for the petroleum industry.

It is convenient in many applications of Darcy's law to introduce commonly used oil-field units. The following is a summary of the more common equations with the conversion factors to convert to oilfield terminology.

Linear Flow: Liquids (or Gases with Volume at Mean Pressure).

Rate in barrels per day:

$$Q = 1.1271 \frac{kA(P_1 - P_2)}{\mu L} \tag{2-30}$$

Rate in cubic feet per day:

$$Q = 6.3230 \frac{kA(P_1 - P_2)}{\mu L} \tag{2-31}$$

where Q is the volume rate of flow, P_1 and P_2 are in pounds per square inch, k is in darcys, μ is in centipoise, A is in square feet, and L is in feet.

Gases at Base Pressure P_b and Average Flowing Temperature T_f.

Linear:

$$Q_b = \frac{3.1615kA(P_1{}^2 - P_2{}^2)}{\mu P_b L} \tag{2-32}$$

Radial:

$$Q_b = \frac{19.88kh(P_e{}^2 - P_w{}^2)}{\mu P_b \ln (r_e/r_w)} \tag{2-33}$$

where Q_b is in cubic feet per day at pressure P_b and flowing temperature T_f, P_b is in psia, μ is in centipoises, L is in feet, k is in darcys, r_e and r_w are in consistent units, A is in square feet, h is in feet, and P_1, P_2, P_e, and P_w are in psia.

Radial Flow: Liquids (or Gases with Volume at Mean Pressure).

Rate in barrels per day:

$$Q = 7.082 \frac{kh(P_e - P_w)}{\mu \ln (r_e/r_w)} \tag{2-34}$$

Rate in cubic feet per day:

$$Q = 39.76 \frac{kh(P_e - P_w)}{\mu \ln (r_e/r_w)} \tag{2-35}$$

where Q is the volume rate of flow, P_e and P_w are in pounds per square inch, k is in darcys, μ is in centipoises, h is in feet, and r_e and r_w are in consistent units.

The above equations describe the flow in the porous medium when the rock is 100 per cent saturated with the flowing fluid. Appropriate modifications will be discussed in later sections to take into account presence of other fluids. Since the above equations describe the flow in the medium, appropriate volume factors must be introduced to account for changes in the fluids due to any decrease in pressure and temperature from that of the medium to standard or stock tank conditions.

Example 2-9 lists various unit conversions to change from the unit of the darcy to other systems of units.

———•◆•———

Example 2-9. Permeability Conversion Factors.

1 darcy = 1,000 millidarcys; 1 millidarcy = 0.001 darcy

$$k = \frac{Q\mu}{(A)(\Delta P)/L}$$

$$1 \text{ darcy} = \frac{(\text{cc/sec})(\text{cp})}{(\text{sq cm})(\text{atm})/\text{cm}}$$

$$= 9.869 \times 10^{-7} \frac{(\text{cc/sec})(\text{cp})}{\text{sq cm}[\text{dyne}/(\text{sq cm})(\text{cm})]}$$

$$= 9.869 \times 10^{-9} \text{ sq cm}$$

$$= 1.062 \times 10^{-11} \text{ sq ft}$$

$$= 7.324 \times 10^{-5} \frac{[\text{cu ft}/(\text{sec})](\text{cp})}{(\text{sq ft})(\text{psi})/\text{ft}} \qquad FTCP$$

$$= 9.679 \times 10^{-4} \frac{[\text{cu ft}/(\text{sec})](\text{cp})}{(\text{sq cm})(\text{cm water})/\text{cm}}$$

$$= 1.127 \frac{[\text{bbl}/(\text{day})](\text{cp})}{(\text{sq ft})(\text{psi})/\text{ft}}$$

$$= 1.424 \times 10^{-2} \frac{[\text{gal}/(\text{min})](\text{cp})}{(\text{sq ft})(\text{ft water})/\text{ft}}$$

———•◆•———

Permeability of Combination Layers

The foregoing flow equations were all derived on the basis of one continuous value of permeability between the inflow and outflow face. It is seldom that rocks are so uniform. Most porous rocks will have space variations of permeability. If the rock system is comprised of distinct layers, blocks, or concentric rings of fixed permeability, the average permeability of the flow system can be determined by one of the several averaging procedures.

Consider the case where the flow system is comprised of layers of porous rock separated from one another by infinitely thin impermeable barriers as shown in Fig. 2-34. The average permeability $\bar{k}$ can be computed as follows:

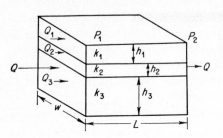

FIG. 2-34. Linear flow, parallel combination of beds.

$$Q_t = Q_1 + Q_2 + Q_3$$

$$h_t = h_1 + h_2 + h_3 = \sum_{j=1}^{3} h_j$$

$$Q_t = \frac{\bar{k}wh_t(P_1 - P_2)}{\mu L}$$

where $wh_t = A$.

$$Q_1 = \frac{k_1wh_1(P_1 - P_2)}{\mu L}$$

$$Q_2 = \frac{k_2wh_2(P_1 - P_2)}{\mu L}$$

$$Q_3 = \frac{k_3wh_3(P_1 - P_2)}{\mu L}$$

Therefore

$$\frac{\bar{k}wh_t(P_1 - P_2)}{\mu L} = \frac{k_1wh_1(P_1 - P_2)}{\mu L} + \frac{k_2wh_2(P_1 - P_2)}{\mu L} + \frac{k_3wh_3(P_1 - P_2)}{\mu L}$$

and

$$\bar{k}h_t \frac{w(P_1 - P_2)}{\mu L} = \frac{w(P_1 - P_2)}{\mu L}(k_1h_1 + k_2h_2 + k_3h_3)$$

or

$$\bar{k}\sum_{j=1}^{3} h_j = \sum_{j=1}^{3} k_jh_j$$

then generalizing,

$$\bar{k} = \frac{\sum_{j=1}^{n} k_jh_j}{\sum_{j=1}^{n} h_j} \qquad (2\text{-}36)$$

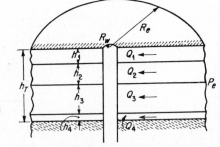

FIG. 2-35. Radial flow, parallel combination of beds.

From Fig. 2-35 it is noted that the same terms appear in the radial-flow network as in the linear system. The only difference in the two systems is the manner of expressing the length over which the pressure drop occurs.

As all these terms are the same in each of the parallel layers, an evaluation of the parallel radial system would yield the same solution as obtained in the linear case.

Example 2-10. Average Permeability of Beds in Parallel. What is the equivalent linear permeability of four parallel beds having equal widths and lengths under the following conditions?

Bed	Pay thickness, ft	Horizontal permeability, millidarcys
1	20	100
2	15	200
3	10	300
4	5	400

$$\bar{k} = \frac{\sum\limits_{j=1}^{n} k_j h_j}{\sum\limits_{j=1}^{n} h_j}$$

so

$$\bar{k} = \frac{100 \times 20 + 200 \times 15 + 300 \times 10 + 400 \times 5}{20 + 15 + 10 + 5} = \frac{10,000}{50}$$

$$= 200 \text{ millidarcys}$$

Another possible combination for flow systems is to have the values of different permeability arranged in series as shown in Fig. 2-36. In case] of linear flow the average series permeability for the total volume can be evaluated as follows:

$$Q_t = Q_1 = Q_2 = Q_3$$

$$P_1 - P_2 = \Delta P_1 + \Delta P_2 + \Delta P_3$$

$$L = L_1 + L_2 + L_3 = \sum_{j=1}^{3} L_j$$

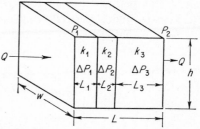

FIG. 2-36. Linear flow, series combination of beds.

$$Q_t = \frac{\bar{k}wh(P_1 - P_2)}{\mu L} \qquad Q_1 = \frac{k_1 wh\, \Delta P_1}{\mu L_1}$$

$$Q_2 = \frac{k_2 wh\, \Delta P_2}{\mu L_2} \qquad Q_3 = \frac{k_3 wh\, \Delta P_3}{\mu L_3}$$

Solving for pressure and substituting for ΔP in the equations above,

$$\frac{Q_t \mu L}{\bar{k} w h} = \frac{Q_1 \mu L_1}{k_1 w h} + \frac{Q_2 \mu L_2}{k_2 w h} + \frac{Q_3 \mu L_3}{k_3 w h}$$

or

$$\frac{L}{\bar{k}} \frac{Q_t \mu}{w h} = \frac{Q_t \mu}{w h} \left(\frac{L_1}{k_1} + \frac{L_2}{k_2} + \frac{L_3}{k_3} \right)$$

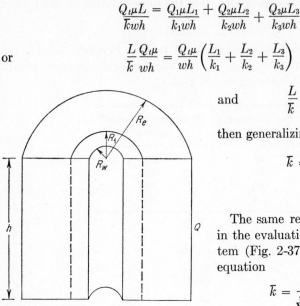

and

$$\frac{L}{\bar{k}} = \sum_{j=1}^{3} \frac{L_j}{k_j}$$

then generalizing,

$$\bar{k} = \frac{L}{\displaystyle\sum_{j=1}^{n} \frac{L_j}{k_j}} \qquad (2\text{-}37)$$

The same reasoning can be used in the evaluation of the radial system (Fig. 2-37) so as to yield the equation

$$\bar{k} = \frac{\log r_e/r_w}{\displaystyle\sum_{j=1}^{n} \frac{\log r_j/r_{j-1}}{k_j}} \qquad (2\text{-}38)$$

Fig. 2-37. Radial flow, series combination of beds.

———•◆•———

Example 2-11. Average Permeability of Beds in Series. What is the equivalent permeability of four beds in series, having equal formation thicknesses under the following conditions: (1) for a linear system and (2) for a radial system if the radius of the penetrating well bore is 6 in. and the radius of effective drainage is 2,000 ft?

Bed	Length of bed, ft	Horizontal permeability, millidarcys
1	250	25
2	250	50
3	500	100
4	1000	200

Assume bed 1 adjacent to the well bore.

Linear system

$$\bar{k} = \frac{L}{\displaystyle\sum_{j=1}^{n} \frac{L_j}{k_j}}$$

$$= \frac{250 + 250 + 500 + 1{,}000}{\dfrac{250}{25} + \dfrac{250}{50} + \dfrac{500}{100} + \dfrac{1{,}000}{200}} = \frac{2{,}000}{10 + 5 + 5 + 5}$$

$$= \frac{2{,}000}{25} = 80 \text{ millidarcys} \qquad (2\text{-}37)$$

Radial system

$$\bar{k} = \frac{\log r_e/r_w}{\displaystyle\sum_{j=1}^{n} \frac{\log r_j/r_{j-1}}{k_j}}$$

$$= \frac{\log 2{,}000/0.5}{\dfrac{\log 250/0.5}{25} + \dfrac{\log 500/250}{50} + \dfrac{\log 1{,}000/500}{100} + \dfrac{\log 2{,}000/1{,}000}{200}}$$
$$= 30.4 \text{ millidarcys} \qquad (2\text{-}38)$$

Channels and Fractures in Parallel

Only the matrix permeability has been discussed in the analysis to this point. In some sand and carbonate reservoirs the formation frequently contains solution channels and natural or artificial fractures. These channels and fractures do not change the permeability of the matrix but do change the effective permeability of the flow network. In order to determine the contribution made by a fracture or channel to the total conductivity of the system, it is necessary to express their conductivity in terms of the darcy.

Channels. Recalling Poiseuille's equation for fluid conductivity of capillary tubes,

$$Q = \frac{\pi r^4 \, \Delta P}{8\mu L} \qquad (2\text{-}3)$$

The total area available to flow is

$$A = \pi r^2$$

So that the equation reduces to

$$Q = A \frac{r^2}{8} \frac{\Delta P}{L\mu}$$

From Darcy's law it is also known that

$$Q = Ak \frac{\Delta P}{L\mu}$$

Equating Darcy's and Poiseuille's equations for fluid flow in a tube,

$$k = \frac{r^2}{8} \tag{2-39}$$

where k and r are in consistent units.

If r is in centimeters, then k in darcys is given by

$$k = \frac{r^2}{8(9.869)(10^{-9})} = 12.50 \times 10^6 r^2$$

where 9.869×10^{-9} is a conversion factor from Example 2-9.

Then if r is in inches,

$$k = 12.50 \times 10^6 (2.54)^2 r^2$$
$$= 80 \times 10^6 r^2 = 20 \times 10^6 d^2$$

where d is the diameter of the opening in inches.

Therefore, the permeability of a circular opening 0.005 in. in radius is 2,000 darcys or 2,000,000 millidarcys.

Consider a cube of reservoir rock 1 ft on the side and having a matrix permeability of 10 millidarcys. If a liquid of 1-centipoise viscosity flows linearly through the rock, under a pressure of gradient 1 psi per ft, the rate of flow will be 0.011271 bbl per day. If a circular opening 0.01 in. in diameter traverses the same rock, then the rate of flow can be considered to be the above value plus the rate of flow Q' through the circular opening.

Then

$$Q' = 1.1271 kA \frac{P}{\mu L}$$

$$= 1.1271(2,000) \frac{\pi d^2}{4(144)} \frac{1}{1(1)}$$

$$= \frac{0.785(2254.2)(0.0001)}{144}$$

$$= \frac{0.785(0.22542)}{144} = 0.00122 \text{ bbl/day}$$

Therefore the combined rate is 0.012491, or an increase of about 11 per cent. If the cube matrix has 1-millidarcy permeability, the increase would be 108 per cent.

Fractures. For flow through slots of fine clearances and unit width Buckingham (see Croft[24]) reports that

$$\Delta P = \frac{12 \mu v L}{h^2} \tag{2-40}$$

where h is the thickness of the slot. By analogy to Darcy's law where

$$\Delta P = \frac{v\mu L}{k}$$

then

$$k = \frac{h^2}{12} \qquad (2\text{-}41)$$

where h is in centimeters and k in darcys. The permeability of the slot is given by

$$k = \frac{h^2}{12(9.869)(10^{-9})} = 84.4 \times 10^5 h^2$$

When h is in inches and k is in darcys,

$$k = 54.4 \times 10^6 h^2$$

The permeability of a fracture 0.01 in. in thickness would be 5,440 darcys or 5,440,000 millidarcys.

Under the same flowing conditions used for the circular opening, a fracture 0.01 in. in thickness across the width of the block would contribute a flow rate Q'', computed as

$$Q'' = 1.1271 \frac{kA\,\Delta P}{\mu L}$$

$$= 1.1271(5,440)\frac{(0.01/12)(1)(1)}{1(1)}$$

$$= 1.1271(5,440)(0.00083)$$

$$= 5.1095 \text{ bbl/day}$$

The combined rate is 5.12077 bbl per day, or an increase of 45,437 per cent. It is obvious that *in situ* fractures and solution cavities contribute substantially to the productivity of any reservoir.

Analogy of Darcy's Law and Other Physical Laws

In using Darcy's law to define fluid flow in porous media it is often found that complex flow systems make a solution practically impossible. In analyzing Darcy's law it was found that it was comparable to Ohm's law for conductance of electrical current and Fourier's equation for conductance of heat in a solid.

Ohm's law as commonly written is

$$I = \frac{E}{r} \qquad (2\text{-}42)$$

where I = current, amp
E = voltage drop, volts
r = resistance of the circuit, ohms

but $$r = \rho \frac{L}{A} \quad \text{or} \quad r = \frac{L}{\sigma A}$$

where ρ = resistivity, ohm-cm
$\sigma = 1/\rho$ = conductivity
L = length of flow path, cm
A = cross-sectional area of conductor, sq cm

Therefore $$I = \frac{AE}{\rho L} \tag{2-43}$$

Comparing to Darcy's law for a linear system

$$Q = \frac{k}{\mu} A \frac{\Delta P}{L} \tag{2-44}$$

note that

$$Q \sim I \quad \frac{k}{\mu} \sim \frac{1}{\rho} = \sigma \quad \frac{\Delta P}{L} \sim \frac{E}{L} \tag{2-45}$$

Using the analogue between fluid and electrical systems it is possible to obtain solutions of complex fluid-flow networks by use of electrical networks. Further analogies of fluid systems with electrical parameters are possible.

The Fourier heat equation can be written as

$$q = k'A \frac{\Delta T}{L} \tag{2-46}$$

where q = rate of heat flow, Btu/hr
A = cross-sectional area, sq ft
ΔT = temperature drop, °F
L = length of conductor, ft
k' = thermal conductivity, Btu/(hr)(ft)(°F)

From (2-44)

$$Q \sim q \quad \frac{k}{\mu} \sim k' \quad \frac{\Delta P}{L} \sim \frac{\Delta T}{L} \tag{2-47}$$

As in the case of Ohm's law, further analogies are possible and will be discussed in later sections.

The above-listed analogies are useful in that many complex problems of both heat and electrical conduction have been solved analytically so that the mathematics can be extended readily to problems of flow through porous media. In addition, many fluid-flow problems involving complex geometry can be solved by appropriate electrical or heat models scaled down in size or time for convenient laboratory performance.

Measurement of Permeability

The permeability of a porous medium can be determined from samples extracted from the formation or by in-place testing. The procedures discussed in this section pertain to the permeability determinations on small samples of media.

Two methods are used to evaluate the permeability of cores. The method most used on clean, fairly uniform formations utilizes small cylindrical samples, perm plugs, approximately ¾ in. in diameter and 1 in. in length. The second method uses full-diameter core samples in lengths of 1 to 1½ ft. The fluids used with either method may be gas or any nonreactive liquid.

Perm Plug Method. As core samples ordinarily contain residual oil and water saturation, it is necessary that the sample be subjected to preparation prior to the determination of the permeability. Perm plugs are drilled from the larger cores parallel to the bedding planes. The perm plugs are dried in an oven or extracted by a soxhlet extractor and then subsequently dried. The residual fluids are thus removed, and the core sample becomes 100 per cent saturated with air. The perm plug is then inserted in a core

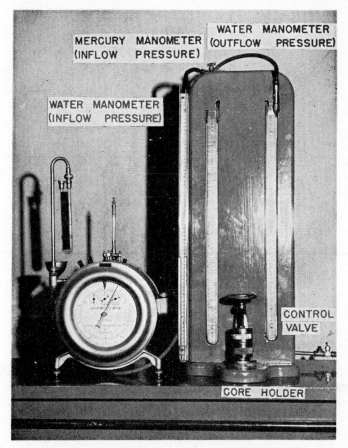

Fig. 2-38. Permeability apparatus. (*From Stevens.*[25])

holder of a permeability device such as illustrated in Fig. 2-38. An appropriate pressure gradient is adjusted across the perm plug, and the rate of flow of air through the plug is observed. The requirement that the permeability be determined for conditions of viscous flow is best satisfied by obtaining data at several flow rates and plotting results as shown in Fig. 2-39 from either Eq. (2-20) or (2-21). For conditions of viscous flow, the

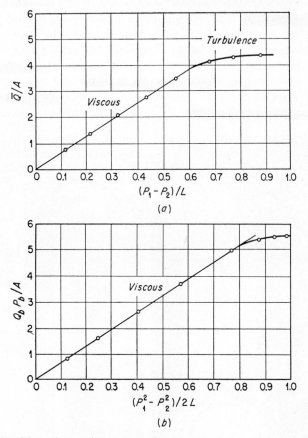

FIG. 2-39. (a) Plot of experimental results for calculation of permeability, from $k = \bar{Q}L/A(P_1 - P_2)$; (b) plot of experimental results for calculation of permeability, from $k = 2Q_bP_bL/A(P_1^2 - P_2^2)$. (*From Stevens*.[25])

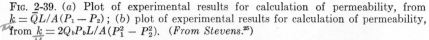

data should plot a straight line, passing through the origin. Turbulence is indicated by curvature of the plotted points. The slope of the straight-line portion of the curve is equal to k/μ, from which the permeability can be computed. To obtain k in darcys, Q must be in cubic centimeters per second, A in square centimeters, P_1 and P_2 in atmospheres, L in centimeters, and μ in centipoises.

A permeameter designed for the determination of the permeability of rocks with either gas or liquid is illustrated in Fig. 2-40. Data are ordi-

FIG. 2-40. Ruska universal permeameter. (A) Rubber stopper; (B) core-holder sleeve; (C) core holder; (D) flow-meter selector valve; (E) pressure-regulating valve; (F) pressure gauge (G) fill nipple; (H) liquid-gas valve; (I) discharge-fill valve; (J) burette. (*From Stevens.*[25])

narily taken from this device at only one flow rate. To assure conditions of viscous flow, it is the lowest possible rate which can be accurately measured.

Example 2-12. Permeability Measurement. 1. The following data were obtained during a routine permeability test. Compute the permeability of this core.

Flow rate = 1,000 cc of air at 1 atm abs and 70°F in 500 sec

Pressure, downstream side of core = 1 atm abs, flowing temperature, 70°F

Viscosity of air at test temperature = 0.02 centipoise

Cross-sectional area of core = 2.0 sq cm

Length of core = 2 cm

Pressure, upstream side of core = 1.45 atm abs

$P_1V_1 = P_2V_2 = \overline{P}\overline{V}$

$$\overline{P} = \frac{P_1 + P_2}{2} = \frac{1.45 + 1}{2} = 1.225$$

$$1 \times 1,000 = 1.225 \overline{V}$$

$$\overline{V} = 815 \text{ cc}$$

$$\overline{Q} = \frac{\overline{V}}{t} = \frac{815}{500} = 1.63$$

$$k = \frac{\overline{Q}}{A} \frac{L}{\Delta P} \mu$$

$$= \frac{1.63(2)(0.02)}{2(0.45)} \times 1,000$$

$$= 72.5 \text{ millidarcys}$$

2. Assuming that the data indicated above were obtained but water was used as the flowing medium, compute the permeability of the core. The viscosity of water at test temperature was 1.0 centipoise.

$$\overline{Q} = \frac{\overline{V}}{t} = \frac{1,000}{500} = 2.0$$

$$k = \frac{\overline{Q}\mu}{A} \frac{L}{\Delta P} = \frac{2(1)(2)}{2(0.45)} \times 1,000 = 4,450 \text{ millidarcys}$$

Whole-core Measurement. The core must be prepared in the same manner as perm plugs. The core is then mounted in special holding devices as shown in Fig. 2-41. The measurements required are the same as for the perm plugs, but the calculations are slightly different.

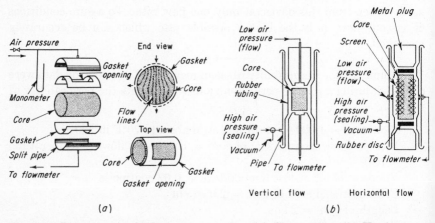

Fig. 2-41. Permeameters for large core samples. (*a*) Clamp type; (*b*) Hassler type. (*From Kelton.*[13])

In the case of the clamp-type permeameter, the geometry of the flow paths is complex and an appropriate shape factor must be applied to the data to compute the permeability of the sample. The shape factor is a function of the core length and the size of the gasket opening.

The values of permeability obtained from large core measurements to those for perm plugs (designated as matrix samples) are compared in Table 2-5. Measurements of permeability on large core samples generally

TABLE 2-5. COMPARISON OF MATRIX AND WHOLE-CORE PERMEABILITY OBSERVATIONS FOR ELLENBURGER LIMESTONE[13]

Core group	1	2	3	4
Maximum whole core permeability	10	409	23	94
Whole core permeability, measured at 90° to flow direction of maximum value	0.6	1.2	10	38
Matrix permeability from perm plugs	0.3	0.2	0.3	3.7

yield better indications of the permeability of limestones than do the small core samples. However, rocks which contain fractures *in situ* frequently separate along the natural planes of weakness when cored. Therefore, the conductivity of such fractures will not be included in the laboratory data. In general, the laboratory measurement of permeability represents a minimum value except in the case of highly argillaceous materials.

Factors Affecting Permeability Measurements

In the techniques of permeability measurement previously discussed, certain precautions must be exercised in order to obtain accurate results. When gas is being used as the measuring fluid, corrections must be made for gas slippage. When liquid is the testing fluid, care must be taken that it does not react with the solids in the core sample. Also corrections can be applied for the change in permeability because of the reduction in confining pressure on the sample.

Effect of Gas Slippage on Permeability Measurements. Klinkenberg[27] has reported variations in permeability as determined using gases as the flowing fluid from that obtained when using nonreactive liquids. These variations were ascribed to slippage, a phenomenon well known with respect to gas flow in capillary tubes. The phenomenon of gas slippage occurs when the diameter of the capillary openings approach the mean free path of the gas.

The mean free path of a gas is a function of the molecular size and the kinetic energy of the gas. Therefore, the "Klinkenberg effect" is a function of the gas with which the permeability of the porous medium is determined. Figure 2-42 is a plot of the permeability of a porous medium as determined at various mean pressures using hydrogen, nitrogen, and carbon dioxide as the flowing fluids. Note that for each gas a straight line is obtained for the

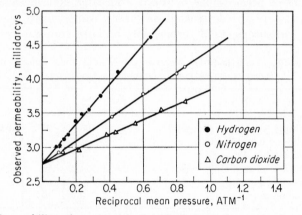

FIG. 2-42. Permeability of core sample L to hydrogen, nitrogen, and carbon dioxide at different pressures. Permeability to isooctane, 2.55 millidarcys. (*From Klinkenberg.*[27])

observed permeability as a function of the reciprocal of the mean pressure of the test. The data obtained with lowest molecular weight gas yield the straight line with greater slope, indicative of a greater slippage effect. All the lines when extrapolated to infinite mean pressure $(1/\overline{P} = 0)$ intercept the permeability axis at a common point. This point is designated k_L, or the equivalent liquid permeability. Klinkenberg and others established that the permeability of a porous medium to a nonreactive homogeneous single-phase liquid was equal to the equivalent liquid permeability.

The linear relationship between the observed permeability and the reciprocal of mean pressure can be expressed as follows:

$$k_L = \frac{k_g}{1 + (b/\overline{P})} = k_g - m\frac{1}{\overline{P}}$$

$$b = \frac{m}{k_L} \tag{2-48}$$

where k_L = permeability of medium to a single liquid phase completely filling the pores of the medium

k_g = permeability of medium to a gas completely filling the pores of the medium

$\overline{P}$ = mean flowing pressure of the gas at which k_g was observed

b = constant for a given gas in a given medium

m = slope of the curve

The constant b in the above equation depends on the mean free path of the gas and the size of the openings in the porous medium. Since permeability is, in effect, a measure of the size openings in a porous medium, it is found that b is a function of permeability. Figure 2-43 represents a corre-

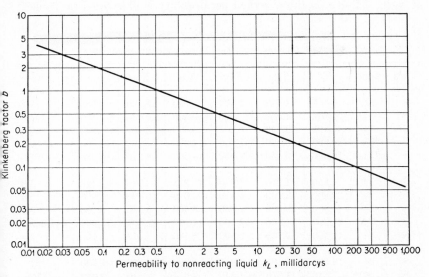

Fig. 2-43. Correlation of Klinkenberg factor b with permeability. (*From American Petroleum Institute.*[19])

lation of b with k_L, determined from measurements made on 175 samples.[19] The factor b increases with decreasing permeability as slippage effects become proportionately greater for smaller openings.

To obtain accurate permeability measurements on small samples requires approximately 12 flow tests. Permeability values should be determined for four flow rates, each at three different values of mean pressure. This procedure permits the obtaining of three values of permeability under viscous flow conditions at three mean pressure values, from which the permeability to liquid can be graphically determined.

Effect of Reactive Liquids on Permeability. While water is commonly considered to be nonreactive in the ordinary sense, the occurrence of swelling clays in many reservoir rock materials results in water being the most frequently occurring reactive liquid in connection with permeability determinations. Reactive liquids alter the internal geometry of the porous medium. This phenomenon does not vitiate Darcy's law but rather results in a new porous medium the permeability of which is determined by the new internal geometry.

The effect of clay swelling in the presence of water is particularly important in connection with the determination of the permeability of the graywackes common to the Gulf Coast and the arkosic sediments of California. The degree of hydration of the clays is a function of the salinity of the water. Permeability changes of 50-fold or more may be noted between that determined with air and that determined with fresh water. The effect of water salinity on the observed permeability is illustrated in Table 2-6

TABLE 2-6. EFFECT OF WATER SALINITY ON PERMEABILITY OF NATURAL CORES[28] (Grains per gallon of chloride ion as shown[a])

Field	Zone	K_a	K_{1000}	K_{500}	K_{300}	K_{200}	K_{100}	K_w
S	34	4,080	1,445	1,380	1,290	1,190	885	17.2
S	34	24,800	11,800	10,600	10,000	9,000	7,400	147
S	34	40,100	23,000	18,600	15,300	13,800	8,200	270
S	34	39,700	20,400	17,600	17,300	17,100	14,300	1,680
S	34	12,000	5,450	4,550	4,600	4,510	3,280	167
S	34	4,850	1,910	1,430	925	736	326	5.0
S	34	22,800	13,600	6,150	4,010	3,490	1,970	19.5
S	34	34,800	23,600	7,800	5,460	5,220	3,860	9.9
S	34	27,000	21,000	15,400	13,100	12,900	10,900	1,030
S	34	12,500	4,750	2,800	1,680	973	157	2.4
S	34	13,600	5,160	4,640	4,200	4,150	2,790	197
S	34	7,640	1,788	1,840	2,010	2,540	2,020	119
S	34	11,100	4,250	2,520	1,500	866	180	6.2
S	34	6,500	2,380	2,080	1,585	1,230	794	4.1
T	36	2,630	2,180	2,140	2,080	2,150	2,010	1,960
T	36	3,340	2,820	2,730	2,700	2,690	2,490	2,460
T	36	2,640	2,040	1,920	1,860	1,860	1,860	1,550
T	36	3,360	2,500	2,400	2,340	2,340	2,280	2,060
T	36	4,020	3,180	2,900	2,860	2,820	2,650	2,460
T	36	3,090	2,080	1,900	1,750	1,630	1,490	1,040

[a] For example, K_a means permeability to air; K_{500} means permeability to 500 grains per gal chloride solution; K_w means permeability to fresh water.

While fresh water may cause the cementation material in a core to swell owing to hydration, it is a reversible process. A highly saline water can be flowed through the core and return the permeability to its original value. The reversibility of the effect of reactive liquids is illustrated in Fig. 2-44.

Care must be taken that laboratory permeability values are corrected to liquid values obtained with water whose salinity corresponds to formation water. An example of the variation of air permeability and formation water permeability is shown in Fig. 2-45. California and Gulf Coast sands will normally exhibit lower formation permeabilities than those measured by air in the laboratory.

Overburden Pressure. When the core is removed from the formation, all the confining forces are removed. The rock matrix is permitted to expand in all directions, partially changing the shapes of the fluid-flow paths inside the core.

Compaction of the core due to overburden pressure may cause as much as a 60 per cent reduction in the permeability of various formations, as shown in Fig. 2-46. It is noted that some formations are much more compressible than others; thus more data are required to develop empirical correlations which will permit the correction of surface permeability for overburden pressures.

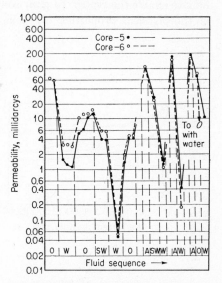

Fig. 2-44. The permeability history of two Stevens sand cores when subjected to and measured with various fluids. (*From Muskat.*[29])

Factors in Evaluation of Permeability from Other Parameters

Permeability, like porosity, is a variable which can be measured for each rock sample. To aid better in the understanding of fluid flow in rocks and possibly to reduce the number of measurements required on rocks, correlations among porosity, permeability, surface area, pore size,

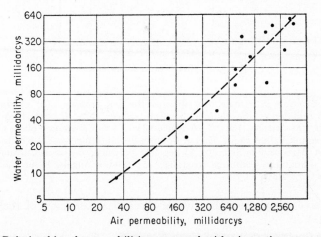

Fig. 2-45. Relationship of permeabilities measured with air to those measured with water having a concentration of 20,000 to 25,000 ppm chloride ion.

and other variables have been made. The reasoning behind some of the
correlations among porosity, permeability, and surface area are presented
here to enable the reader to gain some understanding of the interrelation
of the physical properties of rocks. Although these relations are not

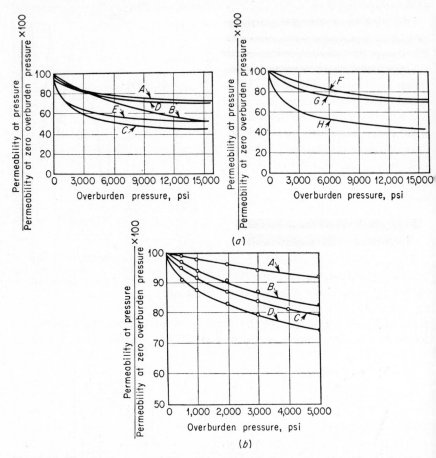

FIG. 2-46. Changes in permeability with overburden pressure. (a) Curve A—Colo-
rado; 3.96 millidarcys; B—Southern California coast, 40.9; C—San Joaquin Valley,
Calif., 45.0; D—Arizona, 4.36; E—Arizona, 632; F—San Joaquin Valley, Calif., 40.5;
G—San Joaquin Valley, Calif., 55.5; H—Southern California coast, 318.8. (b) A—basal
Tuscaloosa, Miss., 229 millidarcys, 15 per cent porosity; B—basal Tuscaloosa, Miss.,
163, 24; C—Southern California coast, 335, 25; D—Los Angeles basin, Calif., 110, 22.
(*From Fatt and Davis.*[30])

quantitative, they are indicative of the interdependence of rock characteristics.

Use of Capillary Tubes for Flow Network. The simplest expression for
the rate of flow in a circular conduit is Poiseuille's equation, which, like

Darcy's law, holds for viscous conditions of flow. It has previously been shown that

$$k = \frac{r^2}{8} \qquad (2\text{-}39)$$

If a porous system is conceived to be a bundle of capillary tubes, then it can be shown that the permeability of the medium depends on the pore size distribution and porosity.

Consider a medium formed of a bundle of capillary tubes with the spaces between the tubes sealed by a cementing material. If the conductors are all of the same size and are arranged in cubic arrangement, then, neglecting wall thickness of the tubes, the number of tubes per unit area is given by

$$\frac{\text{Porosity}}{\text{Area/tube}} = \frac{\pi/4}{\pi r^2} = \frac{1}{4r^2}$$

The rate of flow is then given by

$$Q = \frac{1}{4r^2} \frac{\pi r^4 \, \Delta P}{8\mu L} = \frac{k \, \Delta P}{\mu L}$$

Therefore

$$k = \frac{\pi r^2}{32}$$

If the arrangement of the tubes is the most compact possible, the number of tubes per unit area is given by

$$\frac{\pi}{4 \sin 60°} \div \pi r^2 = \frac{1}{3.46 r^2}$$

where the porosity $\phi = \dfrac{\pi}{4 \sin 60°}$

Therefore

$$k = \frac{\pi r^2}{27.68}$$

Note, therefore, that the permeability for this simple case is a function not only of the pore size but of the arrangement of the tubes. The arrangement of the tubes is defined by the porosity ϕ for these simple cases.

Consider a system comprised of a bundle of capillary tubes of the same radii and length; k, the permeability, can be written as a function of porosity as follows:

$$k = \frac{\phi r^2}{8}$$

The internal surface area per unit of pore volume can be derived as follows:

Surface area $= n2\pi r L$

Pore volume $= n\pi r^2 L$

Therefore $S_p = \dfrac{n2\pi r L}{n\pi r^2 L} = \dfrac{2}{r}$ $\qquad r = \dfrac{2}{S_p}$

where S_p is the internal surface area per unit pore volume. Substituting the above into the preceding expression gives

$$k = \frac{4\phi}{8S_p{}^2} = \frac{1}{2}\frac{\phi}{S_p{}^2}$$

if, for the constant ½, $1/k_z$ is substituted,

$$k = \frac{\phi}{k_z S_p{}^2} \tag{2-49}$$

which is the Kozeny equation where k_z is the Kozeny constant.

Wyllie[31] derived the Kozeny relation from Poiseuille's law as follows (symbols in cgs units):

$$v = \frac{d^2\,\Delta P}{32\mu L}$$

where v = velocity of flow, cm/sec
d = diameter of conduit, cm
ΔP = pressure loss, dynes/sq cm
μ = fluid viscosity, poises
L = average path length, cm

Introducing the concept of mean hydraulic radius m, where

$$m = \frac{\text{volume of conduit}}{\text{area of wetted surface}}$$

for a circular pipe,

$$m = \frac{\pi r^2 L}{2\pi r L} = \frac{d}{4}$$

which when substituted into Poiseuille's laws yields

$$v = \frac{m^2}{2}\frac{\Delta P}{\mu L}$$

Poiseuille's law in this form is applicable to noncircular conduits. Wyllie[31] then suggests that the factor 2 in the denominator of the above expression be generalized and replaced by a shape factor k_0, which takes on values for porous materials ranging from 2.5 to 3.0.

The conduit in a porous medium is conceived by Wyllie to be of a cross-sectional area ϕA, where ϕ is the fractional porosity and A is the cross-sectional area of the porous medium. A further correction must be applied to the length of the conduit, for the average distance traversed by the fluid is greater than the distance L between two mutually perpendicular faces across which flow is occurring. The actual fluid velocity v_a within the pores of the medium is greater than the macroscopic velocity v, such as implied by Q/A, where Q is the volume rate of flow and A is the cross-sectional area of the porous medium. The increased velocity is due to the decreased area actually available for flow and to the increased length of the actual flow path L_a as compared with the length L across the porous medium. Then

$$v_a = \frac{v}{\phi}\frac{L_a}{L}$$

Correcting Poiseuille's law for the actual flow conditions gives

$$\frac{v}{\phi}\frac{L_a}{L} = \frac{m^2}{k_0}\frac{\Delta P}{L_a}$$

or

$$v = \frac{\phi m^2}{k_0}\frac{\Delta P}{\mu L}\left(\frac{L}{L_a}\right)^2$$

In a porous medium, m is equal to the ratio of the pore space per unit volume of the medium ϕ to the surface per unit volume s, so that

$$m = \frac{\phi}{s}$$

The surface per unit volume of the medium s is related to the surface per unit volume of pore space S_p as follows:

$$s = \phi S_p$$

therefore

$$m = \frac{1}{S_p}$$

Then

$$v = \frac{\phi}{k_0 S_p{}^2}\frac{\Delta P}{\mu L}\left(\frac{L}{L_a}\right)^2$$

From Darcy's law

$$v = \frac{k\,\Delta P}{\mu L} \tag{2-14}$$

$$k = \frac{\phi}{k_0 S_p{}^2}\left(\frac{L}{L_a}\right)^2$$

where k is the permeability of the porous medium.

Let

$$(L_a/L)^2 = \tau = \text{tortuosity of the porous medium}$$
$$k_z = k_0\tau = \text{Kozeny constant}$$

then

$$k = \frac{\phi}{k_z S_p{}^2} \tag{2-49}$$

Carman[32] reported that the value of k_z was in all cases about 5. To apply Eq. (2-49) to the calculation of permeability, the porosity ϕ, the surface area per unit pore volume S_p, and the Kozeny constant k_z must be known. Rapoport and Leas[33] have reported a method of calculation of permeability based on Eq. (2-49) and determination of S_p from capillary data. Other investigators have reported discrepancies in the assumption that $k_z = 5$. In consolidated porous media there is little reason to believe k_z to be a constant, but rather, k_z depends on k_0 and τ for the particular medium.

Wyllie[31] and others have reported that k_z can be evaluated from electrical properties of the porous material when saturated with an electrolyte.

FLUID SATURATIONS

In the previous sections of this chapter the storage and conduction capacity of a porous rock were discussed. To the engineer there is yet another important factor to be determined. What is the fluid content of the rock? In most oil-bearing formations it is believed that the rock was completely saturated with water prior to the invasion and trapping of petroleum. The less dense hydrocarbons are considered to migrate to positions of hydrostatic and dynamic equilibrium, thus displacing water from the interstices of the structually high part of the rock. The oil will not displace all the water which originally occupied these pores. Thus, reservoir rocks normally will contain both petroleum hydrocarbons and water (frequently referred to as connate water) occupying the same or adjacent pores. To determine the quantity of hydrocarbons accumulated in a porous rock formation, it is necessary to determine the fluid saturation (oil, water, and gas) of the rock material.

Methods of Determining Fluid Saturation

There are two approaches to the problem of determining the original fluid saturations within a reservoir rock. The direct approach is the selecting of rock samples and measuring the saturations of these samples as they are recovered from the parent formations. The indirect approach is to determine the fluid saturation by measuring some other physical property of the rock. The direct approach is all that will be discussed here. The indirect approach, such as using electric logs or capillary-pressure measurements, will be discussed in later chapters.

Determination of Fluid Saturations from Rock Samples

In determining fluid saturations directly from a sample removed from a reservoir, it is necessary to understand first how these values are measured; second, what these measured values represent; and third, knowing what they represent, how they can be applied.

In order to measure values of original rock saturations there have been essentially three methods devised. These methods involve either the evaporation of the fluids in the rock or the leaching out of the fluids in the rock by extraction with a solvent.

One of the most popular means of measuring the initial saturations is the retort method. This method takes a small rock sample and heats the sample so as to vaporize the water and the oil, which is condensed and collected in a small receiving vessel. An electric retort is shown in Fig.

2-47. The retort method has several disadvantages as far as commercial work is concerned. First in order to remove all the oil, it is necessary to approach temperatures on the order of 1000 to 1100°F. At temperatures of this magnitude the water of crystallization within the rock is driven off,

FIG. 2-47. Retort distillation apparatus. (*From Stevens.*[25])

causing the water-recovery values to be greater than just the interstitial water. An example of such a system is illustrated in Fig. 2-48. Here the water removed in the first 30 min was approximately the interstitial water. As the application of heat was continued, the water of crystallization was removed, amounting to approximately 2 cc of water out of a total recovery of 8 cc. Thus, it is seen that an error of 33 per cent is possible if the water of crystallization is not accounted for. The second error which occurs from retorting samples is that the oil itself when heated to high temperatures has a tendency to crack and coke. This change of a hydrocarbon molecule tends to decrease the liquid volume and also in some cases coats the internal walls of the rock sample itself. The effect of cracking and coking in a retort is shown in Fig. 2-49, wherein 0.4 cc of oil actually in the sample yields about 0.25 cc in the receiving vessel. Thus a fluid correction must be made on all sample data obtained with a retort. Before retorts can be used, calibration curves must be prepared on various gravity fluids to correct for the losses from cracking and coking with the various applied temperatures. Another correction curve can also be obtained which correlates recovered

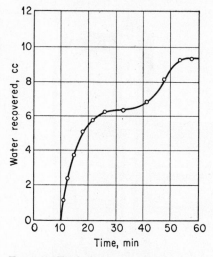

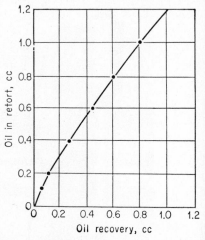

FIG. 2-48. Typical retort calibration curve for water from a Wilcox sand. Altair Field core, depth 8,270 ft, 107 millidarcys, 22.0 per cent porosity. 6.2 cc plateau reading. (*From Emdahl.*[34])

FIG. 2-49. Typical retort calibration curve for oil from a Wilcox sand. Clay Creek Field, 43°API oil, observed trace = 0.1 cc in retort. (*From Emdahl.*[34])

API oil gravity with initial API oil gravity. It is normal for the oil gravity of the recovered liquid to be less than the oil gravity of the liquid originally in the rock sample. These curves can be obtained by running "blank" runs (retorting known volumes of fluids of known properties). The retort is a rapid method for the determination of fluid saturations, and utilizing the corrections yields satisfactory results. It gives both the water and oil volumes, so that the oil and water saturations can be calculated from the following formulas:

$$S_w = \frac{\text{water, cc}}{\text{pore volume, cc}} \qquad (2\text{-}50)$$

$$S_o = \frac{\text{oil, cc}}{\text{pore volume, cc}} \qquad (2\text{-}51)$$

$$S_g = 1 - S_w - S_o \qquad (2\text{-}52)$$

The other method of determining fluid saturation is by extraction with a solvent. Extraction can be accomplished by a modified ASTM method or a centrifuge method. In the standard distillation test the core is placed so that a vapor of toluene, gasoline, or naphtha rises through the core and is condensed to reflux back over the core. This process leaches out the oil and water in the core. The water and extracting fluid are condensed and are collected in a graduated receiving tube. The water settles to the bottom of the receiving tube because of its greater density, and the extracting fluid refluxes back into the main heating vessel. The process is continued until no more water is collected in the receiving tube. The distillation apparatus is shown in Fig. 2-50. The water saturation can be determined directly; i.e.,

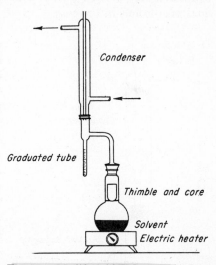

Fig. 2-50. Modified ASTM extraction apparatus.

$$S_w = \frac{\text{water, cc}}{\text{pore volume, cc}}$$

The oil saturation is an indirect determination. It is necessary to note the weight of the core sample prior to extraction. Then, after the core has been cleaned and dried, the sample is again weighed. The oil saturation as a fraction of pore volume is given by

$$S_o = \frac{(\text{wt of wet core, gm} - \text{wt of dry core, gm} - \text{wt of water, gm})}{(\text{pore volume, cc})(\text{density of oil, gm/cc})}$$

(2-53)

The core can be completely cleaned in the ASTM extraction apparatus, or once all water is removed, the remainder of the cleaning can be done in a soxhlet extractor (Fig. 2-51). The mechanics of the soxhlet extractor are essentially the same as the ASTM extraction apparatus except that no receiving vessel is supplied for trapping water. The cleaning solution is continually vaporized and condensed on the core. This action leaches out the oil and water from the core. The ASTM extraction method does less damage to a core sample and results in perhaps the cleanest core of any of the saturation determinations. The core sample is ready for porosity or permeability determinations after this extraction process.

Before permeability and porosity can be measured, it is necessary to clean the core sample in a device similar to the soxhlet extractor or one which uses centrifugal force. Thus, using the ASTM distillation only one additional step is required to obtain information from which to calculate fluid saturations in the core.

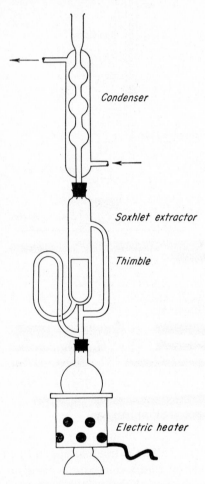

Fig. 2-51. Soxhlet extractor.

Condenser

Soxhlet extractor

Thimble

Electric heater

Another method of determining water saturation is to use a centrifuge (Fig. 2-52). A solvent is injected into the centrifuge just off center. Owing to centrifugal force it is thrown to the outer radii, being forced to pass through the core sample. The solvent removes the water and oil from the core. The outlet fluid is trapped, and the quantity of water in the core is measured. The use of the centrifuge provides a very rapid method because of the high forces which can be applied. At the same time that the water content is determined, the core is cleaned in preparation for the other measurements. The values of water and oil saturation are calculated by using Eqs. (2-50) to (2-53) as for the ASTM extraction method.

There is another procedure for saturation determination which is used with either of the extraction methods. The core as received from the well is placed in a modified mercury porosimeter (Fig. 2-14) wherein the bulk volume and gas volume are measured. The volume of water is determined by one of the extraction methods. The fluid saturations can be calculated from these data.

In connection with all procedures for determination of fluid content, a value of pore volume must be established in order that fluid saturations can be expressed as percentage of pore volume. Any of the porosity procedures previously described can be used. Also the bulk volume and gas volume determined from the mercury porosimeter can be combined with the oil and water volumes obtained from the retort to calculate pore volume, porosity, and fluid saturations.

FIG. 2-52. Centrifugal core cleaning. (*Courtesy of Core Laboratories, Inc.*)

Factors Affecting Fluid Saturations of Cores

The core sample delivered to the laboratory for fluid-saturation determinations was obtained from the ground by rotary, side-wall, or cable-tool coring. In all cases, the fluid content of these samples has been altered by two processes. First, especially in the case of rotary drilling, the formation is under a greater pressure from the mud column in the well than from the fluid in the formation. The differential pressure across the well face causes mud and mud filtrate to invade the formation immediately adjacent to the well surface, thus flushing the formation with mud and its filtrate. As most drilling is done with water-base mud, water filtrate invades the core and displaces some of the oil and perhaps some of the original interstitial water. This displacement process changes the original fluid contents of the in-place rock. Second, as the sample is brought to the surface, the confining pressure of the fluid column is constantly decreasing. The reduction of pressure permits the expansion of the entrapped water, oil, and gas. Gas, having the greater coefficient of expansion, expels oil and water from the core. Thus, the contents of the core at the surface have been changed from those which existed in the formation. The core has been invaded with water, and the contents subsequently subjected to a solution-gas–drive mechanism. As the invasion of the filtrate precedes the core bit, it is not possible to use pressurized core barrels to obtain undisturbed samples.

In the case of drill cuttings, chips, or cores from cable-tool drilling, they also have undergone definite physical changes. If little or no fluid is maintained in the well bore, the formation adjacent to the well surface is depleted owing to pressure reduction. As chips fall into the well, they may or may not be invaded, depending on the fluids in the well bore and the physical properties of the rock. In all probability, fluid will permeate this depleted sample, resulting in flushing. Thus, even cable-tool cores have undergone the same two processes as was noted in the case of rotary coring but in reverse order.

In an attempt to understand better the over-all effect of the physical changes which occur in the core because of flushing and fluid expansion, Kennedy, Van Meter, and Jones[35] undertook a study to simulate rotary coring techniques. In this study a cylindrical sample was used which had a hole drilled in the middle to represent the well bore (Fig. 2-53). Mud under pressure was supplied to the middle hole, allowing filtrate to enter the core sample. The oil and water forced from the core were collected, and the amount was measured at the outer boundary. The values gave the change in saturation caused by the flushing action of the filtrate. The pressure on the core was reduced to atmospheric pressure, and the amount of water and oil that remained in the core was determined. The total effect of both flushing and expansion because of pressure reduction was thus measured.

Schematic illustrations of the changes in saturation resulting from these two processes for oil-base and water-base muds are shown in Fig. 2-54. It is noted that the original flushing action reduced the oil saturation by approximately 14 per cent. The expansion to surface pressure displaced water and additional oil. The final water saturation was greater than the water saturation prior to coring. In coring with an oil-base mud, the filtrate is oil, so that the flushing action did not alter the initial water saturations but did result in replacement of approximately 20 per cent of the initial oil. On pressure depletion a small fraction of the water was expelled, reducing the water saturation from 49.1 to 47.7 per cent. The oil saturation was reduced by both processes from 50.9 to 26.7 per cent. Thus, even when high water saturations are involved, up to approximately 50 per cent, the water-saturation values obtained with oil-base muds may be considered to be representative of the initial water saturations in the reservoir. Hence, it is possible to obtain fairly representative values of in-place water saturations by selecting the fluids with which the core samples are obtained. Kennedy et al. studied cores with permeabilities ranging from 2.3 to 3,040

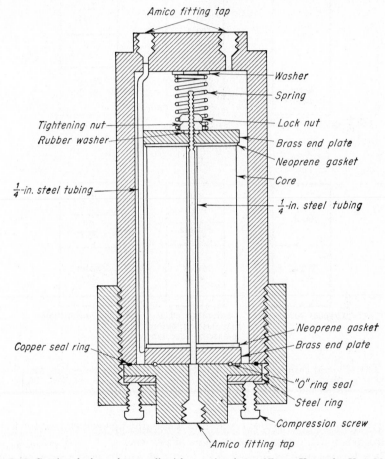

FIG. 2-53. Sectional view of test cell with core in place. (*From Kennedy, Van Meter, and Jones.*[35])

millidarcys and porosities ranging from 5.2 to 21.7 per cent. The drilling fluids used were clay bentonite, lime starch, and oil-base muds with viscosities from 65 to 133 centipoises and water losses from zero for the oil-base mud to 6.8 cc API for the clay bentonite mud.

Kennedy et al. correlated hydrocarbon saturations before and after coring. These correlations are shown in Fig. 2-55. It is noted that for cores of 5- and 10-millidarcy permeability, the initial and final hydrocarbon saturation yields an approximate straight line for initial saturations greater than 15 per cent. Data for cores of from 127- to 3,040-millidarcy permeability were correlated in the same manner as the data for the low-perme-

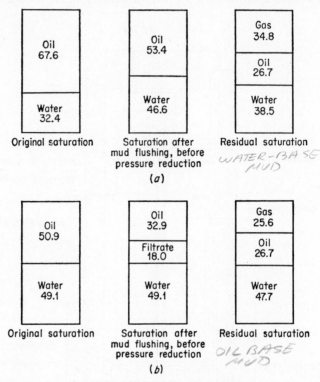

FIG. 2-54. Typical changes in saturation of cores flushed with water-base and oil-base muds. (a) Water-base mud; (b) oil-base mud. (*From Kennedy, Van Meter, and Jones.*[35])

ability samples. These also resulted in a straight-line correlation for initial hydrocarbon saturations greater than 15 per cent.

Correlations such as presented in Fig. 2-55 can be used to correct saturations measured from cores to original conditions. Additional data are required before universal correlations can be established.

Attempts have been made to use tracers in the drilling fluid to determine the amount of water in the core which is due to mud filtrate invasion. The theory was that mud filtrate displaced only oil. Thus, when the core is recovered to the surface, the salt concentration of the core water can be determined. Knowing the salt concentration in the reservoir water and the tracer concentration in the drilling fluid, it was thought possible to calculate the volume of filtrate and reservoir water in the core. A large fraction of the initial reservoir water may have been displaced by the invading filtrate, so the tracer method would give low values of reservoir water saturation.

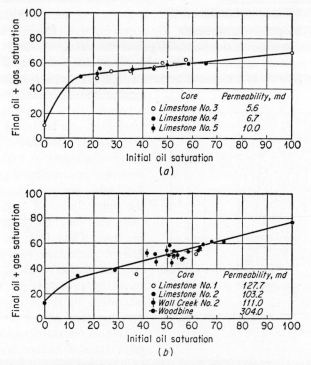

FIG. 2-55. Laboratory determination of fluid saturation of oil-field cores subjected to mud flushing and pressure depletion. (*From Kennedy, Van Meter, and Jones.*[35])

In order to obtain realistic values of fluid saturation it is necessary to choose the proper drilling fluid or resort to correlations similar to that reported by Kennedy et al. or to use indirect methods of saturation determination.

The Use of Core-determined Fluid Saturations

The saturation values obtained directly from rock samples are usually not reliable for determining the quantity of each fluid in the rock. Other uses exist for fluid-saturation determinations from core samples. It has already been shown that water saturations obtained from core samples cut with oil-base mud are essentially reliable. The saturations of cores cut with water-base mud are used to determine the original oil-gas contact, original oil-water contact, and whether a sand is productive of oil or gas.

The determination of contacts is made by carefully studying the residual oil saturations of the cores as a function of depth. In the oil-saturated regions the samples will have essentially a constant value for residual oil saturations, probably 15 per cent or greater. In the gas region the oil saturation is small or vanishes. Thus the depth of the gas-oil contact is defined by a sharp increase in oil saturation. In the water zone, the oil saturation gradually disappears with depth. By observing these changes in oil saturation, it is possible to choose the depth of the water-oil contact.

It is possible to establish a correlation of the water content of cores and permeability from which it can be determined whether a formation will be productive of hydrocarbons. Such a correlation is shown in Fig. 2-56, wherein it can be noted that low-permeability formations with core water saturations as high as 55 per cent may be considered productive. For higher permeability formations the upper limits of water saturations may be slightly less than 50 per cent. Thus, from the investigation of saturation values of cores one can gather that a formation would be productive if the water saturation in the surface samples were less than 50 per cent.

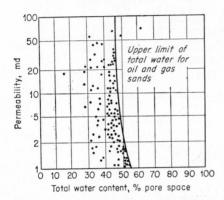

FIG. 2-56. Limiting values of total core water for oil and gas production—water-base muds. (*From Emdahl.*[34])

Another reason for measuring fluid saturations of surface samples is to obtain other correlations such that direct or indirect measurements of other physical properties may also give indications of initial fluid distributions. The measurement of electrical resistivity of the core samples, prior to cleaning, permits correlations of electrical resistivities with other physical properties to aid in electrical log interpretation.

Thus, in summary, it is seen that although fluid-saturation determinations made on core samples at the surface may not give a direct indication of the saturations within the reservoir, they are of value and do yield very useful and necessary information.

ELECTRICAL CONDUCTIVITY OF FLUID-SATURATED ROCKS

Resistivity Relations

Porous rocks are comprised of an aggregate of minerals, rock fragments, and void space. The solids, with the exception of certain clay minerals, are nonconductors. The electrical properties of a rock depend on the geometry of the voids and the fluids with which those voids are filled. The fluids of interest in petroleum reservoirs are oil, gas, and water. Oil and gas are nonconductors. Water is a conductor when it contains dissolved salts. Current is conducted in water by movement of ions and can therefore be termed electrolytic conduction. The resistivity of a material is the reciprocal of conductivity and is commonly used to define the ability of a material to conduct current. The resistivity of a material is defined by the following equation:

$$\rho = \frac{rA}{L}$$ (2-54)

where ρ = resistivity
    r = resistance
    A = cross-sectional area of the conductor
    L = length of the conductor

For electrolytes, ρ is commonly reported in ohm-centimeters, r is expressed in ohms, A in square centimeters, and L in centimeters. In the study of the resistivity of soils and rocks, it has been found that the resistivity can be expressed more conveniently in ohm-meters. To convert to ohm-meters from ohm-centimeters, divide the resistivity in ohm-centimeters by 100. In oil-field practice, the resistivity in ohm-meters is commonly represented by the symbol R with an appropriate subscript to define the conditions to which R applies.

Formation Factor. The most fundamental concept in considering electrical properties of rocks is that of formation factor.

As defined by Archie[36], the formation factor is

$$F = \frac{R_o}{R_w}$$ (2-55)

where R_o is the resistivity of the rock when saturated with water having a resistivity of R_w.

The relationships between the electrical properties and other physical properties of the rock are complex but can be illustrated by the following developments.

Consider a cube of salt water (cube 1, Fig. 2-57) having a cross-sectional area A, a length L, and a resistivity R_w. If an electrical current is caused to flow across the cube through an area A and a length L, the resistance of the cube can be determined. Let this resistance be r_1. Then

$$r_1 = \frac{R_w L}{A}$$

In Fig. 2-57 cube 2 represents a cube of porous rock of the same dimensions of cube 1 and 100 per cent saturated with water of resistivity R_w. Considering the solids to be nonconducting, the electrical flow must then be through the water-filled pores. The cross-sectional area available for conduction is now A_a, actual or effective cross section of the water-filled pores. The path length of current flow is increased to a value L_a, the average length that an ion must traverse in passing through the pore channels.

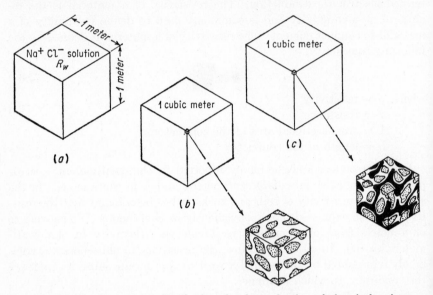

FIG. 2-57. Idealized saturation distributions for determination of electrical resistance. (a) Cube 1; resistivity of a cube of NaCl solution. (b) Cube 2; $F = R_o/R_w$ (basic definition, resistivity of a cube of rock 100 per cent saturated with water). (c) Cube 3; resistivity of a cube of rock—interstices filled with water and hydrocarbons. (*From Winn.*[37])

The resistance r_2 of such a cube can be determined as for the cube of water:

$$r_2 = \frac{R_w L_a}{A_a}$$

By definition,

$$R_o = \frac{r_2 A}{L}$$

Hence

$$R_o = \frac{R_w L_a A}{L A_a}$$

Therefore

$$F = \frac{R_o}{R_w} = \frac{L_a/L}{A_a/A} = \frac{\sqrt{\tau}}{A_a/A} \tag{2-56}$$

where τ is the tortuosity.

Resistivity Index. If the cube of porous rock contains both water and hydrocarbons (Fig. 2-57, cube 3), the water is still the only conductor. The cross-sectional area available for conduction is reduced further to A_a', and the path length changed to L_a'. In a similar manner to the foregoing examples, the resistance of the cube is given by

$$r_3 = \frac{R_w L_a'}{A_a'}$$

The resistivity of a partially water-saturated rock is defined as

$$R_t = \frac{r_3 A}{L}$$

and

$$R_t = \frac{R_w L_a' A}{L A_a'} \qquad (2\text{-}57)$$

The second fundamental notion of electrical properties of porous rocks is that of the resistivity index I:

$$I = \frac{R_t}{R_o} \qquad (2\text{-}58)$$

Therefore

$$I = \frac{A_a/A_a'}{L_a/L_a'} \qquad (2\text{-}59)$$

Both the formation factor and the resistivity index are shown to be functions of effective path length and effective cross-sectional area. It is desirable to relate these quantities with other physical parameters of the rock. To do so requires the use of idealized models of porous systems, as the internal geometry of the pores is too complex to express analytically.

Three idealized representations have been introduced in the literature from which relations have been developed relating F and I with porosity ϕ and tortuosity τ.

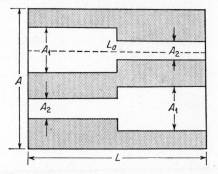

Fig. 2-58. Straight capillary-tube model of porous media. (*After Wyllie and Spangler.*[38])

The first of these models was presented by Wyllie[38] et al. and is as shown in Fig. 2-58. In the model, it is considered that the various pore openings are continuous. The cross-sectional areas of the pore openings vary along their length but in such a manner that the sum of the areas of the pores is constant. L_a in such a model represents the average path length through the pores.

In such a model,

$$A_a = A_1 + A_2 = \phi A$$

Therefore, substituting ϕA for A_a in Eq. (2-56),

$$F = \frac{L_a/L}{\phi A/A} = \frac{L_a}{L} \frac{1}{\phi} \qquad (2\text{-}60)$$

If a hydrocarbon is introduced into the pores, the water saturation S_w can be expressed as a fraction of the pore volume. Presence of the hydrocarbons further reduces the effective cross-sectional area available for flow to A_a', and the average path length is altered to L_a'. Again considering that the cross-sectional area available for flow is the same at each plane in the cube,

$$A_a' = \phi S_w A$$

then substituting $\phi S_w A$ into Eq. (2-59),

$$I = \frac{\phi A / \phi S_w A}{L_a / L_a'} = \frac{L_a'}{L_a} \frac{1}{S_w} \qquad (2\text{-}61)$$

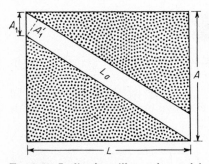

FIG. 2-59. Inclined capillary-tube model of porous media. (*After Cornell and Katz.*[39])

Cornell and Katz[39] have presented a slightly different model as illustrated in Fig. 2-59. In the simplest form of this model, the pores can be considered uniform in cross sections but oriented so that they have an effective length L_a which is greater than L. The cross-sectional area available for flow is once again considered constant at each plane in the model.

The effective cross-sectional area A_a is the area normal to the direction of flow in the pore;

therefore $A_a = A_1'$

but $A_1' = A' \dfrac{L}{L_a}$

and $A_1 = \phi A$

Therefore $A_a = \phi A \dfrac{L}{L_a}$

and substituting in Eq (2-56)

$$F = \frac{L_a / L}{\phi A (L/L_a)/A} = \left(\frac{L_a}{L}\right)^2 \frac{1}{\phi} = \frac{\tau}{\phi} \qquad (2\text{-}62)$$

Following the same reasoning as above and considering a hydrocarbon saturation present,

$$A_a' = A_1''$$

$$A_1'' = A_1 \frac{L}{L_a'}$$

and $\qquad A_1 = \phi S_w A \qquad$ therefore $A'_a = \phi S_w A \dfrac{L}{L'_a}$

then substituting into Eq. (2-59)

$$I = \frac{\phi A (L/L_a)/\phi S_w A (L/L'_a)}{L_a/L'_a}$$

$$= \left(\frac{L'_a}{L_a}\right)^2 \frac{1}{S_w} \tag{2-63}$$

Wyllie and Gardner[40] have recently introduced a third model which is shown in Fig. 2-60. In this model, the cross-sectional area of the pores is

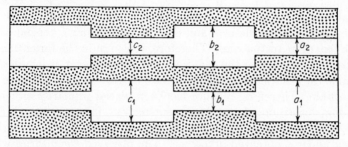

FIG. 2-60. Capillary tube model of porous media. (*After Wyllie and Gardner.*[40])

again considered constant. However, it is conceived that the effective flow cross section is only the net exit area at each plane. Thus the probability that a selected point will fall in a pore opening in one plane is ϕ, that it will fall also in a pore opening in the contiguous plane is $(\phi)^2$,

therefore $\qquad\qquad A_a = (\phi)^2 A$

Substituting in Eq. (2-56),

$$F = \frac{L_a/L}{\phi^2 A/A} = \frac{L_a}{L}\frac{1}{\phi^2}$$

In such a model flow is considered only in matching pore openings such that $L_a = L$.

Therefore $\qquad\qquad F = \dfrac{1}{\phi^2} \tag{2-64}$

If a hydrocarbon saturation is admitted, on the basis of the previous assumptions

$$A'_a = (\phi S_w)^2 A$$

and $\qquad\qquad L'_a = L_a = L$

Substituting in Eq. (2-59),

$$I = \frac{(\phi)^2 A/\phi^2 S_w^2 A}{L_a/L'_a} = \frac{1}{S_w^2} \tag{2-65}$$

From the analysis of the electrical properties of the foregoing models, general relationships between electrical properties and other physical properties of the rock can be deduced. The formation factor has been shown to be some function of the porosity and the internal geometry of the rock system. In particular, it can be stated from examination of Eqs. (2-60), (2-62), and (2-64) that the formation factor can be expressed in the following form:

$$F = C\phi^{-m} \qquad (2\text{-}66)$$

where C is some function of the tortuosity and m is a function of the number of reductions in pore opening sizes or closed-off channels. Since C is a function of the ratio L_a/L, it is suggested that C should be 1 or greater. The value of m has been shown from theory to range from 1 to 2.

Both the formation factor F and the resistivity index I depend on ratios of path length or tortuosities. Therefore, to compute the formation factor or resistivity index from the equations developed above, it is necessary to determine the electrical tortuosity. Direct measurement of the path length is impossible. Therefore, reliance has been placed primarily on empirical correlations based on laboratory measurements. Winsauer[41] et al. devised a method of determining tortuosity by transit time of ions flowing through the rock under a potential difference. The observed tortuosities were believed to be reliable. The data obtained were correlated with the product $F\phi$ as suggested by Eq. (2-62), rearranged as follows:

$$\left(\frac{L_a}{L}\right)^{1.67} = F\phi \qquad (2\text{-}67)$$

The deviation from the theory is believed to be an indication of the greater complexity of the actual pore system than that of the model on which the theory was based.

The dependence of the formation factor on porosity was suggested by Sundberg[42] in 1932. Table 2-7 summarizes Sundberg's computations for

TABLE 2-7. PHYSICAL AND ELECTRICAL CHARACTERISTICS OF SIMPLE
PACKINGS OF SPHERICAL GRAINS OF UNIFORM SIZE[43]

Packing	Porosity, %	Resistivity of packing fully saturated with water of resistivity, ρ_w	
Cubic	47.6		$2.64\rho_w$
Rhombic	39.5	Perpendicular to plane	$4.40\rho_w$
		Parallel of paper	$3.38\rho_w$
Hexagonal	25.9		$5.81\rho_w$

uniform spheres arranged systematically. Archie[36], in 1942, correlated observed formation factors with porosity and permeability. He suggested that the correlation with porosity was the better correlation and that the formation factor could be expressed

$$F = \phi^{-m} \tag{2-68}$$

where ϕ is the fractional porosity and m is the cementation factor. Archie further reported that the cementation factor probably ranged from 1.8 to 2.0 for consolidated sandstones and for clean unconsolidated sands was about 1.3. Figure 2-61 presents the family of curves defined by Eq. (2-68)

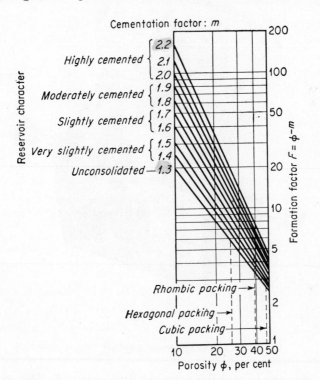

Fig. 2-61. Formation factor F versus per cent porosity for various reservoir characters or cementation classes. (*From Pirson.*[43])

and cementation factors ranging from 1.3 to 2.2. The dashed lines indicate the values computed for systematic packing of uniform spheres.

Measurement of Electrical Resistivity of Rocks

Laboratory measurements of electrical properties of rocks have been made with a variety of devices. The measurements require a knowledge of the dimension of the rock, the fluid saturation of the rock, the resistivity of the water contained in the rock, and a suitable resistivity cell in which to test the sample.

A simple cell is shown in Fig. 2-62. A sample cut to suitable size is placed in the cell and clamped between electrodes. Current is then passed

Fig. 2-62. Core sample resistivity cell. (*From Rust.*[44])

through the sample, and the potential drop observed. The resistance of the sample is computed from Ohm's law:

$$r = \frac{I}{E}$$

(2-69)

and R (the resistivity) is computed from

$$R = \frac{rA}{L}$$

(2-70)

where A is the cross-sectional area of the sample and L is the length of the sample. The saturation conditions of the test can be established at known values prior to measurement or determined by an extraction procedure after measurement.

A second type of cell is shown in Fig. 2-63 which is a combined capillary-pressure and resistivity cell. This device has the advantage that two different tests can be performed simultaneously. The disadvantage is the length of time required for a capillary-pressure test. Capillary-pressure tests are discussed in Chap. 3 of this volume.

Empirical Correlation of Electrical Properties

Archie, as previously mentioned, reported the results of correlating laboratory measurements of formation factor with porosity. He expressed his results in the form

$$F = \phi^{-m}$$

Archie derived from experimental data that $F = \phi^{-1.3}$. Slawinski and Maxwell[46] derived theoretical expressions for the formation factor based on models of unconsolidated spheres. Slawinski stated that for spheres in contact

$$F = \frac{(1.3219 - 0.3219\phi)^2}{\phi}$$

FIG. 2-63. Combined interstitial water and resistivity cell (*From Rust.*[44])

For dispersed spheres, not necessarily in contact, Maxwell states that

$$F = \frac{3 - \phi}{2 - \phi}$$

Wyllie[45] investigated the influence of particle size and cementation on the formation factor of a variety of materials. Unconsolidated materials were packed in tubes, and some were artificially consolidated.

Wyllie's experimental data are compared with the results calculated using Archie's and Slawinski's and Maxwell's expressions in Fig. 2-64. Archie's and Slawinski's equations fit the data reasonably well except for the aggregate of cubes. The data for the cubes fall above the other data as well as above all three lines calculated from the equations. This could possibly be indicative of a greater tortuous path length in such a system.

Observed formation factors for artificially cemented aggregates are shown in Fig. 2-65. It may be noted that cementation results in increased values of formation factor over that observed for uncemented aggregates. Furthermore, the cemented aggregates exhibit a greater change in formation factor with a change in porosity than the unconsolidated aggregates. The curves no longer pass through the point $F = 1$, $\phi = 100$ per cent.

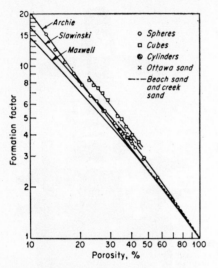

From these data Wyllie concluded that the general form of the relation between formation factor and porosity should be

$$F = C\phi^{-m} \qquad (2\text{-}66)$$

where m is a constant depending on cementation and C a constant controlled by the porosity of the unconsolidated matrix prior to cementation. This is identical with the general form [Eq. (2-66)] deduced theoretically using simple models.

Winsauer[41] et al. reported a similar relationship based on correlations of data from a large number of sandstone cores. This equation, commonly referred to as the Humble relation, is

$$F = 0.62\ \phi^{-2.15} \qquad (2\text{-}71)$$

FIG. 2-64. Formation factor—porosity data of unconsolidated porous aggregates. $\phi = 12.4 - 56$ per cent. (*From Wyllie.[45]*)

In discussing the theory it was stated that C should be greater than 1 and that m should be 2 or less. The discrepancy between theory and experiment must at this time be attributed to limiting the values of C and m to constants for a wide range of porosities.

Improved correlations should result from considering other parameters such as permeability as variables in the relations.

A comparison of suggested relationships between porosity and the formation factor is shown in Fig. 2-66.

Since the formation factor is a function of porosity and some unknown effect of the complex internal geometry, it is suggested that the constants in formulas similar to Eq. (2-71) are functions of the depositional environment and must be determined on each formation to yield the most reliable results. Of the correlations presented in Fig. 2-66, the Humble relation appears to be of the greatest general utility.

Effect of Conductive Solids.   It was pointed out in the opening discussion that clay minerals might act as conductors and contribute to the conductivity of a water-saturated porous rock. Investigations by Wyllie[48] indicate that clays contribute substantially to the conductivity of a rock when the rock is saturated with a low-conductivity water. The effect of water resistivity on the formation factor for sands containing clay minerals is shown in Fig. 2-67. The formation factor for a comparable clean (clay-free) sand is a constant. The formation factor for the clayey sand increases

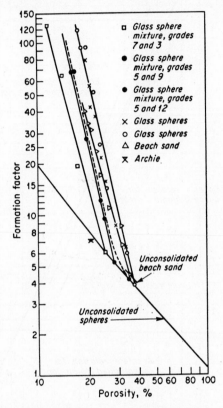

FIG. 2-65. Effect of artificial cementation on the formation factor–porosity relationship of unconsolidated aggregates. (*From Wyllie.*[45])

with decreasing water resistivity and approaches a constant value at a water resistivity of about 0.1 ohm-m. The apparent formation factor F was calculated from the definition of the formation factor and observed values of R_{oa} and R_w. Wyllie proposed that the observed effect of clay minerals was similar to having two electrical circuits in parallel: the conducting clay minerals and the water-filled pores. Thus

$$F_a = \frac{R_{oa}}{R_w} \quad \text{and} \quad \frac{1}{R_{oa}} = \frac{1}{R_c} + \frac{1}{FR_w} \qquad (2\text{-}72)$$

where R_{oa} is the resistivity of a shaly sand when 100 per cent saturated with water of resistivity R_w, R_c is the resistivity due to the clay minerals; FR_w is the resistivity due to the distributed water, and F is the true formation factor of the rock (i.e., the constant value of formation factor approached when the rock contains low-resistivity water).

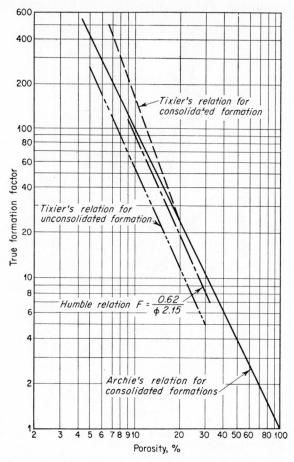

FIG. 2-66. Comparison of various formation factor correlations. (*From Owen.*[47])

The data presented in Fig. 2-68 represent graphically the confirmation of the relationship expressed in Eq. (2-72). The graphs were plotted by deWitte[49] from data presented by Hill and Milburn.[50] The plots are linear and are of the general form

$$\frac{1}{R_{oa}} = m\,\frac{1}{R_w} + b \tag{2-73}$$

where m is the slope of the line and b is the intercept. Comparing Eq. (2-72) with Eq. (2-73), it may be noted that $m = 1/F$ and $b = 1/R_c$. The curve labeled suite 1, No. 40, indicates a clean sand, since the line passes through the origin, therefore

$$b = \frac{1}{R_c} = 0 \qquad \text{Then} \frac{1}{R_{oa}} = m\frac{1}{R_w} = \frac{1}{FR_w} \qquad \text{or} \qquad R_o = FR_w$$

The remaining samples are from shaly sands which have a finite conductivity of the clay minerals as indicated by the intercepts of the lines. The linearity of the plots indicate that $1/R_c$ is a constant independent of R_w. This phenomenon can be explained in terms of the ions adsorbed on the clay. When the clay is hydrated, the adsorbed ions form an ionic conducting path which is closely bound to the clay. The number of adsorbed ions is apparently little changed by the salt concentration of the interstitial water.

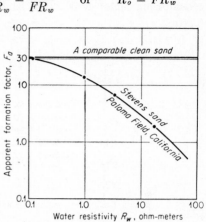

Fig. 2-67. Apparent formation factor versus water resistivity for Stevens sand of Paloma Field, Calif. (*From Winn.*[37])

Equation (2-72) can be rearranged to express the apparent formation factor in terms of R_c and FR_w.

$$R_{oa} = \frac{R_c R_w}{R_w + (R_c/F)} \qquad \text{and} \qquad F_a = \frac{R_c}{R_w + (R_c/F)} \qquad (2\text{-}72)$$

As $R_w \to 0$, $\lim\limits_{R_w \to 0} F_a = \frac{R_c}{R_c/F} = F$. Therefore F_a approaches F as a limit as R_w becomes small. This was observed in Fig. 2-67.

Hill and Milburn[50] presented a somewhat more complex correlation of the formation factor of shaly sands. The correlation was based on measurements on a large number of samples. They state that the contribution of clay minerals to the conductivity of a rock is not a constant as proposed by Patnode[48] and deWitte.[49] Their correlation is as follows:

$$F_a = F_{0.01}100R_w{}^{b\,\log\,(100R_w)} \qquad (2\text{-}74)$$

where $F_{0.01}$ is the formation factor of the rock when saturated with water having a resistivity of 0.01 ohm-m. The quantity b in the exponent is defined as a shaliness factor and was correlated with the cation-exchange capacity. The cation-exchange capacity is related to the clay content of the rock and provides an independent determination of the amount of shale in a rock.

Additional work is required to determine how the effect of clay in a rock can best be evaluated.

The experimental data discussed are largely from measurements on sandstones or similar materials having intergranular porosity. Little data are available on the electrical properties of limestone. Tixier[51] states that a cementation factor m of 2.0 in Archie's formula yields a satisfactory correlation.

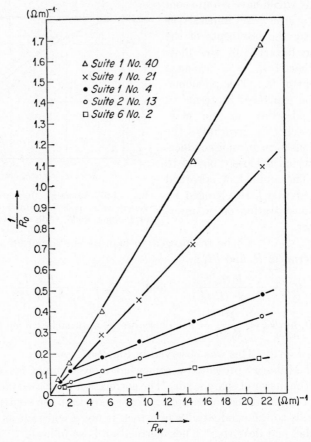

FIG. 2-68. Water-saturated rock conductivity as a function of water conductivity. (*From deWitte.[49]*)

Resistivity of Partially Water-saturated Rocks. A rock containing both water and hydrocarbon has a higher resistivity than the rock when fully saturated with water. The resistivity of partially water-saturated rocks has been shown to be a function of the water saturation S_w.

The resistivity index introduced earlier is a convenient function for correlating experimental data on the resistivity of rocks with water saturation. Equations (2-61), (2-63), and (2-65) indicate that the resistivity index is a function of the water saturation and the path length. From the theoretical developments, the following generalization can be drawn:

$$I = C'S_w^{-n} \tag{2-75}$$

where $I = R_t/R_o$, the resistivity index; C' is some function of tortuosity; and n is the saturation exponent.

Archie compiled and correlated experimental data from Wyckoff,[52] Leverett,[53] Jakosky,[54] and Martin[55] from which he suggested that the data could be represented by

$$I = S_w^{-2} \tag{2-76}$$

Williams[56] studied consolidated sands, the results for which are shown in Fig. 2-69. The solid line was fitted to the data points by the method of least squares. The equation of the best fitting line is

$$I = S_w^{-2.7} \tag{2-77}$$

The dashed line is a plot of Archie's relation [Eq. (2-76)] for comparison.

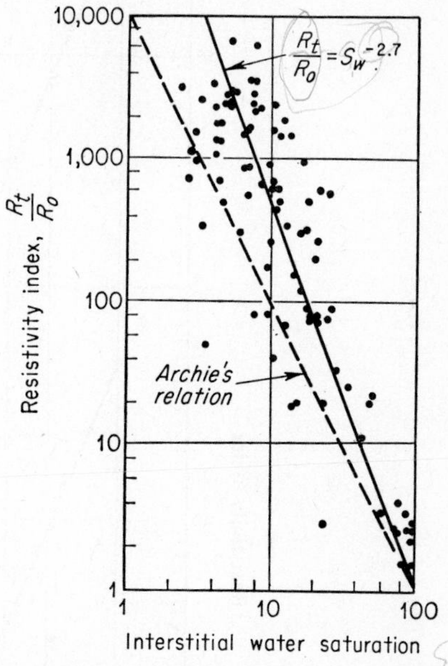

In Fig. 2-70 are presented results obtained by Rust[44] on consolidated samples from Woodbine sand outcrop. The saturation exponent n ranges from 2.31 to 2.40.

Fig. 2-69. Resistivity index versus interstitial water saturation. (*From Williams.*[56])

All the equations fitted to the experimental data have assumed that both C' and n of Eq. (2-75) were constants and furthermore that $C' = 1$. From the theory, it would be expected that C' is a function of saturation and that n would range between 1 and 2. Additional study is required to ascertain the discrepancy between theory and experiment.

Morse[57] et al. presented data showing the effect of fluid distribution on the saturation-resistivity relationship. In Fig. 2-71 curves 2 and 3 are from data by Morse while curve 1 is for Archie's relation. The data for curve 2 are from artificially consolidated sands containing water and air, while those for curve 3 are from the same material containing water and oil. The difference in the results is attributed to the distribution of the fluids within the rock. The material was believed to be water wet in the water-air tests and oil wet in the water-oil tests. The difference in wettability would cause a different distribution of fluids to be established at the same value of water saturation.

Whiting[58] et al. reported tests of the saturation relationship in limestones.

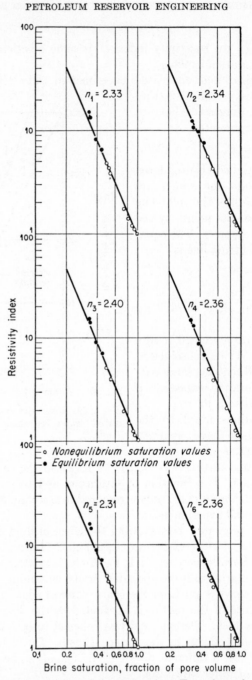

FIG. 2-70. Resistivity index versus brine saturation. Four-electrode method. Woodbine outcrop. Gas permeability 1,130 millidarcys; effective porosity, 38.8 per cent. (*From Rust.*[44])

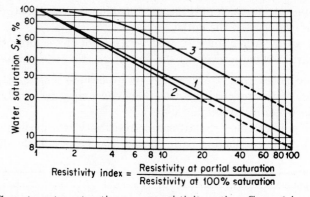

Resistivity index = $\dfrac{\text{Resistivity at partial saturation}}{\text{Resistivity at 100\% saturation}}$

Fig. 2-71. Connate water saturation versus resistivity ratios. Curve 1 is composite of data from Wycoff, Leverett, Jakosky, and Martin; curve 2 is for brine-air displacement results of Morse et al. on artificially consolidated sands; curve 3 is for brine-oil displacement results. (*From Pirson.*[43])

The data are presented in Fig. 2-72. The resistivity ratio plotted is the reciprocal of the resistivity index. The effect of the method of changing the saturation was also investigated in these tests. The trends of the curves are the same as for sandstones. The three methods of changing the saturation in the test specimens were:

1. Dynamic air brine in which the desired water saturation was obtained by flowing air and water simultaneously through the sample

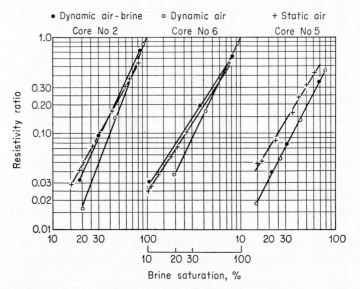

Fig. 2-72. Comparison of electrical resistivity–brine saturation relationships obtained in dynamic and static experiments. (*From Whiting et al.*[58])

2. Dynamic air in which only air was introduced at the inlet, displacing both air and water from the outlet

3. Static air in which air displaced water from the sample through a capillary barrier which prevented the flow of air from the sample

It may be noted that the dynamic air procedure consistently yielded lower values of the resistivity ratio. This effect may be attributed to a difference in water distribution.

Conductive clays affect the saturation-resistivity relationship as shown in Fig. 2-73. The conducting path through the clays is little affected by

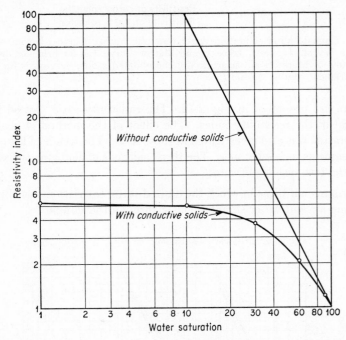

FIG. 2-73. Effect of conductive solids on the resistivity-index saturation relationship in Stevens sandstone core. (*From Patnode and Wyllie.*[48])

the presence of hydrocarbon. Thus as the water saturation is reduced to zero, the resistivity approaches the resistivity of the clay path rather than approaching infinity as in clean sands. The relationship of saturation and resistivity in shaly sands is complex and will not be considered at greater length.

Use of Electrical Parameters in Characterizing Porous Media. In the section on permeability, the Kozeny equation was developed as follows:

$$k = \frac{\phi}{k_0 \tau S_p{}^2} \qquad (2\text{-}49)$$

where k is the permeability, ϕ is the porosity fraction, k_0 is a shape factor, S_p is the internal surface area per unit pore volume, and τ is the Kozeny tortuosity.

By definition,
$$\tau = \left(\frac{L_a}{L}\right)^2$$

and from Eq. (2-60)
$$\tau = F^2\phi^2 \tag{2-78}$$

which on substitution in Eq. (2-49) leads to
$$k = \frac{1}{k_0 F^2 \phi S_p{}^2} \tag{2-79}$$

However, from Eq. (2-62)
$$\tau = F\phi \tag{2-80}$$

which on substitution in Eq. (2-49) leads to
$$k = \frac{1}{k_0 F S_p{}^2} \tag{2-81}$$

The only experimental measurements of L_a/L yielded, as previously stated,
$$\left(\frac{L_a}{L}\right)^{1.67} = F\phi \tag{2-67}$$

or
$$\tau = \left(\frac{L_a}{L}\right)^2 = (F\phi)^{2/1.67} \tag{2-82}$$

which leads on substitution to
$$k = \frac{\phi}{k_0(F\phi)^{1.33}S_p{}^2} \tag{2-83}$$

The quantity
$$S_p = \frac{1}{r_m} \tag{2-84}$$

where r_m is the mean hydraulic radius. Thus, the general form of the relationship can be stated
$$k = \frac{\phi r_m{}^2}{k_0(F\phi)^x} \tag{2-85}$$

or
$$\frac{r_m{}^2}{k} = k_0\frac{(F\phi)^x}{\phi} \tag{2-86}$$

The above relations have been partially verified by experiment.

REFERENCES

1. Slichter, C. S.: Theoretical Investigation of the Motion of Ground Water, *U.S. Geol. Survey 19th Annual Rep.*, 1899.
2. Fraser, H. J., and L. C. Graton: Systematic Packing of Spheres—With Particular

Relation to Porosity and Permeability, *J. Geol.*, November–December, 1935, pp. 785–909.

3. Nanz, Robert H., Jr.: Genesis of Oligocene Sandstone Reservoir, Seeligson Field, Jim Wells and Kleberg Counties, Texas, *Bull. Am. Assoc. Petrol. Geologists*, January, 1954, p. 96.

4. Tickell, F. G., O. E. Mechen, and R. C. McCurdy: Some Studies on the Porosity and Permeability of Rocks, *Trans. AIME*, 1933, pp. 250–260.

5. Nuss, W. F., and R. L. Whiting: Technique for Reproducing Rock Pore Space, *Bull. Am. Assoc. Petrol. Geologists*, November, 1947, p. 2044.

6. Core Laboratories, Inc., *Trade Circ.*

7. Kaye, Emil, and M. L. Freeman, Jr.: A New Type of Porosimeter, *World Oil*, March, 1949, p. 94.

8. Nutting, P. G.: Physical Analysis of Oil Sands, *Bull. Am. Assoc. Petrol. Geologists*, vol. 14, 1930.

9. Russell, W. L.: A Quick Method for Determining Porosity, *Bull. Am. Assoc. Petrol. Geologists*, vol. 10, 1926.

10. Stevens, A. B.: New Device for Determining Porosity by the Gas Expansion Method, *Tech. Publ.* 1061, *Trans. AIME*, May, 1939.

11. Rall, C. G., and D. B. Taliaferro: A Bureau of Mines Method for Determining Porosity: A List of Porosity of Oil Sands, *U.S. Bur. Mines*, September, 1948.

12. Dotson, B. J., R. L. Slobod, P. N. McCreery, and James W. Spurlock: Porosity—Measurement Comparisons by Five Laboratories, *Trans. AIME*, 1951, p. 341.

13. Kelton, Frank C.: Analysis of Fractured Limestone Cores, *Trans. AIME*, 1950, p. 225.

14. Krumbein, W. C., and L. L. Sloss: "Stratigraphy and Sedimentation," p. 218, 1st ed., W. H. Freeman Publishing Company, 1951.

15. Geertsma, J.: The Effect of Fluid Pressure Decline Oil Volumetric Changes of Porous Rocks, *Trans. AIME*, vol. 210, 1957.

16. Fatt, I.: Pore Volume Compressibilities of Sandstone Reservoir Rocks, *J. Petrol. Technol.*, March, 1958.

17. Carpenter, Charles B., and George B. Spencer: Measurements of Compressibility of Consolidated Oil-bearing Sandstones, *U.S. Bur. Mines* Rept. Invest. 3540, October, 1940.

18. Hall, Howard N.: Compressibility of Reservoir Rocks, *Trans. AIME*, 1953, p. 309.

19. "Recommended Practice for Determining Permeability of Porous Media," American Petroleum Institute, Division of Production, September, 1952.

20. Fancher, G. H., J. A. Lewis, and K. B. Barnes: Some Physical Characteristics of Oil Sands, *Penn. State Coll. Bull.* 12, 1933, pp. 65–171.

21. Darcy, H.: "Les fontaines publiques de la ville de Dyon," Victor Dalmont, 1856.

22. Hubbert, M. King: Entrapment of Petroleum under Hydrodynamic Conditions, *Bull. Am. Assoc. Petrol. Geologists*, August, 1953, p. 1954.

23. Muskat, Morris: "Flow of Homogeneous Fluids," p. 287, McGraw-Hill Book Company, Inc., New York, 1937.

24. Croft, H. O.: "Thermodynamics, Fluid Flow and Heat Transmission," p. 129, McGraw-Hill Book Company, Inc., New York, 1938.

25. Stevens, A. B.: "A Laboratory Manual for Petroleum Engineering," p. 308, Exchange Store, A. and M. College of Texas, 1954.

26. Beeson, C. M.: The Kobe Porosimeter and the Oilwell Research Porosimeter, *Trans. AIME*, 1950.

27. Klinkenberg, L. J.: The Permeability of Porous Media to Liquids and Gases, *Drilling and Production Practices*, p. 200, American Petroleum Institute, 1941.

28. Johnston, Norris, and Carrol M. Beeson: Water Permeability of Reservoir Sands, *Trans. AIME*, 1945, p. 292.

29. Muskat, Morris: "Physical Principles of Oil Production," p. 142, McGraw-Hill Book Company, Inc., New York, 1949.

30. Fatt, I., and D. H. Davis: Reduction in Permeability with Overburden Pressure, *Trans. AIME*, 1952, p. 329.

31. Wyllie, M. R. J., and M. B. Spangler: Application of Electrical Resistivity Measurements to Problems of Fluid Flow in Porous Media, *Bull. Am. Assoc. Petrol. Geologists*, February, 1952.

32. Carman, P. C.: *J. Soc. Chem. Ind.*, vols. 57 and 58, 1939.

33. Rapoport, L. A., and W. J. Leas: Relative Permeability to Liquid in Gas-Liquid Systems, *Trans. AIME*, vol. 192, 1951.

34. Emdahl, Ben A.: Core Analysis of Wilcox Sands, *World Oil*, June, 1952.

35. Kennedy, H. T., O. E. Van Meter, and R. G. Jones: Saturation Determination of Rotary Cores, *Petrol. Engr.*, January, 1954.

36. Archie, G. E.: The Electrical Resistivity Log as an Aid in Determining Some Reservoir Characteristics, *Trans. AIME*, 1942.

37. Winn, R. H.: The Fundamentals of Quantitative Analysis of Electric Logs, *Symposium on Formation Evaluation*, AIME, October, 1955.

38. Wyllie, M. R. J., and M. B. Spangler: Application of Electrical Resistivity Measurements to Problem of Fluid Flow in Porous Media, *Bull. Am. Assoc. Petrol. Geologists*, February, 1952, p. 359.

39. Cornell, D., and D. L. Katz: *Ind. Eng. Chem.*, vol. 45, 1953.

40. Wyllie, M. R. J., and G. H. F. Gardner: The Generalized Kozeny-Carman Equation, *World Oil*, March and April, 1958.

41. Winsauer, W. O., H. M. Shearin, P. H. Masson, and M. Williams: Resistivity of Brine-saturated Sands in Relation to Pore Geometry, *Bull. Am. Assoc. Petrol. Geologists*, February, 1952.

42. Sundberg, Karl: Effect of Impregnating Waters on Electrical Conductivity of Soils and Rocks, *Geophysical Prospecting*, AIME, 1932.

43. Pirson, S. J.: "Oil Reservoir Engineering," 2d ed., McGraw-Hill Book Company, Inc., New York, 1958.

44. Rust, C. F.: Electrical Resistivity Measurements on Reservoir Rock Samples by the Two-electrode and Four-electrode Methods, *Trans. AIME*, 1952.

45. Wyllie, M. R. J.: Formation Factors of Unconsolidated Porous Media: Influence of Particle Shape and Effect of Cementation, *Trans. AIME*, 1953.

46. Slawinski, A.: Conductivity of an Electrolyte Containing Dielectric Bodies, *J. chem. phys.*, 1926.

47. Owen, Joe D.: Well Logging Study: Quinduno Field, Roberts County, Texas, *Symposium on Formation Evaluation*, AIME, October, 1955.

48. Patnode, H. W., and M. R. J. Wyllie: The Presence of Conductive Solids in Reservoir Rocks as a Factor in Electric Log Interpretation, *Trans. AIME*, 1950.

49. deWitte, A. J.: Saturation and Porosity from Electric Logs in Shaly Sands, *Oil Gas J.*, Mar. 4, 1957.

50. Hill, H. J., and J. D. Milburn: Effect of Clay and Water Salinity on Electrochemical Behavior of Reservoir Rocks, *Trans. AIME*, 1956.

51. Tixier, M. P.: Porosity Index in Limestone from Electrical Logs, *Oil Gas J.*, 1951.

52. Wyckoff, R. D., and H. G. Botset: Flow of Gas Liquid Mixtures through Sands, *Physics*, 1936, p. 325.

53. Leverett, M. C.: Flow of Oil-Water Mixtures through Unconsolidated Sands, *Trans. AIME*, 1939.

54. Jakosky, J. J., and R. H. Hopper: The Effect of Moisture on the Direct Current Resistivities of Oil Sands and Rocks, *Geophysics*, vol. 2, 1937.

55. Martin, M., G. H. Murray, and W. J. Gillingham: Determination of the Potential Productivity of Oil-bearing Formations by Resistivity Measurements, *Geophysics*, vol. 3, 1938.

56. Williams, Milton: Estimation of Interstitial Water from the Electrical Log, *Trans. AIME*, 1950.

57. Morse, R. A., et al.: Relative Permeability Measurements on Small Core Samples, *Oil Gas J.*, Aug. 23, 1947.

58. Whiting, R. L., E. T. Guerrero, and R. M. Young: Electrical Properties of Limestone Cores, *Oil Gas J.*, July 27, 1953.

CHAPTER 3

PROPERTIES OF POROUS MEDIA
CONTAINING MULTIPLE FLUID SATURATIONS

In preceding chapters the physical properties of reservoir rocks are defined in terms of single-fluid systems. Such a simplified case is seldom found in actual petroleum reservoirs. In petroleum reservoirs two fluids are present, and many times three fluid phases are involved. All the basic definitions must be modified and other definitions added for a complete classification of the properties of a petroleum reservoir.

The simultaneous existence of two or more fluids in a porous rock requires that terms such as capillary pressure, relative permeability, and wettability be defined. When only one fluid exists in the pore spaces, there is only one set of forces to consider, the attraction between the rock and the fluid. When more than one fluid phase is present, there are at least three sets of active forces affecting capillary pressure and wettability. In the preceding chapter, permeability was defined and discussed in terms of a rock saturated with a single fluid. The material which follows amplifies the previous definitions and introduces concepts which are required for multifluid systems. The measurements and use of these various factors also will be discussed.

SURFACE FORCES AND CAPILLARY PRESSURE

In dealing with multiphase systems, it is necessary to consider the effect of the forces acting at the interface when two immiscible fluids are in contact. When these two fluids are liquid and gas, the interface is normally referred to as the liquid surface. All molecules are attracted one to the other in proportion to the product of their masses and inversely as the square of the distance between them. Considering water and oil, fluids commonly found in petroleum reservoirs, it is found that an interfacial tension always exists between the fluids. A water molecule which is remote from the interface is surrounded by other water molecules, thus having a resulting net attractive force on the molecule of zero. However, a molecule at the interface has a force acting upon it from the oil lying immediately above the interface and water molecules lying below the interface.

133

The resulting forces are unbalanced and give rise to interfacial tension. The unbalanced attractive force between the molecules creates a membranelike surface. A certain amount of work is required to move a water molecule from within the body of the liquid through the interface. This work is frequently referred to as the free surface energy of the liquid. Free surface energy, in ergs per square centimeter, may be defined as the work necessary to create a unit area of new surface. The interfacial tension is the force per unit length required to create a new surface. Interfacial tension and surface tension are commonly expressed in dynes per centimeter, which is numerically equal to the surface energy in ergs per square centimeter. Surface tension is measured in the laboratory by standard means such as a tensiometer, the drop method, or other methods which can be found described in physical chemistry texts.

Fundamentals of Surface and Capillary Forces

In dealing with hydrocarbon systems, it is necessary to consider not only the interface between a gas and a liquid but also the forces that are active at the interface between two immiscible liquid phases and between the liquids and solids. The combination of all the active surface forces determines the wettability and capillary pressure of a porous rock.

Wetting. The adhesion tension, which is a function of the interfacial tension, determines which fluid will preferentially wet the solid. A sketch is shown in Fig. 3-1, wherein two liquids, oil and water, are in contact with a solid. By convention, the contact angle theta (θ) is measured through the denser liquid phase and ranges from 0 to 180°. Based on the above convention of expression the adhesion tension is defined in Eq. (3-1).

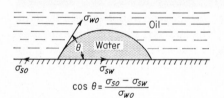

$$\cos \theta = \frac{\sigma_{so} - \sigma_{sw}}{\sigma_{wo}}$$

FIG. 3-1. Equilibrium of forces at a water-oil-solid interface. (*After Benner and Bartell.*[1])

$$A_T = \sigma_{so} - \sigma_{sw} = \sigma_{wo} \cos \theta_{wo} \quad (3\text{-}1)$$

A_T is the adhesion tension, σ_{so} is the interfacial tension between the solid and lighter fluid phase, σ_{sw} is the interfacial tension between the solid and denser phase, and σ_{wo} is the interfacial tension between the fluids.

A positive adhesion tension indicates that the denser phase preferentially wets the solid surface. An adhesion tension of zero indicates that both phases have an equal affinity for the surface. The magnitude of the adhesion tension, as defined by Eq. (3-1), determines the ability of the wetting phase to adhere to the solid and to spread over the surface of the solid. If the adhesion tension value is large or the contact angle θ is small, the denser phase will readily spread and tend to coat the surface. If the contact angle is large, an outside source of energy will be required to cause the denser phase to spread over the surface. The degree of spreading as affected by the contact angle of the system is illustrated in Fig. 3-2, wherein various multiliquid systems are in contact with silica and calcite

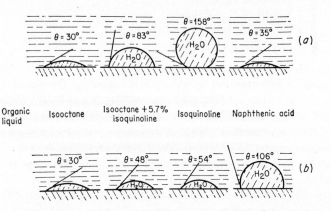

Fig. 3-2. Interfacial contact angles. (a) Silica surface; (b) calcite surface. (*From Benner and Bartell.*[1])

surfaces. It is noted that when water and isooctane are used, the water preferentially wets both the calcite and silica surfaces. When naphthenic acid is used, it is observed that water preferentially wets the silica surface with a contact angle of 35° whereas naphthenic acid preferentially wets the calcite surface with a contact angle of 106°. The other two systems, water–isooctane-plus and water–isoquinoline, yield results similar to the two previously discussed systems. This illustrates the effects that might be expected from varying the mineralogy of the rock and the composition of the two liquid phases. It further illustrates that for an oil-water-solid system, it is possible to have either a water-wet or oil-wet surface, depending on the chemical composition of the fluids and rock.

Rise of Fluids in Capillaries. Consider the case of capillary tubes wherein the internal diameter of the tube is extremely small. If the tube is placed in a large open vessel containing liquid, liquid will rise in the capillary tube above the height of the liquid in the large vessel. This rise in height is due to the attractive forces (adhesion tension) between the tube and the liquid and the small weight represented by the column of liquid in the tube. The adhesion tension is the force tending to pull the liquid up the wall of the tube. The liquid will rise in the tube until the total force acting to pull the liquid upward is balanced by the weight of the column of liquid being supported in the tube.

The total upward force can be expressed as

$$A_T \times 2\pi r = \text{force up} \qquad (3\text{-}2)$$

The weight of the column being supported is

$$\pi r^2 h g \rho = \text{force down} \qquad (3\text{-}2a)$$

where A_T = adhesion tension, dynes/cm
    r = radius of tube, cm
    h = height of liquid column, cm
    ρ = density of liquid in tube, gm/cc
    g = force of gravity, cm/sec^2

Equating these two quantities would yield a force balance such that the total adhesion tension force would be just balancing the gravitational pull on the column of liquid. The pressure existing in the liquid phase beneath the air-liquid interface is less than the pressure which exists in the gaseous phase above the interface. This difference in pressure existing across the interface is referred to as the capillary pressure of the system. This pressure can be calculated on the basis of a U tube, balancing the pressure between the two points.

Figure 3-3 represents the conditions that exist when a capillary tube is immersed in a beaker of water. If the equilibrium height h of the interface

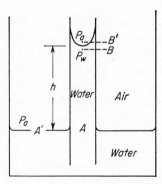

FIG. 3-3. Pressure relations in capillary tubes.

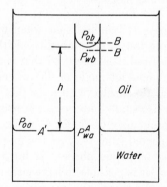

FIG. 3-4. Pressure relations in capillary tubes.

is small, the head of the air is negligible. Therefore, the pressure in the air immediately above the interface is essentially equal to the pressure in the air immediately above the free water level in the large vessel. However, owing to the greater density of the water, the pressure in the water just beneath the interface differs from that at the bottom of the column due to the head of water h. Since the beaker is large compared with the capillary tube, the gas-water interface in the beaker is essentially horizontal. The capillary pressure is zero in a horizontal or plane interface. Therefore the pressure in the water at the bottom of the column is equal to the pressure in the gas at the surface of the water in the large vessel. The pressure in the water at the top of the water column is equal to the pressure in the water at the bottom minus the pressure due to a head of water h. By denoting the pressure in the water at the top by P_w and P_a as the pressure in the gas at both the top and bottom and the pressure in the water at the bottom, from hydrostatics

$$P_a - P_w = \rho_w g h = P_c \tag{3-3}$$

where ρ_w = density of water

g = acceleration due to gravity

h = height of the column of water in the tube above that in the large vessel

Since the pressure in the air at the top is also P_a and by definition the capillary pressure is the pressure difference across an interface, an expression for the height of fluid rise in the tube is obtained by balancing the upward and downward forces.

$$2\pi r A_T = \pi r^2 h g \rho_w$$

$$h = \frac{2\pi r A_T}{\pi r^2 g \rho_w} = \frac{2A_T}{r g \rho_w}$$

by substituting from Eq. (3-1)

$$h = \frac{2\sigma_{wg} \cos \theta_{wg}}{r g \rho_w} \tag{3-4}$$

Substituting the above value for the height in Eq. (3-3), an expression for capillary pressure in terms of the surface forces is obtained.

$$P_c = \frac{2\sigma_{wg} \cos \theta_{wg}}{r} \tag{3-5}$$

Consider the capillary tube immersed in a beaker of water wherein oil is the other fluid rather than air (Fig. 3-4).

Let P_{oa} = pressure in oil at A

P_{ob} = pressure in oil at point B

P_{wa} = pressure in water at point A

P_{wb} = pressure in water at point B

Once again, if the beaker is large, the interface at A is a plane interface and the capillary pressure is zero. Therefore

$$P_{oa} = P_{wa}$$

at the free water level in the beaker. The density of both the oil and the water must be considered in deriving the pressure relationship at point B.

$$P_{ob} = P_{oa} - \rho_o g h$$

$$P_{wb} = P_{wa} - \rho_w g h$$

The pressure difference across the interface is therefore

$$P_{ob} - P_{wb} = (\rho_w - \rho_o)gh = P_c \tag{3-6}$$

Therefore, the capillary pressure must be in equilibrium with gravitational forces if the fluids are in equilibrium and not flowing. The expression of capillary pressure in terms of the surface forces is obtained in the same manner as that for air and water and results in the same expression.

$$P_c = \frac{2\sigma_{wo} \cos \theta_{wo}}{r} \tag{3-7}$$

It is noted in Eq. (3-7) that the capillary pressure is a function of the adhesion tension $(\sigma_{wo} \cos \theta_{wo})$ and inversely proportional to the radius of the capillary tube. Figure 3-5 illustrates the effect of varying the wetting

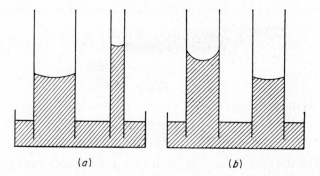

(a)   (b)

FIG. 3-5. Dependence of interfacial curvature upon pore size and contact angle. (a) Same contact angle, different pore size; (b) same pore size, different contact angle.

characteristics of the system and of varying the radius of the capillary tube. If the wetting characteristics remain constant and the radius of the tube is increased, the weight of the water column increases as the square of the radius whereas the magnitude of the adhesion force increases in direct relation to the radius. Therefore, the height of the water column will be decreased proportionally to the increase in the tube radius. This fact is illustrated in Fig. 3-5 wherein it is noted that the smaller the radius of the tube, the higher the water column will rise before an equilibrium system is obtained.

The changes in wetting characteristics are such that the greater the adhesion tension, the greater the equilibrium height obtained. If the only variable is the wetting characteristic of the solid, it is noted that the smaller the contact angle θ, the stronger the adhesion tension and the greater the height to which the liquid column will rise before equilibrium is obtained. This fact is illustrated in Fig. 3-5b, wherein it is noted that for small values of the contact angle, a large height is obtained.

When the radius of the tube or the adhesion tension is changed, the capillary pressure is altered accordingly. From the variation indicated by Fig. 3-5, the following statements can be made: (1) The greater the affinity of the denser phase for the solid, the greater will be the capillary pressure across the interface for a given size tube; (2) the smaller the radius when the wetting characteristics are the same, the greater will be the capillary pressure.

Capillary Pressure in Packings of Uniform Spheres. A general expression for capillary pressure as a function of interfacial tension and curvature of the interface is due to Plateau[2] and is given in Eq. (3-8).

$$P_c = \sigma \left(\frac{1}{R_1} + \frac{1}{R_2} \right) \tag{3-8}$$

where R_1 and R_2 are the principal radii of curvature of the interface and σ is the interfacial tension between the two fluids. The distribution and measurement of these two radii in a porous system are shown in Fig. 3-6. It is noted that these two radii are measured in perpendicular planes. Comparing Eq. (3-8) with the equation for capillary pressure as determined by the capillary tube method, it is found that the mean radius R_m is defined by

$$\frac{1}{R_m} = \left(\frac{1}{R_1} + \frac{1}{R_2} \right) = \frac{2 \cos \theta}{r_t} = \frac{\Delta \rho g h}{\sigma} \tag{3-9}$$

It is practically impossible to measure the values of R_1 and R_2, so they are generally referred to by the mean radius of curvature and empirically determined from other measurements on a porous medium.

The distribution of the liquid in a porous system is dependent upon the wetting characteristics. It is necessary to determine which is the wetting fluid so as to ascertain which fluid occupies the small pore spaces (Fig. 3-6). From packings of spheres, the wetting-phase distribution within a porous system has been described as either funicular or pendular in nature. In funicular distribution, the wetting phase is continuous, completely covering the surface of the solid. The pendular ring is a state of saturation in which the wetting phase is not continuous and the nonwetting phase is in contact with some of the solid surface. The wetting phase occupies the smaller interstices. These distributions are illustrated in Fig. 3-7a and b, the pendular ring distribution in Fig. 3-7a, and the funicular saturation

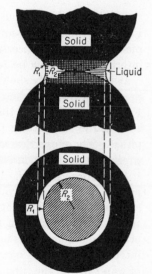

Fig. 3-6. Idealized contact for wetting fluid and spherical grains. (*From Leverett.*[3])

distribution in Fig. 3-7*b*. It is noted in Figs. 3-6 and 3-7 that as the wetting-phase saturation progresses from the funicular to the pendular ring distri-

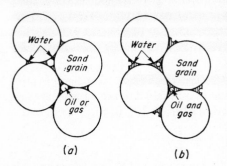

(*a*)   (*b*)

FIG. 3-7. Idealized representation of distribution of wetting and nonwetting fluid phase about intergrain contacts of spheres. (*a*) Pendular-ring distribution; (*b*) funicular distribution. (*From Fancher et al.*[4])

bution, the quantity of the wetting phase decreases and the mean curvature or values of R_1 and R_2 both tend to decrease in size. Referring to Eq. (3-8), it is seen that if R_1 and R_2 both decreased in size, the magnitude of the capillary pressure would in turn have to increase in size. It is therefore possible to express the capillary pressure as a function of rock saturation when two immiscible phases are used within the porous matrix. As will be illustrated later, it is also possible to approximate the pore distribution of this particular system because the capillary pressure would be dependent upon the radii of the various pores for any particular value of saturation. For the capillary pressure to be zero in a porous system with two liquid phases, it would be necessary that R_1 and R_2 be infinitely large or that the interfacial tension σ be zero.

It was previously shown in the case of the capillary tube that the greater pressure is always on the concave side of the interface. It can be seen from Figs. 3-6 and 3-7 that the curvature of the interface is such that the pressure in the nonwetting phase is greater than the pressure in the wetting phase. Therefore the wetting phase in a porous material is at a lower pressure than the nonwetting phase.

Saturation History. To study the effect of saturation history, it is necessary to consider various-size interconnected pores. In the case of a capillary tube of varying diameter, the height to which the fluid will rise in the tube depends on the adhesion tension, fluid density, and variation of tube diameter with height. If pressure is applied to the interface, the interface moves to a new equilibrium position, thus decreasing the volume of water within the tube. This decrease in water volume means a reduction in saturation and is accompanied by an increase in capillary pressure. This fact is illustrated in Fig. 3-8 wherein the capillary pressure would be greater for the small radius of curvature than for the large radius of curvature. This behavior indicates that there is an inverse functional relationship between capillary pressure and the wetting-phase saturation. Also, it indicates that the lower the saturation, the smaller will be the radii of curvature and the wetting-phase material will then exist in the smaller crevices and openings of the system, leaving the large open channels to the nonwetting phase.

Not only is saturation a function of capillary pressure, but it is also a function of the saturation history of the particular pore matrix that is

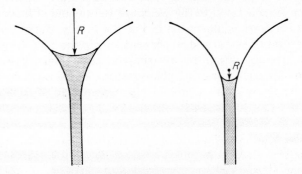

FIG. 3-8. Dependence of interfacial curvature on fluid saturation in a nonuniform pore. Same pore, same contact angle, different fluid saturation. (*From McCardell.*[5])

being considered. For example, in a continuous capillary tube which changes in diameter from small to large to small, as illustrated in Fig. 3-9, the saturation for capillary pressures of equal magnitude depends upon whether the system is initially 100 per cent saturated with a wetting fluid

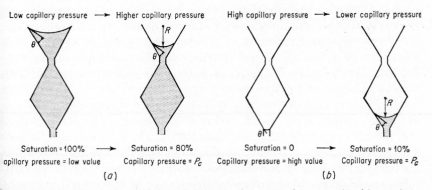

FIG. 3-9. Dependence of equilibrium fluid saturation upon the saturation history in a nonuniform pore. (*a*) Fluid drains; (*b*) fluid imbibes. Same pore, same contact angle, same capillary pressure, different saturation history. (*From McCardell.*[5])

or it is being saturated with the wetting fluid. Forcing the entry of a non-wetting fluid into a tube saturated with a wetting fluid causes the wetting fluid to be displaced to a point such that the capillary pressure across the interface is equal to the applied pressure plus the pressure due to the column of suspended fluid. In the case of Fig. 3-9, the rock is 80 per cent saturated with the wetting phase for the higher value of capillary pressure. Now consider the case where the tube is initially saturated with a nonwetting phase and is immersed in a container filled with a fluid which will preferentially wet the tube. The wetting fluid will be imbibed owing to the adhesion force between the wetting fluid and the surface of the tube until the adhesion force is equal to the weight of the column of fluid. The saturation thus obtained as illustrated in Fig. 3-9 is only 10 per cent. In this example, saturations of 10 and 80 per cent are obtained for identical values of capillary pressure. From this oversimplified example, it is seen that the relationship between the wetting-phase saturation and capillary pressure is dependent on the saturation process. A higher value of saturation for a given capillary pressure would be obtained if the porous system were being desaturated than if the porous system were being resaturated with the wetting-phase fluid.

It is thus seen that the capillary-pressure saturation relationship is dependent upon (1) the size and distribution of the pores, (2) the fluids and solids that are involved, and (3) the history of the saturation process. Thus, in order to use capillary-pressure data properly, these factors must be taken into consideration before the data are actually applied to reservoir calculations.

Laboratory Measurements of Capillary Pressure

The results of a capillary-pressure experiment on an unconsolidated sand pack conducted by Leverett[3] are illustrated in Fig. 3-10. In conducting the experimental work, long tubes filled with sand were saturated with a liquid and suspended vertically. The experiments were performed in such a manner that imbibition and drainage capillary-pressure curves were defined. To obtain the drainage curve, the sand pack was saturated with water and then one end was lowered into a container having a free water level. The water saturation in the tube was then determined at various positions above the free water level in the container. The data obtained are shown in Fig. 3-10 as the drainage curve. The tube was also initially packed dry and then lowered into the water container so that water was imbibed by the sand pack owing to the capillary forces. Again the saturations were measured at various heights above the free water level in the container, and the data are illustrated in Fig. 3-10 as the imbibition curve. Note the difference between the drainage and the imbibition curves as determined by Leverett. The difference in the curves is due to a hysteresis effect which is dependent on the saturation process. Similar data to those reported by Leverett have been obtained by other investigators on different types of systems.

The capillary pressures of Fig. 3-10 are expressed in terms of a dimensionless correlating function. This futionen is equal to the product of the mean radius of curvature, Eq. (3-9), and the square root of the permeability divided by the porosity $(\Delta\rho\, gh/\sigma)(k/\phi)^{1/2}$. The correlating function was proposed so that capillary-pressure data from different sands could be expressed in generalized form. If the correlating function were universally

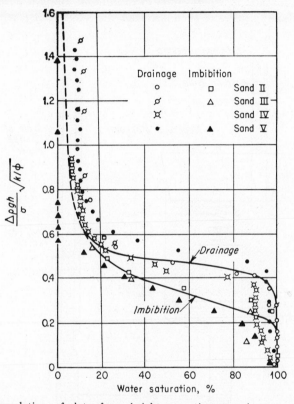

Fig. 3-10. Correlation of data from height-saturation experiments on clean unconsolidated sands. (*From Leverett.*[3])

applicable, one curve would be obtained for all samples. This particular correlating function will be considered in more detail later in the text.

It is not possible to determine the capillary properties of naturally occurring rock materials by a method such as used by Leverett. Core samples of natural materials are necessarily small and are not available in continuous sections of sufficient length for study by the simple drainage method. Therefore other means of measuring capillary pressure have been devised. Essentially five methods of measuring capillary pressure on small core samples are used. These five methods are (1) desaturation or displacement process through a porous diaphragm or membrane (restored state method of Welge[6]), (2) the centrifuge or centrifugal method, (3) the dynamic capillary-pressure method, (4) the mercury-injection method, and (5) the evaporation method.

Porous Diaphragm. The first of these, illustrated in Fig. 3-11, is the displacement cell or diaphragm method. The essential requirement of the diaphragm method is a permeable membrane of uniform pore-size distribution containing pores of such size that the selected displacing fluid will

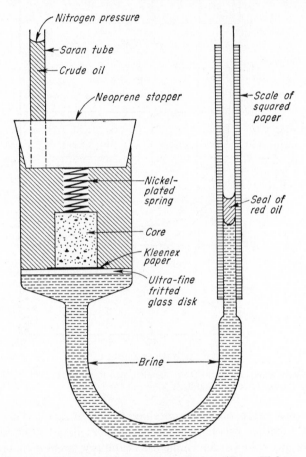

FIG. 3-11. Porous diaphragm capillary-pressure device. (*From Welge and Bruce.*[6])

not penetrate the diaphragm when the pressures applied to the displacing phase are below some selected maximum pressure of investigation. Various materials including fritted glass, porcelain, cellophane, and others have been used successfully as diaphragms. The membrane is saturated with the fluid to be displaced; the test sample is placed on the membrane with some suitable material, such as Kleenex, to aid in establishing contact;

and the test sample is subjected to displacement in a stepwise fashion. Pressure applied to the assembly is increased by small increments. The core is allowed to approach a state of static equilibrium at each pressure level. The saturation of the core is calculated at each point defining the capillary-pressure curve. Any combination of fluids can be used: gas, oil, and/or water. Complete determinations of capillary-pressure curves by the diaphragm method are time-consuming, varying from 10 to 40 days for a single sample, owing to the vanishing pressure differentials causing flow as the core approaches equilibrium at each imposed pressure. As low saturations are approached, the reduction in effective permeability to the displaced phase also contributes to the slow approach to equilibrium. Although most determinations of capillary pressure by the diaphragm method are drainage tests, by suitable modifications, imbibition curves similar to Leverett's can be obtained.

Mercury Injection. The mercury capillary-pressure apparatus (Fig. 3-12) was developed to accelerate the determination of the capillary-pressure–saturations relationship. Mercury is normally a nonwetting fluid.

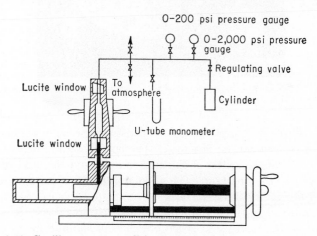

FIG. 3-12. Capillary-pressure cell for mercury injection. (*From Purcell.*[7])

The core sample is inserted in the mercury chamber and evacuated. Mercury is forced in the core under pressure. The volume of mercury injected at each pressure determines the nonwetting-phase saturation. This procedure is continued until the core sample is filled with mercury or the injection pressure reaches some predetermined value. Two important advantages are gained: The time for determination is reduced to a few minutes, and the range of pressure investigation is increased as the limitation of the properties of the diaphragm is removed. Disadvantages are the difference in wetting properties and permanent loss of the core sample.

Centrifuge Method. A third method for determination of capillary properties of reservoir rocks is the centrifuge method[8] illustrated in Fig. 3-13. The high accelerations in the centrifuge increase the field of force on the fluids, subjecting the core, in effect, to an increased gravitational force.

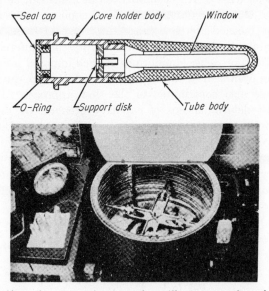

FIG. 3-13. Centrifuge for determination of capillary properties of rocks. (*From Slobod et al.[8]*)

When the sample is rotated at various constant speeds, a complete capillary-pressure curve can be obtained. The speed of rotation is converted into force units in the center of the core sample, and the fluid removed is read visually by the operator. The cited advantage of the method is the increased speed of obtaining the data. A complete curve can be established in a few hours, while the diaphragm method requires days. It is difficult to account for the increase in speed of reaching equilibrium as compared with the diaphragm method, since the same resisting forces appear to be involved in the core.

Dynamic Method. Brown[9] reported the results of determination of capillary-pressure–saturation curves by a dynamic method. Figure 3-14 shows, schematically, the test apparatus. Simultaneous steady-state flow of two fluids is established in the core. By the use of special wetted disks, the pressure of the two fluids in the core is measured and the difference is the capillary pressure. The saturation is varied by regulating the quantity of each fluid entering the core. It is thus possible to obtain a complete capillary-pressure curve.

Comparison of Methods of Measurement. Intuitively, it appears that the diaphragm method (restored state) is superior in that oil and water can be used, therefore more nearly approaching actual wetting conditions. Hence, the diaphragm method is used as the standard against which all other methods are compared.

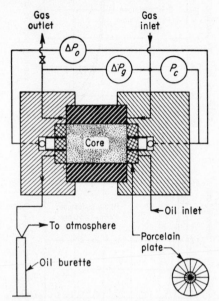

In Fig. 3-15 are presented data of Purcell[7] on capillary properties determined by the diaphragm and mercury-injection methods. Note that the pressure scale for the curves determined by mercury injection is greater by a factor of 5 than that for the curves determined by drainage of water displaced by air in a displacement cell. Purcell assumed that the contact angle for mercury against the rock surfaces was 140° and that of water was 0°.

By analogy to capillary tubes, the ratio of mercury capillary pressure to water-air capillary pressure is

FIG. 3-14. Dynamic capillary-pressure apparatus (Hassler's principle). (*From Brown.[9]*)

$$\frac{P_{cm}}{P_{cw}} = \frac{\sigma_m \cos 140°}{\sigma_w \cos 0°} \tag{3-10}$$

taking σ_m = surface tension of mercury = 480 dynes/cm

σ_w = surface tension of water = 70 dynes/cm

The above ratio

$$\frac{P_{cm}}{P_{cw}} = 5$$

The agreement of the data when corrected by this ratio is good, as shown in Fig. 3-15.

There is some doubt as to the validity of incorporating the contact angle into the ratio of pressures. The geometry of the pores of a rock is complex, and the relationship between the curvature of the interface and the radius of the pore is not necessarily a function of the cosine of the contact angle.

It appears that an equally valid assumption is that the mean curvature of an interface in rock is a unique function of fluid saturation. This assumption permits the ratio of pressure to be defined as follows:

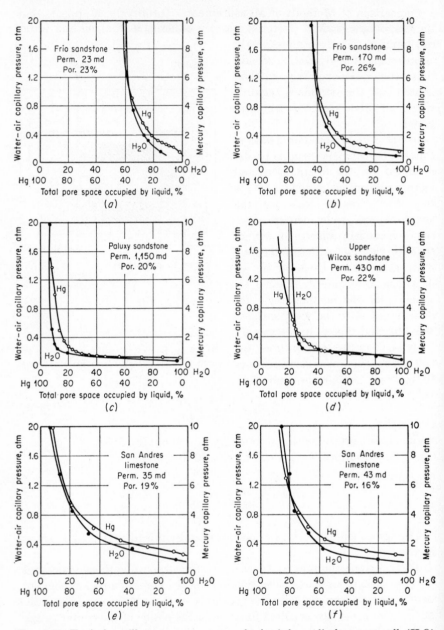

Fig. 3-15. Typical capillary-pressure curves obtained from displacement cell (H₂O) and from mercury apparatus (Hg). (*From Purcell.*[7])

$$\frac{P_{cm}}{P_{cw}} = \frac{\sigma_m}{\sigma_w} = \frac{480}{70} = 6.57 \tag{3-11}$$

With the exception of curves C and D in Fig. 3-15, a superior correlation is established using a conversion factor of 6.57 rather than 5. Brown[9] found that the correlating factor between the diaphragm and mercury-injection methods was not solely a function of interfacial tensions and contact angles. It is noted in Fig. 3-16 that for a sandstone core, the correlating factor is 7.5. For the same fluids it is shown that the correlating factor for a limestone core is 5.8. It appears that a universal conversion factor cannot be defined, as it is different for each type of porous rock.

Good agreement of centrifuge data with those from the diaphragm method was reported by Slobod.[8] A typical curve showing reproducibility between successive determinations with the centrifuge as well as the correlation with data obtained by the diaphragm method is shown in Fig. 3-17. Unlike the mercury-injection method, there is no need of conversion factors to correct for wetting properties. The same fluids are used in the centrifugal and diaphragm methods.

The excellent correlation obtained by Brown[9] between the diaphragm and dynamic methods is illustrated in Fig. 3-18. The dynamic data were obtained by simultaneous steady flow of oil and gas through the porous sample at a predetermined level of pressure difference between the fluids. Care was taken to maintain uniform saturations throughout the core as well as to conduct the test so that a close correspondence to drainage conditions existed.

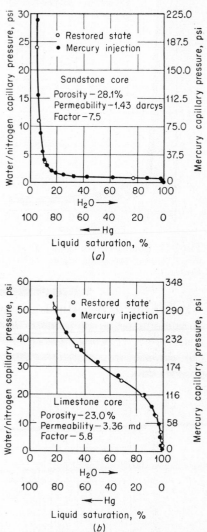

FIG. 3-16. Capillary pressures by restored-state and mercury-injection methods (*From Brown.[9]*)

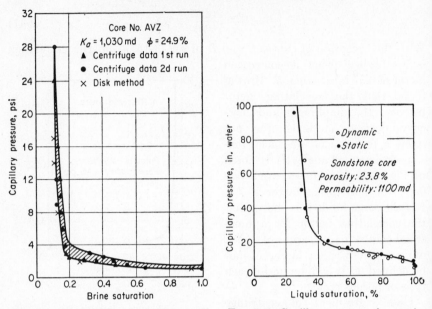

FIG. 3-17. Comparison of centrifuge and disk capillary-pressure data. (*From Brown.*[9])

FIG. 3-18. Capillary pressure by static and dynamic methods. (*From Brown.*[9])

Interstitial-water Saturations

Essentially three methods are available to the reservoir engineer for the determination of connate- or interstitial-water saturations. These methods are (1) coring formations with oil-base or tracer-bearing fluids (2) calculated from electric log analysis and (3) determined from capillary pressure data.

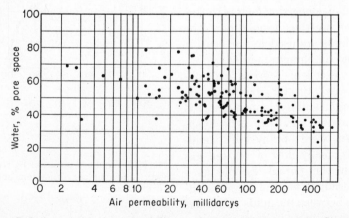

FIG. 3-19. Relation of the air permeability to the water content of the South Coles Levee cores. (*From Gates et al.*[10])

A correlation between water saturation and air permeability for cores obtained with oil-base muds is shown in Fig. 3-19. A general trend of increasing water saturation with decreasing permeability is indicated. It is accepted from field and experimental evidence that the water content de-

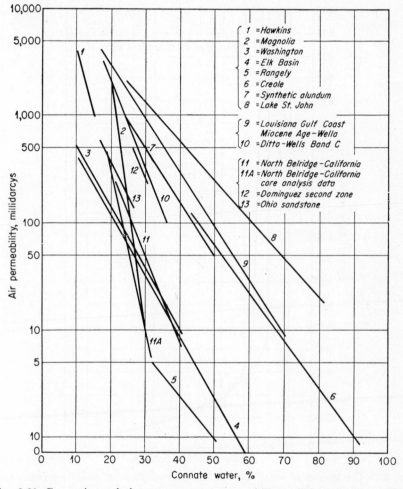

FIG. 3-20. Comparison of the connate-water–permeability relationships for various formations. (*From Welge and Bruce.*[6])

termined from cores cut with oil-base mud reflects closely the water saturation as it exists in a reservoir except in transition zones, where some of the interstitial water is replaced by filtrate or displaced by gas expansion.

In Fig. 3-20 are shown permeability–connate-water relationships reported in the literature for a number of fields and areas. There is no general correlation applicable to all fields. However, an approximately linear correlation between connate water and the logarithm of permeability exists for each individual field. The general trend of the correlation is decreasing connate water with increasing permeability.

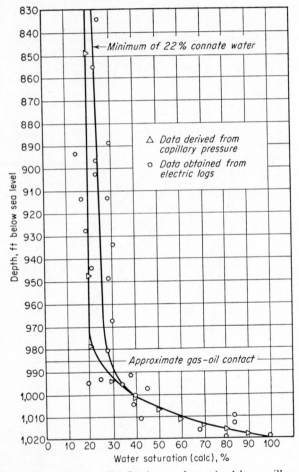

Fɪɢ. 3-21. Comparison of water distribution as determined by capillary-pressure and electric-log data. (*From Owen.*[11])

If capillary-pressure data are to be used for determining fluid saturations, the values obtained should be comparable to those of other methods. Water distributions as determined from electric logs and capillary-pressure data are normally in good agreement. A comparison of these two methods is shown in Fig. 3-21. Shown also is the approximate position of the gas-oil contact as determined from other test data. In the gas-bearing portion of the formation there is no significant variation in water saturation with depth or method of determination. However, in the oil-bearing portion of the rock there is a significant variation in the water saturation with depth. The oil segment is almost wholly in the oil-water transition zone. Variations in water saturations with depth within that zone must be taken into account in order to determine accurately average reservoir connate- or interstitial-water saturations.

Water Saturation from Capillary-pressure Data. Before going into the actual determinations of water distributions from capillary-pressure data, it is best to discuss the basis upon which these determinations are made. In the prior section, using the classic capillary tube, it was shown that

$$P_c = gh(\rho_1 - \rho_2) \tag{3-6}$$

where P_c = capillary pressure

   g = gravitational constant

   h = height above plane of zero capillary pressure between fluids 1 and 2

   ρ_1 = density of heavier fluid

   ρ_2 = density of lighter fluid

It was also shown that $P_c = \sigma\left(\dfrac{1}{R_1} + \dfrac{1}{R_2}\right)$ $\tag{3-8}$

When these equations are put in oil-field terms, the capillary pressure in pounds per square inch can be stated as

$$P_c = \frac{h}{144}(\rho_1 - \rho_2) \tag{3-12}$$

where h is in feet, ρ_1 and ρ_2 are the densities of fluids 1 and 2, respectively, in pounds per cubic feet at the conditions of the capillary pressure.

Converting Laboratory Data. To use laboratory capillary-pressure data it is necessary to convert to reservoir conditions. Laboratory data are obtained with a gas-water or an oil-water system, which does not normally have the same physical properties as the reservoir water, oil, and gas.

There are essentially two techniques, differing only in the initial assumptions, available for correcting laboratory capillary-pressure data to reservoir conditions. As shown previously, by means of the capillary tube, the capillary pressure is expressed as

$$P_c = \frac{2\sigma\cos\theta}{r}$$

Considering a specific case wherein the laboratory values are determined with gas and water, the capillary pressure becomes

$$(P_c)_L = \frac{2\sigma_{wg}\cos\theta_{wg}}{r} \tag{3-5}$$

PETROLEUM RESERVOIR ENGINEERING

where σ_{wg} is the interfacial tension between gas and water used in laboratory tests, and r is the radius of the capillary. The capillary pressure which would exist if reservoir fluids, oil and water, were used in the same capillary would be

$$(P_c)_R = \frac{2\sigma_{wo} \cos \theta_{wo}}{r}$$

where σ_{wo} = interfacial tension between reservoir water and oil at reservoir temperature and pressure

θ_{wo} = contact angle for reservoir water and oil

r = radius of capillary

Comparing the equations for laboratory and reservoir capillary pressure, it is found that the reservoir capillary pressure is

$$(P_c)_R = \frac{\sigma_{wo} \cos \theta_{wo}}{\sigma_{wg} \cos \theta_{wg}} (P_c)_L \tag{3-13}$$

Thus reservoir capillary pressure can be calculated from laboratory capillary pressure when the interfacial tensions and contact angle between oil and water in the reservoir and gas and water in the laboratory are known. This relationship assumes that the saturations as measured in the laboratory remain equal to the saturations in the reservoir so that the height of rise in capillary tubes of equal radii are the same. It is difficult, if not impossible, to determine the exact value of the contact angle for fluids in a porous matrix. The cosine of the contact angle can vary between -1 and 1, which can cause considerable variation in the resulting conversion of laboratory data, and therefore it is often desirable to neglect the contact angle in Eq. (3-13).

A second technique, a relationship neglecting the contact angle, for converting laboratory data to reservoir conditions can be obtained by stating the capillary pressure as

$$P_c = \sigma \left(\frac{1}{R_1} + \frac{1}{R_2} \right) \tag{3-8}$$

where $\quad P_c$ = capillary pressure, dynes/sq cm

σ = interfacial tension, dynes/cm

R_1 and R_2 = principal radii of curvature, cm

If it is assumed that the radii of curvature are uniquely defined by the saturation in the wetting phase for a given displacement process (imbibition or drainage), an equation expressing the capillary pressure as a function of saturation can be written as

$$P_c = \sigma[f(S_w)]$$

where $f(S_w)$ is a function of saturation which can be determined by a laboratory test for laboratory conditions, so that

$$P_{cL} = \sigma_L[f(S_w)]$$

For reservoir conditions

$$P_{cR} = \sigma_R[f(S_w)]$$

therefore

$$f(S_w) = \frac{P_{cL}}{\sigma_L} = \frac{P_{cR}}{\sigma_R}$$

so that

$$P_{cR} = \frac{\sigma_R}{\sigma_L} P_{cL} \qquad (3\text{-}14)$$

Since the interfacial tensions enter as a ratio, pressure in any consistent units can be used together with the interfacial tension in dynes per centimeter.

As was noted in the discussion on mercury-injection tests, the capillary-pressure data obtained with one set of fluids cannot be exactly converted to the basis of another set of fluids by either Eq. (3-13) or (3-14). In the case of mercury-injection tests, Eq. (3-14) yielded the better results. As this relationship is simpler and does not require knowledge of the contact angles under reservoir conditions, it will be used in all future conversion calculations in this text.

Averaging Capillary-pressure Data. As capillary-pressure data are obtained on small core samples which represent an extremely small part of the reservoir, it is necessary to combine all the capillary data to classify a particular reservoir. As would be expected from Fig. 3-20, fluid-saturation–capillary-pressure relationships are affected by the permeability of the sample. It therefore becomes necessary to evaluate the various sets of capillary-pressure data with respect to the permeability of the core sample from which they were obtained.

There are two proposed methods of correlating capillary-pressure data for a reservoir. The first method is that proposed by Leverett[3] wherein a correlating function, commonly called the J function, is used. The second method, which was best illustrated by Guthrie,[12] is a statistical approach to the problem.

The J-function correlating term uses the physical properties of the rock and fluid and is expressed as

$$J(S_w) = \frac{P_c}{\sigma} \left(\frac{k}{\phi}\right)^{\frac{1}{2}} \qquad (3\text{-}15)$$

where P_c = capillary pressure, dynes/sq cm
  σ = interfacial tension, dynes/cm
  k = permeability, sq cm
  ϕ = fractional porosity

Some authors alter the above expression by including the cos θ (where θ is the contact angle) as follows:

$$J(S_w) = \frac{P_c}{\sigma \cos \theta} \left(\frac{k}{\phi}\right)^{1/2} \qquad (3\text{-}16)$$

The inclusion of the cos θ term will not be used herein by the authors, and the J function will be as defined by Eq. (3-15).

The J function was originally proposed as a means of converting all capillary-pressure data to a universal curve. There are significant differences in correlation of the J function with water saturation from formation to formation, so that no universal curve can be obtained.

Correlation of the J function with water saturation for a number of materials is illustrated in Fig. 3-22. Note that there is an independent

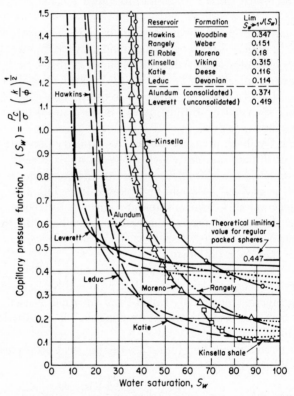

FIG. 3-22. Capillary retention curves. (*From Rose and Bruce.*[13])

correlation for each material considered. Brown[9] considered the J function as a correlating device for capillary-pressure data. In an evaluation of samples from the Edwards formation in the Jourdanton Field he used the expression

$$J(S_w) = \frac{P_c}{\sigma \cos \theta} \left(\frac{k}{\phi}\right)^{1/2}$$

Figure 3-23a shows the correlation obtained for all samples available from the field. There is considerable dispersion of data points, although the trend of the correlation is good. Brown found that the correlations could be improved by dividing materials on a textural basis. The core materials were subdivided into limestone and dolomites, both materials occurring within the productive section of the Edwards formation. The correlation

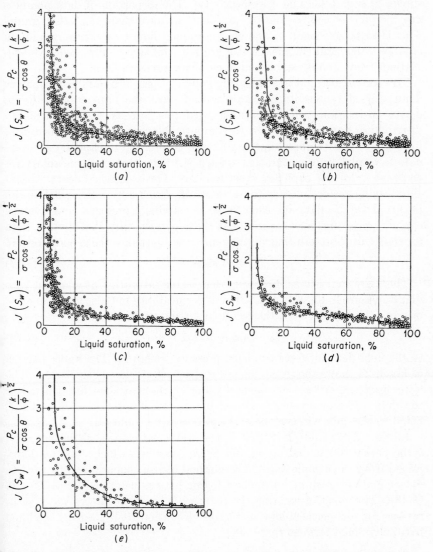

FIG. 3-23. J-function correlation of capillary-pressure data in the Edwards formation, Jourdanton Field. J curve for (a) all cores; (b) limestone cores; (c) dolomite cores; (d) microgranular limestone cores; (e) coarse-grained limestone cores. (*From Brown.*[9])

for the limestone samples is shown in Fig. 3-23b and for the dolomite samples in Fig. 3-23c. The dolomite samples indicate a good correlation, while the limestone samples exhibit a scattering of data in the range of low water saturations. In an attempt to obtain a better correlation, the limestone cores were further subdivided into microgranular and coarse-grain samples. The J curves for microgranular and coarse-grained limestone samples are shown in Fig. 3-23d and e, respectively. The dispersion of data points is greater for the coarse-grained limestone samples than any other group. This is to be expected, as the coarse-grained limestone contains solution cavities, vugs, and channels which are not capillary in size, hence the deviations from trends established in capillary-pressure data.

The second method of evaluating capillary-pressure data is to analyze a number of representative samples and treat the data statistically to derive correlations which, together with the porosity and permeability distribution data, can be used to compute the connate-water saturations for a field. A first approximation for the correlation of capillary-pressure data is to plot water saturation against the logarithm of permeability for constant values of capillary pressure. An approximately linear relationship usually results such as those shown in Fig. 3-20. A straight line can be fitted to the data for each value of capillary pressure, and average capillary-pressure curves computed from permeability distribution data for the field. The resulting straight-line equation takes the general form of

$$S_w = a \log k + C \qquad (3\text{-}17)$$

There are indications, however, that water saturation at constant capillary pressure is not only a function of permeability but also some function of porosity. In Fig. 3-24, the results of fitting an equation

$$S_w = a_1\phi + a_2 \log k + C \qquad (3\text{-}18)$$

to the field data for 5-psi capillary pressure is shown. The upper portion of Fig. 3-24 shows the three-dimensional aspect of such a correlation. The lower part shows lines of constant porosity (light dashed lines) fitted to the data. The heavy black dashed line is the straight line [Eq. (3-17)] fitted to the data wherein the effect of porosity is omitted.

In Eq. (3-17) and (3-18), S_w is the water saturation, ϕ the porosity, k the permeability, and a_1, a_2, and C are constants which must be determined from the sample data. The method of least squares can be used to determine the constants of the best fitting lines as described by (3-17) and (3-18). The effect of ignoring the porosity is to predict lower water saturations for low-permeability materials. Equation (3-18) can be modified to a polynomial form so that

$$S_w = a_1\phi + a_2\phi^2 + a_3 \log k + a_4(\log k)^2 + C \qquad (3\text{-}19)$$

Note the change in three-dimensional aspects when the polynomial form of the correlation is used (Fig. 3-25). In the lower portion of the figure are shown curves of constant porosity resulting from correlations using Eq. (3-19). A better correlation is obtained using Eq. (3-19) rather than Eb.

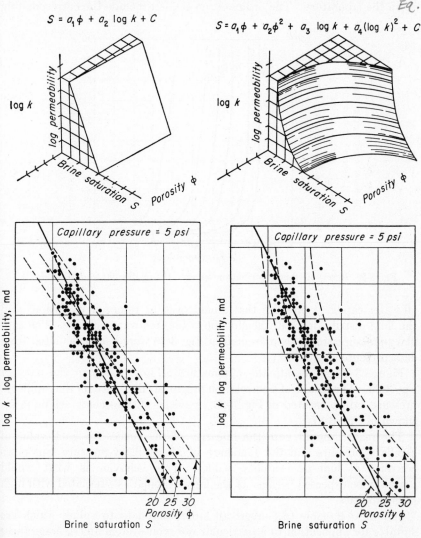

$$S = a_1 \phi + a_2 \log k + C$$

$$S = a_1 \phi + a_2 \phi^2 + a_3 \log k + a_4 (\log k)^2 + C$$

Fig. 3-24. Correlation of capillary-pressure data. (*From Guthrie and Greenburger.*[12])

Fig. 3-25. Correlation of capillary-pressure data. (*From Guthrie and Greenburger.*[12])

(3-17), but for most engineering purposes, with limited data, correlations as implied by the latter equation are satisfactory.

Fluid-distribution curves are reported for several values of permeability, ranging from 10 to 900 millidarcys in Fig. 3-26. These data may be considered also to be capillary-pressure curves. The ordinate on the right reflects values of capillary pressure determined by displacing water with air in the laboratory. The ordinates on the left include the corresponding

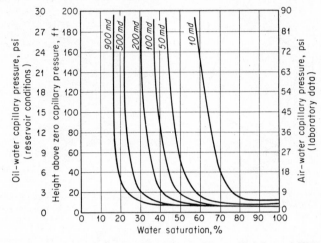

FIG. 3-26. Reservoir fluid-distribution curves. (*From Wright and Wooddy.*[14])

oil-water capillary pressure that would exist at reservoir conditions and the fluid distribution with height above the free water surface. The capillary-pressure ordinates presented in Fig. 3-26 were back-calculated from data presented by Wright.[14]

Figure 3-27 represents the application of Eq. (3-17) to the data of Fig. 3-26. The results of a correlation previously discussed, of the capillary-pressure data presented in Fig. 3-26, by means of the second technique are shown in Fig. 3-27.

The reader should note the linearity of the curves for each value of capillary pressure and the tendency of all capillary-pressure curves to converge at high-permeability values. This behavior is what would normally be expected because of the larger capillaries associated with high permeabilities.

It is now possible to convert all laboratory data to values which are suitable for application to a particular reservoir system and average these values to obtain a reservoir fluid distribution. The capillary-pressure–saturation data have to be converted into height-saturation data to be applicable to the reservoir. Such data are illustrated in Fig. 3-21 wherein capillary-pressure–saturation data are compared with saturation data calculated from electric logs. In Fig. 3-21 it is noted that the water saturation within the oil zone varies from 100 to approximately 24 per cent and the water saturation in the gas zone varies from 24 to 20 per cent. Such a large variation of water saturation in the oil zone could cause a large error in the estimated oil reserve if not properly accounted for. It therefore becomes imperative that the water distribution with height within the oil zone be determined so that a mean water saturation for the oil zone is obtained for use in determining oil reserves.

To convert capillary-pressure–saturation data to height saturation, it is only necessary to rearrange the terms in Eq. (3-12) so as to solve for the height instead of the capillary pressure so that

$$h = \frac{P_c \times 144}{\rho_w - \rho_o}$$

where h = height above free water surface, ft

ρ_w = density of water at reservoir conditions, lb/cu ft

ρ_o = density of oil at reservoir conditions, lb/cu ft

P_c = capillary pressure at some particular saturation for reservoir conditions, which means it must first be converted from laboratory data

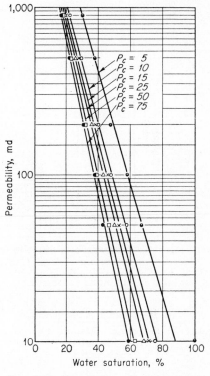

Fig. 3-27. Correlation of water saturation with permeability for various capillary pressures. (*From Wright and Wooddy.*[14])

By use of this equation, it is possible to convert laboratory capillary-pressure data into a water-saturation curve as a function of height as was shown in Fig. 3-21. This type of system, then, would be represented by an idealized fluid distribution as shown in Fig. 3-28. Here, two cores are used for illustrative purposes where core *A* represents a core sample within the oil zone and core *B* represents a core sample within the oil-gas zone. By use of capillary-pressure data, it is possible to calculate the water saturations that exist in cores *A* and *B*.

It is emphasized here that all height measurements are from the free water surface. This surface is not necessarily defined by the level at which only water is produced but is that surface defined by zero capillary pressure. The free water surface is dependent upon the capillary pressure and the relative permeability of the porous system. For this reason, the means of selecting or calculating the proper free water surface will be delayed until after the discussion of relative and effective permeabilities for multi-

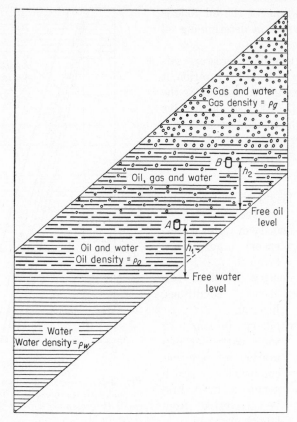

Fig. 3-28. Fluid distribution in a petroleum reservoir. (*From Welge and Bruce.*[6])

phase systems. For the time being, it will be considered that the free water surface can be defined and that all measurements can be made from that surface. In order that the relationship used to convert capillary-pressure data to height above the free water surface be valid, it is necessary that the following conditions be satisfied:

1. The pressure in the oil and water are equal at the free surface.
2. The water and oil columns are continuous and connected throughout the range of the calculations.
3. The system is in static equilibrium.

If any of these three conditions are not satisfied, then the equation for calculating the height of a saturation plane above the free water surface is not valid.

It is possible to determine the range over which the conversion equation is valid from the laboratory data. The wetting phase is said to be discontinuous when the capillary pressure increases without changing its saturation. Referring to Fig. 3-26, it is seen that the wetting phase, water, becomes discontinuous at a height of approximately 130 ft above the free water surface.

By determining the free water surface from coring, electric logs, or drill-stem tests, it is possible to calculate the water saturations as a function of height above the free water surface by using Eq. (3-14).

———•—•—•———

Example 3-1.   Calculation of Height of Saturation Plane from Laboratory Capillary-pressure Data.

If $P_{cL} = 18$ psi for $S_w = 0.35$,

$\sigma_{wo} = 24$ dynes

$\rho_w = 68$ lb/cu ft

$\sigma_{wg} = 72$ dynes

$\rho_o = 53$ lb/cu ft

then, from Eq. (3-14),

$$P_{cR} = 18 \left(\frac{24}{72}\right) = \frac{18}{3} = 6 \text{ psi}$$

$$h = \frac{P_{cR}(144)}{\rho_w - \rho_o} = \frac{6(144)}{68 - 53} = \frac{6(144)}{15} \cong 58 \text{ ft}$$

Thus, a water saturation of 35 per cent exists at a height of 58 ft above the free water surface.

———•—•—•———

To calculate the fluid saturation in the gas zone, it is necessary to consider all three phases: oil, water, and gas. If all three phases are continuous, it can be shown that

$$(P_c)_{wg} = (P_c)_{wo} + (P_c)_{og}$$

where $(P_c)_{wg}$ = capillary pressure at given height above free water surface determined by using water and gas

$(P_c)_{wo}$ = capillary pressure at given height above free water surface using oil and water

$(P_c)_{og}$ = capillary pressure at height above free oil surface using oil and gas

If the wetting phase becomes discontinuous, then the wetting-phase saturation takes on a minimum value, and at all heights above the point of discontinuity the wetting-phase saturation cannot be less than this mini-

mum value. It is then possible to calculate the fluid saturations above the
free oil surface by the following relations:

1. S_w at h, calculated using oil and water as the continuous phases
2. S_t at h, calculated using oil and gas as the continuous phases and
height denoted by the free oil surface
3. $S_g = 1 - S_t$
4. $S_o = S_t - S_w$

Example 3-2. Calculation of Water and Oil Saturation in Gas Zone
from Capillary-pressure Data. Let oil zone thickness $h_o = 70$ ft

$$\sigma_{wg} = 72 \text{ dynes} \qquad \rho_o = 53 \text{ lb/cu ft}$$
$$\sigma_{og} = 50 \text{ dynes} \qquad \rho_w = 68 \text{ lb/cu ft}$$
$$\sigma_{wo} = 25 \text{ dynes} \qquad \rho_g = 7 \text{ lb/cu ft}$$

From Fig. 3-26 for a 900-millidarcy sample

let $P_{cL} = 54$ psi by the method illustrated in Example 3-1
$P_{cR} = 18$ psi
h_{fw} = height above free water level = 120 ft
S_w = 16 per cent at a height of 70 ft or greater (read from curve)

As the oil zone is only 70 ft thick, then the height of 120 ft above the free
water surface must be at least 50 ft into the gas-saturated zone. The first
step is to calculate the total fluid saturation S_t using gas and oil as the continuous phases.

$$h_{fo} = h_{fw} - h_o = 120 - 70 = 50 \text{ ft}$$

$$(P_{cR})_{og} = \frac{h_{fo}}{144} (\rho_o - \rho_g)$$

$$= \frac{50}{144} \times (53 - 7) = \frac{50}{144} \times 46$$

$$= 15.96 \text{ psi}$$

$$P_{cL} = P_{cR} \frac{\sigma_{wg}}{\sigma_{og}} = 15.96 \times \frac{72}{50} = 23 \text{ psi}$$

From Fig. 3-26 for a laboratory capillary pressure of 23 psi, permeability
of 900 millidarcys, the total wetting saturation is

$$S_t = 18 \text{ per cent}$$
therefore $S_o = S_t - S_w = 18 - 16 = 2$ per cent
$$S_g = 100 - S_t = 100 - 18 = 82 \text{ per cent}$$

It must be understood that the relationships used in calculating the fluid
saturations in the gas zone were based upon continuity of all three phases.
As this is not normally the case, it might be expected that saturations

somewhat different from the calculated values might exist. As the capillary pressure for a discontinuous phase could vary from pore to pore, it is impossible to ascertain the exact relationships that should exist. Hence, the preceding method of calculating fluid distributions is not exact but is usually as accurate as the data available for making the computation.

Calculation of Wettability

Wettability of Reservoir Rocks. As mentioned earlier in the discussion of capillary pressure, the curvature of an interface confined in a pore is some function of the contact angle, which, in turn, is a function of the wetting properties of the fluids and the rock surfaces. The degree to which fluids wet a solid surface was shown to depend on the interfacial tensions between the various contacts, fluid-solid and fluid-fluid.

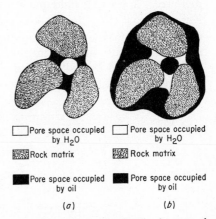

There are two means of expressing the degree of wettability. The first is expressed in terms of the contact angle. A contact angle of zero would indicate complete wetting by the more dense phase, an angle of 90° indicates that neither phase preferentially wets the solid, and an angle of 180° indicates complete wetting by the less dense phase. The contact angle is, therefore, a measure of the relative wetting of a solid by a fluid.

Fig. 3-29. Wetting in idealized pores of reservoir rocks. (a) Oil wet; (b) water wet. (*From Calhoun.*[15])

Another convenient index of wettability is the sessile drop ratio, defined as the ratio of the height of a droplet on a surface to the breadth of the droplet. A sessile drop ratio of 1 indicates complete nonwetting, whereas a ratio of zero indicates complete wetting.

The wettability of reservoir rocks to the fluids present in these rocks is of great importance in that the distribution of the fluids within the interstices is a function of the wettability. Figure 3-29 is an idealized representation of the change in fluid distribution in a given pore due to a change from oil wetting to water

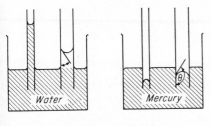

$$P_c = \frac{2\sigma \cos \theta}{r}$$

Fig. 3-30. Illustration of relation between wettability and capillary pressure.

distribution in a given pore due to a change from oil wetting to water wetting. Because of the attractive forces, the wetting fluid tends to occupy the smaller interstices of the rock and the nonwetting fluid occupies the more open channels.

Since reservoir rocks are, for the most part, aggregates of small mineral and rock fragments, it is not possible to determine the wetting properties by direct measurement of contact angles or sessile drop ratios. However, by analogy with the effect of wetting properties on capillary pressure in capillary tubes, an indirect measurement is indicated. In Fig. 3-30 are shown the capillary rise of water in a tube and the capillary depression of mercury. A wetting fluid tends to enter a pore or tube spontaneously while a nonwetting fluid resists entry. It is suggested that the contact angle and some degree of wettability can be calculated from the threshold pressure (pressure just causing nonwetting fluid entry) of a porous system.

Data obtained by Calhoun and Yuster[16] on core samples for the threshold pressure as a function of permeability are reported in Fig. 3-31. The

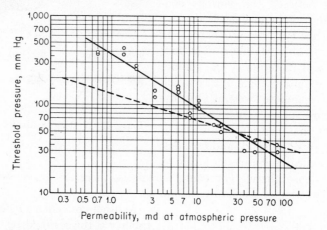

Fig. 3-31. Threshold pressure as a function of permeability and wettability. (- - -) Oil into water; (—) water into oil. (*From Calhoun and Yuster.*[16])

dashed curve is for water entering a rock containing oil, while the solid curve is for oil entering a water-bearing rock. The lower pressures required for water entry are concluded to be indicative of preferential wetting of the core samples by water.

Slobod[17] extended the concept of using threshold pressures as a means of determining the wetting characteristics to compute a wettability number and an apparent contact angle. The wettability number and apparent contact angle are both defined by Eqs. (3-20) and (3-21).

$$\text{Wettability number} = \frac{\cos\theta_{wo}}{\cos\theta_{oa}}\frac{P_{Two}\,\sigma_{oa}}{P_{Toa}\,\sigma_{wo}} \qquad (3\text{-}20)$$

$$\text{Contact angle} = \cos \theta_{wo} = \frac{P_{Two}\,\sigma_{oa}}{P_{Toa}\,\sigma_{wo}} \tag{3-21}$$

where $\cos \theta_{oa} = 1$

$\cos \theta_{wo}$ = contact angle between water and oil in core

$\cos \theta_{oa}$ = contact angle between air and oil in core

P_{Two} = threshold pressure of core for oil to enter when core initially saturated with water

P_{Toa} = threshold pressure of core for air to enter when core initially saturated with oil

σ_{oa} and σ_{wo} = interfacial tensions between air and oil and oil and water, respectively

Table 3-1 summarizes the data presented by Slobod.[17] A wettability number of 1.0 would indicate complete wetting by water; of zero, complete wetting by oil. In general, intermediate wetting is exhibited for the core samples reported on in Table 3-1.

The effects of wettability must be considered in all laboratory determinations of residual oil saturations, capillary pressure, and other similar tests. Important changes in the physical properties of core samples have been noted which are apparently due to changes which occurred in the rock-wetting characteristics during laboratory extraction with solvents.

Some of the unexplained irregularities in capillary pressure and other laboratory core data could possibly be attributed to the uncertainties in the wetting properties of the rock samples and possible changes in these wetting properties due to aging and laboratory procedures.

Pore-size Distribution and Calculation of Permeability from Capillary-pressure Data

In the discussion of permeability in Chap. 2, the analogy between fluid flow in tubes and the permeability of porous media was noted. Furthermore, it was stated that if a porous medium was conceived to be a bundle of capillary tubes, the permeability could be shown to be some function of the porosity or, more rigorously, the arrangement of the tubes. In one form or another all the above relationships connect the fluid-conducting capacity with the volume storage capacity of a flow system.

As the pores of a rock vary in size, the concept of pore-size distribution must be introduced before relationships connecting the permeability and porosity of a porous matrix can be derived.

As noted in the capillary-pressure curves previously presented, the capillary pressure is a function of the fluid properties and of the saturation. For a given rock-fluid system and saturation history, the saturation at a given capillary pressure is some function of the pore geometry. If the porous medium is conceived to be a bundle of capillary tubes of various radii, then the capillary-pressure–saturation curve relates the number and

TABLE 3-1. COMPARISON OF WETTABILITY AMONG CORE SAMPLES OF
DIFFERENT ORIGIN[a,17]

Core No.	Description	Initial desaturation pressure (threshold pressure, psi)		Wettability number[b]	Apparent contact angle, deg
		Air-oil	Oil-water		
BTL	Devonian	6.5	6.1	0.835	33.4
BTN	limestone	6.8	6.2	0.811	35.8
BTO		6.25	6.0	0.854	31.3
BTP		6.4	3.9	0.541	57.2
1588	Yates	0.86	0.32	0.331	70.7
1589	sandstone	0.85	0.3	0.314	71.4
1590		0.85	0.31	0.324	71.1
1591		1.00	0.4	0.356	69.1
1542	Alundum	0.70	0.25	0.318	71.5
1543	(RA 1139)	0.70	0.28	0.356	69.1
1544		0.68	0.4	0.522	58.5
1545	Synthetic	0.67	0.28	0.372	68.2
1592	Clearfork	0.72	0.24	0.297	72.7
1593		0.54	0.32	0.528	58.1
1594	Limestone	1.58	0.32	0.180	79.6
1595		2.90	0.45	0.138	82.1
1620	Tensleep	0.86	0.21	0.217	77.5
1621	sand	0.86	0.21	0.217	77.5
1622		0.68	0.12	0.157	81.0
1623		0.86	0.27	0.280	73.8

[a] Routine extraction with chloroform preceded wettability tests except alundum, which has been regenerated at 1400° F for 3 hr.
[b] Air-oil surface tension = 24.9 dynes/cm. Oil-water interfacial tension = 28.0 dynes/cm.

size of pores penetrated by the nonwetting fluid at a given capillary pressure.

Purcell[7] and Burdine[18] both have reported on computation of permeability from capillary-pressure data obtained by the mercury-penetration method. Purcell utilized the concept of pore-size distribution without evaluating the distributions. He applied the data directly to the computation of permeability. Burdine reported pore-size distributions as well as the results of computation of permeability.

The equation presented by Purcell for the calculation of permeability from the pore properties of a rock is developed as follows:

The minimum capillary pressure required to displace a wetting fluid from or inject a nonwetting fluid into a capillary tube of radius r is given by

$$P_c = \frac{2\sigma \cos \theta}{r} \tag{3-5}$$

The flow rate from a single tube of radius r is given by Poiseuille's law

$$Q = \frac{\pi r^4 \, \Delta P}{8\mu L} \tag{3-22}$$

Since the volume of the capillary is $V = \pi r^2 L$, substituting the volume in Eq. (3-22) above gives

$$Q = \frac{V r^2 \, \Delta P}{8\mu L^2}$$

Solving Eq. (3-5) for r and substituting yield

$$Q = \frac{(\sigma \cos \theta)^2 V \, \Delta P}{2\mu L^2 (P_c)^2}$$

If the porous medium is conceived to be comprised of n capillary tubes of equal length but random radii, the total rate of flow is given by

$$Q_t = \frac{(\sigma \cos \theta)^2 \, \Delta P}{2\mu L^2} \sum_{i=1}^{i=n} \frac{V_i}{(P_c)_i^2} \tag{3-23}$$

From Darcy's law of fluid flow in porous media

$$Q_t = \frac{kA \, \Delta P}{\mu L} \tag{3-24}$$

Combining Eqs. (3-23) and (3-24) a relation for permeability as a function of pore volume and capillary pressure is obtained, Eq. (3-25):

$$k = \frac{(\sigma \cos \theta)^2}{2AL} \sum_{i=1}^{i=n} \frac{V_i}{(P_c)_i^2} \tag{3-25}$$

The volume V_i of each capillary can be expressed as a fraction S_i of the total void volume V_T of the system, so that

$$\frac{V_i}{V_T} = S_i$$

Since AL is the bulk volume of the system and ϕ is the fractional porosity,

$$\phi = \frac{V_T}{AL}$$

Substituting in Eq. (3-25),

$$k = \frac{(\sigma \cos \theta)^2}{2} \phi \sum_{i=1}^{i=n} \frac{S_i}{(P_c)_i^2} \tag{3-26}$$

To account for the deviation of the actual pore space from the simple geometry used in the derivation, Purcell introduced a lithology factor λ into the final equation. Introducing conversion factors and generalizing, Eq. (3-26) reduces to

$$k = 10.24(\sigma \cos \theta)^2 \phi \lambda \int_{S=0}^{S=1} \frac{dS}{(P_c)^2} \qquad (3\text{-}27)$$

where k = permeability, millidarcys
  ϕ = fractional porosity
  S = fraction of total pore space occupied by liquid injected or forced out of sample
  P_c = capillary pressure, psi
  σ = interfacial tension, dynes/cm
  θ = contact angle

Purcell assumed that the contact angle for mercury was 140° and that the interfacial tension of mercury was 480 dynes/cm. Therefore, using mercury capillary-pressure data, Eq. (3-27) further reduces to

$$k = 14{,}260\phi \lambda \int_{S=0}^{S=1} \frac{dS}{(P_c)^2} \qquad (3\text{-}28)$$

To evaluate Eq. (3-28) the integral is found by reading values of P_c from the capillary-pressure curve at various saturations, calculating values of $1/(P_c)^2$, and plotting these values as a function of the corresponding values of saturation which existed on the original capillary-pressure curves. The value of the integral is the area under the curve $1/(P_c)^2$ (see Fig. 3-32).

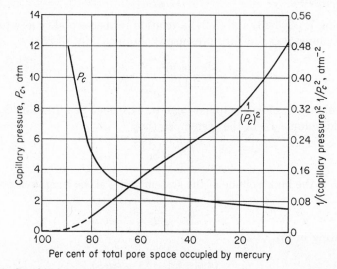

Fig. 3-32. Graphical presentation of capillary-pressure data for calculating permeability. (*After Purcell.*[7])

TABLE 3-2. OBSERVED VALUES OF LITHOLOGY FACTOR[7]

Sample no.	Factor λ [Eq. (3-28)] required to make calculated and observed permeabilities identical	Permeability calculated from Eq. (3-28) using an average λ of 0.216	Observed air permeability, millidarcys
	Upper Wilcox Formation		
1	0.085	3.04	1.2
2[a]	0.122	21.2	12.0
3	0.168	17.3	13.4
4	0.149	53.5	36.9
5	0.200	61.9	57.4
6	0.165	91.6	70.3
7	0.257	92.3	110
8	0.256	97.5	116
9	0.191	163	144
10	0.107	680	336
11	0.216	430	430
12	0.273	348	439
13	0.276	388	496
14[a]	0.185	902	772
15	0.282	816	1,070
16	0.363	865	1,459
	Paluxy Formation		
17		0.003	<0.1
18		0.10	<0.1
19	0.182	42.2	35.7
20	0.158	54.9	40.2
21	0.231	172	184
22	0.276	183	235
23	0.215	308	307
24	0.163	422	320
25	0.284	383	506
26	0.272	502	634
27	0.338	734	1,150
	Av 0.216		

[a] "Cuttings."

Table 3-2 presents Purcell's summary of observed and computed permeabilities as well as computed values for the lithology factor, λ. Figure 3-33 shows the correlation obtained between observed and computed data using an average lithology factor of 0.216. Good agreement is indicated between calculated and measured values of permeability. The calculated values are higher than observed at low permeabilities and lower than observed at high values of permeability.

Burdine[18] adapted the method of Ritter and Drake[19] to the determination of pore-size distribution of reservoir rocks and also presented a method of

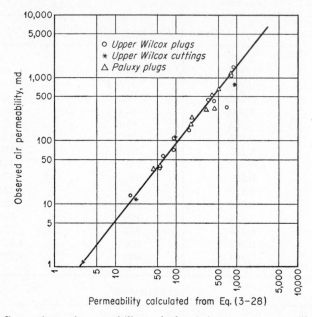

Permeability calculated from Eq. (3-28)

FIG. 3-33. Comparison of permeability calculated from mercury capillary-pressure data to the observed permeability; (○) upper Wilcox—plugs; (*) upper Wilcox—cuttings; (△) Paluxy—plugs. (*From Purcell.*[7])

calculating permeability from this distribution. The method involves injection of mercury into an evacuated core sample, thus obtaining a mercury capillary-pressure curve. The equations presented by Burdine for calculating the pore size and rock permeability are derived as follows:

A distribution function is defined as $D(r_i)$, so that

$$dV = D(r_i)\,dr$$

where dV is the total volume of all pores having a radius between r_i and $r_i - dr$. The quantity $D(r_i)$ can be computed from the mercury capillary-pressure data by using the following two equations:

$$P_{ci}r_i = 2\sigma \cos \theta \qquad (3\text{-}5)$$

and

$$D(r_i) = \frac{P_{ci}}{r_i}\frac{dS_m}{dP_c} \qquad (3\text{-}29)$$

where P_{ci} = capillary pressure

  r_i = pore entry radius

  σ = interfacial tension

  θ = contact angle

  S_m = mercury saturation, per cent of pore volume

The distribution function can be evaluated by graphically taking slopes of the mercury capillary-pressure curve at different values of mercury saturation, computing the pore radius from the capillary pressure corresponding to the point at which the slope was taken by means of Eq. (3-5), and evaluating Eq. (3-29) for the distribution function $D(r_i)$.

A typical mercury capillary-pressure curve and the corresponding distribution curve are presented in Fig. 3-34a and b. The area under the distribution curve to a given radius is the fraction of the volume having pores larger than the given radius.

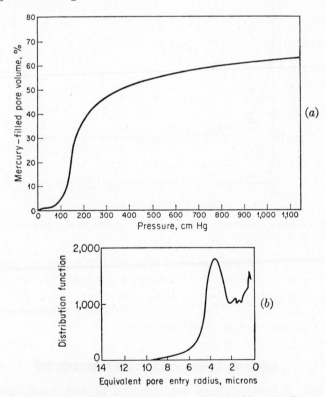

Fig. 3-34. (a) Mercury capillary-pressure curve. Sample No. 173 C, permeability 20.1 millidarcys, porosity 14.3 per cent. (*From Burdine et al.*[18]) (b) Equivalent pore entry radius relation with distribution function. Sample No. 173 C, permeability 20.1 millidarcys, porosity 14.3 per cent. (*From Burdine et al.*[18])

The permeability equation developed by Burdine[18] is based on an analogy to a bundle of capillary tubes.

$$k = \frac{100\phi}{8(9.87 \times 10^{-7})} \sum_{i=1}^{i=n} \frac{\Delta S_m \, \bar{r}_i^4}{x_i^2 \bar{r}_i^2} \qquad (3\text{-}30)$$

where r_i = pore entry radius, cm
ΔS_m = incremental change in mercury saturation
ϕ = fractional porosity
x_i^2 = factor to account for more complex geometry of system and termed dividing factor

Empirically determined values of the dividing factor as a function of permeability are presented in Fig. 3-35.

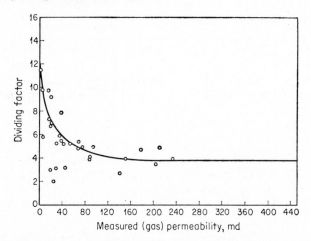

FIG. 3-35. Dividing factor correlation with measured (gas) permeability. (*From Burdine et al.[18]*)

The various equations presented here are better known examples and can be used to yield reasonable estimates of permeability. The chief value of these relationships, however, is in computing relative permeabilities. The procedure of calculating relative-permeability relations will be presented in the next section of this work.

EFFECTIVE AND RELATIVE PERMEABILITY

In Chap. 2, permeabilities were referred to rock conditions where a single-phase fluid saturation was considered. Darcy's law, as originally formulated and developed in Chap. 2, was considered to apply when the porous medium was fully saturated with a homogeneous, single-phase fluid. In petroleum reservoirs, however, the rocks are usually saturated with two or more fluids, such as interstitial water, oil, and gas. It is necessary to generalize Darcy's law by introducing the concept of effective permeability to describe the simultaneous flow of more than one fluid. In the definition of effective permeability each fluid phase is considered to be completely independent of the other fluids in the flow network. The fluids are considered immiscible, so that Darcy's law can be applied to each individually. Thus, Darcy's law can be restated as follows:

$$v_{os} = -\frac{k_o}{\mu_o}\left(\frac{dP_o}{ds} - \rho_o g\,\frac{dz}{ds}\right)$$

$$v_{gs} = -\frac{k_g}{\mu_g}\left(\frac{dP_g}{ds} - \rho_g g\,\frac{dz}{ds}\right)$$

$$v_{ws} = -\frac{k_w}{\mu_w}\left(\frac{dP_w}{ds} - \rho_w g\,\frac{dz}{ds}\right)$$

In the above equations, the subscripts o, g, and w refer to oil, gas, and water, respectively.

The effective permeability is a relative measure of the conductance of the porous medium for one fluid phase when the medium is saturated with more than one fluid. This definition of effective permeability implies that the medium can have a distinct and measurable conductance to each phase present in the medium.

Experimentation has established that effective permeability is a function of the prevailing fluid saturation, the rock-wetting characteristics, and the geometry of the pores of the rock. It becomes necessary, therefore, to specify the fluid saturation when stating the effective permeability of any particular fluid in a given medium. The effective permeability is stated as some numerical value at some given saturation conditions. Just as k is the accepted symbol for permeability, k_o, k_w, and k_g are the accepted symbols for the effective permeability to oil, water, and gas, respectively. The saturations, if known, should be specified to define completely the conditions at which a given effective permeability exists. Unlike the previously defined permeability, many values of effective permeability now exist, one for each particular condition of fluid saturation. Symbolically, $k_{o(60,13)}$ is the effective permeability of the medium to oil when the fluid saturations are 60 per cent oil, 13 per cent water, and 27 per cent gas. The saturation succession given above, that is, oil and water, is always followed. The gas saturation is understood to be the difference of the sum of oil and water saturations from 100 per cent.

Effective permeabilities are normally measured directly in the laboratory on small core samples. However, owing to the many possible combinations of saturation for a single medium, laboratory data are usually summarized and reported as relative permeability. Relative permeability is defined as

the ratio of the effective permeability of a fluid at a given value of saturation to the effective permeability of that fluid at 100 per cent saturation.
It is normally assumed that the effective permeability is the same for all
fluids at 100 per cent saturation, this permeability being denoted as the
permeability of the porous medium. Thus, relative permeability can be
expressed symbolically as

$$k_{ro(50,30)} = \frac{k_{o(50,30)}}{k}$$

$$k_{rw(50,30)} = \frac{k_{w(50,30)}}{k}$$

$$k_{rg(50,30)} = \frac{k_{g(50,30)}}{k}$$

which are the relative permeabilities to oil, water, and gas, respectively,
when the medium is saturated with 50 per cent oil, 30 per cent water, and
20 per cent gas, and k is the permeability at 100 per cent saturation of one
of the fluid phases.

Laboratory Investigations of Relative Permeability

The first experimental data based on the concept of a generalized set of
equations for Darcy's law were recorded by Wyckoff and Botset.[21] The
results of their work are shown by curves 1 in Fig. 3-36, which represent
the relative permeabilities for water and gas in an unconsolidated sand
pack. The fluids used in obtaining these data were water and carbon dioxide, where water was the wetting fluid. The curve labeled k_{rw} denotes
the relative permeabilities to water, while that labeled k_{rg} denotes the relative permeabilities to gas. The trends which are presented in this figure
have been substantiated by many investigations since the original work by
Wyckoff and Botset. The k_{rw} curve is typical of the trend of relative-
permeability curves for the wetting phase in a porous system regardless
of whether that phase is oil or water. The relative permeability to the
wetting phase is characterized by a rapid decline in value for small decreases in an original high saturation of that particular phase. The relative
permeability for the wetting phase normally approaches zero or vanishes
at saturations of the wetting phase greater than zero. Likewise, the k_{rg}
curve is typical of the relative permeability to a nonwetting phase, whether
that phase is gas, oil, or water.

The principal characterizing features which can be gained by the study
of relative-permeability curves are indicated in Fig. 3-36. The first of
these characteristics is commonly called the point of equilibrium saturation and is denoted by point A. Equilibrium saturation is that value at
which the nonwetting phase becomes mobile. This saturation may vary
between zero and 15 per cent nonwetting-phase saturation.

The second feature is the rapid rise in the relative permeability of the nonwetting phase for very small increases in nonwetting-phase saturations above the equilibrium saturation. The third general characteristic is the attainment of a nonwetting-phase relative permeability of nearly 100 per cent at nonwetting-phase saturations much less than 100 per cent.

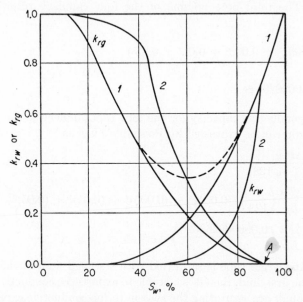

FIG. 3-36. Relative permeability to gas and water. (*a*) Unconsolidated sand; (*b*) consolidated sand. (*After Botset.*[22])

Some indication of the distribution of the fluid within the porous medium can be deduced from a study of relative-permeability data. The rapid decline in relative permeability to the wetting phase indicates that the larger pores or larger flow paths are occupied first by the nonwetting fluid. As the saturation of the nonwetting phase increases, the average pore size saturated with wetting fluid becomes successively smaller. This is confirmed by the rapid rise in the relative permeability to the nonwetting phase. In other words, at a saturation above the so-called equilibrium saturation, the nonwetting fluid occupies larger pores than does the wetting fluid. The attainment of 100 per cent relative permeability to the nonwetting phase at saturations of less than 100 per cent indicates that a portion of the available pore space, even though interconnected, contributes little to the fluid-conductive capacity of the porous medium.

In Chap. 2, it was shown that the pore space is proportional to the square of the diameter of the pore openings and the fluid-conductive capacity is proportional to the diameter of the pore openings to the fourth power.

Using the relations of Chap. 2 it is possible to use capillary tubes to illustrate the effective reduction in permeability caused by the introduction of a second fluid.

Example 3-3. Effect of Saturation on Fluid Conductance. Consider four capillary tubes of length L and diameters of 0.001, 0.005, 0.01, and 0.05 cm. The total pore volume of the four capillary tubes would be given by

$$\frac{\pi L}{4} (0.001^2 + 0.005^2 + 0.01^2 + 0.05^2)$$

$$\frac{\pi L}{4} (0.002626) \text{ cc}$$

The total conductive capacity for the four tubes under the same imposed pressure drop can be expressed by Poiseuille's law as

$$Q_t = \frac{\Delta P}{L} \frac{\pi}{\mu} \frac{d^4}{128}$$

$$Q_t = \frac{\Delta P \pi}{L\mu} \left(\frac{1}{128}\right) [(0.001)^4 + (0.005)^4 + (0.01)^4 + (0.05)^4]$$

$$Q_t = \frac{\Delta P \pi}{L\mu} \left(\frac{0.000006260626}{128}\right)$$

Now if the larger tube is saturated with a second fluid of the same viscosity as the first fluid, then it is possible to express the conductive capacity when two fluids are saturating the system to the conductive capacity when only one fluid saturates the system. Thus, it is seen that the ratios of the conductive capacities are

$$\frac{Q_2}{Q_t} = \frac{0.000,006,25}{0.000,006,260,626} = 0.9983$$

and from Darcy's law

$$\frac{Q_2}{Q_t} = \frac{k_{2(S_2,S_1)}}{k_{t(0,S_1)}} = 0.9983$$

$$\frac{Q_1}{Q_t} = \frac{k_{1(S_2,S_1)}}{k_{t(0,S_1)}} = 0.0017$$

$$S_2 = \frac{0.0025}{0.002626} = 0.952$$

S_2 = saturation of second fluid phase

S_1 = saturation of first fluid phase

The results of Example 3-3 are extreme, in that the sizes chosen for the capillary tubes vary over an extreme range. The wetting-phase saturation S_1, was changed by 95 per cent, and the effective permeability to this phase was decreased by 99.8 per cent. The relative-permeability values for the two fluids in Example 3-3 sum up to 1; that is, $(Q_1/Q_t) + (Q_2/Q_t) = 1$. This behavior is not true in actual porous systems. It would not be the case in this example if the minute film which would wet the surface were considered. This film would decrease the diameter of the larger tube, thus reducing the flow capacity for the second fluid, and yet the film itself would contribute no flow capacity to the wetting fluid. Thus, the total fluid capacity of the tubes would be decreased. This is a rather normal feature of most relative-permeability curves, where it is found that the total of all values of relative permeability seldom add up to 1.

As most reservoirs are comprised of consolidated porous media, Botset[22] subsequently reported results of similar relative-permeability tests conducted on consolidated sandstone. These tests were performed with water and carbon dioxide, and the results are indicated in Fig. 3-36 as curves 2. Again, water was the wetting fluid and carbon dioxide the nonwetting phase. Note the similarity of the curves for the consolidated and unconsolidated cores. Both cores give the same general results, the differences being in the slopes of the curves and water-saturation value at which the relative permeability to water vanishes. It is noted that the relative permeability to water vanishes at a much greater wetting-phase saturation for the consolidated core. This difference in flow behavior indicates that the relative permeability of a pore system is dependent in some fashion upon the pore geometry of that system.

The average results of 26 tests on relative-permeability for Permian dolomites by Bulnes and Fitting[23] are shown in Fig. 3-37. The reader will note that the general trend and shape of these curves on a Permian dolomite are essentially the same as those found for consolidated and unconsolidated sandstones. The same three characteristic points are noted: (1) The wetting-phase saturation declines very rapidly for small increases in nonwetting-phase saturations; (2) all the cores indicate an equilibrium gas, nonwetting-phase saturation somewhere between 10 and 30 per cent; (3) the relative permeability to the wetting phase, which in this case is oil, tends to vanish at saturations between 20 and 40 per cent. Thus, it seems that the trends are very similar to those obtained for sandstones, indicating that materials with intergranular porosity possess similar relative-permeability saturation characteristics.

It would be expected that data obtained on small core samples of fractured or vugular material would give very erratic relative-permeability results, which would differ from those obtained for rocks with intergranular porosity. The behavior of fractured or vugular material should more closely approximate that of Example 3-3, where the conductive capacity is in the fractures and the pore volume is largely in the matrix of the system. Such a conductance-volume relationship should give a very different relative-permeability curve from those indicated by Fig. 3-37 for three types of rocks.

All the tests previously discussed were conducted with two-fluid systems, one of which was always gas. Leverett[24] investigated a two-fluid system in which the fluids used were water and oil. He systematically investigated the effect of fluid viscosity, pressure gradients, and interfacial tension on the relative-permeability behavior of porous systems. He concluded from

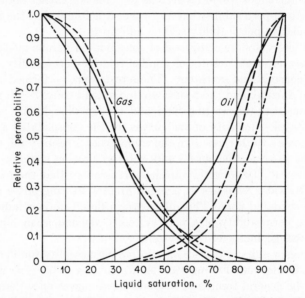

Fig. 3-37. Relative permeability to gas and oil for West Texas dolomites. (———) Wasson Field data; (——— — — ———). Slaughter Field data; (– – –) average results of 26 cores from three West Texas Permian dolomites. (*From Bulnes and Fitting.*[23])

the data, which are shown in Fig. 3-38, that relative permeability was substantially independent of the fluid viscosity but was some function of pore-size distribution, displacement pressure, pressure gradient, and fluid saturations. Subsequent work to that of Leverett has indicated that the displacement pressure and pressure gradient are parameters which are peculiar to laboratory measurements. These parameters have been given the notation "end effects." As these parameters are essentially properties of laboratory measurements, the means of measuring relative permeability in the laboratory must take them into consideration. End effects will be discussed in more detail in the section covering the laboratory measurement of relative permeability.

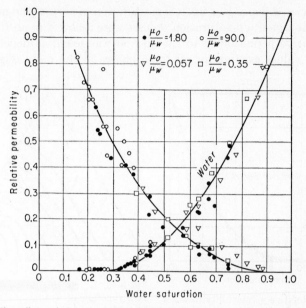

Fig. 3-38. The effect of the viscosity ratio (oil to water) on the relative permeabilities in a 100- to 200-mesh sand. (*After Leverett.*[24])

Three-phase Relative Permeability

As was mentioned previously, there are many instances when, not two fluids, but three fluids exist in the rock simultaneously. Thus two-phase relative-permeability data had to be amplified and extended for three-phase systems. Leverett[25] in 1941 reported results of steady-state flow tests on unconsolidated sand where three phases, oil, water, and gas, were used. From these data, one of the comprehensive studies recorded on three-phase relative permeability, the basic concepts for three-phase fluid flow were established.

The principal results of the work of Leverett are illustrated in Figs. 3-39 through 3-42. The fluids used by Leverett were nitrogen, kerosene, and brine. The relative permeability to the wetting phase, water, was found to correlate closely with the data of Wyckoff and Botset[21] (Fig. 3-39) and to be a unique function of the wetting-phase saturation. The fact that the relative permeability to the wetting phase depends on the saturation of the wetting phase alone can be rationalized from the data. The wetting phase occupies the portions of the pore space adjacent to the sand grains, thus occupying the smaller pore openings. Therefore, at a given level of wetting-phase saturation, the same portion of the pores are occupied by the wetting phase irrespective of the saturation of the other two phases.

This behavior is dependent upon the saturation history of the porous medium and is true if a desaturation process were followed at all times.

The relative permeability to gas and oil was found to depend on the saturation values existing for all three phases in the rock. Figures 3-40

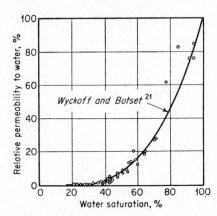

FIG. 3-39. Relative permeability to water as a function of water saturation. (*From Leverett and Lewis.*[25])

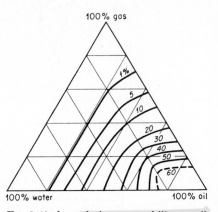

FIG. 3-40. k_{ro}, relative permeability to oil as a function of saturation. Curves are lines of constant permeability as per cent of relative permeability. (*From Leverett and Lewis.*[25])

and 3-41 show the relative-permeability data for gas and oil in a three-phase system. The data are plotted on triangular diagrams to define the saturation condition of the rock. The relative-permeability data are plotted as lines of constant-percentage relative permeability.

The dependency of the oil relative permeability on the saturations of the other phases can be established by the following reasoning: The oil phase has a greater tendency than the gas to wet the solid. In addition, the interfacial tension between water and oil is less than that between water and gas. The oil occupies portions of the rock adjacent to the water or pores that are dimensionally between those occupied by the water and the gas. At lower water saturations the oil occupies more of the smaller pores. The extended flow path length caused by this phenomenon accounts for the change in relative permeability to oil at constant oil saturations and varying water saturations.

For an oil saturation of 60 per cent and a water saturation of 40 per cent, the relative permeability to oil as read from Fig. 3-40 is approximately 34 per cent. For the same oil saturation and a water saturation of 20 per cent, it is noted that the relative permeability to oil increases to approximately 38 per cent. For a water saturation of zero, the relative permeability to oil is approximately 18 per cent. Thus, it is seen that by changing the water and gas saturation the flow characteristics of the oil are changed so that the oil assumes more tortuous paths.

The variation of the gas relative permeability at constant gas saturations to the saturations of the other phases is indicated in Fig. 3-41. The reason for the particular behavior indicated is not definite, as other investigators indicate that the relative permeability to gas should be a unique function of gas saturation. Other studies of three-phase flow systems will be necessary to establish definitely the relative permeability for gas in three-phase systems. It would be expected that when gas is in a system, it is the fluid least likely to wet the surface of the rock and, therefore, should take on a property which is dependent only upon the total fluid saturations of the other two phases. The other phases, oil and water, should occupy the smaller pore openings and wet the surface of the rock. Therefore, the gas phase should be dependent only upon the total liquid saturations and independent of how much of that total is composed of either phase.

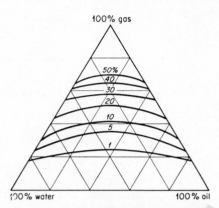

FIG. 3-41. k_{rg}, relative permeability to gas as a function of saturation. Curves are lines of constant permeability as per cent of relative permeability. (*From Leverett and Lewis.*[25])

It is noted from a study of Leverett's data that the saturation region in which simultaneous flow of all three phases occurs is quite small. The regions in which single-phase, two-phase, and three-phase fluids flow will normally occur are indicated in Fig. 3-42. For gas saturations in excess of 35 per cent, essentially only gas is flowing in the system. For gas saturations between 18 and 35 per cent and for water saturations less than approximately 40 per cent, oil and gas are both flowing. For water saturations between 18 and 85 per cent and oil saturations ranging from 15 to 82 per cent, where the gas saturation does not exceed approximately 15 per cent, only oil and water are flowing. The region of three-phase flow is extremely small and essentially centers around the region of 20 per cent gas, 30 per cent oil, and 50 per cent water saturation. This region is illustrated in Fig. 3-42 by the "hatched" area. The single-phase flow regions are illustrated by the shaded area, and the two-phase flow regions are illustrated by the white area. From these data it is evident that in most cases two-phase relative-permeability curves are quite satisfactory. For immobile water saturations it is possible to define gas and oil relative-permeability curves using two-phase techniques. Also in this two-phase flow region, the curves obtained for two-phase systems, gas and liquid, are essentially the same as would be obtained if a third immobile phase is present. The relative permeability to the mobile liquid phase is essentially dependent on

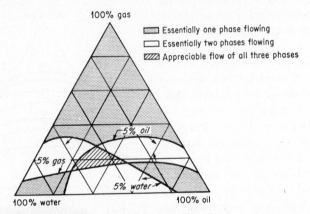

FIG. 3-42. Approximate limits of saturations giving 5 per cent or more of all components in flow stream. Fluids: nitrogen, kerosene, brine. Arrows point to increasing fraction of respective components in stream. (*From Leverett and Lewis.*[25])

the total liquid saturation. Thus the relative permeability to oil at 60 per cent oil and 20 per cent water and that at 80 per cent oil and zero per cent water are not greatly different, as illustrated by Fig. 3-40. This behavior is attributed to the fact that the smaller saturation values of the wetting phase contribute little to the fluid conductance of the porous matrix.

Measurement of Relative-permeability Data

There are essentially four means by which relative-permeability data can be obtained. They are (1) direct measurement in the laboratory by a steady-state fluid flow process, (2) measurement in the laboratory by a displacement or a pseudo-unsteady-state process, (3) calculations of relative-permeability data from capillary-pressure data, and (4) calculations from field-performance data. The methods most used are the laboratory steady-state flow and displacement processes.

There are numerous steady-state methods which can be used in the laboratory to measure relative permeability, but essentially, all of them depend upon the following technique. A small core sample is chosen and prepared for the test. It is mounted either in lucite or in a pressurized rubber sleeve. Either the flow system is designed for a high rate of flow and large pressure differential, or each end of the sample is suitably prepared with porous disks and test sections to minimize end effects. The phases oil and gas, oil and water, or gas and water which are to be used in the test are introduced simultaneously at the inlet end through different piping systems. Most tests are started with the core sample at 100 per cent saturation in the wetting phase, and the tests are known as desaturation tests. The two fluids are introduced at a predetermined fluid ratio and are flowed through the core until the produced ratio is equal to the injected ratio. At this time, the core system is considered to be in a steady-state flow condition and the existing saturations are considered to be stable.

The saturation of the various fluids are determined in one of three fashions: (1) Electrodes have been inserted in the test section, and the saturations are determined by measurement of the core resistivity; (2) the core section is removed and weighed to determine the saturation conditions; or (3) a volumetric balance is maintained of all fluids injected and produced from the sample. Once the saturation has been measured by one of the above methods, the relative permeability of the two phases at these saturation conditions can be calculated. The injected ratio is increased, removing more of the wetting phase, until once again the system is flowing in steady-state condition. The process is continually repeated until a complete relative-permeability curve is obtained.

An alternate method is to use the resaturation process where the test section is originally 100 per cent saturated by the nonwetting phase. In this method the injection ratios start out at high nonwetting-phase values and decline to 100 per cent wetting phase. The results obtained using the desaturation and resaturation processes illustrate a hysteresis effect of the same type discussed earlier in connection with capillary-pressure curves.

Some of the equipment and results obtained using the steady-state process

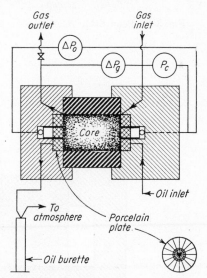

Fig. 3-43. Relative-permeability apparatus (Hassler's principle). *From Osoba et al.*[26])

are illustrated in Figs. 3-43 to 3-49. Four of the apparatus which have been developed for testing small core samples are shown in Figs. 3-43 to 3-46. The four apparatus illustrated represent the Hassler method, Penn State method, Hafford method, and dispersed-feed method. In order to eliminate end effects, porous material has been placed in contact with the outflow face of the test section. In the Hafford apparatus (Fig. 3-45)

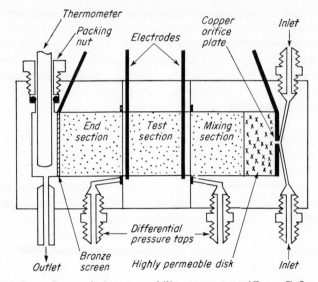

FIG. 3-44. Penn State relative-permeability apparatus. (*From Geffen et al.*[27])

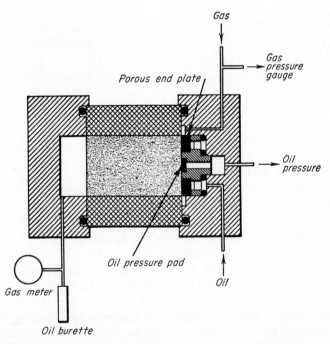

FIG. 3-45. Hafford relative-permeability apparatus. (*From Richardson et al.*[28])

and in the dispersed-feed apparatus (Fig. 3-46), end effects are materially reduced by maintaining a high rate of flow through the test section. All the apparatus depend on the same flow mechanism and are different only in the manner in which they introduce the two fluids and in the manner in which they adjust for end effects. As some of the steady-state relative-permeability measuring devices depend on the rate of flow or pressure drop

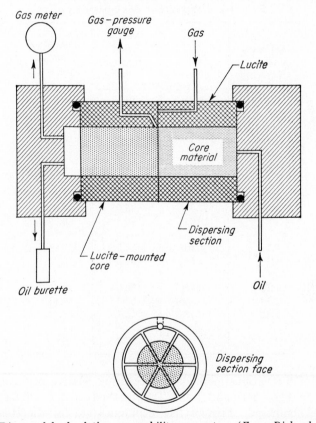

FIG. 3-46. Dispersed-feed relative-permeability apparatus. (*From Richardson et al.*[28])

to eliminate end effects, it is necessary to determine the effect of rate on the accuracy of the measurements.

Table 3-3 presents data on the effect of pressure gradient on relative-permeability measurements. It is evident that the relative permeability is essentially independent of the pressure gradient providing the gradient is maintained sufficiently high in the laboratory to eliminate end effects. For example, the water-oil relative-permeability ratio when the oil-water injection ratio is $\frac{1}{10}$ is essentially constant for pressure gradients ranging

from 2 to 16 psi per in. The same is true for other ratios. In the case of
the 100:1 oil-water ratio, when the pressure gradient gets below 0.678 psi
per in., the permeability ratio of oil to water increases, so that at low-
pressure differentials it is apparent that end effects are becoming a domi-
nate factor in the control of the flow mechanism.

TABLE 3-3. EFFECTS OF PRESSURE GRADIENT ON
RELATIVE PERMEABILITY MEASUREMENTS[27]

Oil-water flow ratio	k_o/k_w	Average brine saturation, %	Pressure gradient, psi/in.
Infinite	Infinite	14.2	2.04
100:1	150.6	36.9	4.71
100:1	149.4	37.2	9.16
100:1	149.4	37.3	1.27
100:1	152.3	37.7	0.678
10:1	15.76	47.3	14.34
10:1	15.36	45.9	4.93
10:1	15.04	46.3	0.994
1:1	1.488	53.5	1.24
1:1	1.510	52.9	3.05
1:10	0.1507	56.0	16.47
1:10	0.1507	55.6	8.14
1:10	0.1537	55.1	2.43
0	0	57.7	15.91

Results of the various methods of measuring relative permeability are
compared in Figs. 3-47 and 3-48. Two methods and six different techniques are compared. Five of the techniques are for the steady-state
processes already discussed; the sixth is for the unsteady-state process,
which is discussed briefly in the text that follows. All the methods tend
to check closely and indicate that as long as proper precautions are taken
to remove errors inherent in each laboratory measuring device, the curves
obtained should closely represent values which would exist in the reservoir
for a duplicate saturation distribution.

Displacement Process of Measuring Relative Permeability. The gas-
drive displacement technique of testing is essentially a nonsteady-state
flow process. The sample is originally 100 per cent saturated with the
wetting phase, and instead of both gas and liquid being injected, only gas

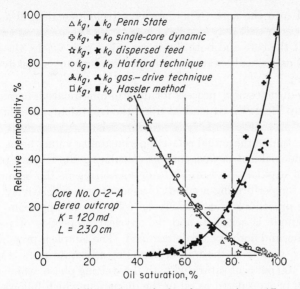

FIG. 3-47. Relative permeability—six methods, short section. (*From Richardson et al.*[28])

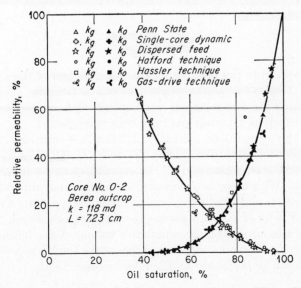

FIG. 3-48. Relative permeability—six methods, long section. (*From Richardson et al.*[28])

is injected into the core. The gas-drive technique then is a nonsteady state process in that only one fluid is entering the core and two fluids are leaving. If the core and both fluids are taken as a whole, the process can be treated as a steady-state volume process but not a steady-state mass flow system.

The gas-displacement process for determining relative permeabilities as proposed by Welge[29] is dependent upon the frontal advance fluid-flow concept. This procedure actually determines relative-permeability ratios and is dependent upon the actual relative-permeability values being determined by some independent means, such as calculating the relative permeability to oil from capillary-pressure data or measuring in the laboratory. The procedure for performing a gas-displacement test is relatively simple and fast. The procedure is essentially as follows: An approximately homogeneous sample is selected, and its physical properties of permeability, bulk volume, and porosity are determined. The sample is properly mounted in a holder, similar to those used in the steady-state tests (Figs. 3-43 to 3-46), and 100 per cent saturated with the wetting phase, which is normally oil and will be considered as oil in the discussion which follows. The sample is desaturated by injecting gas at one end and producing both oil and gas at the other end of the small sample. In the calculations of the data obtained from such a test, there are essentially three necessary conditions or assumptions which must be satisfied. First, the pressure drop across the core sample must be large enough to make any capillary end effects negligible. Second, the gas saturation can be described at a mean value of pressure defined as

$$\bar{P} = \frac{P_i + P_o}{2} \tag{3-31}$$

where P_i represents the pressure at the injection end of the core and P_o represents the pressure at the production end of the core. Third, flow is horizontal, and the core sample is small enough and the test time is short, so that all effects of gravitational forces can be neglected.

If these three conditions are satisfied, then it is necessary to measure only the following quantities during the test: (1) the cumulative gas injected as a function of time and (2) the cumulative oil produced as a function of time. With these two measured quantities and with the pressures at the injection and production ends remaining constant, the relative-permeability ratio of gas with respect to oil can be calculated.

From the measured data, the cumulative volume of gas injected in terms of mean pressure is expressed as a multiple of the total pore volume of the sample and can be calculated by the following equation:

$$(G_i)_{pv} = \frac{2G_{ic}P_i}{LA\phi(P_i + P_o)} \tag{3-32}$$

where $(G_i)_{pv}$ = cumulative injected gas expressed as pore volumes

$LA\phi$ = total pore volume of sample

G_{ic} = cumulative gas injected expressed at pressure P_i

P_i and P_o are as previously defined

A quantity known as $S_{g(av)}$ may be calculated simply by dividing the cumulative oil produced by the pore volume of the sample. Both the gas injected and oil produced have been measured with respect to time and therefore can be cross plotted so that a plot of $S_{g(av)}$ as a function of the cumulative pore volumes of gas injected can be obtained. The slope of the $S_{g(av)}$-injected gas curve represents the fraction of the total outflow volume from the sample that is oil at any given time, which defines the following equation:

$$f_o = \frac{d(S_g)_{av}}{d(G_i)_{pv}} \tag{3-33}$$

where f_o represents the fraction of the total outflow that is oil. The relative-permeability ratio of gas to oil can be calculated from the following equation:

$$\frac{k_g}{k_o} = \frac{1 - f_o}{f_o(\mu_g/\mu_o)} \tag{3-34}$$

where k_g/k_o = relative permeability ratio of gas to oil

μ_g = viscosity of gas

μ_o = viscosity of oil

f_o = fractional flow of the oil as previously defined

This particular value of the relative-permeability ratio applies at the gas saturation at the outflow face. The gas saturation at the outflow face is expressed by the following equation:

$$(S_g)_o = (S_g)_{av} - (G_i)_{pv} f_o \tag{3-35}$$

Hence the relative-permeability ratios are obtained as a function of saturation by solving the above series of equations.

In order to determine the actual value of the relative permeability to either gas or oil, it is necessary that one or the other be independently determined. It would be possible to measure or else to calculate the other functions by one of the means previously discussed in this chapter.

The gas-displacement method has several advantages in that it can be performed with a small amount of equipment and can be performed rapidly and on relatively small core samples. It has the disadvantage in that it cannot determine relative-permeability ratios at low values of gas saturation except by the use of high-viscosity oils. The equations do not apply until such time as gas is being produced at the outflow end of the core. Depending upon the permeability of the core and the pore geometry, the gas saturation at the outflow face at the time gas is initially produced may vary between 2 and 15 per cent, depending upon the viscosities of the gas and oil in the system. The lower the viscosity of the oil, the greater will be the gas saturation at the outflow face at the time that gas production begins. Thus, by using oil samples of various viscosities, it would be possible to define the relative-permeability ratio over the entire saturation range in which two-phase flow might occur.

End effects are not important in the gas-displacement method of measuring relative-permeability ratios because of the high-pressure drops involved. The magnitude of capillary-pressure end effects are extremely small compared with the imposed flow gradient and are neglected. Hence, the equipment necessary for counteracting this phenomenon in the laboratory is eliminated.

Field Determination of Relative-permeability Ratios. The third means of determining relative permeability has the same drawback as the displacement process in that the data obtained are actually determined as relative-permeability ratios. This process is a calculating procedure utilizing field data to calculate the relative-permeability ratio of gas to oil. If Darcy's equation were written for gas and oil flow, both phases being considered to be independent of each other in the flow system, the relative permeability ratio could be defined by the following equation:

$$\frac{Q_g}{Q_o} = \frac{A(k_g/\mu_g)(\Delta P_g/\Delta L)}{A(k_o/\mu_g)(\Delta P_o/\Delta L)} \tag{3-36}$$

If the volumes of flow are expressed in the above equation as Q_g and Q_o at reservoir conditions, and if it is assumed that the pressure drop in the gas system is the same as the pressure drop in the oil system, then the relative-permeability ratio can be expressed in terms of surface volumes by Eq. (3-37).

$$\frac{k_g}{k_o} = \left(\frac{Q_g}{Q_o}\frac{\mu_g}{\mu_o}\right) = \frac{B_g}{5.61B_o}\frac{\mu_g}{\mu_o}(R_p - R_s) \tag{3-37}$$

where B_g represents the formation volume factor of the gas expressed as reservoir cubic feet of gas per standard cubic foot of gas and B_o represents the oil-formation volume factor expressed as reservoir barrels of oil per stock-tank barrel of oil. R_p represents the producing gas-oil ratio, and R_s represents the solution-gas-oil ratio, both expressed as standard cubic foot per stock-tank barrel. μ_g is the gas viscosity and μ_o is the oil viscosity at reservoir conditions of pressure and temperature.

The normal procedure is to use field average gas-oil ratios, which are normally the most accurate values obtainable. The field average gas-oil ratios are preferably obtained from gas-plant production figures. The solution ratio R_s is dependent upon the reservoir pressure and is taken at the average reservoir pressure which is in existence at the time the produced gas-oil ratio is determined. B_g and B_o are also taken at this particular pressure as well as μ_g and μ_o. The saturation at which this particular value of relative-permeability ratio applies must be calculated from field production data.

As oil saturation is dependent upon the type of reservoir performance and reservoir performance has not been previously discussed in the text, further discussion of this particular means of determining relative-permeability ratios will be delayed.

End Effects. During the discussion of the three means of measuring effective and relative permeability, there was a continued mention of end effects and of the fact that in many of the techniques developed special precautions had been taken to eliminate these effects from the laboratory measurement. End effects arise from the saturation discontinuity existing at the outflow face of a porous medium when mounted for a flow test. The fluids flowing through the core are discharged into a region void of the porous medium. Therefore, at the outflow face, all the fluids exist at the same pressure, whereas immediately within the pores of the rock at the outflow face, capillary-pressure conditions require that the saturation of the wetting phase approach 100 per cent. There is, then, a saturation gradient established in the wetting phase of the flow system.

The theory of "end effect" can be developed as follows:

From Darcy's law for a linear system and the concept of effective permeability, when more than one phase is present in a rock,

$$-dP_{wt} = \frac{Q_{wt}\mu_{wt}\,dL}{k_{wt}A} \qquad (3\text{-}38)$$

$$-dP_{nwt} = \frac{Q_{nwt}\mu_{nwt}\,dL}{k_{nwt}A} \qquad (3\text{-}39)$$

$$dP_c = dP_{nwt} - dP_{wt} \qquad (3\text{-}40)$$

where the subscripts nwt and wt refer to the nonwetting and wetting fluids, respectively, other symbols as previously defined.

Combining Eqs. (3-38), (3-39) and (3-40) it is found that

$$\frac{dP_c}{dL} = \left(\frac{Q_{wt}\mu_{wt}}{k_{wt}} - \frac{Q_{nwt}\mu_{nwt}}{k_{nwt}}\right)\frac{1}{A} \qquad (3\text{-}41)$$

where dP_c/dL is the capillary-pressure gradient within the core. Since capillary pressure has been shown to be a function of saturation and in any displacement system the saturation can be expressed as a function of length, it is possible to write

$$\frac{dP_c}{dL} = \frac{dP_c}{dS_{wt}}\frac{dS_{wt}}{dL}$$

where S_{wt} is the wetting-phase saturation. Eq. (3-41) then becomes

$$\frac{dS_{wt}}{dL} = \frac{1}{A}\left(\frac{Q_{wt}\mu_{wt}}{k_{wt}} - \frac{Q_{nwt}\mu_{nwt}}{k_{nwt}}\right)\frac{1}{dP_c/dS_{wt}} \qquad (3\text{-}42)$$

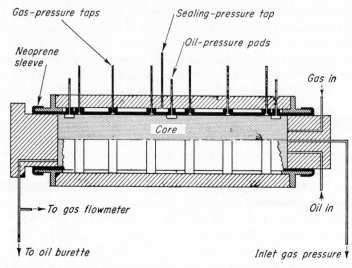

FIG. 3-49. Long-core relative-permeability apparatus. (*From Richardson et al.*[28])

where dS_{wt}/dL is the change in wetting-phase saturation with length. The saturation in the wetting phase at the outflow face is that corresponding to the equilibrium saturation in the nonwetting phase (essentially 100 per cent saturation of the wetting phase). The saturation gradient dS_{wt}/dL

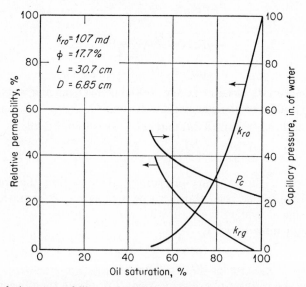

FIG. 3-50. Relative-permeability and capillary-pressure saturation relations for Berea outcrop sand. (*From Richardson et al.*[28])

within a flow system can be determined by graphical integration of Eq. (3-42) using capillary-pressure and relative-permeability data.

The desire to eliminate end effects has influenced the design of most apparatus developed for the determination of the relative permeability of small core samples. Devices to mix the flowing fluids adequately prior to entry into the core have also received considerable attention.

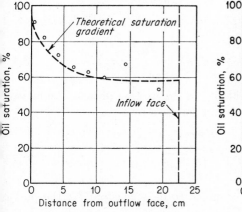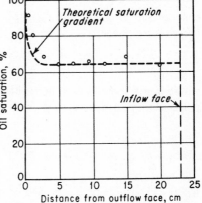

FIG. 3-51. Comparison of experimental and theoretical saturation gradients due to boundary effects. ($q_g = 0.15$ cc/sec, $q_o = 0.000336$ cc/sec). (*From Richardson et al.*[28])

FIG. 3-52. Comparison of experimental and theoretical saturation gradients due to boundary effects. ($q_g = 0.80$ cc/sec, $q_o = 0.00288$ cc/sec). (*From Richardson et al.*[28])

To determine the magnitude of end effects, Richardson[28] studied saturation gradients in a long core apparatus (Fig. 3-49). The test apparatus was designed to determine the pressure in each of the flowing phases at different positions along the core. The relative-permeability relationships were determined for different rates of flow and pressure gradients across the core. In addition, capillary-pressure characteristics were measured. Figure 3-50 presents the results of these tests as well as the physical properties of the core. The relative-permeability data shown are for conditions eliminating end effects and therefore are correct.

Figures 3-51 and 3-52 illustrate both the computed and measured saturation gradients measured by Richardson on the long core shown in Fig. 3-49. Note the good correspondence of theory and experimental results and the reduction of end effect resulting from using higher rates of flow (Fig. 3-52).

Calculation of Relative Permeability from Capillary-pressure Data. In the discussion of capillary pressure, several equations were presented for the calculation of permeability from capillary-pressure data.

The relation developed by Purcell[7] [Eq. (3-27)] can be readily adapted to the computation of wetting-phase relative permeability. The relative permeability is, by definition, the ratio of the effective permeability at a given saturation to the permeability of the medium.

From Eq. (3-27)

$$k = 10.24(\sigma \cos \theta)^2 \lambda \phi \int_{S=0}^{S=1} \frac{dS}{(P_c)^2} \qquad (3\text{-}27)$$

generalizing and considering capillary-pressure data for displacement of the wetting phase,

$$k_{wt} = 10.24(\sigma \cos \theta)^2 \lambda \phi \int_{S=0}^{S=S_{wt}} \frac{dS}{(P_c)^2} \qquad (3\text{-}43)$$

where k_{wt} is the effective permeability to the wetting phase. The relative permeability to the wetting phase is given then by

$$\frac{k_{wt}}{k} = k_{rwt} = \frac{\int_{S=0}^{S=S_{wt}} dS/(P_c)^2}{\int_{S=0}^{S=1} dS/(P_c)^2} \qquad (3\text{-}44)$$

where the lithology factor λ is assumed to be a constant for the porous medium.

The effective permeability to the nonwetting phase (k_{rwt}) can be calculated in a similar fashion as in Eq. (3-43) by assuming that the nonwetting phase is contained in tubes or pores, free of the wetting phase of radius as defined by the capillary-pressure relation in the wetting-phase saturation interval $S_{wt} = S_{wt}$ to $S_{wt} = 1$.

$$k_{nwt} = 10.24(\sigma \cos \theta)^2 \lambda \phi \int_{S=S_{wt}}^{S=1} \frac{dS}{(P_c)^2} \qquad (3\text{-}45)$$

The relative permeability to the nonwetting phase (k_{rnwt}) is given by

$$\frac{k_{nwt}}{k} = k_{rnwt} = \frac{\int_{S=S_{wt}}^{S=1} dS/(P_c)^2}{\int_{S=0}^{S=1} dS/(P_c)^2} \qquad (3\text{-}46)$$

Rapoport and Leas[20] presented two equations for relative permeability to the wetting phase. These are based on surface energy relationships and the Kozeny equation. The equations are presented as defining limiting values, minimum and maximum, for relative permeability.

For $k_{rwt \text{ (min)}}$

$$k_{rwt \text{ (min)}} = \left(\frac{S_{wt} - S_m}{1 - S_m}\right)^3 \frac{\int_1^{S_{wt}} P_c \, ds}{\int_1^{S_m} P_c \, ds + \int_1^{S_{wt}} P_c \, ds} \qquad (3\text{-}47)$$

where S_m is the minimum irreducible saturation of the wetting phase from a drainage capillary-pressure curve and S_{wt} is the saturation in the wetting phase for which $k_{rwt\ (min)}$ is evaluated.

For $k_{rwt\ (max)}$

$$k_{rwt\ (max)} = \frac{\left(\dfrac{S_{wt} - S_m}{1 - S_m}\right)^3 \left(\displaystyle\int_1^{S_m} P_c\, dS\right)^2}{\left[\displaystyle\int_{S_{wt}}^{S_m} P_c\, dS \frac{\displaystyle\int_1^{S_{wt}} P_c\, dS}{\dfrac{1 - \phi}{S_{wt}\phi}\left(\dfrac{P_c(S_{wt} - S_m)}{\displaystyle\int_{S_{wt}}^{S_m} P_c\, dS}\right)^2}\right]^2} \tag{3-48}$$

Symbols are as previously defined.

Fatt and Dykstra,[30] following the basic method of Purcell for calculating permeability, developed an expression for relative permeability considering that the lithology factor λ was a function of saturation. The lithology factor is essentially a correction for deviation of the path length from the length of the porous medium. Fatt and Dykstra assumed that the deviation of the path length was a function of the radius of the conducting pores, so that

$$\lambda = \frac{a}{r^b} \tag{3-49}$$

where r is the radius of a pore and a and b are constants for the material.

The equation for relative permeability for the wetting phase (k_{rwt}) then becomes

$$k_{rwt} = \frac{\displaystyle\int_0^{S_{wt}} dS/P_c^{2(1+b)}}{\displaystyle\int_0^1 dS/P_c^{2(1+b)}} \tag{3-50}$$

which is of the same form as Eq. (3-44), where b is a correction for deviation of the flow paths from straight tubes.

Fatt and Dykstra further assumed that $b = \frac{1}{2}$, thus reducing (3-50) to

$$k_{rwt} = \frac{\displaystyle\int_0^{S_{wt}} dS/(P_c)^3}{\displaystyle\int_0^1 dS/(P_c)^3} \tag{3-51}$$

Figure 3-53 presents experimental data and computed curves based on Eqs. (3-44) and (3-51). There are significant differences in the computed and observed data.

Burdine[18] reported equations for computing relative-permeability curves for both the wetting and nonwetting phases. His equations can be shown to reduce to a form similar to that of Purcell for permeability. Burdine's contribution is principally in handling tortuosity.

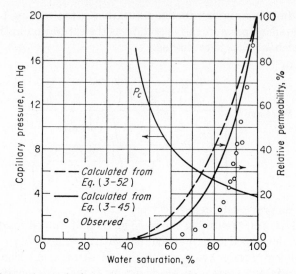

FIG. 3-53. Comparison of observed and calculated relative permeabilities. (*From Fatt and Dykstra.*[30])

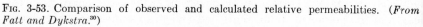

Example 3-4.   Example Calculation of Wetting-phase Relative Permeability.[30]

$$k_{rwt} = \frac{\int_0^{S_{wt}} dS/P_c^3}{\int_0^1 dS/P_c^3}$$

S, %	P_c, cm Hg	$1/P_c^3$, (cm Hg)$^{-3}$	Area from 0 to S, sq in.	k_{rwt}, %
100	3.8	18.2×10^{-3}	4.15	100
95	4.1	14.5	3.13	75
90	4.4	11.7	2.31	56
85	4.8	9.0	1.68	40
80	5.3	6.7	1.19	29
75	5.8	5.1	0.82	20
70	6.4	3.8	0.54	13
65	7.1	2.8	0.34	8.2
60	8.0	2.0	0.20	4.8
55	9.4	1.2	0.10	2.4
50	11.6	0.6	0.04	1.0
45	15.0	0.3×10^{-3}	0.01	0.2

Defining the tortuosity factor for a pore as λ_i when the porous medium is saturated with only one fluid and λ_{wti} for the wetting-phase tortuosity factor when two phases are present, a tortuosity ratio can be defined as

$$\lambda_{rwti} = \frac{\lambda_i}{\lambda_{wti}}$$

Then
$$k_{rwt} = \frac{\int_0^{S_{wt}} (\lambda_{rwti})^2 \, dS/(\lambda_i)^2 (P_c)_i^2}{\int_{S=0}^{S=1} dS/(\lambda_i)^2 (P_c)_i^2} \tag{3-52}$$

If λ_i is a constant for the porous medium and λ_{rwti} depends only on the final saturation, then

$$k_{rwt} = (\lambda_{rwt})^2 \frac{\int_{S=0}^{S=S_{wt}} dS/(P_c)_i^2}{\int_{S=0}^{S=1} (P_c)_i^2} \tag{3-53}$$

In similar fashion, the relative permeability to the nonwetting phase can be expressed utilizing a nonwetting-phase tortuosity ratio λ_{rnwti}

$$k_{rnwt} = (\lambda_{rnwt})^2 \frac{\int_{S=S_{wt}}^{S=1} dS/(P_c)_i^2}{\int_{S=0}^{S=1} dS/(P_c)^2} \tag{3-54}$$

Burdine has shown that

$$\lambda_{rwt} = \frac{S_{wt} - S_m}{1 - S_m} \tag{3-55}$$

where S_m is the minimum wetting-phase saturation from a capillary-pressure curve. The nonwetting-phase tortuosity can be approximated by a straight-line function also and is given by

$$\lambda_{rnwt} = \frac{S_{nwt} - S_e}{(1 - S_m) - S_e} \tag{3-56}$$

where S_e is the equilibrium saturation to the nonwetting phase. Figure 3-54 illustrates the agreement attained between experiment and theory. The calculated curve in Fig. 3-54 was based on equations essentially as given above, although Burdine expressed them in quite different form.

Factors Affecting Relative Permeability. In discussing relative-permeability measurement, reference was made to wetting-phase and nonwetting-phase relative permeabilities. If all materials wet the porous medium to the same degree, then the data taken with an oil-gas system would be the same as the data taken with a water-gas system or a water-oil system. As all fluids do not wet a solid to the same degree and as water may not be the fluid which preferentially we .s the rock, investigations have been made of the effect rock wettability has on relative-permeability data for oil-brine systems. The relative-permeability values are affected by the change in the fluid distribution brought about by different wetting characteristics. Figure 3-55 illustrates changes in relative permeability caused by different wetting characteristics. Curves 1 and 2 are indicative of a water-wet system, while curve 3 indicates that the system is preferentially oil-wet. It is noted that when the rock is preferentially water-wet, the water loses its mobility at a higher value of water saturation than when the rock is preferentially oil-wet. This fact would indicate that the oil is

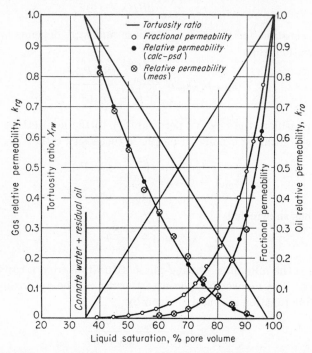

FIG. 3-54. Relative-permeability curves. Sample No. 1374 Atlantic Refining Co. Gas permeability, 72 millidarcys. Effective porosity, 21.9 per cent. Fluids: gas, oil, and connate water. Data: measurements by Atlantic Refining Co. Calculation from mercury-injection data. (*From Burdine et al.*[18])

retained in some of the smaller, more tortuous paths within the medium. It thus becomes necessary to classify reservoir rocks as being oil-wet, water-wet, or intermediate so as to define relative permeability properly.

As was discussed for capillary-pressure data, there is also a saturation-history effect for relative permeability. The effect of saturation history on relative permeability is illustrated in Figs. 3-56 and 3-57. If the rock sample is initially saturated with the wetting phase and relative-permeability data are obtained by decreasing the wetting-phase saturation while flowing onwetting and wetting fluids simultaneously in the core, the process is classified as drainage or desaturation. If the data are obtained by increasing the saturation of the wetting phase, the process is termed imbibition or resaturation. This nomenclature is consistent with that used in connection with capillary pressure. The process used in obtaining relative-permeability data in the laboratory must correspond to the reservoir process to which these data shall be applied.

The difference in the two processes of measuring relative permeability can be seen by observing Figs. 3-56 and 3-57. It is noted that the imbibition technique causes the nonwetting phase (oil) to lose its mobility at

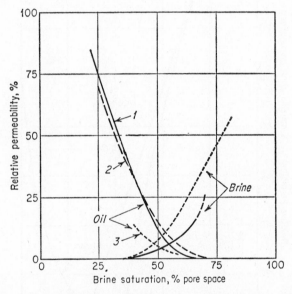

FIG. 3-55. Effect of wettability on flow behavior. Nellie Bly sandstone. (*From Geffen et al.*[27])

higher values of saturation than does the drainage technique. The two methods have similar effects on the wetting-phase (water) curve. The drainage method causes the wetting phase to lose its mobility at higher values of wetting-phase saturation than does the imbibition method.

In the discussion of both porosity and permeability, it was noted that overburden pressure affected the values obtained. Fatt[31] reported, as shown in Fig. 3-58, that overburden pressure did not affect the relative-permeability data although the effective permeabilities are altered.

In Figs. 3-59 and 3-60 are presented relative-permeability and permeability-ratio data for typical tests of various media. These data indicate the effect of pore configuration and pore-size distribution on relative per-meability. The curves should not be interpreted as representative of the types of media but should be viewed as illustrations of the effects of pore geometry on relative permeability. It is apparent then that universal per-meability curves cannot be established. Rather, each reservoir rock (each

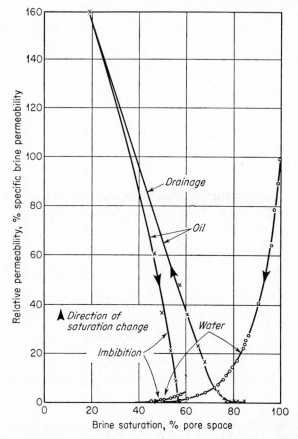

Fig. 3-56. Oil-water flow characteristics. Nellie Bly sandstone, permeability 143 millidarcys, porosity 26.1 per cent. (*From Geffen.*[27])

sample, for that matter) has a characteristic pore geometry which affects relative permeability.

In Fig. 3-61 is shown the effect of connate-water saturation on the gas-oil permeability ratio. In general, low water saturations do not appreci-ably affect the permeability ratio, simply because the water occupies space which does not contribute substantially to the flow capacity of the rock.

Use of Effective- and Relative-permeability Data

Relative-permeability data are essential to all flow work in the field of reservoir engineering. Just a few of its uses will be mentioned here. Other applications of relative-permeability data will be illustrated in the remainder of the text.

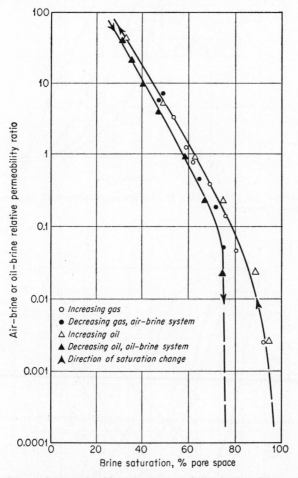

o *Increasing gas*
● *Decreasing gas, air-brine system*
△ *Increasing oil*
▲ *Decreasing oil, oil-brine system*
▲ *Direction of saturation change*

FIG. 3-57. Effect of saturation history on permeability ratios. (*From Geffen.*[27])

Determination of Free Water Surface. As was mentioned in the discussion on capillary pressure, it is necessary to determine the free water surface (level) in order to calculate fluid distribution properly. From the relative-permeability curves which have been presented, it should have become apparent that the point of 100 per cent water flow is not necessarily the

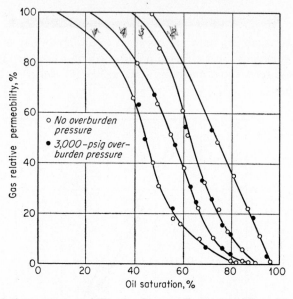

Fig. 3-58. Relative gas permeability in the gas-oil system with and without overburden pressure. (*From Fatt.*[31])

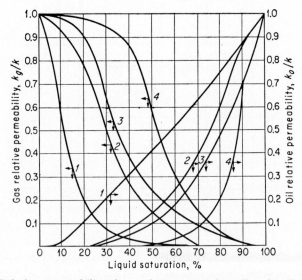

Fig. 3-59. Relative permeability for various types of media. (1) Capillary tube (*Martinelli, Putnam, and Lochart, Trans. Am. Inst. Chem. Engrs., Aug.* 25, 1946); (2) dolomite (Bulnes and Fitting[23]); (3) unconsolidated sand (Botset[22]); (4) consolidated sand Botset[22]).

point of 100 per cent water saturation. Knutsen[32] recognized the fact that two water tables exist. These two water tables can be defined as: *100% water Saturation*

A 1. The free water or zero-capillary-pressure level
B 2. The level below which fluid production is 100 per cent water *100% water Production*

These two definitions for the water table in an oil reservoir are illustrated in Fig. 3-62.

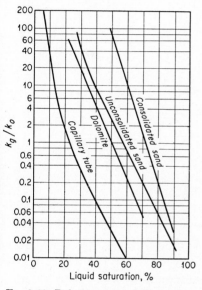

Fig. 3-60. Relative permeability ratios for various types of media. Capillary tube (*Martinelli, Putnam, and Lochart, Trans. Am. Inst. Chem. Engrs., Aug. 25, 1946*); dolomite (Bulnes and Fitting[23]); unconsolidated sand (Botset[22]); consolidated sand (Botset[22]).

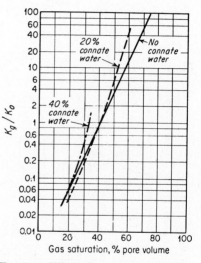

Fig. 3-61. Effect of connate water upon the k_g/k_o value. (*From Calhoun.*[15])

Note that the water table by definition 2 rises as the permeability in the formation decreases while the water table by definition 1 is a horizontal surface, providing static conditions prevail in the reservoir. From a production engineering standpoint, a contact defined as the highest point of water production is useful. From a reservoir engineering standpoint, a contact defined by zero capillary pressure is a more appropriate definition.

The actual location of water tables 1 and 2 can be determined by the use of electric logs, drill-stem tests, and relative-permeability and capillary-pressure data. From electric logs and drill-stem test data, it is possible to determine the depth at which 100 per cent water flow occurs or the point of zero oil permeability. From relative-permeability data the engineer can

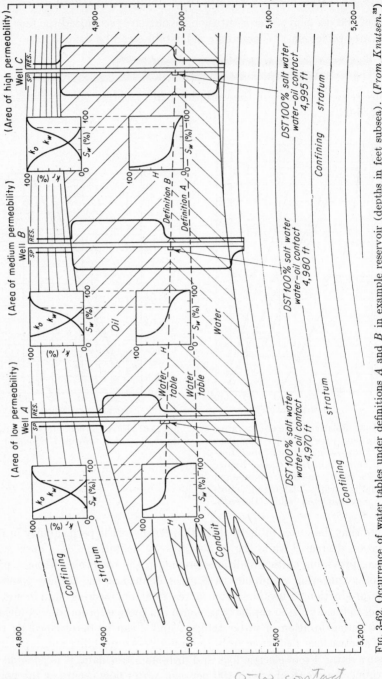

FIG. 3-62. Occurrence of water tables under definitions A and B in example reservoir (depths in feet subsea). (*From Knutsen.*[33])

O-w contact
Tilted due
To change
in Permeability

determine what the fluid saturations must be at the point of zero oil permeability. When the fluid saturations determined from well test data and relative-permeability curves are used, the capillary pressure can be determined and the height above the free water surface or zone of 100 per cent water saturation can be calculated. The above procedure is illustrated in Fig. 3-62. The existence of two distinct water tables materially aids in explaining the occurrence of some "tilted water tables." It is noted that the more permeable the formation, the more closely the pore structure approaches a supercapillary system and the smaller the divergence between the two water tables.

Aid in Evaluating Drill-stem and Production Tests. As indicated in the preceding discussion on the location of the water table, it is not necessary

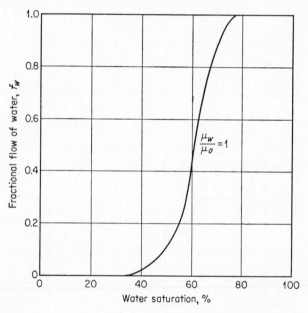

FIG. 3-63. Effect of water saturation on fractional flow of water through a homogeneous medium.

that a formation be 100 per cent water-saturated in order for that formation to produce 100 per cent water. By use of the fractional flow equation*

$$f_w = \frac{1}{1 + (k_o/k_w)(\mu_w/\mu_o)}$$

where μ_w/μ_o = water-to-oil viscosity ratio
k_o/k_w = oil-to-water relative-permeability ratio
f_w = fraction of total production which is water

* This equation is developed in a companion volume.

and relative-permeability data, the engineer can calculate the height above the free water surface at which both water and oil can be produced. Using the resultant solution of the fractional flow equation shown in Fig. 3-63 and the capillary-pressure data shown in Figs. 3-26 and 3-27, the height of the two-phase producing interval was determined for permeabilities from 10 to 900 millidarcys. The results of these calculations are shown in Fig. 3-64.

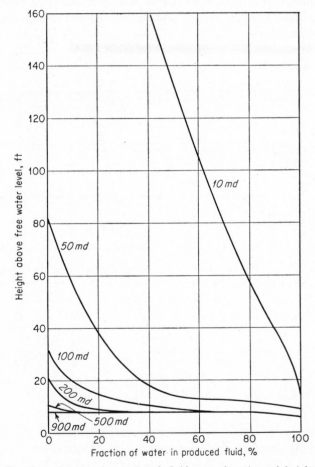

Fig. 3-64. Fraction of water in produced fluid as a function of height above the free water level (using capillary-pressure data shown in Fig. 3-27).

It is noted from the fractional flow curve (Fig. 3-63) that water flows at oil saturation as high as 65 per cent. Thus, from the capillary-pressure curve (Fig. 3-26), it is determined that water would be produced on a drill-stem test 10 ft above the free water surface for a formation with a permeability of 900 millidarcys. Water would be produced 200 ft above the free water surface for a reservoir with a permeability of 100 millidarcys. From the results of these calculations, it is noted that the lower the reservoir permeability, the longer the region of two-phase production or transition zone. It should be pointed out that the same relative-permeability curve was used in determining the fractional flow for all the different permeability samples. The use of one set of relative-permeability data exaggerates the effect of the low-permeability samples but does not change the general trend of the data. The capillary-pressure data used in the calculations for Fig. 3-64 are for a particular set of core samples. Another set of core samples would change the relative magnitude of the curves but would not change the indicated trend.

Other Uses of Relative-permeability Data. There are numerous other uses of relative-permeability data most of which will be discussed in detail later in conjunction with applications to particular problems. A few of these applications are

1. Determination of residual fluid saturations
2. Fractional flow and frontal advance calculations to determine the fluid distributions
3. Making future predictions for all types of oil reservoirs where two-phase flow is involved

REFERENCES

1. Benner, F. C., and F. E. Bartell: The Effect of Polar Impurities upon Capillary and Surface Phenomena in Petroleum Production, *Drilling and Production Practices*, American Petroleum Institute, 1941.
2. Plateau, J. A. F.: Experimental and Theoretical Research on the Figures of Equilibrium of a Liquid Mass Withdrawn from the Action of Gravity, *Smith Inst. Ann. Repts.*, 1863–1866.
3. Leverett, M. C.: Capillary Behavior in Porous Solids, *Trans. AIME*, 1941.
4. Fancher, G. H., J. A. Lewis, and K. B. Barnes: Penn. State Coll. Mineral Ind. *Expt. Sta. Bull.* 12, 1933.
5. McCardell, W. M.: A Review of the Physical Basis for the Use of the J-function, Eighth Oil Recovery Conference, Texas Petroleum Research Committee, 1955.
6. Welge, H. J., and W. A. Bruce: The Restored State Method for Determination of Oil in Place and Connate Water, *Drilling and Production Practices*, American Petroleum Institute, 1947.
7. Purcell, W. R.: Capillary Pressures—Their Measurement Using Mercury and the Calculation of Permeability Therefrom, *Trans. AIME*, 1949.
8. Slobod, R. L., Adele Chambers, and W. L. Prehn, Jr.: Use of Centrifuge for Determining Connate Water, Residual Oil and Capillary Pressure Curves of Small Core Samples, *Trans. AIME*, 1951.
9. Brown, Harry W.: Capillary Pressure Investigations, *Trans. AIME*, 1951.
10. Gates, George L., Frank C. Morris, and W. Hodge Caraway: Effect of Oil-base Drilling Fluid Filtrate on Analysis of Core from South Coles Levee, California and Rangely, Colorado Field, *U.S. Bur. Mines Rept. Invest.* 4716, August, 1950.

210   PETROLEUM RESERVOIR ENGINEERING

11. Owen, J. F.: Electric Logging in the Quinduno Field, Roberts County, Texas, Symposium on Formation Evaluation, AIME, October, 1955.

12. Guthrie, R. K., and Martin H. Greenburger: "The Use of Multiple Correlation Analyses for Interpreting Petroleum Engineering Data," presented at the Spring Meeting of the S. W. District Division of Production, New Orleans, La., Mar. 9–11, 1955.

13. Rose, Walter, and W. A. Bruce: Evaluation of Capillary Characters in Petroleum Reservoir Rock, *Trans. AIME*, 1949.

14. Wright, H. T., Jr., and L. D. Wooddy, Jr.: Formation Evaluation of the Borregas and Seeligson Field, Brooks and Jim Wells County, Texas, Symposium on Formation Evaluation, AIME, October, 1955.

15. Calhoun, John C., Jr.: "Fundamentals of Reservoir Engineering," University of Oklahoma Press, Norman, Okla., 1953.

16. Calhoun, J. C., and S. T. Yuster: Effect of Pressure Gradients and Saturations on Recovery in Water Flooding, *Proc. 8th Tech. Conf. on Petrol. Production*, Pennsylvania State College, 1944.

17. Slobod, R. L., and H. A. Blum: Method for Determining Wettability of Reservoir Rocks, *Trans. AIME*, 1952.

18. Burdine, N. T., L. S. Gournay, and P. O. Reicherty: Pore Size Distribution of Reservoir Rocks, *Trans. AIME*, 1950.

19. Ritter, H. L., and L. C. Drake: Pore Size Distribution in Porous Materials, *Ind. Eng. Chem.*, December, 1945.

20. Rapoport, L. A., and W. J. Leas: Relative Permeability to Liquid in Liquid-Gas Systems, *Trans. AIME*, 1951.

21. Wyckoff, R. D., and H. G. Botset: Flow of Gas Liquid Mixtures through Sands, *Physics*, 1936.

22. Botset, H. G.: Flow of Gas Liquid Mixtures through Consolidated Sand, *Trans. AIME*, vol. 136, 1940.

23. Bulnes, A. C., and R. U. Fitting, Jr.: An Introductory Discussion of the Reservoir Performance of Limestone Formations, *Trans. AIME*, vol. 160, 1945.

24. Leverett, M. C.: Flow of Oil-Water Mixtures through Unconsolidated Sands, *Trans. AIME*, 1939.

25. Leverett, M. C., and W. B. Lewis: Steady Flow of Gas-Oil-Water Mixtures through Unconsolidated Sands, *Trans. AIME*, 1941.

26. Osoba, J. S., J. G. Richardson, J. K. Kerver, J. A. Hafford, and P. M. Blair: Laboratory Measurements of Relative Permeability, *Trans. AIME*, 1951.

27. Geffen, T. M., W. W. Owens, D. R. Parrish, and R. A. Morse: Experimental Investigation of Factors Affecting Laboratory Relative Permeability Measurements, *Trans. AIME*, 1951.

28. Richardson, J. G., J. K. Kerver, J. A. Hafford, and J. S. Osoba: Laboratory Determinations of Relative Permeability, *Trans. AIME*, 1952.

29. Welge, H. J.: Simplified Method for Computing Oil Recoveries by Gas or Water Drive, *Trans. AIME*, vol. 195, 1952.

30. Fatt, I., and H. Dykstra: Relative Permeability Studies, *Trans. AIME*, 1951.

31. Fatt, I.: Effect of Overburden Pressure on Relative Permeability, *Trans. AIME*, 1953.

32. Knutsen, Carroll F.: Definition of Water Table, *Am. Assoc. Petrol. Geologists*, vol. 38, pt. 2, 1954.

CHAPTER 4

FUNDAMENTALS OF THE BEHAVIOR
OF HYDROCARBON FLUIDS

INTRODUCTION

In Chap. 1 of this work, a definition of an oil and gas reservoir by Uren[1] was presented. It can be noted from the definition that an oil-gas reservoir is defined not only by the rocks in which the fluids are contained but by the fluids themselves. Oil and gas are naturally existing hydrocarbon mixtures quite complex in chemical composition which exist at elevated temperatures and pressures in the reservoir. On production and capture of hydrocarbons at the surface, the temperature and pressure of the mixture are reduced. The state of the hydrocarbon mixture at the surface conditions depends upon the composition of the hydrocarbon fluid as produced from the well and upon the pressure and temperature at which it is captured. Furthermore, the fluid remaining in the reservoir at any stage of depletion undergoes physical changes as the pressure is reduced by producing quantities of oil or gas from that reservoir. It is necessary to study the physical properties of these naturally existing hydrocarbons and in particular, their variation with pressure and temperature. Knowledge of the physical properties enables the engineer to evaluate the yield in terms of standard volumes of gas and stock-tank barrels of liquid that may be obtained upon production to the surface of a unit volume of reservoir fluid. These data are necessary in estimating the performance of the reservoir. The complexity of the naturally occurring hydrocarbon mixtures makes it necessary, in general, to rely upon empirical data obtained from laboratory tests.

The chemical composition of hydrocarbon gases can be specified through heptanes readily. The chemical composition of a crude oil, however, is much more difficult to evaluate because a large proportion of the oil is comprised of hydrocarbons heavier than heptanes. In this chapter the fundamental physical behavior of hydrocarbons will be discussed together with the laboratory tests and calculation procedures that are available in evaluating physical properties of hydrocarbon mixtures.

In addition to the hydrocarbons that occur in the reservoir, the prop-

211

erties of the associated waters are of interest to the engineer, as this water occupies space in the reservoir, contributes energy to the production of oil, and may be produced with oil and gas. Succeeding chapters will deal with the applications of these physical properties of hydrocarbon fluids and with the properties of formation waters.

In order to study properly the physical properties of fluids, it is first necessary to gain some understanding of simple systems. A single-component hydrocarbon can be obtained only after extensive processing and does not exist in natural occurrences. However, the physical properties of a single-component hydrocarbon and its behavior when subjected to changes in pressure and temperature are qualitatively similar to those of more complex systems. It is convenient, therefore, to introduce the basic definitions and to review the concepts of thermodynamics and physical chemistry in terms of a single hydrocarbon.

Physical properties of interest to an engineer ordinarily are defined in terms of the pressure and temperature at which a hydrocarbon exists. Fluids in general are classified as gases, vapors, or liquids. It should be pointed out that these particular words convey ideas only when conditions of pressure and temperature are specified. A material may exist as a gas or as a liquid, depending upon the pressure and temperature to which that material is subjected. Vapor is defined in the dictionary as any substance in the gaseous state which, under ordinary conditions, is usually a liquid or solid. By ordinary conditions are meant atmospheric conditions of pressure and temperature. In dealing with hydrocarbons it is convenient to think of the words gas and vapor as being synonymous.

As in other fluid systems, a hydrocarbon system may be homogeneous or heterogeneous. In a homogeneous system, all parts of the system have the same physical and chemical properties. A heterogeneous system is one in which the physical and chemical properties are not the same throughout.

A heterogeneous system is comprised of phases. A phase is defined by Daniels[2] as "a definite part of a system which is homogeneous throughout and physically separated from other phases by distinct boundaries." An example of a heterogeneous system is that of water, ice, and water vapor in which three phases are present. The degree of dispersion does not enter in consideration of the number of phases. In the example cited the ice is a single phase whether it exists in one piece or several.

BASIC CONCEPTS OF PHASE BEHAVIOR

Single-component Systems

Ethane is the hydrocarbon chosen for an example of a single-component system. Ethane is ordinarily thought of as being a gas and quite properly (from the dictionary definition), as at ordinary conditions ethane exists as a gas. At other than ordinary conditions, ethane may exist as a liquid as is shown in Fig. 4-1. The curve plotted is a vapor-pressure curve. At the conditions of pressure and temperature specified by the curve two phases, liquid and vapor, co-exist in equilibrium. At any condition of pressure and temperature which does not fall on the line, only one phase exists, i.e., either liquid or gas. As labeled on the figure, it can be seen that ethane exists in the liquid state at those pressures lying above the vapor-pressure curve for the appropriate temperatures. It can be noted further that ethane is a vapor or gas at those pressures lying below the vapor-pressure curve for the same values of temperature.

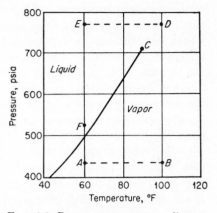

Fig. 4-1. Pressure-temperature diagram of ethane. (*From Standing.*[3])

The boiling characteristics of water are well known. At ordinary conditions of pressure and temperature water boils at 212°F; that is, the material can be transformed from the liquid phase to the vapor phase at a constant temperature and pressure. Although the properties of the liquid and vapor are not changing, the energy for the total system is being increased as more of the material is converted from the low-energy liquid state to the high-energy gaseous state. By analogy, at 636 psia, ethane may be considered to have a boiling temperature of 80°F. Thus, the vapor-pressure curve may also be considered to be the locus of the boiling temperatures of the substance. The boiling temperature of a material is a function of pressure.

In Fig. 4-1 the point C, where the curve terminates, is known as the critical point of the system. For a single-component system, the critical point may be defined as the highest value of pressure and temperature at which two phases can coexist. This definition will not hold for multicomponent systems, and for that reason a different definition will be given here which is applicable for the most complex systems. The classical definition of the critical point is *that state of pressure and temperature at which the intensive properties of the gas and liquid phases are continuously identical.*

The state of a substance is therefore determined by the pressure and temperature at which that substance exists. Consider point A of Fig. 4-1. At a temperature of 60°F and a pressure of approximately 430 psia, ethane exists in the gaseous state. The material has all the properties ordinarily associated with a gas. If the pressure is increased at a constant temperature to a pressure of about 495 psia, a point on the vapor-pressure curve, both the vapor and liquid states may coexist. If the pressure is increased further to point F, ethane is in the liquid state and all the physical properties ordinarily associated with a liquid are exhibited by ethane at that condition. In proceeding from point A to point F, the condensation of a liquid from the ethane vapor can be observed if the compression is conducted in a visual cell. Therefore, the formation of a liquid phase can be readily identified. If, however, a different path from point A to point F is followed by appropriate changes of pressure and temperature, the transition from gas to liquid is not perceptible. For example, proceed at a constant pressure from point A to point B at a temperature of 100°F. At point B ethane still exists in the vapor phase and has properties associated with the vapor phase. If, then, the material is compressed to about 770 psia, point D, there is a continuous change in the material. The material at point D should be referred to as a fluid rather than as a vapor or a liquid. If the material is cooled from point D at 100°F to point E at 60°F while holding the pressure constant, the change in physical properties is continuous and no phase discontinuities can be observed. When the pressure at a constant temperature is dropped to point F, a substance is obtained which is readily identified as having the physical properties ordinarily associated with liquids. Yet in tracing out the complete path it was noted that no discontinuities were observed along that path. In other words, by selecting a particular path of temperature and pressure change it is possible to pass from the vapor state to the liquid state without observing a discontinuity or the formation of a second phase.

Another means of illustrating the behavior of a single-component system is shown in Fig. 4-2 in which pressure and specific volume are the independent parameters. Here the two-phase region is more readily illustrated. Point C is the critical point. It may be noted that the critical temperature for ethane is slightly above 90°F and the critical pressure about 710 psia. The dashed curve defines the two-phase region. The solid curves are lines of equal temperature (isotherms). That portion of the dashed curve to the left of the critical point, the dashed line going through point A and then continuing to the pressure of about 400 psia at a specific volume of about 0.095 cu ft per lb, is the bubble-point curve. That portion of the dashed line to the right of the critical point C and extending through points A' and H is the dew-point curve. Within the region enclosed by the bubble-point and dew-point curves is the two-phase region within which vapor and liquid coexist in equilibrium. At point A on the bubble-point curve ethane has the properties of a liquid, having a specific volume of 0.0516 cu ft per lb. Point A' on the dew-point curve is at the same pressure and temperature as point A, but the specific volume is 0.138 cu ft per lb, which is of the order of magnitude commonly associated with a vapor. Within the two-phase region at point A_2 exists a material having a specific volume of 0.08 cu ft per lb. The material existing at the conditions specified by A_2 consists of two distinct homogeneous phases. One phase has the density specified by A', and the other that specified by A. The specific volume at A_2

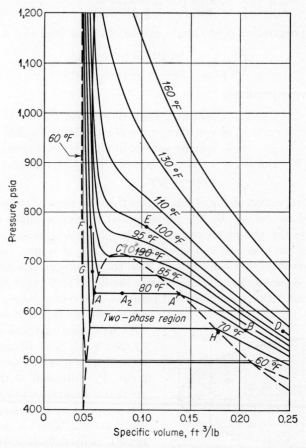

Fig. 4-2. Phase behavior of pure ethane. (*From Brown et al.*[4])

is determined by the relative quantity of liquid A and vapor A' present.

At any condition within the two-phase region, the relative amounts of vapor and liquid coexisting in equilibrium can be determined from the specific volume of the composite of the gas and the liquid. A volume balance can be written as

$$v_{com}(W_{com}) = v_L(W_L) + v_g(W_g)$$

and a weight balance as

$$W_{com} = W_L + W_g$$

Combining the balances,

$$v_{\text{com}}(W_L + W_g) = v_L(W_L) + v_g(W_g)$$

which can be reduced to an expression for the weight ratio of liquid to gas.

$$\left(\frac{W_L}{W_g}\right)_{\text{com}} = \frac{v_g - v_{\text{com}}}{v_{\text{com}} - v_L}$$

A graphical expression of the above can be obtained from Fig. 4-2 as

$$\left(\frac{W_L}{W_g}\right)_{A_2} = \frac{\text{length from } A_2 \text{ to } A'}{\text{length from } A \text{ to } A_2}$$

In the above expressions,

$$v_{\text{com}} = \text{specific volume of composite}$$
$$v_L = \text{specific volume of equilibrium liquid}$$
$$v_g = \text{specific volume of equilibrium vapor}$$
$$W_{com}, W_L, \text{ and } W_g = \text{weights of composite, liquid, and gas,}$$
$$\text{respectively}$$

Further inspection of Fig. 4-2 indicates that the state of the ethane can be specified completely by the pressure and specific volume. However, only outside the two-phase region can the physical state be completely specified by the pressure and temperature. This, of course, corresponds to the observation that can be made from the vapor-pressure curve.

Figure 4-3 shows the relation between the liquid and vapor densities of ethane. Considering the material at point A_2, in the two-phase region, the density of the liquid in equilibrium with its coexisting vapor is specified by point A. The density of the coexisting vapor is specified by point A'. Note that at the critical point the density, one of the intensive properties of ethane, becomes single-valued. This is to be expected, as by definition, the critical point is that point at which the vapor and liquid phases become continuously identical. In other words, as the critical point is approached along the dew-point curve, the density of the vapor progressively increases. As the critical point is approached along the bubble-point curve, the density of the liquid continually decreases, approaching the same value as the vapor at the critical point.

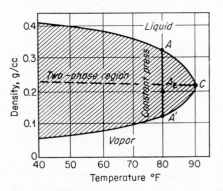

FIG. 4-3. Saturated liquid and vapor densities of ethane. (*From Brown et al.*[4])

For pure components, the critical point can be determined by application of the principle of average composite densities. A straight line is obtained by plotting the arithmetic average of the densities of the vapor and liquid defined at the pressures and temperatures of the vapor-pressure curve. This line passes through the critical point as illustrated by the dashed line of Fig. 4-3. The critical point is defined by the convergence of the vapor, liquid, and average composite density curves.

Binary Systems

When a second component is added to a hydrocarbon system, the phase behavior becomes more complex. This increase in complexity is caused by the introduction of another variable, composition, to the system.

The effect of this variable can be noted by contrasting the pressure-temperature curve plotted in Fig. 4-1 with that of Fig. 4-4. For a single-

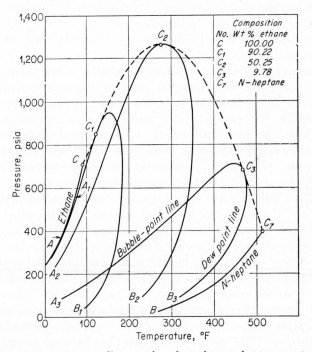

FIG. 4-4. Pressure-temperature diagram for the ethane–n-heptane system. (*From Brown et al.[4]*)

component system the vapor-pressure curve represents the trace of both the bubble-point and dew-point curves on the pressure-temperature plane. For a binary or other multicomponent system, the bubble-point and dew-point lines no longer coincide. The traces of the bubble-point and dew-point curves for the multicomponent system form a phase diagram. Furthermore, for each possible composition a distinct phase diagram exists.

The behavior of ethane–n-heptane mixtures is illustrated in Fig. 4-4. On the left of the figure, terminating at the point C, is the vapor-pressure curve of pure ethane, and on the extreme right of the figure, terminating in point C_7, is the vapor-pressure curve for pure n-heptane. Between the vapor-pressure curves of the pure constituents lie the phase diagrams of the various binary mixtures of the ethane–n-heptane system. The points labeled C are the critical points of the mixtures as defined by the respective subscripts in the legend of the figure. The dashed curve is the locus of critical points for the ethane–n-heptane system. Points C_1, C_2, and C_3 are the critical points of mixtures containing 90.22, 50.25, and 9.8 wt % ethane respectively.

The phase diagram for the mixture containing 90 wt % heptane consists of the bubble-point curve A_3–C_3 and the dew-point curve $C_3 - B_3$. Within the region enclosed by these curves exists the two-phase region. Above and to the left of the bubble-point curve $A_3 - C_3$, the mixture exists as a liquid. Below and to the right of the dew-point curve, the mixture exists as a gas.

If the composition of the mixture is changed, the phase diagram and the two-phase region are shifted on the pressure-temperature plane. This is illustrated by the phase diagrams terminating at C_1 and C_2, respectively. Comparing the phase diagrams, it is noted that with large weight percentages of ethane, the critical point of the mixture lies to the left of the maximum pressure and maximum temperature at which two phases can coexist in equilibrium. When the composition of the mixture is evenly distributed by weight, it is noted that the critical point, point C_2, is located approximately at the highest value of pressure. In the case where heptane comprises nearly 90 per cent by weight of the mixture, the critical pressure point C_3 is now shifted to the right of the highest pressure at which two phases can coexist in equilibrium. From Fig. 4-4 also can be noted that the dew-point or bubble-point line, depending on whether the heaviest or lightest constituent is predominant in the mixture, will approach the vapor-pressure line of the major constituent. Also, as one constituent becomes more predominant, the critical temperature and pressure of the mixture tend to approach those values which define the critical point of the major pure constituent. The critical temperature lies between the critical temperatures of the lightest and heaviest constituents present in the mixture. The critical pressure of the system will always be greater than the critical pressure of any constituent in the system except when one constituent is so predominant that the critical properties of the mixture approach the critical properties of that one component. As the composition of the mixture becomes more evenly distributed between the constituents, the two-phase region increases in size, whereas when one constituent becomes predominant, the two-phase region tends to shrink in size.

The study of other binary mixtures of hydrocarbons provides a valuable insight into the effect of the diversity of volatility and molecular weight on the behavior of hydrocarbon mixtures. In Fig. 4-5 are presented the vapor-pressure curves of several paraffin hydrocarbons together with the critical loci of various binary mixtures of these hydrocarbons.

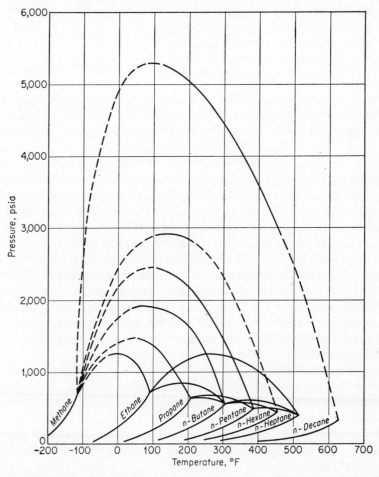

FIG. 4-5. Illustrating effect of divergence in composition on critical point loci. (*From Brown et al.*[4])

If the mixture is composed of two constituents which are quite different in volatility and molecular weight, the critical locus encompasses a wide range of temperature and pressure. For example, the critical locus (Fig. 4-5) of the methane–*n*-decane system exhibits a maximum critical pressure of approximately 5,250 psia. The critical temperatures of the system lie between the critical temperature of methane and of *n*-decane.

The systematic change in the critical loci with increasing diversity of the properties of the constituents is illustrated in Fig. 4-5 by the critical loci of the various methane mixtures. The maximum critical pressures for the various mixtures are as follows:

Methane-ethane system, 1,250 psia
Methane-propane system, 1,450 psia
Methane–n-butane system, 1,950 psia
Methane–n-pentane system, 2,450 psia
Methane–n-hexane system, 2,900 psia
Methane–n-decane system, 5,250 psia

For constituents very similar in molecular structure, such as propane and n-pentane or n-butane and n-heptane, the loci of the critical points form nearly flat curves connecting the critical points of the constituents.

Multicomponent Systems

Naturally occurring hydrocarbon systems are composed of a wide range of constituents. These constituents include not only the paraffin series of hydrocarbons but components from various other series. The phase behavior of a hydrocarbon mixture is dependent on the composition of the mixture as well as the properties of the individual constituents.

A phase diagram for a multicomponent system is shown in Fig. 4-6. Before discussing the significance of the diagram, a number of important physical concepts associated with phase diagrams must be defined.

Critical point (*C* of Fig. 4-6). That state of pressure and temperature at which the intensive properties of each phase are identical.

Critical temperature (*C*). The temperature at the critical point.

Critical pressure (*C*). The pressure at the critical point.

Intensive properties. Those properties that are independent of the amount of material under consideration.

Extensive properties. Those properties that are directly proportional to the amount of material under consideration.

Bubble-point curve. The locus of the points of pressure and temperature at which the first bubble of gas is formed in passing from the liquid to the two-phase region.

Dew-point curve. The locus of the points of pressure and temperature at which the first droplet of liquid is formed in passing from the vapor to the two-phase region.

Two-phase region. That region enclosed by the bubble-point line and dew-point line wherein gas and liquid coexist in equilibrium.

Cricondentherm (*M*). The highest temperature at which a liquid and vapor can coexist in equilibrium.

Cricondenbar (*T̃*). The highest pressure at which the liquid and vapor can coexist in equilibrium. (Some authorities name this point the Crivaporbar.)

Retrograde region (shaded area). Any region where condensation or vaporization occurs in reverse to conventional behavior, i.e., *retrograde condensation* in which liquid is condensed upon either lowering the pres-

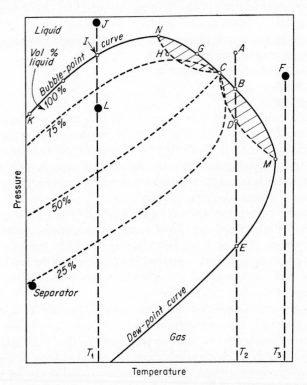

Fɪɢ. 4-6. Phase diagram to illustrate nomenclature of retrograde condensation.

sure at constant temperature (lines *A*, *B*, *D*) or increasing the temperature at constant pressure (lines *H*, *G*, *A*) and *retrograde vaporization* in which vapor is formed upon decreasing the temperature at constant pressure (lines *A*, *G*, *H*) or increasing the pressure at constant temperature (lines *D*, *B*, *A*).

Iso vol lines (quality lines). The loci of points of equal liquid volume per cent within the two-phase region.

Several important observations can be made from a study of Fig. 4-6. The bubble-point curve coincides with the dew-point curve at the critical point. The bubble-point curve represents 100 per cent liquid by volume, and the dew-point curve represents zero per cent liquid by volume. The shaded areas represent regions of retrograde phenomena. The region defined by points C, B, M, D is the region of isothermal retrograde condensation.

In oil-field nomenclature multicomponent systems are broadly classified as oils or gases. These broad classifications are further subdivided depending on the changes in state of the hydrocarbon mixture in the reservoir and the surface yield of hydrocarbon fluids.

Gases. Systems which exist in the gaseous state in the reservoir are classified as gases and subdivided into

1. Condensate or retrograde gases
2. Wet gases
3. Dry gases

The phase diagram and prevailing reservoir conditions determine the classification of the system. If a reservoir contains a mixture having a phase diagram such as that of Fig. 4-6, the reservoir temperature is between the critical temperature and the cricondentherm and the initial reservoir pressure is equal to or greater than the dew-point pressure. Then the reservoir contains a condensate gas.

A reservoir temperature and an initial reservoir pressure corresponding to point B of Fig. 4-6 are typical of reservoir conditions associated with condensate gases. The hydrocarbon mixture originally exists as a single-phase dew-point gas. Fluids produced from the reservoir are brought to the surface and are separated at separator conditions such as shown on the diagram.

Two phenomena associated with the production are of interest. As fluids are produced, the reservoir pressure declines and isothermal retrograde condensation occurs in the reservoir, since the pressure decline occurs along the path B–D. The produced fluid is subjected to both pressure decline and temperature decline. Liquid is accumulated in the separator as a result of normal condensation associated with a decline in temperature.

A portion of the phase diagram of a natural-gas–natural-gasoline mixture is presented in Fig. 4-7. The diagram is more typical of a condensate gas than that of Fig. 4-6. Although the range of investigation did not define the cricondentherm, it is obvious that both the cricondenbar and cricondentherm exist at higher temperatures than the critical temperature. This is commonly true of condensate gases. The region of isothermal retrograde condensation is larger for such a material than for that shown in Fig. 4-6.

Another important observation which can be made from both phase diagrams discussed is that the reservoir temperature must be between the critical temperature and the cricondentherm for the fluid to be a condensate gas. The relative position of the critical point is determined by the amount of light hydrocarbons in the mixture. As was mentioned previously, when the light hydrocarbons comprise a large weight percentage of the total mixture, the critical temperature of the mixture will approach the critical temperature of the lightest constituent. Comparing the phase diagrams of Figs. 4-6 and 4-7, it can be observed that the natural-gas–natural-gasoline mixture contains larger amounts of light constituents.

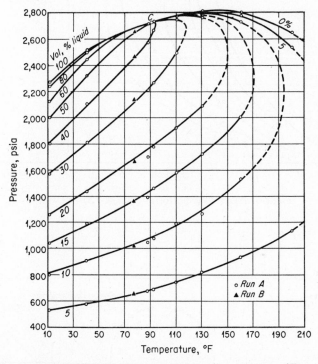

FIG. 4-7. Phase diagram for natural-gas–natural-gasoline mixture. (*From Katz and Kurata.*[5])

The critical temperature of the natural-gas–natural-gasoline mixture is such that if the mixture were to be accumulated in a reservoir at moderate depth (reservoir temperature 100 to 200°F), the fluid would behave as a condensate gas.

A wet gas normally is comprised of a lesser percentage of heavy components than is a condensate gas. As a result, the phase diagram is somewhat less broad and the critical point is shifted to lower temperatures. Furthermore, the reservoir temperature exceeds the cricondentherm as is shown on the phase diagram of Fig. 4-8. Thus, in the reservoir the fluid is at all times in a single phase. The term "wet" is derived from the fact

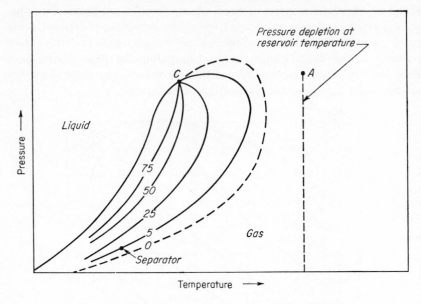

FIG. 4-8. Phase diagram for a wet gas. (*After Clark.*[6])

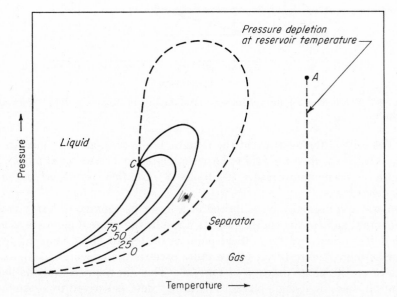

FIG. 4-9. Phase diagram for a dry gas. (*After Clark.*[6])

that the separator conditions lie in the two-phase region and a liquid phase is condensed in the separator. Wet gases ordinarily differ from condensate gases in the following respects:

1. Isothermal retrograde condensation does not occur in the reservoir during pressure depletion.
2. Separator liquid yields are lower than for a condensate.
3. Less heavy components are present in the mixture.

A dry gas is comprised largely of methane and ethane with small percentages of heavier components. A phase diagram of a dry gas, together with typical reservoir and operating conditions, is shown in Fig. 4-9. The separator conditions as well as the reservoir conditions lie in the single-phase region. Hydrocarbon liquid is not condensed from the mixture either in the reservoir or at the surface. Dry gases may contain water vapor, which will condense. "Dry" in this instance means free of hydrocarbon liquids, not necessarily free of water.

Oils. Hydrocarbon mixtures which exist in the liquid state at reservoir conditions are commonly classified as crude oils and subdivided on the basis of liquid yield at the surface into low- and high-shrinkage oils.

A phase diagram for a low-shrinkage oil is shown in Fig. 4-10. Two characteristics are apparent. The critical point lies to the right of the

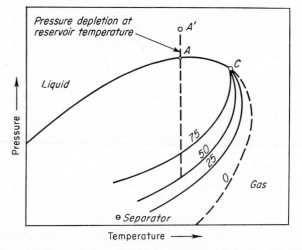

Fig. 4-10. Phase diagram for a low-shrinkage oil. (*After Clark.*[6])

cricondenbar, and the quality lines (volume per cent liquid) are closely spaced near the dew-point curve. Furthermore, at atmospheric pressure and reservoir temperature, the mixture is in the two-phase region. At separator conditions, substantial liquid recoveries are obtained even though the liquid volume per cent is quite low. This phenomenon is caused by the great expansion of the gas phase at low pressures. The characteristics of the phase diagram indicate relatively large amounts of heavy constituents in the mixture.

Oil is frequently further classified, depending upon initial reservoir conditions, as either a saturated or undersaturated oil. If the initial reservoir conditions correspond to A in Fig. 4-10, on the bubble-point curve, the oil is said to be saturated.

As can be observed from the diagram, on an infinitesimal pressure decline, gas is evolved from a saturated oil. If, however, the initial conditions correspond to A', above the bubble-point curve, the oil is said to be undersaturated. An undersaturated oil must be subjected to a substantial pressure decline before gas is evolved; in fact, the pressure must decline from A' to A in the present example before a free gas phase is formed.

A high-shrinkage oil contains more of the lighter hydrocarbons than does a low-shrinkage oil. The reservoir temperature is ordinarily closer to the critical temperature, and the quality lines are less closely grouped near the dew-point curve.

A typical phase diagram for a high-shrinkage oil is presented in Fig. 4-11. Substantially lower amounts of liquid are obtained on pressure decline

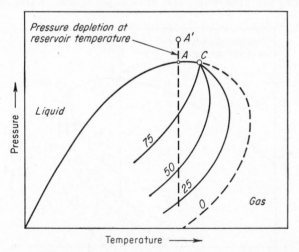

Fig. 4-11. Phase diagram for a high-shrinkage oil. (*After Clark.*[6])

both in the reservoir and at the surface. A high-shrinkage oil may be either saturated (point A) or undersaturated (point A').

The various classifications of hydrocarbon fluids may be characterized by composition, API gravity of liquid yields, and gas-liquid ratios as well as by phase diagrams. In fact, usually only a partial phase diagram is obtained on laboratory analysis of reservoir fluids. This consists of determining the behavior along the isotherm corresponding to reservoir temperature and a group of tests at various separator conditions.

Characteristic analyses of typical reservoir fluids are presented in Table 4-1. These data are to be considered typical but not necessarily characteristic of the classes. A wide range of possible compositions and reservoir conditions exist for naturally occurring hydrocarbon accumulations. Therefore, each reservoir fluid presents a different problem in analysis and classification.

The gas-liquid ratio that is initially produced and the API gravity of the produced liquid are indicative in general of the classification of the reservoir fluid.

Dry gas is indicated by lack of condensed fluids at separator conditions. Wet gases are usually indicated by gas-liquid ratios of 60,000 to 100,000 cu ft per bbl with liquid gravities higher than 60°API. Condensate gases yield gas-liquid ratios of 8,000 to 70,000 cu ft per bbl with liquid gravities between 50 and 60°API.

TABLE 4-1. COMPARISON OF COMPOSITION AND LIQUID YIELDS OF
HYDROCARBON RESERVOIR FLUIDS[7]

Fluid	Dry gas, mole %	Condensate gas, mole %	Low-shrinkage oil, mole %
Methane	91.32	87.07	57.83
Ethane	4.43	4.39	2.75
Propane	2.12	2.29	1.93
Putanes	1.36	1.74	1.60
Pentanes	0.42	0.83	1.15
Hexanes	0.15	0.60	1.59
Heptanes and heavier	0.20	3.08	33.15

	Gravity of liquid, °API	Gas-liquid ratio, cu ft per bbl	Reservoir conditions		Dew point, psi	Bubble point, psi
			Temperature, °F	Pressure, psi		
Gas:						
Wet gas	65	67,000	160	1,700		
Condensate gas	55	18,500	203	4,810	4,470	
Oil:						
Low shrinkage	38	900	211	4,750		4,600
High shrinkage	58	2,700	203	4,700		3,855

Low-shrinkage oils usually yield gas-liquid ratios of 1,000 cu ft per bbl or less with API gravities of the produced liquid 45° or less. High-shrinkage oils yield gas-liquid ratios from 1,000 to 8,000 cu ft per bbl with liquid gravities of 45 to 60°API. Many fluids can be classified only after a detailed study of reservoir-fluid samples.

In Chap. 1 it was stated that gas occurred at reservoir conditions as dissolved gas, associated free gas, or nonassociated free gas. Referring to Fig. 4-6 and considering that the phase diagram shown represents the phase diagram of the total mass of hydrocarbon accumulated in a given reservoir, the dependence of the state of the hydrocarbon system on reservoir conditions can be shown.

If the reservoir temperature is T and the initial reservoir pressure corresponds to point I, the reservoir contains a single-phase bubble-point liquid or, in oil-field terminology, a saturated oil. At a reservoir pressure corresponding to point J, the reservoir contains a single-phase liquid which is several hundred pounds per square inch above the bubble-point pressure—an undersaturated oil. On production of the fluid to the separator, the two-phase region is entered and both vapor and liquid are obtained. Thus, the fluid at both conditions can be considered to contain dissolved gas.

At initial reservoir conditions, corresponding to point L, the reservoir contains two phases, gas and liquid. Thus, in oil-field terminology the

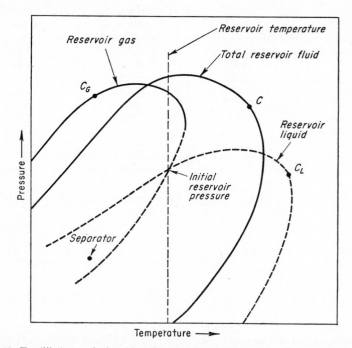

Fig. 4-12. Equilibrium relationship of reservoir containing an oil accumulation with a gas cap. (C_G, critical point for gas; C_L, critical point for liquid; C, critical point for system.)

reservoir consists of an oil accumulation with a gas cap. The gas-cap gas is classified as associated free gas and is usually in equilibrium with the contiguous oil accumulation. Thus, the gas is at its dew point and the liquid at its bubble point. This is shown schematically in Fig. 4-12, in which the phase diagrams for the equilibrium phases are superimposed on the original phase diagram for the total mass of hydrocarbon in the reservoir. The equilibrium gas contains larger percentages of light hydrocarbons and lesser percentages of heavy hydrocarbons than does the equilibrium liquid. The gas-cap gas may be dry, wet, or condensate, depending on the composition and phase diagram of the gas. That shown in Fig. 4-12 is a wet gas.

If the reservoir temperature corresponds to T_2 (Fig. 4-6) and the reservoir pressure to that for point B, then as previously mentioned the reservoir fluid is a condensate gas. The gas is saturated, or at the dew point. However, if initial reservoir conditions correspond to point A, the reservoir contains a single-phase gas which is several hundred pounds per square inch above the dew-point pressure and is undersaturated.

For a reservoir temperature of T_3, such as at point F, the reservoir contains a wet gas.

From the foregoing discussion it can be observed that hydrocarbon mixtures may exist in either the gaseous or liquid state, depending on the reservoir and operating conditions to which they are subjected. The qualitative concepts presented may be of aid in developing quantitative analyses.

PROPERTIES OF THE GASEOUS STATE

Ideal Gas Laws

Gas is defined as a fluid which has neither independent shape or volume but expands to fill completely the vessel in which it is contained. In order to define the thermodynamic properties of gas it is necessary to review some of the basic laws. One of the first laws, perhaps the best known, is Boyle's law.

Boyle's law states that at a constant temperature, the pressure of a gas is inversely proportional to the volume of the gas. Stated in equation form

$$V \propto \frac{1}{P} \quad \text{or} \quad PV = C' \quad \text{for } T = \text{constant}$$

where P is pressure, V is volume, T is temperature.

A second law of equal importance, credited to Charles and Gay-Lussac, is that at a constant pressure, the volume of a perfect gas varies directly with the temperature; also at a constant volume, the pressure varies directly with the temperature. Expressing the laws in equation form,

$$P = KT \quad \text{for } V = \text{constant}$$
$$V = K'T \quad \text{for } P = \text{constant}$$

C', K, and K' are constants of proportionality dependent upon the units used for volume, pressure, and temperature and the nature and weight of gas present. The pressure and temperature in Boyle's and Charles's laws are in absolute units.

From the above laws it is possible to derive the general or perfect gas law. In brief it can be shown that by using only one part of Charles's law and Boyle's law, it is possible to write for a unit weight of gas

$$P_1 v_1 = P_2 v_n \quad \text{at } T_1 \text{ from Boyle's law}$$

$$\frac{v_n}{T_1} = \frac{v_2}{T_2} \quad \text{at } P_2 \text{ from Charles's law}$$

where v_n is the specific volume at P_2 and T_1, v_2 is the specific volume at P_2 and T_2. Then v_1 is the specific volume at P_1 and T_1.

$$v_n = \frac{P_1 v_1}{P_2} = \frac{v_2 T_1}{T_2}$$

From which

$$\frac{P_1 v_1}{T_1} = \frac{P_2 v_2}{T_2} = R' \tag{4-1}$$

Since the conditions were chosen at random, it can also be shown that

$$\frac{P_1 v_1}{T_1} = \frac{P_3 v_3}{T_3} = R'$$

where conditions P_3, v_3, and T_3 represent any other state of pressure, volume, and temperature. Therefore, the above can be generalized as

$$Pv = R'T \tag{4-2}$$

where R' is the gas constant, a constant of proportionality dependent upon the type of gas and the units used. The gas constant for a number of gases has been determined from experimental data. For methane at atmospheric pressure and $32°F$, $R' = 96.2$ ft lb per (lb)(°R), with P in pounds per square foot absolute, v in cubic feet per pound, and T in degrees Rankine.

If both sides of Eq. (4-2) are multiplied by w, the weight of gas, then

$$Pwv = wR'T$$
or
$$PV = wR'T \tag{4-3}$$

where V is the volume of gas having a weight w.

Equations (4-1) and (4-2) are equations of state for perfect or ideal gases. A more useful form of these equations can be obtained by introducing Avogadro's law: *All ideal gases at a given pressure and temperature have the same number of molecules for a given volume.*

One mole of a material is a quantity of that material whose mass, in the unit system selected, is numerically equal to the molecular weight. Furthermore, 1 mole of any substance contains the same number of molecules as 1 mole of any other substance. Thus, 1 mole of any gas will occupy the same volume at a given pressure and temperature.

For 1 mole of gas,

$$w = M$$

where M is the molecular weight. Substituting for w in Eq. (4-3),

$$PV = MR'T$$

or

$$V = \frac{MR'T}{P}$$

From Avogadro's law, for a given pressure and temperature, V must have the same value for all ideal gases. Thus, MR' must be a constant. This fact has been verified by experiment on gases such as oxygen, helium, and nitrogen. The product MR' is called the universal gas constant and designated by the symbol R.

In the general case $nM = w$ and the ideal gas equation of state can be stated as

$$PV = nRT \tag{4-4}$$

where n is the number of moles of gas present.

The universal gas constant is considered to be independent of the type gas, but as it has dimensions, the numerical value depends on the system of units used. The dimensions of R are energy units per mole divided by absolute temperature. The numerical value of R was obtained from measurement of the specific volume, at 14.7 psia and 32°F, of air, oxygen, nitrogen, helium, and hydrogen.

In the English system of units, with pressure in pounds per square foot, volume in cubic feet, n in number of pound moles, and T in degrees Rankine,

$$R = 1{,}544 \text{ ft-lb/(mole)(°F)}$$

and if the pressure is in pounds per square inch, other units as above,

$$R = \frac{1{,}544}{144} = 10.72 \text{ cu ft-lb/(mole)(sq in.)(°R)}$$

In Table 4-2 are listed values of R for various unit systems.

The physical properties density and specific volume can be defined from the equation of state as follows:

$$\text{Density } \rho = \frac{w}{V} = \frac{PM}{RT} \tag{4-5}$$

and

$$\text{Specific volume } v = \frac{V}{w} = \frac{RT}{PM} \tag{4-6}$$

TABLE 4-2. VALUE OF THE UNIVERSAL GAS CONSTANT R FOR VARIOUS UNITS

Pressure	Volume	Temperature, deg	n	R
Atm	Cc	Kelvin	Gm-moles	82.057
Atm	Liters	Kelvin	Gm-moles	0.082054
Atm	Cu ft	Rankine	Lb-moles	0.7302
Psi	Cu ft	Rankine	Lb-moles	10.72
Psf	Cu ft	Rankine	Lb-moles	1,544

thus the density and specific volume of gases are functions of pressure, temperature, and molecular weight. For a particular gas the conditions of pressure and temperature must be specified to define the density or specific volume.

Example 4-1. Calculation of Density from Ideal Gas Equation of State. Find the density of methane at 0 psig and 60°F when atmospheric pressure is 14.7 psia.

$$\text{Methane density} = \frac{14.7(16)}{10.72(520)} = 0.04122$$

Density of methane at 60°F and 14.7 psia = 0.04122 lb/cu ft

Now find the density of methane at 50 psig and 32°F when atmospheric pressure is 14.7 psia.

$$\text{Methane density} = \frac{64.7(16)}{10.72(492)} = 0.1962$$

Density of methane at 60°F and 64.7 psia = 0.1962 lb/cu ft

Kinetic Theory. Another method by which the equation of state for ideal gases can be derived is by use of the kinetic theory of gases. There are two essential features in the theory: one is that matter is made up of small particles called molecules; second, heat energy in matter is a manifestation of molecular motion. The first is easily understood; the second just states that when heat is added to a gas, it is transferred into energy of motion by increasing the velocity of the gas molecules.

In the derivation it is necessary to make three limiting assumptions:

1. The volume occupied by the molecules is insignificant with respect to the volume occupied by the total mass.

2. All collisions of molecules are perfectly elastic.

3. There are no attractive or repulsive forces between the molecules or the containing wall.

Using the afore-mentioned assumptions, the equation of state can be derived. Consider a cube of dimension L. If it is assumed that one-third of the molecules in the container are traveling normal to each pair of parallel sides, the impact pressure on the walls of the vessel can be determined. The time required for one molecule to travel from one side of the vessel to the other is

$$\frac{\text{Distance}}{\text{Velocity}} = \frac{L}{\bar{v}} = t$$

The time required to make a round trip, which is the time between impacts on each wall, is

$$\frac{2L}{\bar{v}} = 2t = \bar{t}$$

The number of impacts per unit time is the reciprocal of the time per impact, so that the impacts per unit time can be expressed as

$$I = \frac{\bar{v}}{2L} = \frac{1}{\bar{t}}$$

Since force is defined as the time rate of change of momentum and momentum is defined as mass times velocity, the total change in momentum per molecule per impact is

$$m\bar{v} - (-m\bar{v}) = 2m\bar{v}$$

Therefore, the change in momentum per second per molecule is the product of the number of impacts per unit time and the change in momentum per impact:

$$(2m\bar{v}) \frac{\bar{v}}{2L} = \frac{m\bar{v}^2}{L}$$

To obtain the total rate of change in momentum at one wall it is necessary to consider all the molecules which are hitting the wall. The total change in momentum per second is force.

$$\text{Force} = \frac{n'}{3} \frac{m\bar{v}^2}{L}$$

where n' is the number of molecules. Since force equals the product of pressure and area, the pressure on the wall is

$$P = \frac{n'}{3} \frac{m\bar{v}^2}{L} \div L^2 = \frac{2n'}{3L^3} \left(\frac{1}{2} m\bar{v}^2 \right)$$

as

$$L^3 = \text{volume}$$

$$PV = \frac{2n'}{3} \left(\frac{1}{2} m\bar{v}^2 \right)$$

It has already been said that heat energy is manifested in molecular motion; hence it can be written that

$$\frac{1}{2} m\bar{v}^2 = \text{kinetic energy} \propto T$$

or

$$\frac{1}{2} m\bar{v}^2 = KT$$

where K is a constant of proportionality; that is, as the temperature is increased, so is the kinetic energy of the molecules. Then

$$PV = \frac{2}{3} Kn'T$$

It can be shown that K is a constant independent of the gas. If T remains constant and the number of molecules n' remains constant, then

$$PV = \text{constant}$$

which is Boyle's law.

Charles's law can also be obtained from the equation of state by arbitrarily letting the pressure or volume be constant.

$$\text{Let } PV = \frac{n'}{A}\left(\frac{2}{3} KA\right)T \qquad \text{since } \frac{n'}{A} = n$$

$$PV = nRT \tag{4-4}$$

where A is Avogadro's number of molecules per mole, n is the number of moles of gas in the vessel, and $\frac{2}{3}KA = R$. As K and A are both constants which are independent of the gas, then R is a constant independent of the gas.

Behavior of Natural Gases. So far, in the cases of both Boyle's and Charles's laws and the kinetic theory, the assumption has been made that the gas involved is perfect. Realizing that no gas obeys the perfect gas laws, many attempts have been made to correct the perfect gas law and make its application more general. One of the better known equations is van der Waals's equation of state for a pure substance. It should be remembered that the kinetic theory assumed that there were no attractive forces existing between molecules, which is known to be incorrect. At low pressure the molecules are so far apart that the attractive force is nearly zero, but at high pressure the molecules are close together and the attractive force becomes an important factor. The pressure that is measured by gauges is the impact pressure, but the internal pressure P_i is greater than the impact pressure by the amount of energy that is expended in overcoming the intermolecular force. In order to express the true internal pressure, it is necessary to add a factor P' to the measured pressure to correct for the attractive force. Therefore, $P_i = (P + P')$ should be used in the equation of state. P' can be shown to be proportional to n^2/v^2 where n is the number of moles in the volume, V. Thus

$$P' = \frac{n^2}{V^2} a$$

where a is a constant dependent on the type of gas in the system.

The volume occupied by the molecules also was neglected in the kinetic theory. The actual "free space" available for compression is less than the total volume by the amount of space occupied by the molecules. If 1 mole of molecules of a pure gas occupies a volume of b, then the free space available to change would be

$$V' = V - nb$$

where b is a constant for the particular gas. The equation of state can be written in the form developed by van der Waals as

$$\left(P + \frac{n^2 a}{V^2}\right)(V - nb) = nRT \qquad (4\text{-}7)$$

or for 1 mole of gas

$$\left(P + \frac{a}{V^2}\right)(V - b) = RT \qquad (4\text{-}8)$$

TABLE 4-3. VAN DER WAALS CONSTANTS FOR SELECTED GASES[2]

Gas	Formula	a*	b†
Carbon dioxide	CO_2	3.59	0.0427
Ethane	C_2H_6	5.49	0.0638
Hydrogen	H_2	0.244	0.0266
Methane	CH_4	2.25	0.0428
Nitrogen	N_2	1.39	0.0391
Oxygen	O_2	1.36	0.0318

* a in atm-liter^2 mole^{-2}.
† b in liter mole^{-1}.

The perfect gas law holds rather well for the so-called "permanent" gases in the low-pressure ranges. Van der Waals's equation applies over a greater pressure range for the same gases. The increased range of pressure does not include pressures of the order of the reservoir pressures encountered in many oil and gas fields.

Furthermore, the equation contains two arbitrary constants which depend on the properties of the gas and limit the usefulness of van der Waals's equation in describing the behavior of mixtures of gases. To a large extent the constants a and b depend on the size of the molecules. Therefore, in a mixture of gases it is evident that the attractive forces depend on the proportion of the various-size molecules present. No adequate method has been devised for evaluating the effect on a and b of the molecular interaction in mixtures.

The Beattie-Bridgeman[8] equation of state, involving five arbitrary constants, describes the behavior of pure substances with considerable precision. It, too, is limited in application for mixtures of gases. To apply this equation to mixtures requires the simultaneous solutions of equations comparable to the number of constituents in the mixture.

The Beattie-Bridgeman equation of state for a pure substance is as follows:

$$P = \frac{RT}{V^2} V + B_0\left(1 - \frac{b}{V}\right)1 - \frac{c}{VT^3} - \frac{A_0}{V^2}\left(1 - \frac{a}{V}\right) \qquad (4\text{-}9)$$

where P = pressure,
  T = temperature, °K
  V = volume, liters per mole

and A_0, B_0, a, b, c are constants empirically defined for each pure gas.

The Benedict, Webb, Rubin[9] equation of state for a mixture of hydrocarbons

$$P = RT\rho_m + \left(BRT - A - \frac{C}{T^2}\right)\rho_m{}^2 + (bRT - a)\rho_m{}^3 +$$

$$a\alpha\rho_m{}^6 + \frac{c\rho_m{}^3}{T^2}[(1 + \gamma\rho_m{}^2)e^{-\gamma\rho_m{}^2}] \qquad (4\text{-}10)$$

where $B = \Sigma y_i B_i$
  $A = (\Sigma y_i A_i{}^{1/2})^2$
  $C = (\Sigma y_i C_i{}^{1/2})^2$
  $b = (\Sigma y_i b_i{}^{1/3})^3$
  $a = (\Sigma y_i a_i{}^{1/3})^3$
  $c = [\Sigma y_i c_i{}^{1/3}]^3$
  $\alpha = [\Sigma y_i \alpha_i{}^{1/3}]^3$
  $\gamma = [\Sigma y_i \gamma_i{}^{1/2}]^2$

and B_i, A_i, C_i, b_i, a_i, c_i, α_i, and γ_i are constants defined empirically for the ith component; ρ_m is the molal density; P is the absolute pressure; R is the universal gas constant; e is the base of the natural logarithm; and T is the absolute temperature.

The petroleum industry has adopted the concept of a compressibility factor Z for describing the behavior of mixtures of gases at moderate to high pressure. The compressibility factor Z is simply a correction factor to the perfect gas law; that is,

$$PV = ZnRT \qquad (4\text{-}11)$$

and, by definition,

$$Z = \frac{PV}{P_0V_0} = \frac{PV}{nRT} \qquad (4\text{-}12)$$

P_0 and V_0 are the pressure and volume, respectively, which would be calculated from the perfect gas law.

Compressibility factors, determined experimentally and correlated with pressure and temperature, are presented in Figs. 4-13 to 4-16 for methane, ethane, propane, and n-butane, respectively. The dashed curve on each of the illustrations represents the boundary of the two-phase region. Com-

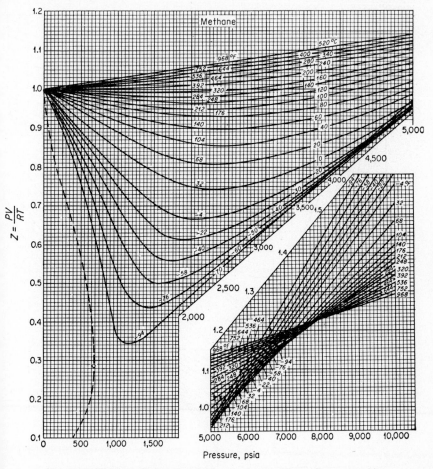

Fig. 4-13. Compressibility factors for methane. (*From Brown et al.*[4])

pressibility factors are defined only in the single-phase region. Several general characteristics of compressibility curves are apparent from a study of the figures. The temperature isotherms have distinct minimums which vanish as the temperature is increased. The compressibility factor decreases with decreasing temperature except in the high-pressure range, where a reversal of trend occurs (see inset in each figure). The minimums in the isotherms become more pronounced as the molecular weight of the gas increases.

Charts such as Figs. 4-13 through 4-16 can be prepared from experimental data for a particular gas. Compressibility factors determined from such charts can be used together with Eq. (4-11) to calculate the volume of the gas.

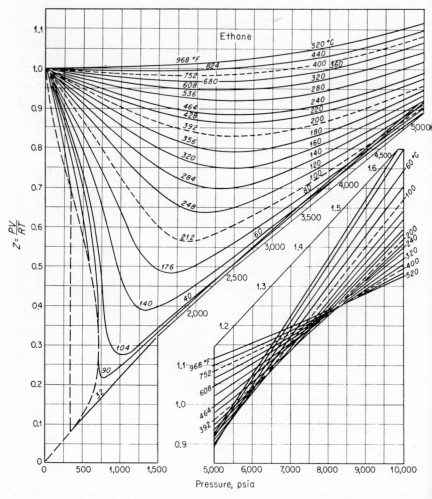

FIG. 4-14. Compressibility factors for ethane. (*From Brown et al.*[4])

Densities of the gas can be determined from the same experimental data from which the compressibility factors were calculated. For pure gases, charts correlating the density with pressure and temperature are perhaps as useful as correlations of compressibility factors. In Figs. 4-17 through 4-20 are presented density correlations of methane, ethane, propane, and normal butane. The two-phase regions are outlined by dashed curves on

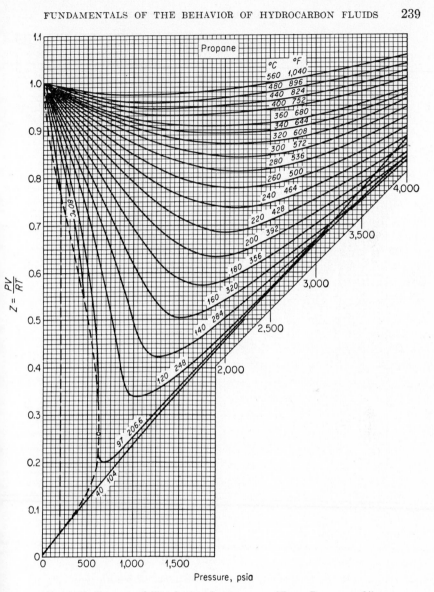

FIG. 4-15. Compressibility factors for propane. (*From Brown et al.*[4])

the charts for ethane, propane, and normal butane. It may be noted that the density of the gas approaches that of a liquid at elevated pressure.

Mixtures of Gases. Natural gases are mixtures of hydrocarbons which, as stated earlier, may be characterized by composition. The composition of any mixture may be reported in terms of per cent by weight, per cent by volume, or mole per cent. For gases, according to Avogadro's law, mole and volume per cent are identical, since 1 mole of any gas occupies the

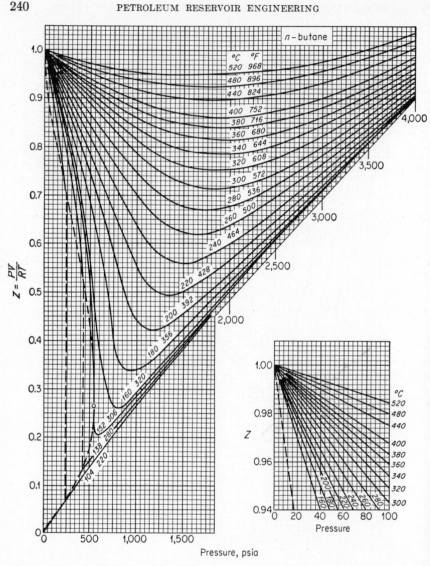

FIG. 4-16. Compressibility factors for *n*-butane. (*From Brown et al.*[4])

same volume at the same conditions of pressure and temperature (provided that the conditions of pressure and temperature are sufficiently close to atmospheric conditions). It will be recalled that a pound-mole of a substance is a quantity of material having a mass in pounds numerically equal to the molecular weight of the substance.

On occasion, analyses are reported on a weight basis, and for computational purposes, it is desirable to convert the analysis to a mole basis. An

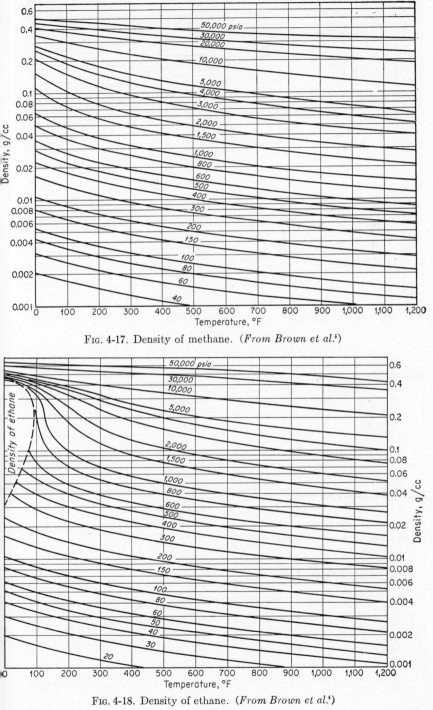

Fig. 4-17. Density of methane. (*From Brown et al.*[4])

Fig. 4-18. Density of ethane. (*From Brown et al.*[4])

241

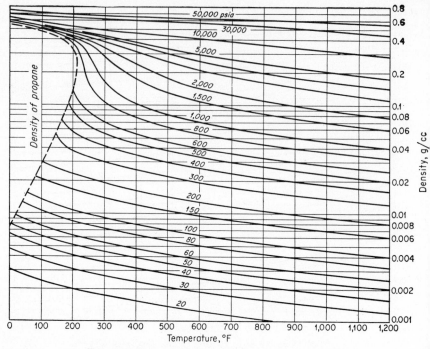

Density of propane

50,000 psia
10,000 30,000
5,000
2,000
1,500
1,000
800
600
500
400
300
200
150
100
80
60
50
40
30
20

0.8
0.6
0.4
0.2
0.1
0.08
0.06
0.04
0.02
0.01
0.008
0.006
0.004
0.002
0.001

Density, g/cc

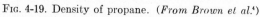
Temperature, °F

FIG. 4-19. Density of propane. (*From Brown et al.*[4])

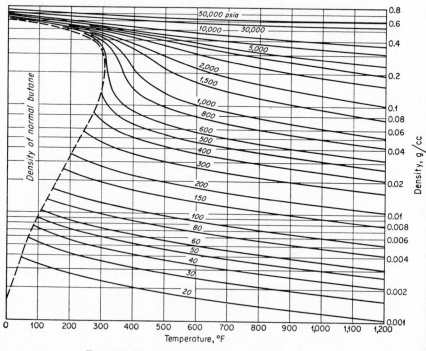

Density of normal butane

50,000 psia
10,000   30,000
5,000
2,000
1,500
1,000
800
600
500
400
300
200
150
100
80
60
50
40
30
20

0.8
0.6
0.4
0.2
0.1
0.08
0.06
0.04
0.02
0.01
0.008
0.006
0.004
0.002
0.001

Density, g/cc

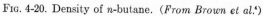
Temperature, °F

FIG. 4-20. Density of *n*-butane. (*From Brown et al.*[4])

example calculation of the mole analysis of a gas when given the weight analysis is given in Example 4-2.

Example 4-2. Converting from a Weight per Cent Analysis to Mole per Cent for a Hydrocarbon Gas.

(1) Component	(2) Weight %	(3) Molecular weight	(4) (2) ÷ (3) Moles/100 lb	Mole %
C_1	30	16.04	1.87	48.2
C_2	40	30.07	1.33	34.3
C_3	30	44.09	0.68	17.5
			3.88	100.0

For many purposes the physical constants of pure hydrocarbons are required. These constants are tabulated in Table 4-4 for the paraffin hydrocarbons through normal heptane.

The equation of state (4-11) can be applied to a mixture of gases by means of Amagat's law of partial volume. Amagat's law states that the total volume of a gaseous mixture is equal to the sum of the volumes of the individual components of the mixture, all at the same temperature and pressure; i.e.,

$$V = V_1 + V_2 + V_3 + \cdots + V_n = \sum_{i=1}^{n} V_i$$

where

$$y_i V = V_i$$

y_i is the mole fraction of the ith component in the gaseous phase, V is the volume of the system, and V_i is the partial volume of the ith component.

The application of this principle to calculation of the volume of a gaseous mixture is given in Example 4-3.

Example 4-3. Gas Volume Calculated Using Partial Volumes. Determination of the volume at 1,000 psia and 104°F occupied by 1,000 scf of gas whose composition is given below:

Gas analysis	Mole fraction y_i	V_i, scf	Z at 1,000 psia and 104°F	$\dfrac{P_a T_s}{P_s T_a} V_i$	V_i at 1,000 psia and 104°F
Methane	0.70	700	0.918[a]	11.13	10.217
Ethane	0.20	200	0.274[b]	3.18	0.871
Propane	0.10	100	0.236[c]	1.59	0.375
	1.000			15.90	11.463

[a] From Fig. 4-13.   [b] From Fig. 4-14.   [c] From Fig. 4-15.

TABLE 4-4. PHYSICAL CONSTANTS OF HYDROCARBONS

(Values from CNGA and NGAA. Modified for standard conditions of 60°F and 14.65 psia)

	Methane	Etnane	Propane	Iso-butane	Normal butane	Iso-pentane	Normal pentane	Normal hexane	Normal heptane	Heptanes plus C_{7+}
Molecular formula	CH_4	C_2H_6	C_3H_8	$i\text{-}C_4H_{10}$	$n\text{-}C_4H_{10}$	$i\text{-}C_5H_{12}$	$n\text{-}C_5H_{12}$	$n\text{-}C_6H_{14}$	$n\text{-}C_7H_{16}$	
Molecular weight	16.042	30.068	44.094	58.120	58.120	72.146	72.146	86.172	100.198	M
Critical temp, °F abs	343	550	666	733	765	830	847	914	972	
Critical pressure, psia	673	708	617	530	551	482	485	434	397	
Critical density, lb/gal	1.351	1.695	1.888	1.945	1.891	1.955	1.935	1.958	1.957	
Critical volume, (cu ft/lb-mole)	1.586	2.371	3.123	3.990	4.130	4.930	4.98	5.88	6.84	
SG gas (air = 1.00)	0.554	1.038	1.522	2.006	2.006	2.491	2.491	2.975	3.459	$M/28.966$
Cu ft/lb gas	23.73	12.66	8.63	6.55	6.55	5.28	5.28	4.42	3.80	$380.69/M$
Cu ft gas/gal of liquid	59.18	39.43	36.56	30.75	31.89	27.50	27.76	24.46	21.77	$3170\ S/M$
Lb Mcf of gas	42.14	78.98	115.83	152.67	152.67	189.51	189.51	226.36	263.20	$2.6268\ M$
Gal liquid per Mcf of gas	16.87	25.36	27.38	32.57	31.41	36.41	36.07	40.94	45.94	$0.3155\ M/S$
Bbl liquid/MMcf of gas	401.6	603.9	651.9	775.4	747.5	867.0	858.8	974.8	1094	$7.511\ M/S$
SG liquid (60/60)	0.3	0.374	0.508	0.563	0.584	0.625	0.631	0.664	0.688	S
Lb/gal liquid at 60°F	2.5	3.11	4.23	4.69	4.86	5.20	5.25	5.53	5.73	$8.327\ S$
Lb/bbl liquid at 60°F	105	130.6	177.7	197.0	204.1	218.4	220.5	232.3	240.7	$349.7\ S$
Lb/cu ft liquid at 60°F	18.70	23.26	31.64	35.08	36.35	38.90	39.27	41.36	42.86	$62.29\ S$
Cu ft liquid/lb at 60°F	0.0535	0.0430	0.0316	0.0285	0.0275	0.0257	0.0254	0.0242	0.0233	$0.01605/S$
Gal/lb-mole at 60°F	6.4	9.64	10.41	12.38	11.94	13.84	13.71	15.57	17.47	$0.120\ M/S$

Constants Used in Calculations

0°F = 459.58°R
Density of water at 60°F = 8.327 lb per gal
Molecular weight of air = 28.966

1 cu ft = 28.316 liters
1 cu ft = 7.480 gal
1 gal = 3,785.53 ml

Volume at 1,000 psia and 104°F calculated as if the mixture were a perfect gas is 15.90 cu ft. Actual volume occupied by gas at 1,000 psia and 104°F is 11.463 cu ft.

$$Z_{mix} = \frac{11.463}{15.90} = 0.721$$

The partial volume method of determining the volumes of a gaseous mixture is tedious when the analysis is known and cannot be used when the analysis is unknown. For these reasons generalized compressibility factors have been determined for methane-rich natural gases and, through the work of Kay[10] and others, correlated with reduced pressure and temperature.

The law of corresponding states provides the theoretical basis for correlations utilizing reduced pressures and temperatures. Sage and Lacey[11] state this law as follows: "The ratio of the value of any intensive property to the value of that property at the critical state is related to the ratios of the prevailing absolute temperature and pressure to the critical temperature and pressure by the same function for all similar substances." Thus, the reduced temperature is the ratio of the prevailing absolute temperature to the critical temperature and the reduced pressure is the ratio of the prevailing absolute pressure to the critical pressure. Other reduced properties can be defined in a like manner.

The law of corresponding states can be derived from van der Waals's equation of state. In expanded form, Eq. (4-9) is

$$PV^3 - V^2(bP + RT) + a(V - b) = 0 \qquad (4\text{-}9)$$

which is a cubic equation. A typical solution of Eq. (4-9) at constant temperature is shown by the dashed 70°F isotherm on Fig. 4-21. Within the two-phase region the equation indicates a continuous transition of physical properties from gas to liquid. This is untrue as shown by experimental data (solid line on Fig. 4-21). The 90°F (critical temperature) isotherm is tangent to the phase envelope at the critical point. Furthermore, the slope of the curve at that point is zero. The isotherm has an inflection point at the critical point.

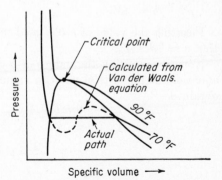

FIG. 4-21. Characteristic isotherms calculated from van der Waals's equation of state for a pure substance.

Mathematically these two situations correspond to zero values of the first and second derivatives of pressure with respect to volume.

That is, the change in pressure with a small change in volume is zero, $(\partial P/\partial V)_{cp} = 0$, and the change in slope with the change in volume is zero, $(\partial^2 P/\partial V^2)_{cp} = 0$.

Solving van der Waals's equation for pressure yields for 1 mole

$$P = -\frac{a}{V^2} + \frac{RT}{V - b} \qquad (4\text{-}13)$$

Then to evaluate a, b, and R at the critical point of the material, it becomes necessary only to obtain the first and second derivatives with respect to volume of Eq. (4-13) and set them equal to zero. Therefore,

$$\left(\frac{\partial P}{\partial V}\right)_{cp} = \left(\frac{+2a}{V_c^3} - \frac{RT_c}{(V_c - b)^2}\right)_{cp} = 0$$

$$a = \frac{RT_c V_c^3}{2(V_c - b)^2}$$

also

$$\left(\frac{\partial^2 P}{\partial V^2}\right)_{cp} = \left(-\frac{6a}{V_c^4} + \frac{2RT}{(V_c - b)^3}\right)_{cp} = 0$$

$$a = \frac{V_c^4}{3}\frac{RT_c}{(V_c - b)^3}$$

Equating the two equations for a and solving for b in terms of T_c, P_c, V_c, and R,

$$\frac{V_c^3}{2}\frac{RT}{(V_c - b)^2} = \frac{V_c^4}{3}\frac{RT_c}{(V_c - b)^3}$$

$$V_c - b = \tfrac{2}{3}V_c$$

$$b = \tfrac{1}{3}V_c$$

Inserting the value of b obtained in the first equation for a it is found that $a = \tfrac{9}{8}RT_cV_c$.

Now, by inserting the values of a and b into van der Waals's original equation, at the critical point, a value for R in terms of P_c, V_c, and T_c is obtained

$$\left(P_c + \frac{9}{8}\frac{RT_cV_c}{V_c^2}\right)\left(V_c - \frac{1}{3}V_c\right) = RT_c$$

$$\frac{2}{3}P_cV_c + \frac{3}{4}RT_c = RT_c$$

$$R = \frac{8}{3}\frac{P_cV_c}{T_c} \qquad (4\text{-}14)$$

from which

$$a = 3P_cV_c^2$$

$$b = \frac{1}{3}V_c$$

By using the values of a, b, and R stated in terms of the critical properties of pressure, volume, and temperature of the gas involved, it is possible to rewrite van der Waals's equation in the following form:

$$\left(P + \frac{3P_c V_c^2}{V^2}\right)\left(V - \frac{1}{3} V_c\right) = \frac{8}{3}\frac{P_c V_c}{T_c} T \qquad (4\text{-}15)$$

Defining the reduced properties of a material as follows:

$$\text{Reduced pressure} \quad = P_r = \frac{P}{P_c}$$

$$\text{Reduced volume} \quad = V_r = \frac{V}{V_c} \qquad (4\text{-}16)$$

$$\text{Reduced temperature} = T_r = \frac{T}{T_c}$$

and substituting in Eq. (4-15) for the pressure temperature and volume, obtain

$$\left(P_c P_r + \frac{3P_c}{V_r^2}\right)\left(V_r V_c - \frac{1}{3} V_c\right) = \frac{8}{3} P_c V_c T_r$$

which reduces to

$$\left(P_r + \frac{3}{V_r^2}\right)(3V_r - 1) = 8T_r \qquad (4\text{-}17)$$

The above equation contains only reduced values of pressure, volume, and temperature. Equation (4-17) is commonly referred to as van der Waals's reduced equation of state. From the above, it is possible to state that regardless of the gas, as long as it is pure, if the reduced pressures and temperatures are equal, then the reduced volumes must be equal. This confirms the law of corresponding states as previously stated.

For real gases, the law of corresponding states does not hold over wide ranges of pressure, but the agreement is close enough to permit the use of reduced properties as the basis for correlating experimentally determined compressibility factors. The correlation of compressibility factors with reduced pressure and temperature of selected pure hydrocarbons is shown in Fig. 4-22.

Kay[10] introduced the concept of pseudo-critical and pseudo-reduced properties for treating mixtures of natural gas. The pseudo-critical properties are obtained by applying Amagat's law of partial volumes for mixtures to the critical properties of the individual constituents of the mixture. The pseudo-critical pressure and pseudo-critical temperature are defined mathematically:

$$_p P_c = \sum_{i=1}^{n} y_i P_{ci} \qquad \text{and} \qquad _p T_c = \sum_{i=1}^{n} y_i T_{ci}$$

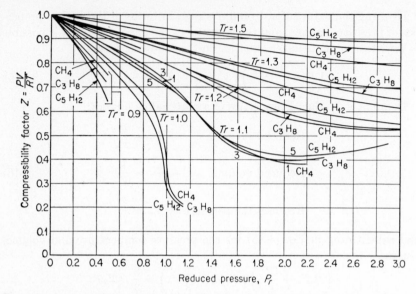

FIG. 4-22. Compressibility factor as a function of reduced pressure for a series of reduced temperatures. (*From Sage and Lacey.*[11])

where $_pP_c$ = pseudo-critical pressure
$_pT_c$ = pseudo-critical temperature
P_{ci} and T_{ci} = critical pressure and temperature respectively of ith component
y_i = mole fraction of ith component in mixture
n = number of components

The pseudo-reduced properties are defined as the ratio of the property of the mixture to the pseudo-critical property of the mixture. Thus, the pseudo-reduced pressure and pseudo-reduced temperature are defined as

$$_pP_r = \frac{P}{_pP_c} \quad \text{and} \quad _pP_r = \frac{T}{_pT_c}$$

Compressibility factors, experimentally determined, for a large number of natural gases were correlated with pseudo-reduced pressure and temperature. The results of these correlations are presented in Figs. 4-23 to 4-25. Since most natural gases contain large amounts of methane, the correlations for natural gases are superior to the correlations between individual hydrocarbons such as previously shown in Fig. 4-22. The petroleum industry has universally adopted the charts of Figs. 4-23 to 4-25 to determine the compressibility factor for use in the equation

$$PV = ZnRT \tag{4-11}$$

Thus, if any analysis of the gaseous mixture is available, the pseudo-critical properties can be calculated and compressibility factors determined for use in the generalized equation of state.

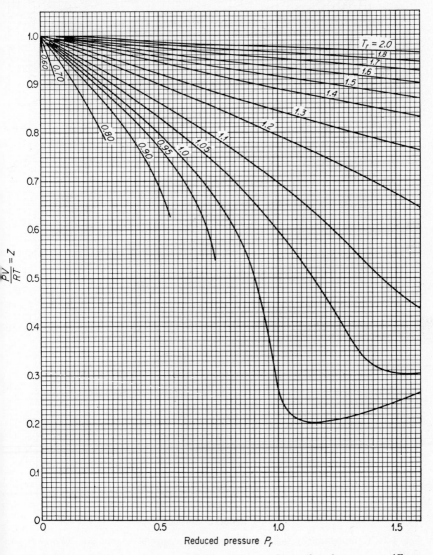

FIG. 4-23. Generalized plot of compressibility factors at low reduced pressures. (*From Brown et al.*[4])

The conversion of an analysis on a weight basis to that on a mole basis was shown in Example 4-2. The mole method of calculation can be applied to the determination of other properties of gaseous mixtures. One of the

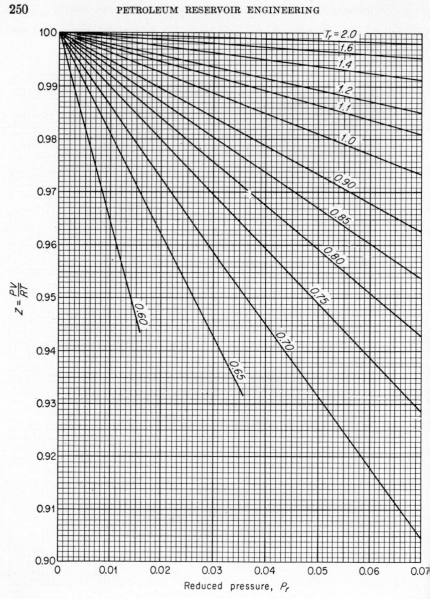

FIG. 4-24. Compressibility factors for gases near atmospheric pressure. (*From Brown et al.[4]*)

properties frequently of interest is the average molecular weight. The average molecular weight $\overline{M}$ is defined mathematically as

$$\overline{M} = \sum_{i=1}^{n} y_i M_i \qquad (4\text{-}18)$$

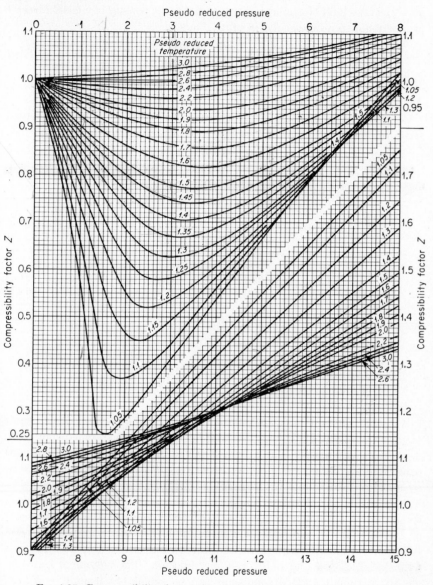

FIG. 4-25. Compressibility factors for natural gases. (*From Brown et al.*[4])

where M_i is the molecular weight of the ith component and y_i is the mole
fraction of the ith component in the mixture.

The gas gravity can be calculated from the average molecular weight.
By definition, the gas gravity is the ratio of the density of the gas to the
density of air, both densities defined at atmospheric conditions. Since the

ideal gas laws hold at atmospheric conditions, the density of a gas is directly
proportional to its molecular weight. Thus

$$G = \frac{\overline{M}}{M_a} = \frac{\overline{M}}{28.96}$$ (4-19)

where G is the gas gravity, $\overline{M}$ is the average molecular weight, and M_a is
the molecular weight of air.

Calculations of the average molecular weight, gas gravity, pseudo-critical
pressure, and pseudo-critical temperature are illustrated in Example 4-4.

Example 4-4. Calculation of Molecular Weight, Gas Gravity, and
Pseudo-critical Properties for Hydrocarbon Gas.

(1) Component	(2) Molecular weight	(3) Mole fraction	(4) P_c	(5) T_c	(2) × (3)	(3) × (4)	(3) × (5)
C_1	16.04	0.70	673	343	11.23	471.1	240.1
C_2	30.07	0.20	708	550	6.01	141.6	110.0
C_3	44.09	0.10	617	666	4.41	61.7	66.6
					21.65	674.4	416.7

Molecular weight of mixture = 21.65
Pseudo-critical pressure = 674.4 psia
Pseudo-critical temperature = 416.7°F
Gas gravity = 21.65/28.96, or 0.749

A useful constant in gas calculations is the volume occupied by 1 mole
of gas at standard conditions. Standard conditions are specified conditions
of pressure and temperature at which gas volumes are reported. Standard
conditions are defined, in part, by usage and, in some oil-producing states,
by statute. The values of pressure and temperature chosen are approximately average atmospheric conditions. In scientific work, the standard
temperature is frequently taken as 20°F (68°F) and the standard pressure
as 14.696 psia. In the petroleum industry 60°F is the standard temperature. The standard pressure varies from state to state. For example, it is
14.65 psia in Texas and 15.025 psia in Louisiana. The volume of 1 mole of
gas for various standard conditions can be calculated from $V = RT/P$,
assuming a perfect gas. Volumes for 1 mole of gas for several commonly
used standard conditions are given in Table 4-5.

It is frequently necessary in engineering calculations to compute the
density or volume of a gas at elevated pressures and temperatures. Various
methods for calculating these and other parameters are illustrated in Example 4-5. The treatment of the material as a perfect gas (method 1) is

TABLE 4-5. VOLUME OF 1 MOLE OF GAS AT VARIOUS STANDARD CONDITIONS

Temp, °F	Pressure, psia	Volume of 1 mole, cu ft
68 (20°C)	14.696	385.51
60	14.65	380.69
60	15.025	371.35
60	14.4	387.47
60	14.7	379.56

included to emphasize the deviation of real gases from the perfect gas law. The volume of gas at 1,000 psia and 104°F as calculated from the perfect gas law is in error by more than 20 per cent.

Method 2, treatment of the gas using additive volumes and compressibility factors of individual components, was illustrated in a slightly different form in Example 4-1.

Method 3, treatment of the gas using additive volumes and densities of the individual components, is simply a variation of method 2, as the density curves were plotted from the same data from which the compressibility curves were plotted. The values obtained in Example 4-5 from these two methods are in very close agreement, certainly within the accuracy of reading the charts.

Both methods 2 and 3 are limited in applicability by the lack of data on heavier hydrocarbons which have two-phase regions extending across temperatures of interest in oil-field applications.

The use of the pseudo-reduced concepts and the compressibility-factor chart for natural gases is illustrated in method 4 of Example 4-5. The values obtained differ about 10 per cent from the values from methods 2 and 3. The difference can be attributed largely to the composition of the gas selected. The generalized charts include data from gases having heavier components and higher concentration of methane present. Therefore, method 4 is more applicable to gases having more methane and a greater variation in other hydrocarbons than the gas in the example.

***

Example 4-5. Methods of Calculating Volume Relations for Mixtures of Real Hydrocarbon Gases. Calculation of specific volume, density, compressibility factor, and the volume occupied at 1,000 psia and 104°F for 1,000 cu ft at 14.65 psia and 60°F of a gas having the composition given in Example 4-3:

1. Treated as a perfect gas having an av mol wt $\overline{M} = 21.65$ (from Example 4-4).

   a. Compressibility factor of the mixture $Z = 1.000$ (definition of perfect gas)

$b.$ Specific volume $v = \dfrac{RT}{MP} = \dfrac{10.72(564)}{21.65(1,000)} = 0.2795$ cu ft/lb

$c.$ Density $\rho = \dfrac{1}{v} = \dfrac{MP}{RT} = \dfrac{21.65(1,000)}{10.72(564)} = \dfrac{1}{0.2795} = 3.578$ lb/cu ft

$d.$ Volume of 1,000 scf at 1,000 psia and 104°F (V)

$$V = \frac{P_s T}{P T_s} V_s = \frac{14.65(564)}{1,000(520)}(1,000) = 15.89 \text{ cu ft}$$

2. Treated as a real gas using additive volumes and compressibility factors of individual components from Figs. 4-13 to 4-15.

   $a.$ Compressibility factor of the mixture (Z),

$$\begin{aligned} Z &= Z_1(y_1) + Z_2(y_2) + Z_3 y_3 \\ &= 0.918(0.7) + 0.274(0.2) + 0.236(0.1) \\ &= 0.721 \end{aligned}$$

   $b.$ Specific volume $v = \dfrac{ZRT}{MP} \dfrac{0.721(10.72)(564)}{21.65(1,000)} = 0.2015$ cu ft/lb

   $c.$ Density $\rho = \dfrac{1}{v} \dfrac{MP}{ZRT} = \dfrac{21.65(1,000)}{0.721(10.72)(564)} = \dfrac{1}{0.2015} = 4.963$ lb/cu ft

   $d.$ Volume of 1,000 scf at 1,000 psia and 60°F (V)

$$\frac{P_s T Z}{P T_s Z_s} V_s = \frac{14.65(564)(0.721)}{1,000(520)(1.000)} 1,000 = 11.46 \text{ cu ft}$$

3. Treated as a real gas using additive volumes and densities determined from Figs. 4-17 to 4-19.

   $a.$ Compressibility factor (not directly determined)

   $b.$ Specific volume $v = \dfrac{\displaystyle\sum_{i=1}^{n} M_i y_i / \rho_i}{M}$

$\rho_1 = 0.046$ gm/cc $= 2.916$ lb/cu ft
$\rho_2 = 0.300$ gm/cc $= 19.020$ lb/cu ft
$\rho_3 = 0.490$ gm/cc $= 31.066$ lb/cu ft

Therefore $\qquad v = \dfrac{\dfrac{16.04(0.7)}{2.916} + \dfrac{30.07(0.2)}{19.020} + \dfrac{44.09(0.1)}{31.066}}{21.65}$

$$= \frac{4.309}{21.65} = 0.1990 \text{ cu ft/lb}$$

   $c.$ Density $\rho = \dfrac{M}{\displaystyle\sum_{i=1}^{n} M_i y_i / \rho_i} = \dfrac{1}{v} = \dfrac{1}{0.1990} = 5.03$ lb/cu ft *d.* Volume of 1,000 scf at 1,000 psia and 60°F (V)

$$V = wv$$

where w is the weight of the gas.

$$w = \overline{M}\, \frac{V_s}{V_m} = 21.65\, \frac{1,000}{380.69} = 56.9 \text{ lb}$$

where V_m is the molal volume.

Therefore $V = 56.9(0.1990) = 11.30$ cu ft

**. Treated as a real gas using pseudo-reduced properties and Fig. 4-25.

a. Compressibility factor (Z)

$$_pP_c = 674.4 \qquad \text{and} \qquad _pT_c = 416.7 \quad \text{(from Example 4-4)}$$

Therefore $$_pP_r = \frac{P}{_pP_c} = \frac{1,000}{674.4} = 1.483$$

$$_pT_r = \frac{T}{_pT_c} = \frac{564}{416.7} = 1.353$$

Therefore from Fig. 4-25

$$Z = 0.798$$

b. Specific volume $v = \dfrac{ZRT}{\overline{M}P}\, \dfrac{0.798(10.72)(564)}{21.65(1,000)} = 0.2231$ cu ft/lb

c. Density $\rho = \dfrac{\overline{M}P}{ZRT} = \dfrac{1}{v} = \dfrac{1}{0.2231} = 4.48$ lb/cu ft

d. Volume of 1,000 scf at 1,000 psia and 60°F (V)

$$V = \frac{P_s T Z}{P T_s Z_s}\, V_s = \frac{14.65(564)(0.798)}{1,000(520)(1.000)}\,(1,000) = 12.68 \text{ cu ft}$$

The composition of natural gases is usually reported through the heptanes-plus fraction. The heptanes-plus fraction as obtained from a fractional distillation is a liquid residue which contains heptanes and heavier hydrocarbons. In the laboratory analysis, two properties, molecular weight and specific gravity, of the heptanes plus are determined. These properties have been correlated with pseudo-critical pressure and temperature of the heptanes-plus fraction. The results of these correlations are presented in Figs. 4-26 and 4-27, which differ only in that the specific gravity is used in in Fig. 4-26 and API gravity is used in Fig. 4-27. These correlations, together with the physical constants from Table 4-4, enable the engineer to calculate pseudo-reduced properties of natural gases for which conventional analyses are available.

In many instances analyses are not available. Therefore, correlations with gas gravity of pseudo-critical properties of natural gases have been

developed. These correlations are presented in Figs. 4-28 to 4-30. Th
data in Figs. 4-28 and 4-29 are for gases from particular fields. Thes
figures are included to illustrate the fit of the data points to the correlations
Figure 4-30 presents generalized correlations which include data from botl
of the previously mentioned figures as well as additional data. The mis
cellaneous gas curves cover all natural gases other than those having larg
concentrations of intermediates (propane, butane, and pentane). The con

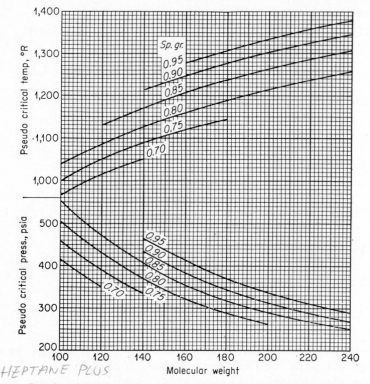

HEPTANE PLUS

FIG. 4-26. Pseudo-critical properties of hydrocarbon liquids. (*From Matthews et al.*[12])

densate well fluid curves should be used for gases having large fractions o
intermediates.

The gas gravity can be readily determined in the field by the Ac-M
balance method or similar techniques. Thus compressibility data can b
obtained even though a gas analysis is not available. The accuracy of th
correlations of reduced properties with gas gravity can be ascertained by
comparing the pseudo-critical pressure and temperature calculated ir
Example 4-4 with those determined from Fig. 4-30. The gas gravity (from
Example 4-4) is 0.75. The pseudo-critical pressure and pseudo-critica
temperature are

$_pP_c$ (from miscellaneous gases, Fig. 4-30) = 665 psia
$_pP_c$ (from Example 4-4) = 674.4
$_pT_c$ (from miscellaneous gases, Fig. 4-30) = 405°F
$_pT_c$ (from Example 4-4) = 416.7°R

The difference is about 1.5 per cent. A gas having a wider range of composition would yield a closer check if the appropriate curves are selected.

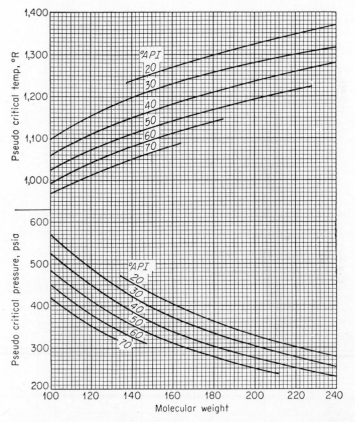

FIG. 4-27. Pseudo-critical properties of hydrocarbon liquids. (*From Matthews et al.*[12])

The petroleum engineer is primarily interested in volume calculations for gaseous mixtures. The volume of reservoir space occupied by a unit volume of gas at standard conditions is defined as the gas-formation volume factor B_g. The gas-formation volume factor has the units of volume per volume and, therefore, is dimensionless. In equation form,

$$B_g = \frac{ZRT}{V_m P} \tag{4-20}$$

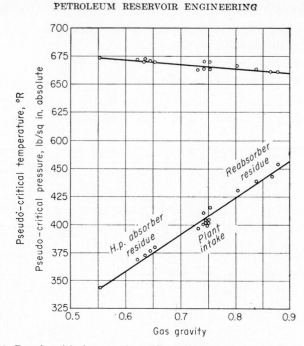

FIG. 4-28. Pseudo-critical properties of Grapeland gases. (*From Brown et al.*[4])

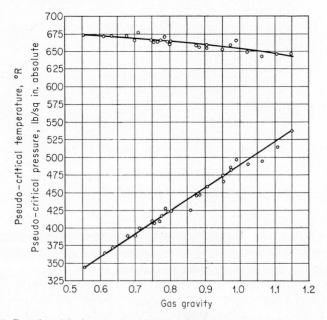

FIG. 4-29. Pseudo-critical properties of Oklahoma City gases. (*From Brown et al.*[4])

where Z = compressibility factor
      R = universal gas constant
      T = reservoir temperature, °F
      P = reservoir pressure, psia
     V_m = molal volume as defined for the particular set of standard
           conditions desired

If a reservoir contains a dry or wet gas, the composition of the gas in the reservoir will remain constant over the producing life of the reservoir.

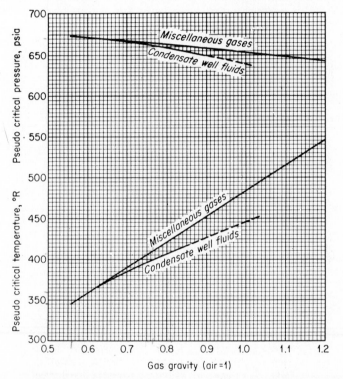

FIG. 4-30. Pseudo-critical properties of natural gases. (*From Brown et al.*[4])

During the depletion history the reservoir pressure ordinarily will decline. As the reservoir temperature is constant, B_g can be defined as a function of pressure. Thus, for a particular gas reservoir and a particular set of standard conditions,

$$B_g = C\left(\frac{Z}{P}\right) \qquad \text{where } C = \frac{RT}{V_m}$$

The calculation of B_g for a particular gas reservoir is illustrated in Example 4-6.

Example 4-6. Calculation of Gas-formation Volume Factor from Gas Analysis. A reservoir having an initial pressure of 3,500 psia and a temperature of 140°F contains a dry gas having the composition listed below. The gas-formation volume factor B_g is desired as a function of pressure.

Component	(1) Mole fraction	(2) Critical temp, °R	(3) Critical pressure, psia	(1) × (2)	(1) × (3)
Methane	0.8686	343	673	297.93	584.57
Ethane	0.0609	550	708	33.50	43.12
Propane	0.0261	666	617	17.38	16.10
Isobutane	0.0048	733	530	3.52	2.54
n-Butane	0.0077	765	551	5.89	4.24
Isopentane	0.0031	830	482	2.57	1.49
n-Pentane	0.0022	847	485	1.86	1.07
Hexanes	0.0038	914	434	3.47	1.65
Heptanes plus	0.0228	1118^a	415^a	25.49	9.46
				$_pT_c = 391.61$	$_pP_c = 664.24$

^{a} From Fig. 4-26 and properties of heptanes-plus fraction.

$M_{C_{7+}} = 128$

$SG_{C_{7+}} = 0.8195$

$T = 600°R$ $\quad _pT_r = \dfrac{600}{391.61} = 1.5321$

$B_g = C\dfrac{Z}{P} \quad C = \dfrac{RT}{V_m} = \dfrac{10.72(600)}{380.69} = 16.895$

Pressure, psia	$_pP_r$	Z	$\dfrac{Z}{P}$	B_g
100	0.151	0.985	0.009850	0.166415
400	0.602	0.950	0.002375	0.040125
700	1.054	0.918	0.001314	0.022200
1,000	1.505	0.885	0.000885	0.014952
1,500	2.258	0.823	0.000549	0.009275
2,000	3.011	0.795	0.000398	0.006724
2,500	3.764	0.790	0.000316	0.005339
3,000	4.516	0.805	0.000268	0.004528
3,500	5.259	0.835	0.000239	0.004038

Impurities in Natural Gas

Mixtures of hydrocarbons were considered in the preceding discussions. Natural gases frequently contain materials other than hydrocarbons such as nitrogen, carbon dioxide, and hydrogen sulfide. Hydrocarbon gases are classified as "sweet" or "sour" depending on the hydrogen sulfide content. Both sweet and sour gases may contain nitrogen, carbon dioxide, or both. In local areas, natural gases contain small percentages of rare gases such as helium.

The common occurrence of small percentages of nitrogen and carbon dioxide is in part considered in the correlations previously cited, as many of the natural-gas mixtures used in developing the data contained small percentages of these materials. To facilitate the handling of these impurities, their critical constants are presented in Table 4-6.

TABLE 4-6. CRITICAL CONSTANTS FOR COMMON NONHYDROCARBON
CONSTITUENTS OF NATURAL GASES

Compound	Formula	Molecular weight	Critical temp, °R	Critical pressure, psia
Carbon dioxide	CO_2	44.01	548	1,073
Helium	He	4.00	9.4	33
Hydrogen sulfide	H_2S	34.08	673	1,306
Oxygen	O_2	32.00	278	731
Nitrogen	N_2	28.02	227	492
Water	H_2O	18.016	1165	3,206

Nitrogen. Eilerts and coworkers [13] investigated the effect of nitrogen content on the volumetric behavior of natural gases. They state that the compressibility factor of nitrogen–natural-hydrocarbon mixtures determined by assuming the validity of the law of corresponding states has an error of less than 1 per cent if the nitrogen concentration is 10 mole % or less. If the gas contains 20 mole % or more, the error in compressibility factor (assuming corresponding states) may be greater than 3 per cent.

Standing[3] states that for carbon dioxide concentrations of 2 mole % or less, the law of corresponding states yields satisfactory results. Olds, Sage, and Lacey[14] indicated that errors in compressibility factors as great as 5 per cent can be obtained from application of the theorem of corresponding states at carbon dioxide concentrations of 4 mole %.

Eilerts and coworkers[13] proposed a procedure for calculating the effect of nitrogen on the compressibility factor. They defined an additive compressibility factor Z_a as follows:

$$Z_a = Z_n y_n + (1 - y_n) Z_g \qquad (4\text{-}21)$$

where Z_n is the compressibility factor of the nitrogen (Fig. 4-31), y_n is the mole fraction of nitrogen in the mixture, and Z_g is the compressibility factor of the hydrocarbon fraction of the mixture (Fig. 4-25). The true compressibility factor Z of the mixture is then defined by

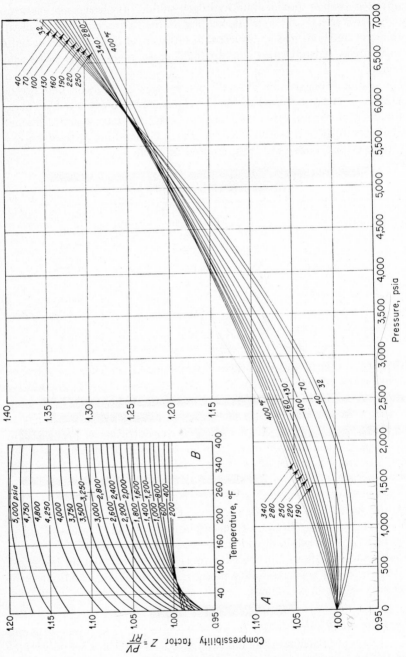

FIG. 4-31. Compressibility factors Z for nitrogen. (*From Eilerts et al.*[13])

262

TABLE 4-7. ADDITIVE VOLUME CORRECTION FACTORS FOR GASES
CONTAINING NITROGEN

Properties of mixture containing 7.907 mole % (11.350 mass %)
added nitrogen
Additive volume correction factors C at indicated temperature, °F

Pressure, psia	70	100	130	160	190	220	250	280
5,000	1.0010	1.0046	1.0049	1.0038	1.0034	1.0037	1.0044	1.0034
4,750	1.0011	1.0047	1.0054	1.0048	1.0048	1.0046	1.0044	1.0035
4,500	1.0010	1.0055	1.0059	1.0050	1.0053	1.0053	1.0049	1.0038
4,250	1.0014	1.0053	1.0061	1.0056	1.0059	1.0053	1.0052	1.0040
4,000	1.0016	1.0063	1.0068	1.0060	1.0062	1.0053	1.0054	1.0043
3,750	1.0032	1.0073	1.0076	1.0066	1.0065	1.0057	1.0058	1.0047
3,500	1.0056	1.0083	1.0083	1.0078	1.0070	1.0059	1.0055	1.0049
3,250	1.0073	1.0101	1.0094	1.0083	1.0070	1.0057	1.0054	1.0048
3,000	1.0093	1.0109	1.0101	1.0085	1.0074	1.0061	1.0051	1.0050
2,800	1.0115	1.0118	1.0110	1.0086	1.0073	1.0064	1.0054	1.0050
2,600	1.0124	1.0122	1.0108	1.0086	1.0076	1.0065	1.0055	1.0054
2,400	1.0143	1.0124	1.0103	1.0086	1.0080	1.0066	1.0055	1.0056
2,200	1.0165	1.0131	1.0087	1.0086	1.0081	1.0066	1.0055	1.0055
2,000	1.0162	1.0128	1.0101	1.0085	1.0081	1.0068	1.0052	1.0055
1,800	1.0168	1.0135	1.0093	1.0079	1.0073	1.0064	1.0049	1.0050
1,600	1.0150	1.0108	1.0079	1.0069	1.0062	1.0053	1.0040	1.0038
1,400	1.0116	1.0096	1.0064	1.0056	1.0047	1.0038	1.0027	1.0019
14.4	1.0000	1.0000	1.0000	1.0000	1.0000	1.0000	1.0000	1.0000

Properties of mixture containing 18.280 mole % (25.013 mass %)
added nitrogen
Additive volume correction factors C at indicated temperature, °F

Pressure, psia	70	100	130	160	190	220	250	280
5,000	1.0035	1.0079	1.0088	1.0086	1.0089	1.0093	1.0092	1.0071
4,750	1.0048	1.0079	1.0101	1.0095	1.0096	1.0098	1.0095	1.0071
4,500	1.0055	1.0090	1.0107	1.0104	1.0104	1.0106	1.0103	1.0075
4,250	1.0065	1.0106	1.0116	1.0115	1.0111	1.0111	1.0107	1.0079
4,000	1.0076	1.0120	1.0134	1.0127	1.0122	1.0116	1.0112	1.0088
3,750	1.0103	1.0137	1.0149	1.0135	1.0132	1.0121	1.0115	1.0092
3,500	1.0136	1.0158	1.0161	1.0146	1.0139	1.0123	1.0113	1.0096
3,250	1.0173	1.0182	1.0176	1.0159	1.0142	1.0125	1.0113	1.0094
3,000	1.0201	1.0204	1.0190	1.0163	1.0146	1.0129	1.0113	1.0096
2,800	1.0236	1.0222	1.0201	1.0166	1.0147	1.0134	1.0114	1.0092
2,600	1.0264	1.0233	1.0199	1.0173	1.0152	1.0131	1.0111	1.0092
2,400	1.0288	1.0240	1.0199	1.0175	1.0153	1.0125	1.0107	1.0093
2,200	1.0314	1.0248	1.0202	1.0170	1.0147	1.0117	1.0102	1.0090
2,000	1.0319	1.0250	1.0195	1.0160	1.0140	1.0115	1.0098	1.0087
1,800	1.0322	1.0239	1.0177	1.0147	1.0132	1.0108	1.0091	1.0079
1,600	1.0284	1.0199	1.0157	1.0134	1.0117	1.0092	1.0076	
1,400	1.0235	1.0172	1.0134	1.0109	1.0094			
14.4	1.0000	1.0000	1.0000	1.0000	1.0000	1.0000	1.0000	1.0000

$$Z = CZ_a \tag{4-22}$$

where C is a correction factor depending on the concentration of nitrogen, the temperature, and the pressure.

The additive volume correction factor C ranges from 1.00 to 1.04 for temperatures and pressures ranging from 70 to 280°F and 14.4 to 5,000 psia, respectively. Values of C as a function of pressure and temperature for two different concentrations of nitrogen are tabulated in Table 4-7. Inspection of the data indicates that a maximum value of C occurs over a small pressure range for each of the temperatures tabulated. These maxi-

TABLE 4-8. MAXIMUM ADDITIVE VOLUME CORRECTION FACTORS[13]

Temp, °F	18,280 mole % nitrogen		7,907 mole % nitrogen	
	Occurs at pressure, psia	Value of C	Occurs at pressure, psia	Value of C
70	1,800	1.0322	2,200	1.0165
100	2,000	1.0250	1,800	1.0135
130	2,200	1.0202	2,800	1.0110
160	2,400	1.0175	2,200–2,800	1.0086
190	2,400	1.0153	2,000–2,200	1.0081
220	2,800	1.0134	2,000	1.0068
250	3,750	1.0115	2,200–2,600	1.0055
280	3,500	1.0096	2,400	1.0056

mums are indicative of the error associated with using the additive volume compressibility factor Z_a without correction. The maximum value of C and the pressure at which that maximum occurs are tabulated in Table 4-8 as a function of temperature. In general, less than 2 per cent error in the additive compressibility factor is possible at temperatures in excess of 130°F.

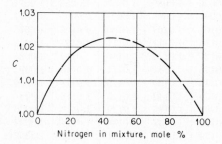

FIG. 4-32. Variation of additive volume-correction factor at a constant pressure and temperature due to nitrogen concentration. (*After Eilerts et al.*[13])

The effect of composition on the additive volume correction factor is illustrated in Fig. 4-32. Using Eilerts's data and procedure, four points are known for the value of C, at a constant temperature and pressure, as a function of the nitrogen concentration. By definition of the correction factor, C must be 1.0000 at 0 and 100 mole % nitrogen in the mixture. Values of C are

vailable from Table 4-7 for 7.907 and 18.280 mole % nitrogen. The
urve of Fig. 4-32 was constructed from this type of information. The
ashed portion represents the extrapolation necessary. The maximum
ʳas estimated from a similar curve presented by Eilerts,s which included
ata up to 56 mole % nitrogen. The maximum of that curve occurred
t about 46 mole % nitrogen.

The computation of the compressibility factor by Eilerts' method and
y the law of corresponding states is presented in Example 4-7 for a natural
as containing 10 mole % nitrogen. The Z factors compare as follows:

Z (Eilerts's method) = 0.892
Z (corresponding states) = 0.885
Difference = 0.77 per cent

Example 4-7. Calculation of the Compressibility Factor for a Natural
ias Containing Nitrogen.

. Eilerts' method

$$({}_pT_r)_g = \frac{T}{({}_pT_c)_g} = \frac{620}{367.25} = 1.69; \; ({}_pP_r)_g = \frac{P}{({}_pP_c)_g} = \frac{3,000}{670.61} = 4.47$$

$Z_g = 0.86$ (from Fig. 4-25)
$Z_n = 1.0855$ (from Fig. 4-31)
$Z_a = y_nZ_n + (1 - y_n)Z_g = 0.10000(1.0855) + 0.90000(0.86)$
$\quad = 0.10855 + 0.77400 = 0.88255$

Additive volume-correction factor (from Fig. 4-32)

$C = 1.0104$
$Z = CZ_a = 1.0104(0.88255) = 0.89173$

. Treating by corresponding states

$$({}_pT_r)_{\text{mix}} = \frac{T}{({}_pT_c)_{\text{mix}}} = \frac{620}{353.22}$$

$$({}_pP_r)_{\text{mix}} = \frac{P}{({}_pP_c)_{\text{mix}}} = \frac{3,000}{652.74} = (Z)_{\text{mix}} = 0.885 \text{ (from Fig. 4-25)}$$

Eilerts calculated a compressibility factor of 0.885 for the same gas composition reported in Example 4-7. The difference in the Z factor calculated
by Eilerts and that obtained in Example 4-7 resulted from slightly different critical properties and reading of the gas-compressibility-factor curve.
It then becomes apparent that if the engineer exercises due care, he can
have about a 1 per cent error regardless of the calculation method used.

Carbon Dioxide. The effect of carbon dioxide on compressibility factors
of hydrocarbon mixtures has not been so extensively investigated as has
hat of nitrogen. Sage and Lacey[15,16] have presented data on the binary

Example 4-7. (*Continued.*)

Component	(1) Mole fraction y_i	(2) Mole fraction y_i (of hydrocarbon fraction)	(3) T_c, °R	(4) P_c, psia	(5) $y_i T_c$ (2) × (3)	(6) $y_i P_c$ (2) × (4)	(7) $y_i T_c$ (1) × (3)	(8) $y_i P_c$ (1) × (4)
Nitrogen	0.10000	0.00000	227	492			22.70	49.20
Methane	0.82394	0.91549	343	673	314.01	616.12	282.61	554.51
Ethane	0.04324	0.04805	550	708	26.43	34.02	23.78	30.61
Propane	0.01766	0.01962	666	617	13.07	12.11	11.76	10.90
Isobutane	0.00443	0.00492	733	530	3.61	2.61	3.25	2.35
n-Butane	0.00378	0.00420	765	551	3.21	2.31	2.89	2.08
Isopentane	0.00225	0.00250	830	482	2.08	1.21	1.87	1.08
n-Pentane	0.00090	0.00100	847	485	0.85	0.49	0.76	0.44
Hexanes	0.00164	0.00182	914	434	1.66	0.79	1.50	0.71
Heptane plus	0.00216	0.00240	972	397	2.33	0.95	2.10	0.86

Temperature = 160°F; pressure = 3,000 psia.

$(_pT_c)_g = 367.25$ $(_pP_c)_g = 670.61$ $(_pT_c)_{mix} = 353.22$ $(_pP_c)_{mix} = 652.74$

$(_pT_c)_g = 367.25$

$(_pT_c = 367.25$

$(_pP_c =)$

$(_pT_{c})_{mix}$

systems methane–carbon dioxide and ethane–carbon dioxide. They suggested that a computation procedure based on partial residual volumes utilizing their basic data be used to correct volume calculations for impurities. The partial residual volume method is not presented in this text, but a carbon dioxide compressibility-factor chart developed from the data of Sage and Lacey is presented in Fig. 4-33. This chart can be used to cal-

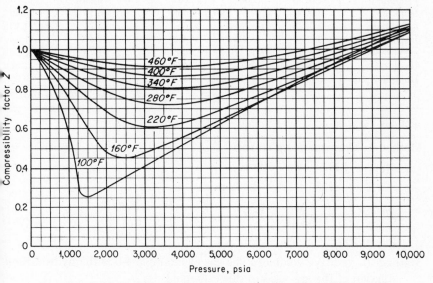

Fig. 4-33. Compressibility factor for carbon dioxide. (*From Olds et al.*[14])

culate, by the method of additive volumes, the compressibility factor for gases containing carbon dioxide. The additive volume compressibility factor is defined as follows:

$$Z_a = y_{CO_2}(Z_{CO_2}) + (1 - y_{CO_2})(Z_g) \qquad (4\text{-}21)$$

where y_{CO_2} = mole fraction of CO_2 in mixture
Z_{CO_2} = compressibility factor of pure CO_2 (from Fig. 4-33)
Z_g = compressibility factor of hydrocarbon fraction

This definition is analogous to that used when nitrogen was the impurity.

The compressibility factors for a gas containing 2.36 mole % carbon dioxide and having a hydrocarbon distribution equivalent to the gas of Example 4-6 are calculated by two methods in Example 4-8.

Example 4-8. Calculation of the Compressibility Factor for a Natural Gas Containing Carbon Dioxide.

Component	(1) Mole fraction y_i	(2) T_c, °R	(3) P_c, psia	(4) y_iT_c (1) × (2)	(5) y_iP_c (1) × (3)
Carbon dioxide	0.0236	548	1073	12.93	25.32
Methane	0.8481	343	673	290.90	570.77
Ethane	0.0595	550	708	32.73	42.13
Propane	0.0255	666	617	16.98	15.73
Isobutane	0.0047	733	530	3.45	2.49
n-Butane	0.0075	765	551	5.74	4.13
Isopentane	0.0030	830	482	2.49	1.45
n-Pentane	0.0021	847	485	1.78	1.02
Hexanes	0.0037	914	434	3.38	1.61
Heptanes plus	0.0223	1118^a	415^a	24.93	9.25
				$_pT_c = 395.31$	$_pP_c = 673.90$

^{a} From Fig. 4-26.

Method 1. Corresponding states

$M_{C7+} = 128$ $T = 600°R$

$SG_{C7+} = .8195$ $P = 3,000$ psia

$$_pT_r = \frac{T}{_pT_c} = \frac{600}{395.31} = 1.52 \qquad _pP_r = \frac{P}{_pP_r} = \frac{3,000}{673.90} = 4.45$$

$Z = 0.795$ (from Fig. 4-25)

Method 2. Additive volumes (The gas of Example 4-6 is the hydrocarbon fraction of the gas used in this example.)

Therefore

$Z_g = 0.805$

$Z = y_{CO_2}(Z_{CO_2}) + (1 - y_{CO_2})(Z_g) = 0.0236(0.453^*) + (0.9764)(0.805)$
$= 0.011 + 0.786 = 0.797$

The validity of the two methods presented in Example 4-8 are compared in Table 4-9 with observed data and with results from the residual volume method of calculation. All three methods of computation yield values which are about 2 per cent below observed data. As the accuracy of the three methods is comparable, the method of corresponding states is preferred because of its simplicity of application. If this method is applied with care, the ratios of observed Z to calculated Z reported in Table 4-9 can be used to adjust calculated values of Z for greater accuracy. For gases containing more than 4 mole % carbon dioxide, the deviations computed from observed values may be greater than indicated in this comparison. Laboratory tests should be made on such materials if accuracy greater than about 4 per cent is required.

* From Fig. 4-33.

TABLE 4-9. COMPARISON OF OBSERVED AND CALCULATED COMPRESSIBILITY
FACTORS FOR GAS CONTAINING CARBON DIOXIDE

Temp, °F	Pressure, psia	Z_{CO_2}	Observed,[a] Z	Calculated compressibility factors					
				Method of residual vol[a]		Method of corresponding states		Additive vol method	
				Z_3	Z/Z_3	Z_1	Z/Z_1	Z_2	Z/Z_2
100	500	0.841	0.928	0.916	1.01	0.915	1.01	0.919	1.01
	1,000	0.580	0.860	0.844	1.02	0.847	1.02	0.842	1.02
	2,000	0.299	0.775	0.753	1.03	0.750	1.03	0.747	1.04
	3,000	0.411	0.787	0.758	1.04	0.760	1.04	0.763	1.03
160	500	0.889	0.951	0.941	1.01	0.945	1.01	0.944	1.01
	1,000	0.759	0.908	0.893	1.02	0.895	1.01	0.894	1.02
	2,000	0.479	0.856	0.836	1.02	0.830	1.03	0.830	1.03
	3,000	0.479	0.860	0.836	1.03	0.830	1.04	0.830	1.04
220	500	0.921	0.967	0.957	1.01	0.962	1.01	0.962	1.01
	1,000	0.838	0.940	0.926	1.02	0.928	1.01	0.929	1.01
	2,000	0.684	0.910	0.889	1.02	0.888	1.02	0.886	1.03
	3,000	0.607	0.916	0.882	1.04	0.885	1.04	0.886	1.03

Gas analysis from which the above data were obtained

Component	Mole fraction[a]
Methane	0.83845
Ethane	0.06366
Propane	0.03744
Isobutane	0.00390
n-Butane	0.00951
Isopentane	0.00160
n-Pentane	0.00110
Hexanes plus[b]	0.00230
Carbon dioxide	0.04204

[a] Reported by Sage and Lacey.[15]
[b] Estimated mole wt, 95, and estimated SG, 0.68.

If both carbon dioxide and nitrogen are present in small concentration,
the method of corresponding states applied to the whole mixture of gases
will yield satisfactory results. For low concentrations of carbon dioxide and
moderate concentrations of nitrogen, Eilerts's[13] method is recommended.
The carbon dioxide can be treated in the method of corresponding states
along with nitrogen and the hydrocarbon gas. The final correction of the

additive of compressibility factor can be made as if carbon dioxide were a part of the hydrocarbon system; thus C corrects for nitrogen only.

$$Z_a = Z_{CO_2}(y_{CO_2}) + Z_n(y_n) + (1 - y_{CO_2} - y_n)Z_g$$
$$Z = C_n Z_a \tag{4-23}$$

Hydrogen Sulfide. Hydrogen sulfide is another impurity frequently present in natural gas. A hydrocarbon gas is termed a sour gas if it contains 1 grain of H_2S per 100 cu ft. Sour gases are corrosive and, if H_2S is in sufficient concentration, toxic. The hydrogen sulfide concentration must be reduced to specified limits prior to sales to transmission lines. The removal of H_2S can yield valuable by-products in the form of sulfur and sulfuric acid.

Few data are available on the effect of hydrogen sulfide on the compressibility factor of natural-gas mixtures. For small concentrations the critical constants for hydrogen sulfide can be used in the calculation of the pseudocritical properties of the mixture. Figure 4-25 can then be used in the normal manner. Insufficient data are available to determine a method for calculating compressibility factors for mixtures containing substantial amounts of hydrogen sulfide. In such instances Z factors should be determined by laboratory tests. For low concentrations, the additive volume method can be used in the same manner as when nitrogen and carbon dioxide were present. Compressibility factors for H_2S are presented in Fig. 4-34.

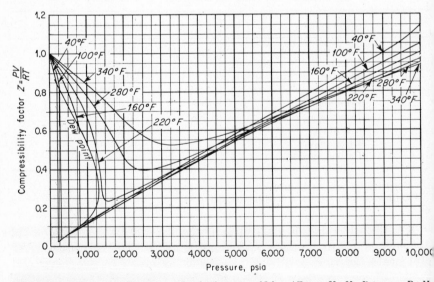

FIG. 4-34. Compressibility factor for hydrogen sulfide. (*From H. H. Reamer, B. H. Sage, and W. N. Lacey, "Volumetric Behavior of Hydrogen Sulfide," Ind. Eng. Chem., vol. 42, no. 1, p. 140, January, 1950.*)

Water Vapor. Water vapor, not ordinarily reported in a routine gas analysis, is a common impurity in natural gas. Gas samples are ordinarily dried before analysis; therefore, the water content of the gas is not determined.

Water has the highest critical temperature and pressure of any of the constituents commonly found in natural gases. The critical temperature of water is 1165°R or 705°F, which is much higher than the temperatures reported from wells drilled to 20,000 ft or more. Reservoir pressures are well in excess of the saturation pressure of water at prevailing reservoir temperatures.

The water content of natural gases can be estimated from the vapor pressure of water (see Table 4-10) at the prevailing temperature.

Dalton's law of partial pressures states that the total pressure of a confined mixture of gas is equal to the sum of the partial pressures of the individual constituents each taken alone in the same volume. Expressed mathematically,

$$P = \sum_{i=1}^{m} P_i \tag{4-24}$$

where P is the pressure of the mixture and P_i is the partial pressure of the ith constituent.

It follows from this statement that the volume fraction of a constituent in the vapor phase is given by

$$y_i = \frac{P_i}{P} \tag{4-25}$$

where y_i is the volume (or mole) fraction of the constituent in the vapor phase.

The partial pressure of water in the vapor phase is its vapor pressure at the prevailing temperature. Thus

$$y_w = \frac{P_{vw}}{P} \tag{4-26}$$

The laboratory analysis can be corrected for water content as follows:

$$(y_i)_c = (1 - y_w)(y_i)_L \tag{4-27}$$

where $(y_i)_c$ is the corrected mole fraction of any constituent other than water and $(y_i)_L$ is the mole fraction of that constituent from the laboratory analysis.

The water content of a gas is usually expressed in pounds of water per million standard cubic feet of hydrocarbon gas. This can be calculated from the mole fraction of water in the corrected analysis.

The weight of water per mole of the mixture is given by

$$y_w(M_w)$$

TABLE 4-10. VAPOR PRESSURE FOR WATER[17]

Temperature, °F	Pressure, psia
60	0.2561
70	0.3628
80	0.5067
90	0.6980
100	0.9487
110	1.274
120	1.692
130	2.221
140	2.887
150	3.716
160	4.739
170	5.990
180	7.510
190	9.336
200	11.525
210	14.123
220	17.188
230	20.78
240	24.97
250	29.82
260	35.43
270	41.85
280	49.20
290	57.55
300	67.01

where M_w is the molecular weight of water. The weight of water per mole of hydrocarbon gas is, then,

$$\frac{y_w(M_w)}{1 - y_w}$$

The number of pound moles of gas per million standard cubic feet (MMscf) at 14.65 psia and 60°F is

$$\frac{10^6 \text{ scf}}{380.69}$$

Therefore the pounds of water per MMscf of hydrocarbon gas W_{hc} is

$$W_{hc} = \frac{y_w(M_w)10^6}{(1 - y_w)380.69} \tag{4-28}$$

or substituting for y_w,

$$W_{hc} = \frac{P_{vw}M_w10^6}{(P - P_{vw})380.69}$$

which on reducing constant terms yields

$$W_{hc} = \frac{P_{vw}(4.73 \times 10^4)}{P - P_{vw}}$$ (4-29)

A sample calculation of a corrected analysis and the water content of a gas is presented in Example 4-9.

———•◦•———

Example 4-9. Correction of Gas Analysis for Water Vapor and Calculation of Water Content of Gas. USE $CHARTS$ $P274-5$

Components	Gas analysis from laboratory, mole fraction	Analysis corrected for water vapor, mole fraction
C_1	0.85	0.8488
C_2	0.07	0.0699
C_3	0.04	0.0399
C_4	0.03	0.0300
C_5	0.01	0.0100
H_2O		0.0014
	1.000	1.0000

Determination of mole fraction of water in vapor phase when gas is water-saturated at a reservoir temperature of 140°F and a reservoir pressure of 2,000 psia:

$$y_w = \frac{P_{vw}}{P} = \frac{2.887*}{2,000} = 0.0014435$$

Water content at 140°F and 2,000 psia, in pounds per MMcf.

$$W_{hc} = \frac{P_{vw}(4.73 \times 10^4)}{P - P_{vw}} = \frac{2.89(4.73 \times 10^4)}{2,000 - 2.89} = \frac{13,6697}{1997.11} = 68.45 \text{ lb/MMscf}$$

———•◦•———

A number of investigators found that at higher pressures and temperatures, the simple relations presented above yielded low water contents. This may be expected, as in the theory presented no provision was made for the attraction of the hydrocarbon molecules for the water molecules. McCarthy, Boyd, and Reid[18] summarized the available data in charts which are presented in Figs. 4-35 and 4-36.

The water content in pounds per MMscf can be estimated from the chart if the pressure and temperature are known. The results obtained from the chart and by calculations as illustrated in Example 4-9 are compared in Table 4-11. The charts are based on experimental data and have been verified by use in field applications. The engineer should choose to use the

* From Table 4-10.

charts when they are available. Not only are the data more accurate but the charts are easier to use.

Hydrates.   Water vapor is usually not considered in volume calculations as the concentrations are quite low. However, transmission lines require

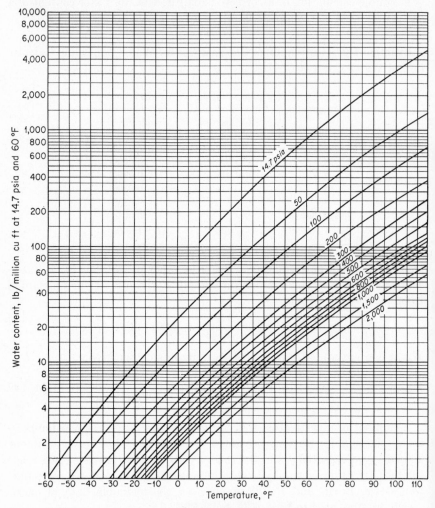

FIG. 4-35. Water-vapor content of natural gas at saturation. Temperature range: −60 to 110°F. Pressure range: 50 to 2,000 psia. (*From McCarthy et al.*[18])

gas to be dehydrated to avoid problems associated with hydrate formation. A natural-gas hydrate is a solid solution of water and natural gas with a "freezing" point which depends on the gas composition, the available water, the pressure, and the temperature. It is believed that the presence of free water is required for hydrate formation.

Since all reservoirs are believed to contain connate water, it is generally assumed that all mixtures which exist as a gas phase in the reservoir are saturated with water vapor. The amount of water contained in such a gas is determined by the prevailing reservoir temperature and pressure (see

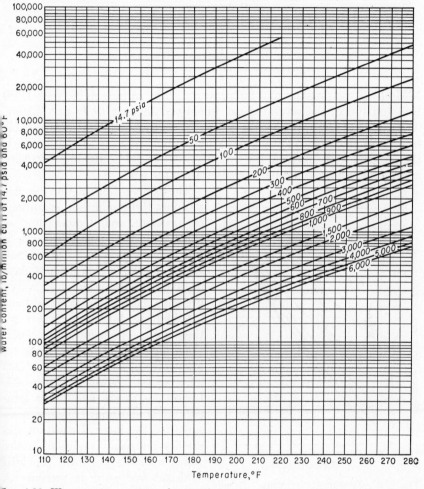

FIG. 4-36. Water-vapor content of natural gas at saturation. Temperature range: 110 to 280°F. Pressure range: 50 to 6,000 psia. (*From McCarthy et al.*[18])

Figs. 4-35 and 4-36). When the gas is produced to the surface, the temperature and pressure are reduced. The decrease in temperature tends to decrease the weight of water which can be maintained in the vapor state, while the decrease in pressure tends to increase the weight of water in a saturated gas. For example, consider a gas existing at 2,000 psia and 140°F at reservoir conditions. The water content (from Fig. 4-36) is 108 lb per

TABLE 4-11. WATER-VAPOR CONTENTS OF GAS
(Comparison of calculated and chart values, temperature 140°F)

Pressure, psi	Water content, lb/MMscf (from Fig. 4-36)	Calculated
2,000	108	68.45
1,000	174.0	137.09
200	830.0	693.51

MMscf. If the pressure and temperature of the surface choke were 1,000 psia and 100°F, the saturated gas would contain only 61 lb per MMscf. Thus, 47 lb of free water would be present for each million standard cubic feet of gas, and one condition for hydrate formation would be satisfied. If the pressure were 500 psia, the water content would be 108 lb per MMscf and no free water would be condensed from the gas.

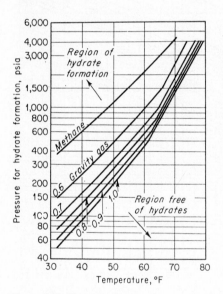

FIG. 4-37. Pressure-temperature curves for predicting hydrate formation. (*From Katz.*[20])

The composition of the gas determines the equilibrium conditions of pressure and temperature at which a hydrate can form even if free water is present. Carson and Katz[19] developed experimental data on hydrate equilibrium, and Katz[26] presented correlations of conditions for hydrate formation based on these data. The equilibrium conditions are dependent on pressure, temperature, and composition (gas gravity) as shown in Fig. 4-37. Conditions to the left (lower temperatures) and above (higher pressures) of the lines of constant gas gravity represent conditions under which hydrates can form provided sufficient free water is present. A 0.6-gravity gas at 1,000 psia and 60°F would be in the region of hydrate formation. If the pressure were reduced to 800 psia while the temperature were maintained at 60°F, the 0.6-gravity gas would be free of hydrate formation.

Another aspect of the problem of hydrates is the cooling associated with expanding a gas. The Joule-Thompson effect results from expansion across

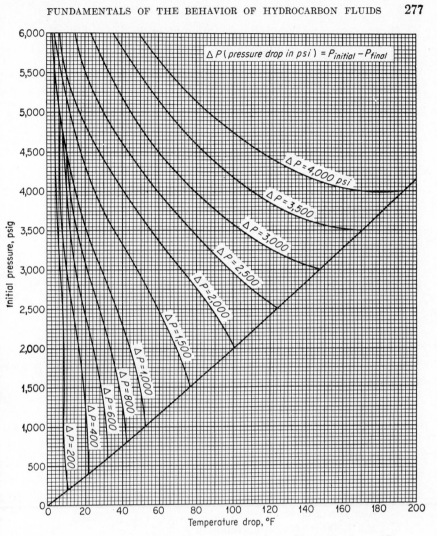

FIG. 4-38. Temperature change due to pressure drop. (*From Vondy et al.*[21])

a choke. The temperature drop associated with given pressure drops can be estimated from the chart (Fig. 4-38) presented by Vondy.[21]

A number of possible situations can be analyzed readily by using data from Figs. 4-35 through 4-38. Solutions for some of the more common problems are illustrated in Example 4-10.

Example 4-10. Determination of Hydrate-formation Conditions. A 0.7-gravity natural gas exists in the reservoir at 180°F and 3,000 psia.

1. How much water does the gas contain at reservoir conditions?

190 lb/MMscf (from Fig. 4-36)

2. What is the surface temperature of the gas if the well head pressure and temperature are 2,000 psia and 120°F for a pressure drop across the choke of
   a. 600 psi?

$\Delta t = 24°F$ (from Fig. 4-38)

Therefore, surface temperature = 96°F
   b. 1,500 psi?

$\Delta t = 69°F$ (from Fig. 4-38)

Therefore, surface temperature = 51°F

3. How much free water is present under conditions expressed in 2
   a. at the well head?

Gas contains 66 lb/MMscf (Fig. 4-36)

Therefore, free water = 190 − 66 = 124 lb/MMscf
   b. at the outlet of the choke if the pressure drop across choke is 600 psia?

Gas contains 45 lb/MMscf (Fig. 4-35)

Therefore, free water = 190 − 45 = 145 lb/MMscf
   c. at the outlet of the choke if the pressure drop across choke is 1,500 psia?

Gas contains 22 lb/MMscf (Fig. 4-35)

Therefore, free water = 190 − 22 = 168 lb/MMscf

4. Will hydrates be formed under conditions expressed in 2
   a. for a 600-psi pressure drop across choke?

Pressure = 1,400 psia, temperature = 96°F

Therefore (from Fig. 4-37), no hydrate will form even though free water is available
   b. for a 1,500-psi pressure drop across choke?

Pressure = 500 psia, temperature = 51°F

Therefore (from Fig. 4-37), hydrate will form, since conditions fall in hydrate region for a 0.7-gravity gas

Viscosity of Gases

The petroleum engineer deals continuously with problems of fluid flow. Oil-field fluids flow through the reservoir rock to the well bore, through the well bore to the surface, and thence to the separator and stock tank. The viscosity of a fluid is required to solve these many flow problems. Viscosity is defined loosely as the internal resistance of the fluid to flow.

A more rigorous definition can be developed mathematically. In Fig. 4-39 two layers of area A within a fluid separated by distance dy are shown to be in motion. The upper layer has a velocity $v + dv$, and the lower layer a velocity v. Because of the friction between the molecules of the fluid, a force F is required in the upper layer to maintain the difference in velocity dv between the layers.

It has been found experimentally that

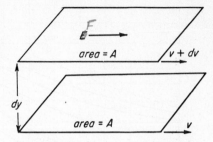

$$\frac{F}{A} \propto \frac{dv}{dy}$$

Fig. 4-39. Two layers of fluid in relative motion.

or

$$\frac{F}{A} = \mu \frac{dv}{dy} \tag{4-30}$$

where μ is a constant of proportionality, by definition the viscosity.

The units of viscosity can be readily determined from rearrangement of Eq. (4-30).

$$\mu = \frac{F/A}{dv/dy} \tag{4-31}$$

In the mass units of the cgs system,

let
$$F = 1 \text{ dyne}$$
$$A = 1 \text{ sq cm}$$
$$dv = 1 \text{ cm/sec}$$
$$dy = 1 \text{ cm}$$

Thus
$$\mu = \frac{1 \text{ dyne/sq cm}}{1 \text{ cm/(sec)/(cm)}} = \frac{1 \text{ dyne/sec}}{\text{sq cm}}$$

but $1 \text{ dyne} = \text{gm}/(\text{cm})/(\text{sec}^2)$

Therefore $\mu = 1 \text{ gm}/(\text{cm})(\text{sec}) = 1 \text{ poise}$

The poise is a large unit for hydrocarbon fluids under normal operating conditions. Therefore, viscosities are usually reported in centipoises. One poise equals 100 centipoises.

The viscosity of gases are difficult to measure accurately, particularly at elevated temperature and pressure. The engineer ordinarily must rely on correlations developed by careful experimentation rather than on limited laboratory measurements on the fluids at hand.

Bicher and Katz[22] presented the first correlations for hydrocarbon gases at elevated temperature and pressure. They found the viscosity to be a function of the temperature, pressure, and molecular weight (gas gravity)

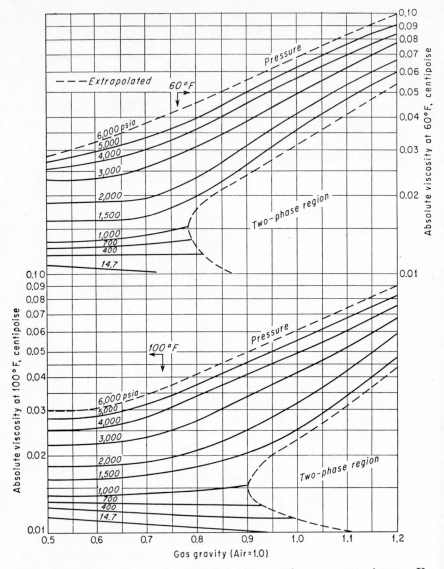

Fɪɢ. 4-40. Viscosity of natural gases. Data from methane-propane mixtures. Use only for samples containing less than 5 per cent nitrogen. Deviation, 5.8 per cent. (*From Bicher and Katz.[22]*)

of the gas. Correlation charts for four temperatures (60, 100, 200, and 300°F) are presented in Figs. 4-40 and 4-41. The charts are based on data from methane-propane mixtures. Katz stated that the correlations should be used for gases containing less than 5 per cent nitrogen. The average deviation of the data and the correlations is 5.8 per cent.

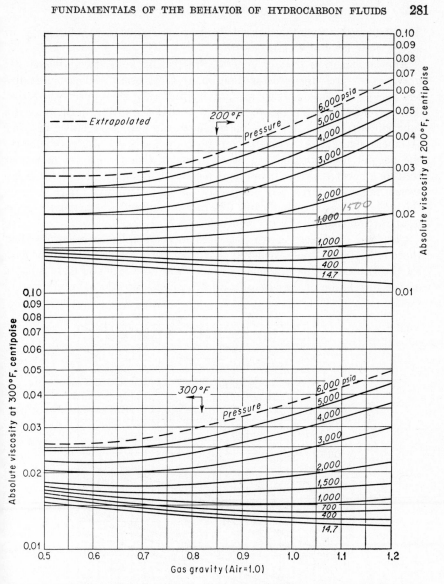

FIG. 4-41. Viscosity of natural gases. Data from methane-propane mixtures. Use only for samples containing less than 5 per cent nitrogen. Deviation, 5.8 per cent. (*From Bicher and Katz.*[22])

Several trends typical of gas viscosity can be noted from examination of Figs. 4-40 and 4-41. As the temperature is increased, the kinetic energy of the molecules increase. More collisions occur between the molecules; thus the viscosity is increased by an increase in temperature. At a constant temperature, an increase in pressure causes an increase in viscosity.

The distance between the molecules is decreased; thus more collisions occur at the same level of kinetic energy.

Another trend is that of generally increasing viscosity at higher gas gravities for pressures greater than about 1,000 psia. At lower pressures the trend is reversed, higher gas gravity materials have lower viscosities. These trends are temperature-sensitive in that the inversion pressure is a function of temperature. This phenomenon can also be explained in terms of kinetic energy. At a given temperature (same level of kinetic energy) the heavier molecules have a lesser velocity; therefore, fewer molecular collisions occur. As the pressure is increased, however, the distance between molecules is reduced sufficiently that the attractive forces between the molecules become significant. The heavier molecules have greater forces of attraction than the lighter molecules.

Carr and coworkers[23] have presented more complete correlations than those of Katz. It is also believed that their correlations yield better data for the viscosity of natural-gas mixtures. The correlation of Carr is based on the correlation of the viscosity ratio μ/μ_1 with pseudo-reduced pressure

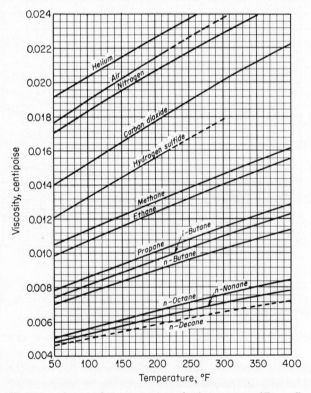

FIG. 4-42. Viscosity of natural gases at atmospheric pressure. (*From Carr et al.*[23])

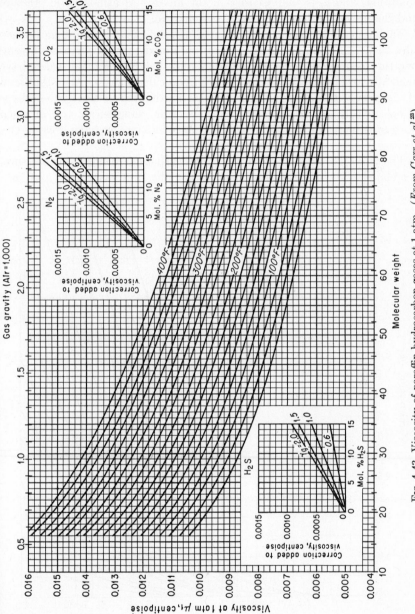

FIG. 4-43. Viscosity of paraffin hydrocarbon gases at 1 atm. (*From Carr et al.*[23])

and temperature, where μ is the viscosity of the mixture at the prevailing conditions and μ_1 is the viscosity of the mixture at atmospheric pressure and the prevailing temperature of the mixture.

The viscosity-temperature relation for several gases at atmospheric pressure is presented in Fig. 4-42. It can be noted that an increase in temperature results in increased viscosity and that hydrocarbon gases have generally lower viscosities than the nonhydrocarbon gases. Also, the trend of viscosity with molecular weight which is exhibited by the hydrocarbons is not reflected in the nonhydrocarbons. For example, air and nitrogen both are greater in molecular weight than methane yet have greater viscosities.

The viscosity relations of paraffin hydrocarbons at atmospheric pressure, correlated with temperature, molecular weight, and gas gravity, are pre-

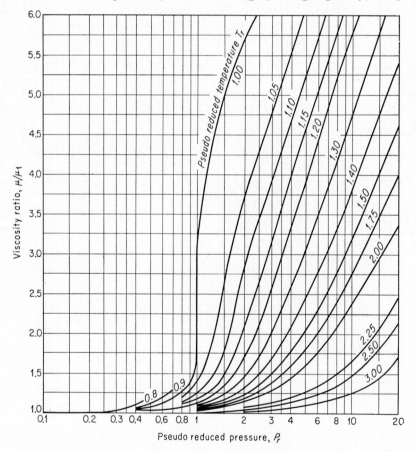

FIG. 4-44. Viscosity ratio versus pseudo-reduced pressure. (*From Carr et al.*[23])

sented in Fig. 4-43. The insert charts in the figure provide means of correcting the viscosity for the presence of nonhydrocarbon components.

Data for mixtures of gases read from Fig. 4-43 were compared by Carr with those calculated from a mixture rule proposed by Herning and Zipperer.[24] The viscosity μ_m of a mixture of gases is given by

$$\mu_m = \frac{\sum_{i=1}^{m} \mu_i y_i M_i}{\sum_{i=1}^{m} y_i M_i} \tag{4-32}$$

where μ_m = viscosity of mixture
μ_i = viscosity of ith component
M_i = molecular weight of ith component
y_i = mole fraction of ith component in mixture

The values obtained compared closely for all mixtures studied.

The correlations of viscosity ratio with pseudo-reduced properties are presented in Figs. 4-44 and 4-45. The pseudo-critical properties of mixtures can be calculated from gas analyses or read from Fig. 4-30 as previously explained. A sample calculation of gas viscosity by two methods

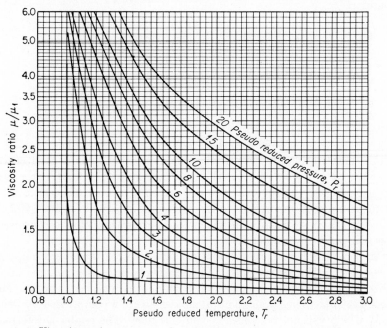

FIG. 4-45. Viscosity ratio versus pseudo-reduced temperature. (*From Carr et al.*[23])

is presented in Example 4-11. The deviation of the values determined was 3 per cent.

- - - • • • - - -

Example 4-11. Calculation of Gas Viscosity. A natural gas having a gas gravity of 0.9080 exists at a pressure of 3,010 psia and a temperature of 224°F. What is the viscosity of the gas?

Method 1. (from Figs. 4-40 and 4-41). The reservoir temperature lies be‑ tween the 200 and the 300°F chart of Fig. 4-41. Interpolation is required.

At 200°F, $\mu = 0.024$ cp
At 300°F, $\mu = 0.022$ cp

Therefore,

$$\mu \text{ at } 224°F = 0.024 - \frac{24}{100}(0.002) = 0.0235 \text{ cp}$$

Method 2. (from Figs. 4-43 to 4-45).

Mol wt $= 28.96G = 26.33$

Therefore,

$\mu_1 = 0.0119$ (from Fig. 4-43)
$_pT_c = 454$ and $_pP_c = 657$ (from Fig. 4-30)

Therefore,

$$_pT_r = \frac{684}{454} = 1.507 \text{ and } _pP_r = \frac{3,010}{657} = 4.581$$

$$\frac{\mu}{\mu_1} = 1.92 \text{ (from Fig. 4-44)}$$

$$\mu = \frac{\mu}{\mu_1}\mu_1 = 1.92(0.0119) = 0.0228 \text{ cp}$$

- - - • • • - - -

Summary of Properties of Gases

The quantitative analysis of the volumetric behavior of hydrocarbon gases has been developed in some detail. Methods of estimating the water content and the viscosity of gases have been presented. These properties are of paramount importance to the petroleum-reservoir engineer and pro‑ vide him the basic data with which to describe the behavior of gases in reservoir production processes.

Problems of gas compression and gas flow in pipes, which are frequently of concern to the petroleum production engineer, often require thermo‑ dynamic properties such as specific heats, enthalpy, and entropy. These properties and their applications alone constitute a comprehensive field of study which will not be considered in this text. The reader is referred to texts on thermodynamics and to the literature for information in this area of study.

PROPERTIES OF THE LIQUID STATE

A liquid is a fluid without independent shape but having a definite volume for a fixed mass under given conditions. Thus, a liquid will assume the shape of but not necessarily fill a vessel within which it is confined. A liquid seeks its own level and is incapable of supporting shearing stresses.

In the discussion of the properties of gases, the kinetic theory was introduced to describe theoretically the behavior of molecules in the gaseous state. Inherent in the treatment of gases was the assumption that the distance between molecules was great enough so that the attractive forces between molecules were negligible. In the case of liquids, the distance between molecules is much less and the force of attraction between molecules is substantial. The proximity of the molecules and the force of attraction between molecules in the liquid state result in substantial differences in the physical properties of liquids as compared with gases. Liquids have less fluidity (greater viscosity) and less volumetric changes with changes in temperature and pressure than do gases.

In general terms, liquids are frequently conceived to be either condensed vapors or melted solids. These concepts are useful in visualizing many processes to which substances are subjected. The continuity of the various concepts can be illustrated by the generalized phase diagram for a one-component system presented in Fig. 4-46. The curves represent conditions

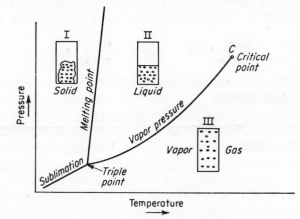

Fig. 4-46. Generalized phase diagram for a pure substance. (*Adapted from Daniels.*[2])

of pressure and temperature at which two phases can exist in equilibrium. At one point, the triple point, three phases can exist in equilibrium. Along the sublimation curve, solid and vapor coexist. The melting-point and vapor-pressure curves indicate coexistence of solid-liquid and liquid-vapor phases, respectively. Insets I, II, and III represent containers of equal size containing a fixed mass of the material in the state prescribed by the region within which the inset is located. Characteristic of the solid phase is the definite shape of the mass of substance, independent of the shape of the confining vessel. The liquid assumes the shape of the vessel but fills only a portion equal to the definite volume of liquid corresponding to the mass of material and the given conditions of pressure and temperature. The molecules are relatively close together in both the solid and liquid states. The vapor, inset III, occupies the entire vessel and assumes the shape of the vessel. The distance between molecules is substantially greater than for either solid or liquid.

Volumetric Behavior of Liquids

The volumetric behavior of liquids as a function of pressure and temperature has been studied intensively in many fields of science. Nevertheless, Daniels[2] states "the theory of liquids is in a much less satisfactory state than the theories of gases and crystals, but important progress is being made in our understanding of the structure of liquids." The state of progress is such that a simple generalized equation of state has not been developed for liquids.

As in the case of gases, the petroleum engineer is primarily concerned with the change in volume of liquids with a change in pressure and temperature. The state changes discussed in this section will be confined to those in which the material remains wholly within the liquid phase.

An ideal liquid is defined mathematically by the following partial derivatives:

$$\left(\frac{\partial V}{\partial T}\right)_P = 0 \quad \text{and} \quad \left(\frac{\partial V}{\partial P}\right)_T = 0 \tag{4-33}$$

where $(\partial V/\partial T)_P$ is the isobaric thermal expansion and $(\partial V/\partial P)_T$ is the isothermal compressibility. The relations expressed in Eq. (4-33) state that changes in volume with pressure and temperature are zero for an ideal liquid. Furthermore, an ideal liquid is conceived to have no internal friction between molecules; consequently, the viscosity of an ideal liquid is zero. Thus, the concept of an ideal liquid is of little quantitative value in predicting the volumetric behavior of liquids but does provide an insight into the behavior of liquids. The volume of real liquids does change with variations in pressure and temperature, but these changes are so small compared with gas that they are represented as a fractional change of some standard or reference volume.

A comparison of the relative changes in volume of liquid and gases can be obtained from inspection of Fig. 4-2. For the conditions specified by the segment of the 80°F isotherm between points A and B, ethane is in the gaseous state. For the conditions specified by the segment between points A and F, ethane is in the liquid state. The isothermal compressibility

$(\partial V/\partial P)_T$ can be represented in finite form as $(\Delta V/\Delta P)_T$. In the liquid state,

$$\Delta V = 0.005 \text{ cu ft/lb} \quad \text{and} \quad \Delta P \cong 135 \text{ psi}$$

or

$$\left(\frac{\Delta V}{\Delta P}\right)_T = 0.0000371 \text{ cu ft/(lb)(psi)}$$

in the gaseous state,

$$\Delta V \cong 0.067 \text{ cu ft/lb} \quad \text{and} \quad \Delta P \cong 75 \text{ psi}$$

or

$$\left(\frac{\Delta V}{\Delta P}\right)_T = 0.000894 \text{ cu ft/(lb)(psi)}$$

The gas is 24.1 times as compressible as the liquid in the ranges considered.

The isobaric thermal expansion can also be expressed in finite form as $(\Delta V/\Delta T)_P$. At a pressure of 700 psia and temperatures between 60 and 85°F, ethane is in the liquid state. The average isobaric thermal expansion is approximately

$$\frac{0.013}{25} = 0.00052 \text{ cu ft/(lb)(°F)}$$

Also at 700 psia and for temperatures between 100 and 160°F, ethane is a gas. The average isobaric thermal expansion is about

$$\frac{0.079}{60} = 0.00131 \text{ cu ft/(lb)(°F)}$$

The thermal expansion of the gas is about 2.5 times that of the liquid.

The effect of pressure on the specific volume of propane is shown in Fig. 4-47a. The curves are for constant temperatures and are called isotherms. The slope of an isotherm is the isothermal compressibility. The compressibility is always negative, indicating that increasing pressures result in smaller volumes. The magnitude of the isothermal compressibility decreases with increasing pressure, whereas it increases with increasing temperature. Thus, the effect of pressure variations are greater at low pressure and high temperature. The change in slope with pressure can be stated mathematically as the second partial derivative.

$$\frac{\partial(\partial V/\partial P)_T}{\partial P} = \left(\frac{\partial^2 V}{\partial P^2}\right)_T = \text{positive number} \quad (4\text{-}34)$$

Plots of the compressibility and change in compressibility are shown in Fig. 4-47b and c. It is noted from these curves that the greater the pressure, the smaller the effect of changes in pressure on the compressibility of the fluid.

At low temperatures, the isothermal compressibility is very nearly constant. For example, the 70°F isotherm in Fig. 4-47a is essentially a straight

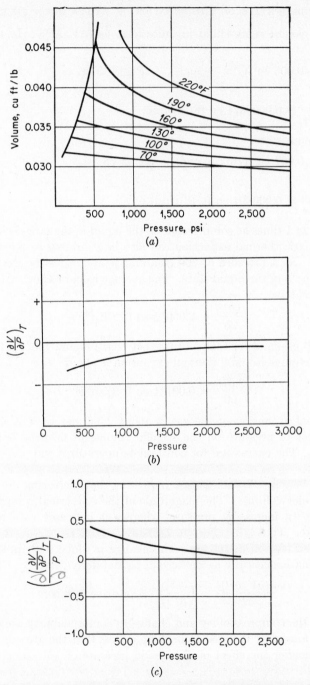

Fig. 4-47. (a) Effect of pressure upon the volume of liquid propane. (*From Sage and Lacey.*[11]) (b) Isothermal compressibility of liquid propane at 160°F. (c) The change in the isothermal compressibility of liquid propane at 160°F.

ne. Hence, the variation in the specific volume of the fluids at 70°F can
e represented (in the range 100 to 2,500 psi) by

$$v = v_0 + bP$$

here v_0 is the intercept with the y axis at zero pressure and b is the average
lope or isothermal compressibility. The volume rather than the specific
olume is usually the desired quantity, so that

$$V = wv \qquad V_0 = wv_0$$

here w is the weight of liquid. Thus

$$V = wv_0 + wv_0 \frac{bP}{v_0}$$

r

$$V = V_0 \left(1 + \frac{b}{v_0} P\right) \tag{4-35}$$

he familiar coefficient of compressibility c is then stated as

$$c = \frac{b}{v_0}$$

Equation (4-35) can also be expressed in the form

$$V = V_0[1 - c_T(P - P_0)] \tag{4-36}$$

here c_T = isothermal coefficient of compressibility at temperature T

V_0 = reference volume at temperature T and pressure P_0

V = volume at temperature T and pressure P

A definition of c which is satisfied whether or not the isotherms are
traight lines is

$$c = -\frac{1}{V} \left(\frac{\partial V}{\partial P}\right)_T \tag{4-37}$$

he coefficient of compressibility as defined by Eq. (4-37) is a point func-
ion and can be computed from the slope of isothermal specific volume
urve for each value of pressure. The minus sign arises from the mathe-
natical convention with respect to slope. As the slope of the curves are
lways negative, v is always a positive number and it is desired that c be
, positive number; then c must be defined as the negative of the slope
livided by the specific volume.

The coefficient of compressibility of a fluid is the reciprocal of the bulk
nodulus. The bulk modulus is analogous to the modulus of elasticity of a
olid. That is,

$$\frac{1}{c} = \left(\frac{\text{stress}}{\text{strain}}\right)_T$$

he stress in a fluid is the change in pressure from some reference pressure
P_0, and the strain is the change in volume per unit volume at the reference
onditions. Therefore

$$\frac{1}{c} = \frac{P - P_0}{\Delta V / V_0} \qquad (4\text{-}38$$

The bulk modulus has the dimensions of pressure, while the coefficient o
compressibility has the dimensions of reciprocal pressure.

The isobaric variation in specific volume of liquid propane is shown in
Fig. 4-48a. The isobaric thermal expansion $(\partial V / \partial T)_P$ is the slope of th
isobars (lines of constant pressure). The slope of the isobars (Fig. 4-48b

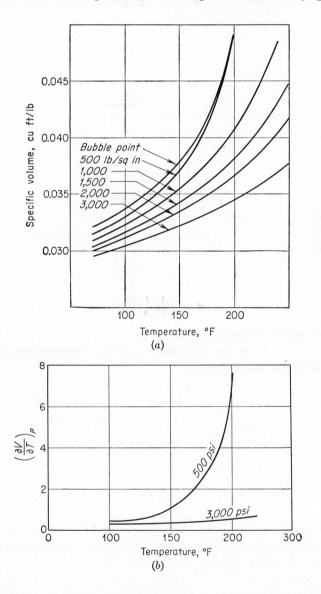

(a)

(b)

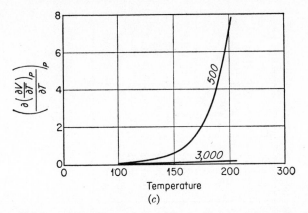

Temperature

(c)

Fig. 4-48. (a) Volume of liquid propane. (*From Sage and Lacey.*[11]) (b) Isobaric expansion of liquid propane. (c) Variation of isobaric expansion of liquid propane.

increases with increasing temperature and decreases with increasing pressure. The slope is at all times positive, indicating that an increase in temperature increases the volume of the fluid. Stated analytically, the change of slope with temperature is the second partial derivative.

$$\frac{\partial(\partial V/\partial T)_P}{\partial T} = \left(\frac{\partial^2 V}{\partial T^2}\right)_P = \text{positive number} \qquad (4\text{-}39)$$

Results of Eq. (4-39) (Fig. 4-48c) confirm the observation that the isobaric expansion increases with increasing temperature.

The coefficient of isobaric thermal expansion β is defined as

$$\beta = \frac{1}{V}\left(\frac{\partial V}{\partial T}\right)_P \qquad (4\text{-}40)$$

The isobars of Fig. 4-48a are sensibly straight over a part of the range of pressure and temperature. Therefore, a relation between the volume at a reference temperature and that at another temperature can be written as

$$V = V_0[1 + \beta_p(T - T_0)] \qquad (4\text{-}41)$$

where β_p = isobaric coefficient of expansion at pressure P

V_0 = reference volume at pressure P and temperature T_0
V = volume at pressure P and temperature T

It is frequently convenient to define the volume relations of liquids in terms of specific volumes or densities.

For thermal expansion

$$v = v_0[1 + \beta_p(T - T_0)] \qquad (4\text{-}42)$$

and
$$\rho = \rho_0[1 - \beta_p(T - T_0)] \qquad (4\text{-}43)$$

For compressibility,

$$v = v_0[1 - c_T(P - P_0)] \qquad (4\text{-}44)$$

TABLE 4-12. PHYSICAL PROPERTIES OF SELECTED PARAFFIN HYDROCARBONS WHICH ARE LIQUIDS AT ATMOSPHERIC CONDITIONS[25]

Hydrocarbon	Mole wt	Boiling point at 14.696, psia	Vapor pressure, psia at 100°F	Critical pressure, psia	Critical temp, °R	Critical specific volume, cu ft/lb	Liquid density, lb/cu ft[b]	Gas density, lb/cu ft[b]	Isobaric coefficient of expansion, 1/°F	Viscosity at 68°F, cp[f]	Heat of vaporization, Btu/lb
n-Pentane	72.146	96.9	15.57	489.5	845.9	0.0690	39.29	0.19	0.00089[c]	} 0.240	153.8
Isopentane	72.146	82.1	20.44	483.0	830.0	0.0685	38.90	0.19			145.9
n-Hexane	86.172	155.7	4.96	439.7	914.5	0.0685	41.35	0.23	0.00078[c]	0.326	144.2
n-Heptane	100.198	209.2	1.62	396.9	972.6	0.0682	42.85	0.26	0.00068[d]	0.416	136.2
n-Octane	114.224	258.2	0.54	362.1	1025.2	0.0682	44.01	0.30	0.00065[d]	0.542	131.9
n-Nonane	128.250	303.4	0.18	345	1073	0.0673	44.94	0.34	0.00062[e]	0.620[a]	126.9
n-Decane	142.276	345.2	0.07	320	1115	0.0671	45.74	0.38	0.00059[e]	0.770	120.2

[a] At 72°F.
[b] At 14.696 psia and 60°F.
[c] At atmospheric pressure and temperature range 50–100°F.
[d] At atmospheric pressure and temperature range 32–86°F.
[e] At atmospheric pressure and temperature range 32–112°F.
[f] From Eshbach.[26]

and $$\rho = \rho_0[1 + c_T(P - P_0)] \tag{4-45}$$

The isotherms of Fig. 4-47a can be fitted by an exponential relation over a greater range of pressure than that for which Eq. (4-45) is valid. Therefore, a relation for density can be expressed as

$$\rho = \rho_0 e^{c(P)} \qquad \text{or} \qquad \rho = \rho_0 e^{c(P - P_0)} \tag{4-46}$$

The coefficients c and β are functions of temperature, pressure, and liquid composition. Although sensibly constant over certain ranges of pressure or temperature, these coefficients must be determined experimentally or from correlations based on experiment.

The more important physical constants of common hydrocarbons which exist as liquids at atmospheric conditions are presented in Table 4-12.

Density of Hydrocarbon Mixtures in the Liquid State

The stock-tank liquids obtained from production of petroleum are complex mixtures of hydrocarbons. The density (or API gravity) of such a mixture is readily measured by means of hydrometers or other devices. Frequently the engineer must estimate the density of a mixture from an analysis which has been computed from other properties of the system.

Typical analyses of both a stock-tank liquid and a reservoir liquid are presented in Table 4-13. The stock-tank liquid is comprised largely of propanes and heavier components, while reservoir liquid contains appreciable quantities of methane and ethane.

TABLE 4-13. TYPICAL LIQUID ANALYSES

Component	Stock-tank liquid, mole fraction[a]	Reservoir liquid, mole fraction[b]
Methane	0.0019	0.3396
Ethane	0.0098	0.0646
Propane	0.0531	0.0987
Butanes	0.0544	0.0434
Pentanes	0.0555	0.0320
Hexanes	0.0570	0.0300
Heptanes plus	0.7681[c]	0.3917[c]
	1.0000	1.0000

[a] Exists at 14.7 psia and 60°F.
[b] Exists at 3,614 psia and 220°F.
[c] Properties of heptanes-plus fraction: Mol wt = 263; density = 55.28 lb/cu ft at 14.7 psia and 60°F.

Methane and ethane are gases at ordinary pressures and temperatures. Because of the volatility of these mat_rials, the densities of methane and ethane depend on the composition of the heavier fractions of the liquid.

Heavier molecules have a greater force of attraction, so that methane or ethane molecules are, in effect, compacted to a greater extent in the presence of heavy molecules than in the presence of light molecules.

Standing and Katz[27] presented correlations of the apparent density of methane and ethane with the density of the system (see Fig. 4-49). Based on these data and assuming that propanes and heavier components followed the rule of additive volumes, they developed a method for com-

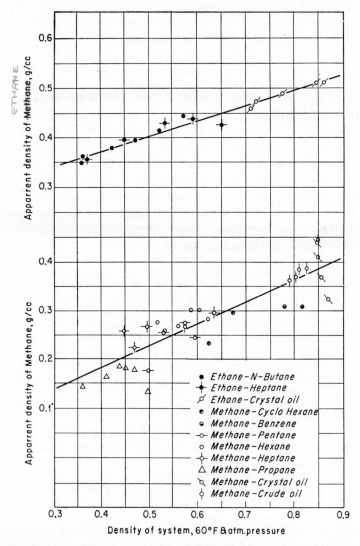

FIG. 4-49. Variation of apparent density of methane and ethane with density of the system. (*Standing and Katz.*[27])

uting the density of mixtures of hydrocarbons. The density of a system containing methane and ethane is correlated (Fig. 4-50) with the density of the propanes-plus fraction, the weight per cent ethane in the ethanes plus, and the weight per cent methane in the system.

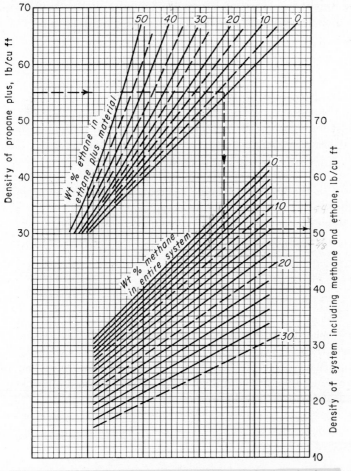

FIG. 4-50. Pseudo-liquid density of systems containing methane and ethane. (*From Standing.*[3])

For stock-tank liquids which have low concentrations of methane and ethane, the additive volume method of calculating density is satisfactory. The method can be developed as follows:

The weight in pounds of a component in 1 lb-mole of a mixture is equal to the product of the molecular weight and the mole fraction of that component in the mixture. Therefore,

$$W_i = x_i M_i \qquad (4\text{-}47$$

where W_i = weight of ith component in 1 lb-mole of mixture
x_i = mole fraction of ith component in mixture
M_i = molecular weight of ith component

The weight in pounds of 1 lb-mole of mixture is simply the sum of th
weights of the components. Thus

$$W_m = \sum_{i=1}^{m} x_i M_i \qquad (4\text{-}48$$

where W_m is the weight in pounds of 1 lb-mole of the mixture.

The volume of a component in a mixture is the product of the weigh
of that component in the mixture and the specific volume of that compo
nent at the prevailing condition of pressure and temperature. Thus

$$V_i = x_i M_i v_i = W_i v_i \qquad (4\text{-}49$$

where V_i is the volume of the ith component in 1 lb-mole of mixture an(
v_i is the specific volume of the ith component.

Applying the concept of additive volumes,

$$V_m = \Sigma V_i$$

and the density of the mixture is given by

$$\rho_m = \frac{W_m}{V_m} \qquad (4\text{-}50$$

The molecular weight and specific volume at 14.65 psia and 60°F can b
obtained from Table 4-4 for the lighter hydrocarbons, methane througl
heptane. The molecular weight and specific gravity of the heptanes-plu:
fraction are normally determined in the laboratory and reported as a par
of the fractional fluid analysis.

The method of Standing and Katz assumes that the propanes-plus frac
tions can be treated by additive volumes. The above-cited equations mus
be modified as follows:

$$W_{\text{C3+}} = \sum_{i=3}^{m} x_i M_i \qquad (4\text{-}51$$

and

$$V_{\text{C3+}} = \sum_{i=3}^{m} V_i \qquad (4\text{-}52$$

where the summations are taken over the C_3 (propane) and heavier com
ponents. Two additional definitions must be stated in mathematical forn

use the corrective charts for methane and ethane. The weight per cent
ethane in the ethanes plus is

$$(\text{wt } \% \text{ C}_2)_{C2+} = \frac{\text{wt of C}_2}{\text{wt of C}_{2+}} = \frac{x_2 M_2}{\displaystyle\sum_{i=2}^{m} x_i M_i} \tag{4-53}$$

and the weight per cent methane in the system is

$$(\text{wt } \% \text{ C}_1)_{C1+} = \frac{\text{wt of C}_1}{\text{wt of C}_{1+}} = \frac{x_1 M_1}{\displaystyle\sum_{i=1}^{m} x_i M_i} \tag{4-54}$$

The calculation of the density, at atmospheric pressure and 60°F, of the
stock-tank liquid from Table 4-13 is given in Example 4-12. The density
is calculated both by additive volumes and by the method of Standing and
Katz. It can be noted from Example 4-12 that the results from the two
methods differ only by 0.02 per cent. This result is because of the small
percentage of methane and ethane in the mixture.

Example 4-12.   Calculation of Liquid Density from Stock-tank Liquid
Analysis.

(1)	(2)	(3)	(4)	(5)	(6)
Component	Mole fraction in liquid phase x_i	Mole wt M_i	Relative weight, lb/mole, $x_i M_i$ (2) × (3)	Liquid density, lb/cu ft at 60°F and 14.65 psia	Liquid volume, cu ft/mole (4) ÷ (5)
Methane, C_1	0.0019	16.04	.0305	(18.70)	(0.0016)
Ethane, C_2	0.0098	30.07	.2947	(23.26)	(0.0127)
Propane, C_3	0.0531	44.09	2.3412	31.64	0.0740
Butanes, C_4	0.0544	58.12	3.1617	35.71[b]	0.0885
Pentanes, C_5	0.0555	72.15	4.0043	39.08[b]	0.1025
Hexanes, C_6	0.0570	86.17	4.9203	41.36	0.1190
Heptanes plus, C_{7+}	0.7681	263[a]	202.0366	55.28[a]	3.6548
Total			216.7893		4.0388
					(4.0531)

[a] From Table 4-13.
[b] Average of iso and normal.
Column 5 from Table 4-4.

Method 1. Additive volumes

$$\text{Density of } C_{1+} = \frac{216.7893}{4.0531} = 53.49 \text{ lb/cu ft}$$

Method 2. Standing and Katz apparent density correlation

$$\text{Density of } C_{3+} = \frac{\sum\limits_{i=3}^{m} x_i M_i}{\sum\limits_{i=3}^{m} \frac{x_i M_i}{\rho_i}} = \frac{(216.7893 - 0.3252) \text{ lb/mole}}{4.0388 \text{ cu ft/mole}}$$

$$= 53.59 \text{ lb/cu ft}$$

$$\text{Wt } \% \ C_2 \text{ in } C_{2+} = \frac{x_2 M_2}{\sum\limits_{i=2}^{m} x_i M_i} 100 = \frac{(0.2947) \text{ lb/mole } (100)}{(216.7893) - (0.0305) \text{ lb/mole}}$$

$$= 0.136$$

$$\text{Wt } \% \ C_1 \text{ in } C_{1+} = \frac{x_1 M_1}{\sum\limits_{i=1}^{m} x_i M_i} 100 = \frac{(0.0305) \text{ lb/mole } (100)}{(216.7893) \text{ lb/mole}} = 0.014$$

Density of C_{1+} = 53.5 lb/cu ft (from Fig. 4-50)

Compressibility of Liquids:

Data on the isothermal compressibility of pure hydrocarbons have not been compiled. However, extensive data are available in the literature on the effect of pressure on the specific volume at 60°F of pure hydrocarbon and hydrocarbon mixtures. Standing and Katz correlated the available data in the form of density-correction curves. The curves (Fig. 4-51) enable the engineer to correct a known density at 14.7 psia and 60°F to a density a a desired pressure and also at 60°F. The density corrections from Fig. 4-5 are adequate for most engineering calculations.

As has been discussed previously in this chapter, many reservoirs contain undersaturated oils. An undersaturated oil is a compressed liquid in the pressure range above the bubble-point pressure. The coefficient of isothermal compressibility can be used to describe the volume change of such materials above the bubble-point pressure.

Trube[28] has reported a method of calculating the compressibility of undersaturated liquids based on pseudo-reduced properties. The pseudo reduced compressibility c_r is defined:

$$c_r = c(_p P_c) \tag{4-55}$$

Since $$P_r = \frac{P}{_p P_c}$$

$$c_r P_r = cP = c_{11} P_1 = c_{12} P_2 = c_{13} P_3 \tag{4-56}$$

Therefore if the pressure temperature and critical properties of a liquid

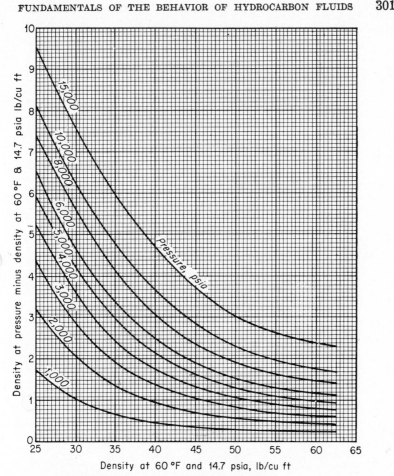

FIG. 4-51. Density correction for compressibility of liquids. (*From Standing.*[3])

are known, the compressibility can be determined from Eq. (4-56) and the correlation of c_r with pseudo-reduced temperature and pressure presented in Fig. 4-52.

At the higher reduced pressures an appreciable segment of the constant reduced-temperature lines have a constant slope. An equation for the straight-line segment has the form

$$\log c_r = n \log P_r + \log a$$

or

$$c_r = aP_r{}^n \qquad (4\text{-}57)$$

where n is the slope of the line and a is the intercept of the straight-line segment at $P_r = 1$. Substituting from Eq. (4-57) for c_r in Eq. (4-56)

$$(aP_r{}^n)P_r = cP$$

or

$$aP_r{}^{n+1} = cP$$

from which can be obtained

$$\frac{a}{P_c^{n+1}} = cP^{-n} \qquad (4\text{-}58)$$

For a particular liquid,

$$\frac{a}{P_c^{n+1}} = \text{constant} = A = cP^{-n}$$

The slope n is negative. The compressibility c is thus shown to decrease with increasing pressure. Furthermore, the slopes of the straight-line seg-

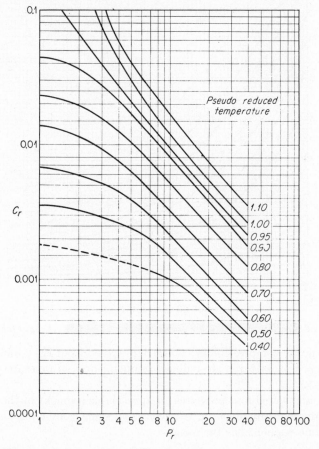

Fig. 4-52. Pseudo-reduced compressibility of undersaturated hydrocarbon liquids. (*From Trube.*[28])

ments of Fig. 4-52 are very close to minus one (-1). If it is assumed that $n = -1$, then

$$A = cP = c_1 P_1 = c_2 P_2 \qquad (4\text{-}59)$$

The pseudo-reduced properties can be calculated from the liquid analysis in the manner described previously for gases. These values can be used together with Fig. 4-52 to calculate the compressibility of the fluid. For many liquids adequate estimates of critical properties are not available. Trube also developed procedures for estimating the critical properties of reservoir liquids.

Thermal Expansion of Liquids. The most frequent application of the thermal expansion of liquids is in correcting the volume and density of stock-tank liquids to 60°F, the standard temperature. The Bureau of Standards published in *Circular* C-410[29] extensive tables for this purpose. Abridged volume-correction tables are presented in the ASTM[30] and Tag[31] manuals. Coefficients of thermal expansion at atmospheric pressure are listed in Table 4-12 for the paraffin hydrocarbons pentane through decane. For the more complex stock-tank oils, the coefficients of thermal expansion

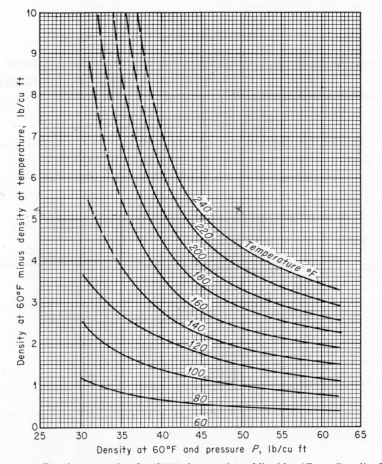

FIG. 4-53. Density correction for thermal expansion of liquids. (*From Standing.*[3])

are correlated with the API gravity of the oil. In Table 4-14 are listed the average coefficients of thermal expansion used in preparing abridged volume-correction tables for oils.

TABLE 4-14. AVERAGE COEFFICIENTS OF THERMAL EXPANSION FOR CRUDE OILS AT ATMOSPHERIC PRESSURE[30]

Group No.	Coefficient of thermal expansion at 60°F	Corresponding gravity, °API	Gravity range of group, °API
0	0.00035	6	Up to 14.9
1	0.00040	22	15–34.9
2	0.00050	44	35–50.9
3	0.00060	58	51–63.9
4	0.00070	72	64–78.9
5	0.00080	86	79–88.9
6	0.00085	91	89–93.9
7	0.00090	97	94–99.9

For correction of liquid densities for thermal expansion, Fig. 4-53 can be used. The chart yields satisfactory results except in the low-density and high-temperature regions. This chart, like Fig. 4-51, is based on a large number of observations of specific volumes of hydrocarbons. The density at 60°F is the only composition parameter.

The calculation of the density of a liquid at reservoir conditions from a fractional analysis requires data on both compressibility and thermal expansion. The procedure for such a calculation is shown in Example 4-13, using the reservoir liquid of Table 4-13.

Example 4-13. Calculation of Density of Reservoir Liquid. Reservoir conditions = 3,614 psia, 220°F.

(1) Component	(2) Reservoir fluid analysis Z_i	(3) Mol wt	(4) Relative weights, lb/mole (2) × (3)	(5) Density at 60°F and 14.7 psia, lb/cu ft[a]	(6) Liquid volume, cu ft/mole (4) ÷ (5)
C_1	0.3396	16.04	5.4472		
C_2	0.0646	30.07	1.9425		
C_3	0.0987	44.09	4.3517	31.64	0.1375
C_4	0.0434	58.12	2.5224	35.71[b]	0.0706
C_5	0.032	72.15	2.3088	39.08[b]	0.0591
C_6	0.03	86.17	2.5851	41.36	0.0625
C_{7+}	0.3917	263	103.0171	55.28[c]	1.8636
Total			122.1748		2.1933

[a] From Table 4-4.   [b] Average iso and normal.   [c] From Table 4-13.

$$\text{Density of } C_{3+} = \frac{\sum\limits_{i=3}^{n} x_i M_i}{\sum\limits_{i=3}^{n} \dfrac{x_i M_i}{\rho_i}} = \frac{(122.1748 - 7.3897) \text{ lb/mole}}{2.193 \text{ cu ft/mole}}$$

$$= 52.34 \text{ lb/cu ft}$$

$$\text{Wt \% } C_2 \text{ in } C_{2+} = \frac{x_2 M_2}{\sum\limits_{i=2}^{i=n} x_i M_i} = \frac{(1.9425) \text{ lb/mole } (100)}{(122.1748 - 5.4472) \text{ lb/mole}}$$

$$= 1.66 \text{ per cent}$$

$$\text{Wt \% } C_1 \text{ in total} = \frac{x_1 M_1}{\sum\limits_{i=1}^{n} x_i M_i} = \frac{(5.4472) \text{ lb/mole } (100)}{122.1748 \text{ lb/mole}} = 4.46 \text{ per cent}$$

Density of liquid including C_1 and C_2 at 60°F and 14.7 psia = 49.4 lb/cu ft (from Fig. 4-50)

Pressure correction from 14.7 to 3614 psia = 1.0 (from Fig. 4-51)
Therefore,

$$\rho_{C_{1+}} = 49.4 + 1.0 = 50.4 \text{ lb/cu ft at 60°F and 3,614 psia}$$

Temperature correction from 60 to 220°F at 3,614 psia = 3.78 (from Fig. 4-53)
Therefore,

$$\rho_{C_{1+}} = 50.4 - 3.78 = 46.62 \text{ lb/cu ft at 220°F and 3,614 psia}$$

Surface Tension

The surface tension at the interface between a liquid phase and an equilibrium vapor phase is a function of pressure, temperature, and the compositions of the phases.

For pure substances only the temperature or pressure must be specified to define the surface tension. As a tension exists only if two phases are present, then either temperature or pressure can be specified and the other will be defined by the vapor-pressure curve. The surface tensions of various pure paraffin hydrocarbons are shown as a function of temperature in Fig. 4-54. The value of zero surface tension occurs at the critical point of the hydrocarbon.

The surface tensions of mixtures of hydrocarbons have been investigated experimentally by Katz,[33] who, from the experimental data, developed a

procedure for calculating surface tension. The method is based on th
parachor and the equation proposed by Sugden[34] relating the surface ten
sion to the properties of the liquid and vapor phases.

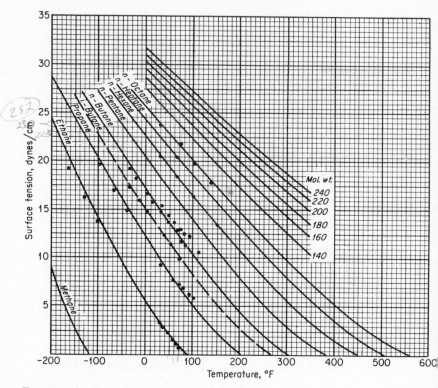

FIG. 4-54. Surface tension of paraffin hydrocarbons. (*From Katz and Saltman.*[32])

For a pure material,

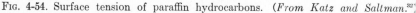

$$\sigma^{1/4} = P_{ch}\frac{\rho_L - \rho_v}{M}\tag{4-60}$$

where P_{ch} = parachor
     ρ_L = density of the liquid phase, gm/cc
     ρ_v = density of the vapor phase, gm/cc
     σ = surface tension, dynes/cm
     M = molecular weight

Parachors for pure hydrocarbons, nitrogen, and carbon dioxide are given
in Table 4-15. A correlation of the parachor with molecular weight is presented in Fig. 4-55. For a mixture the surface tension is defined by the
following relation:

$$\sigma^{\frac{1}{4}} = \sum_{i=1}^{m} P_{chi} \left(x_i \frac{\rho_L}{M_L} - y_i \frac{\rho_v}{M_v} \right) \qquad (4\text{-}61)$$

where $\quad P_{chi} =$ parachor of ith component

$\quad x_i$ and $y_i =$ mole fractions of ith component in liquid and vapor phases, respectively

$\quad \rho_L$ and $M_L =$ density and molecular weight, respectively, of liquid phase

$\quad \rho_v$ and $M_v =$ density and molecular weight, respectively, of vapor phase

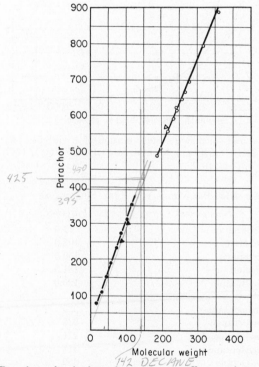

FIG. 4-55. Parachors for hydrocarbons ●, n-paraffins; ○, heptanes plus of Ref. 7; ▲, gasolines; △, crude oil. (*From Katz et al.*[33])

The densities ρ_L and ρ_v must be in grams per cubic centimeter to use in Eq. (4-61) with the parachors of Table 4-15 or Fig. 4-55.

The calculation of surface tension is presented in Example 4-14. In Fig. 4-56 are presented comparisons of experimental and calculated surface tensions of mixtures of hydrocarbons.

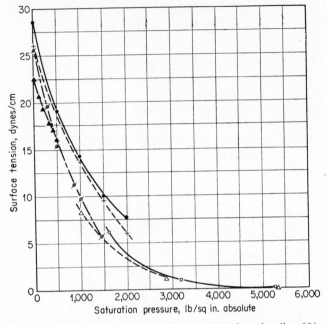

Fig. 4-56. Experimental and calculated surface tension of crude oils. ($\bullet$) (+) (ϕ) experimental data at 88°F; ($\blacktriangle$), experimental data at 95°F; ($\circ$), ($\triangle$), calculated data at 120°F. (*After Katz et al.*[33])

TABLE 4-15. PARACHORS FOR PURE SUBSTANCES

Component	Parachor
Methane	77.0
Ethane	108.0
Propane	150.3
Isobutane	181.5
n-Butane	190.0
Isopentane	225
n-Pentane	232
n-Hexane	271
n-Heptane	311
n-Octane	352
Nitrogen (in *n*-heptane)	41.0
Carbon dioxide	78.0

Example 4-14. Calculation of Surface Tension of Crude Oil Containing Dissolved Gases. Equilibrium phases at 1,744 psia and 200°F:

Mol wt of liquid = 119.9*
Mol wt of vapor = 20.48*

Density of liquid $= 44.429$ lb/cu ft*
Density of vapor $= 5.803$ lb/cu ft*
$\rho_L = 44.429$ lb/cu ft $= 0.729$ gm/cc
$\rho_v = 5.803$ lb/cu ft $= 0.093$ gm/cc

$$\sigma^{1/4} = \sum_{i=1}^{n} P_{chi}\left(x_i\frac{\rho_L}{M_L} - y_i\frac{\rho_v}{M_v}\right)$$

$$\frac{\rho_L}{M_L} = \frac{0.729}{119.9} = 0.005938 \qquad \frac{\rho_v}{M_v} = \frac{0.093}{20.48} = 0.004541$$

(1) Component	(2) Mole fraction liquid x_i	(3) Mole fraction vapor y_i	(4) $x_i\frac{\rho_L}{M_L}$	(5) $y_i\frac{\rho_v}{M_v}$	(6) $(4)-(5)$	(7) P_{chi} parachor	(8) $(6)\times(7)$
C_1	0.2752	0.8424	0.001634	0.003825	-0.002191	77.0	-0.169
C_2	0.0657	0.0752	0.000390	0.000341	$+0.000054$	108.0	$+0.006$
C_3	0.0665	0.0405	0.000395	0.000184	$+0.000211$	150.3	$+0.032$
C_4	0.0584	0.0217	0.000347	0.000099	$+0.000248$	190.0	$+0.047$
C_5	0.0454	0.0095	0.000270	0.000043	$+0.000227$	229.0	$+0.052$
C_6	0.0432	0.0052	0.000257	0.000024	$+0.000233$	271.0	$+0.063$
C_{7+}	0.4456	0.0055	0.002645	0.000025[a]	$+0.002620$	575.0[a]	$+1.511$
	1.0000	1.0000					1.542

[a] C_{7+} mol wt $= 225$.
* Calculated from fluid analyses by methods previously discussed.
$\sigma^{1/4} = 1.542$. $\sigma = 5.650$ dynes/cm surface tension at 1,744 psia and 200°F.

————•••————

Viscosity of Liquids

The definitions of viscosity presented in the discussion of gases hold for liquids. In contrast to gases, liquids decrease in viscosity with increasing temperature. This is generally attributed to the increased distance between molecules due to thermal expansion of the liquid. The viscosity relations of paraffin hydrocarbons at atmospheric pressure are presented in Fig. 4-57. Liquid viscosity increases with molecular weight and decreases with temperature.

The viscosity μ in centipoises divided by the density ρ in grams per cubic centimeter is defined as the kinematic viscosity in centistokes. In mathematical notation

$$v = \frac{\mu}{\rho} \tag{4-62}$$

where v is the kinematic viscosity.

The kinematic viscosity exhibits a linear trend when plotted as a function of temperature on a special chart available from the American Society of Testing Materials (ASTM). Thus, observations at only two temperatures enable the investigator to determine the viscosity at other temperatures.

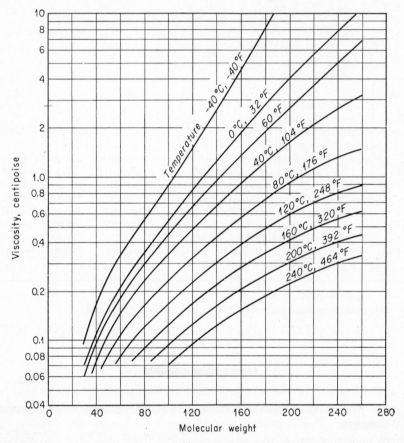

FIG. 4-57. Viscosity of paraffin rydrocarbon liquids at atmospheric pressure. (*From Brown.*[35])

The kinematic viscosity of liquids at atmospheric pressure is usually determined by means of a modified Ostwald (Fig. 4-58a) or other capillary-tube viscosimeter. The viscosity of a liquid at elevated pressure is ordinarily determined with a rolling-ball viscosimeter such as illustrated in Fig. 4-58b.

Hydrocarbon liquids are nonpolar and, therefore, obey closely the rule of additive fluidity. Fluidity is the reciprocal of viscosity. Therefore, the rule of additive fluidity can be written mathematically as

$$\frac{1}{\mu_m} = \Sigma \frac{(\text{vol } \%)_i}{\mu_i} \div 100 \tag{4-63}$$

where μ_m = viscosity of mixture

$(\text{vol } \%)_i$ = vol % of ith component in mixture

μ_i = viscosity of ith component

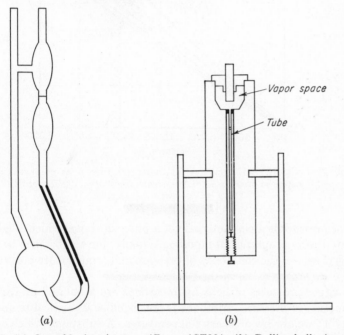

(a)   (b)

FIG. 4-58. (a) Ostwald viscosimeter. (*From ASTM.*) (b) Rolling-ball viscosimeter.

The effect of pressure on the viscosity of a liquid is shown in Fig. 4-59. It will be noted that with an increase in pressure the viscosity is increased. This effect will occur in all hydrocarbon systems if the system is compressed above the bubble-point pressure. If two phases are present during the compression, lighter constituents will enter the liquid phase, thus causing the viscosity to decrease. This effect will be discussed in connection with laboratory measurements in Chap. 5.

Viscosity correlates closely with the density or API gravity of the liquid. However, discussion of the means of estimating the viscosity of liquids will be deferred to Chap. 5, as the existing correlations require concepts yet to be presented.

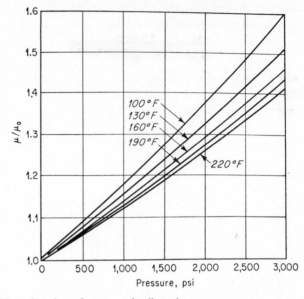

FIG. 4-59. The viscosity of a crystal oil under pressure μ as compared with its viscosity at atmospheric pressure μ_o. (*After Sage, Sherborne, and Lacey.*[36])

Vapor Pressure

Vapor pressure is a characteristic of a pure substance which is generally associated with properties of liquids. Actually, for a substance to exhibit a vapor pressure, the conditions of pressure and temperature must be such that the substance exists in the two-phase region.

The vapor pressures of light hydrocarbons are tabulated in Table 4-16 as a function of pressure. Methane is not listed as its critical temperature is $-117°F$, well below ordinary temperatures. A substance which exists at a pressure and temperature corresponding to a point on the vapor-pressure curve may exist as a liquid, a gas, or a mixture of the two phases depending on the total energy of the system. The change of energy levels is the subject of thermodynamics and will not be considered here. From energy concepts, however, it can be shown that the logarithm of the vapor pressure of a substance is linear with the reciprocal of the temperature.

A more common method for graphically presenting vapor-pressure data of hydrocarbons is the Cox chart (Fig. 4-60). The pressure scale is logarithmic, and the temperature scale is determined by drawing an arbitrary line on the chart for the vapor pressure of water. At each pressure point on the line, a temperature corresponding to the vapor pressure of water is defined. The vapor pressures for hydrocarbons plotted on such a chart are linear and tend to converge to a common point. This property is useful in extrapolating vapor pressures of pure substances to temperatures

TABLE 4-16. VAPOR PRESSURE OF THE LIGHTER HYDROCARBONS, PSI[37]

Temp., °F	Hydrocarbon							Temp, °C
	Ethane	Pro-pane	Iso-butane	Butane	Iso-pentane	Pentane	Hexane	
32	348	69.6	25.9	14.9	5.0	3.5	.87	0.0
34	358	71.4	27.1	15.5	5.3	3.7	.92	1.1
36	367	73.4	28.3	16.1	5.6	3.9	.98	2.2
38	377	75.3	29.6	16.7	5.9	4.1	1.04	3.3
40	387	77.4	30.9	17.4	6.2	4.3	1.10	4.4
42		79.6	32.2	18.1	6.5	4.5	1.16	5.6
44		82.0	33.5	18.9	6.8	4.7	1.22	6.7
46		84.5	34.8	19.7	7.1	4.9	1.29	7.8
48		87.1	36.1	20.5	7.4	5.1	1.36	8.9
50		89.8	37.4	21.4	7.7	5.4	1.43	10.0
52		92.7	38.9	22.3	8.0	5.7	1.50	11.1
54		95.8	40.0	23.2	8.3	6.0	1.58	12.2
56		99.0	41.3	24.1	8.6	6.3	1.66	13.3
58		102.2	42.6	25.1	8.9	6.6	1.74	14.4
60		105.5	43.9	26.1	9.2	6.9	1.83	15.6
62		108.9	45.2	27.1	9.6	7.2	1.93	16.7
64		112.4	46.5	28.1	10.0	7.5	2.03	17.8
66		115.9	47.8	29.1	10.4	7.8	2.14	18.9
68		119.4	49.2	30.1	10.8	8.1	2.25	20.0
70		122.9	50.7	31.1	11.2	8.4	2.37	21.1
72		126.4	52.2	32.1	11.6	8.7	2.49	22.2
74		130.0	53.7	33.1	12.1	9.0	2.62	23.3
76		133.6	55.3	34.2	12.6	9.4	2.76	24.4
78		137.3	56.9	35.3	13.1	9.8	2.90	25.6
80		141.1	58.6	36.4	13.7	10.2	3.05	26.7
82		145.0	60.3	37.5	14.3	10.6	3.21	27.8
84		149.0	62.1	38.7	14.8	11.0	3.37	28.9
86		153.1	63.9	39.9	15.5	11.5	3.53	30.0
88	Critical	157.3	65.8	41.2	16.2	12.0	3.70	31.1
90	temp	161.5	67.7	42.5	16.9	12.5	3.87	32.2
92		165.8	69.7	43.8	17.6	13.0	4.05	33.3
94		170.2	71.7	45.2	18.3	13.5	4.23	34.4
96		174.7	73.7	46.7	19.0	14.1	4.42	35.6
98		179.2	75.8	48.3	19.7	14.7	4.62	36.7
100		183.7	77.9	49.9	20.4	15.3	4.83	37.8
102		188.3	80.1	51.5	21.1	15.9	5.05	38.9
104		193.0	82.3	53.2	21.8	16.5	5.28	40.0
106		197.8	84.6	55.0	22.6	17.2	5.52	41.1
108		202.7	86.9	56.9	23.4	17.9	5.77	42.2
110		207.7	89.3	58.9	24.3	18.6	6.03	43.3
112		212.7	91.7	61.0	25.2	19.3	6.29	44.4
114		217.8	94.2	63.2	26.1	20.0	6.56	45.6
116		223.0	96.7	65.5	27.1	20.8	6.84	46.7
118		228.3	99.3	67.8	28.1	21.6	7.13	47.8
120		233.7	101.9	70.1	29.2	22.4	7.43	48.9

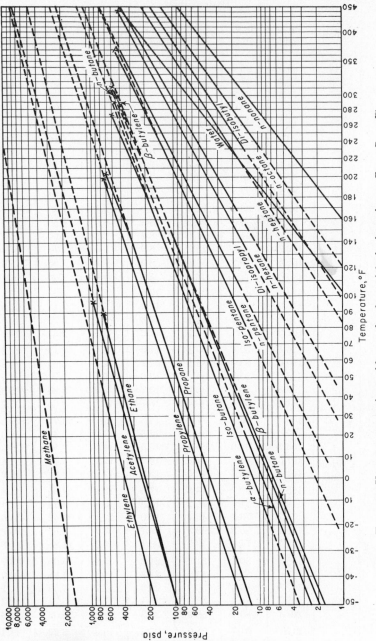

Fig. 4-60. Vapor-pressure chart of low-molecular-weight hydrocarbons. *From Perry.*[38])

314

above their critical temperature. This extrapolation is required in certain calculations of properties of mixtures.

Mixtures of hydrocarbons do not exhibit a true vapor pressure, as the bubble-point and dew-point curves do not coincide. But in evaluating tank storage problems, the concept of vapor pressure is useful. It can be used as a measure of the losses of liquid petroleum by evaporation and to estimate the internal pressures which the tanks must withstand.

Vapor pressure may be considered as a measure of the volatility of a pure substance in a mixture and as such will be discussed in the next section of this chapter.

PROPERTIES OF TWO-PHASE SYSTEMS

The qualitative phase behavior of hydrocarbon systems and the quantitative behavior of the vapor and liquid states have been discussed. The quantitative analysis of two-phase systems will be discussed in the remainder of this chapter. Two-phase, gas and liquid, systems exist at conditions of pressure and temperature within the region enclosed by the bubble-point and dew-point curves.

The quantitative analysis of two-phase systems involves the determination of the mole fractions of gas and liquid present at a given condition and the computation of the composition of the coexisting phases. The physical properties of the gas and liquid phases within this two-phase region can then be computed by the procedures previously described.

As was shown in Fig. 4-6, if a fluid existing at its bubble point is subjected to a pressure decline at constant temperature, two phases are formed. The proportion of the gas phase increases, with decreasing pressure, because of two phenomena. The more volatile constituents escape from the liquid phase and enter the gas phase. Also, the gas formed at higher pressures expands and occupies a larger volume.

The idealized volume relations during such a process are shown in Fig. 4-61. A portion of the gas volume V_g is shown to be gas released between

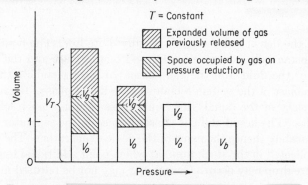

FIG. 4-61. Idealized volume relation in the two-phase region.

pressure increments, and a portion is shown to be the expanded volume of gas previously released. The volume of liquid phase present decreases, or shrinks. The total volume of the system increases, as a mass of gas represents a larger volume than an equal mass of liquid. Fluid samples are frequently subjected to laboratory tests, conducted at reservoir temperature, which duplicate the process illustrated in Fig. 4-61. The results of the laboratory test are referred to as the pressure-volume relation for the fluid.

If a sample of a bubble-point fluid is brought to separator conditions, the fluid enters the two-phase region but at a temperature much lower than reservoir temperature (see Fig. 4-61). The idealized volume relations for expansion of a fluid from bubble point to separator conditions are shown in Fig. 4-62. Large volumes of gas are formed at the separator con-

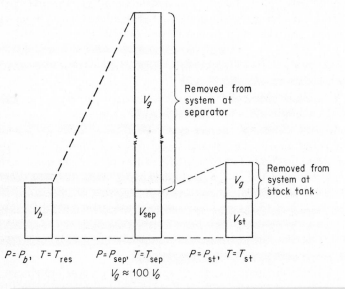
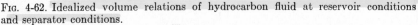

FIG. 4-62. Idealized volume relations of hydrocarbon fluid at reservoir conditions and separator conditions.

ditions, as the density of the gas is usually quite low. The liquid volume shrinks substantially because of decreased temperature and the escape of a portion of the fluid into the gas phase. In most instances the greater part of the mass of the system remains in the liquid phase. Because of the greater density of the liquid, this liquid mass occupies much less volume than the gas. The liquid and gas phases formed in the separator are withdrawn separately (hence the name of the device, separator). The separator liquid is collected in the stock tank, at which point additional temperature and pressure drop may occur. Gas may or may not be released in quantity in the stock tank depending on the relation of separator conditions to stock-tank conditions.

If a given mass of fluid is subjected to the processes described above, then the following definitions can be made:

1. The volume of liquid V_o at reservoir temperature and the prevailing pressure divided by the volume of oil collected in the stock tank V_{st} is defined as the oil formation volume factor B_o.

2. The total volume of the system V_t divided by the volume of stock-tank oil is defined as the total formation volume factor B_t.

In equation form,

$$B_o = \frac{V_o}{V_{st}}$$

$$B_t = \frac{V_t}{V_{st}} \tag{4-64}$$

and $V_t = V_o + V_g$ (see Fig. 4-61).

The two volume factors defined above, together with the gas volume factor B_g, enable the petroleum-reservoir engineer to describe the volume changes in fluid systems. The volume factors B_o and B_t can be computed from equilibrium calculations or measured in the laboratory. The theory of equilibrium calculations will be developed in this chapter, while the laboratory processes will be discussed in Chap. 5.

Equilibrium Relations

If a fluid is subjected to a process such that in the two-phase region the gas and liquid phases remain in contact, equilibrium will be maintained between the phases. One of the first means of evaluating the equilibrium behavior of multicomponent two-phase systems was devised by combining Dalton's and Raoult's laws.

Dalton's law is defined by Eqs. (4-24) and (4-25):

$$P = \sum_{i=1}^{n} P_i \tag{4-24}$$

and

$$y_i = \frac{P_i}{P} \tag{4-25}$$

The statement of Raoult's law is "the partial pressure exerted by a constituent of the liquid phase is equal to the vapor pressure of that constituent times the mole fraction of that constituent in the liquid phase." That is,

$$P_i = x_i P_{vi} \tag{4-65}$$

where P_i = partial pressure of ith component
      x_i = mole fraction of ith component in liquid phase
      P_{vi} = vapor pressure of ith component

When a gas and a liquid phase are in equilibrium, the partial pressure exerted by a constituent in the gaseous phase must be equal to the partial pressure exerted by that same constituent in the liquid phase. Therefore, it is possible to equate the partial pressure calculated by Dalton's and Raoult's laws.

$$y_i P = x_i P_{vi}$$

or
$$\frac{y_i}{x_i} = \frac{P_{vi}}{P} = K_i \tag{4-66}$$

where K_i is defined as the equilibrium ratio of the ith component at pressure P and some temperature T.

Certain limitations are placed on the above definition by Raoult's law and Dalton's law. It is evident that any pure constituent has a vapor pressure only up to its critical temperature. Beyond this temperature there is no true vapor-pressure curve for a pure constituent. The vapor-pressure curves of such constituents can be extrapolated to higher temperatures. However, this practice does not yield satisfactory results in most cases. Since the critical temperature of methane is $-117°F$, it becomes evident then that some other means must be obtained for the determination of equilibrium ratios if they are to be used for hydrocarbon mixtures.

Lewis[39] introduced the concept of fugacity for the extrapolation or correction of vapor pressures to use in equilibrium calculations. The fugacity of a component is a thermodynamic quantity defined in terms of the change in free energy in passing from one state to another state. Dodge and Newton[40] have shown that the fugacity of a component in a phase of a mixture is equal to the fugacity of that component in the same phase in the pure state and at the same conditions of pressure and temperature as the mixture times the mole fraction that component represents of the mixture. Thus

$$(f_{gi})_{\text{mix}} = y_i (f_{gi})_{\text{pure}} \tag{4-67}$$

and
$$(f_{Li})_{\text{mix}} = x_i (f_{Li})_{\text{pure}} \tag{4-68}$$

where the subscript "mix" refers to the mixture and "pure" to the component in the pure state, f_{gi} is the fugacity of a component in the gas phase, and f_{Li} is the fugacity of that component in the liquid phase. The fugacity of a component in the gas phase is equal to the fugacity of that component in the liquid phase if the phases are in equilibrium. Thus

$$y_i (f_{gi})_{\text{pure}} = x_i (f_{Li})_{\text{pure}}$$

or, dropping the subscripts pure,

$$\frac{y_i}{x_i} = \frac{f_{Li}}{f_{gi}} = K_i \tag{4-69}$$

The fugacity as defined above is also limited by the critical temperatures of many hydrocarbons. However, the fugacities can be extrapolated to cover a wider range of conditions than the vapor-pressure relations. Even this is not sufficient accurately to define equilibrium ratios in hydrocarbon systems over a wide range of pressure and temperature. The attractive

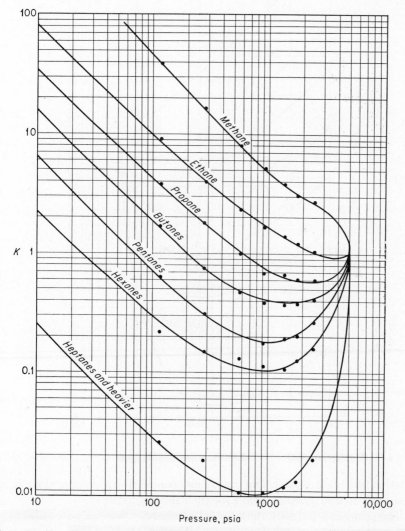

Pressure, psia

FIG. 4-63. Equilibrium ratios at 200°F for a low-shrinkage oil. (*From Katz and Hachmuth.*[41])

force between the molecules affect the vapor-liquid equilibria. Therefore, at higher pressure and temperatures the composition of the mixture affects the equilibrium ratio. As both Raoult's and Dalton's laws and the fugacitv

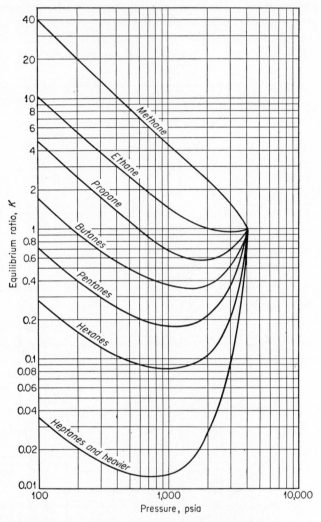

FIG. 4-64. Equilibrium ratios at 200°F for a condensate fluid. (*From Roland et al.*[42])

relation calculate equilibrium ratios independently of composition, it is necessary to find another means for their determination.

The petroleum industry relies on experimentally determined equilibrium ratios (sometimes called equilibrium constants or K values). Katz and Hachmuth[41] presented equilibrium-ratio data for low-shrinkage oils (Fig.

4-63), and Roland, Smith, and Kaveler[42] presented data for condensate fluids (Fig. 4-64).

Development of Equations for Calculating Equilibrium Relations

From equilibrium concepts and appropriate equilibrium ratios, the composition of the equilibrium gas and liquid phases and the fraction of the mixture existing in each phase can be calculated.

In deriving the various equations describing phase equilibrium, it is convenient to define a group of terms and symbols as follows:

n = total moles of both gas and liquid present at any state of pressure and temperature

L = total moles of material within liquid phase at conditions of n

V = total moles in vapor phase at same conditions as L

z_i = mole fraction of ith constituent in mixture

x_i = mole fraction of ith constituent in liquid phase

y_i = mole fraction of ith constituent in vapor phase

By definition,

$$n = L + V \tag{4-70}$$

That is, the total number of moles of composite is equal to the moles in the liquid phase plus the moles in the vapor phase.

From which, by definition of the mole fraction,

$$z_i(n) = x_i(L) + y_i(V) \tag{4-71}$$

This states simply that the moles of the ith constituent in the total must be equal to the sum of the moles of the ith constituent in the vapor and liquid phases.

A third equation by definition of the equilibrium ratio is

$$K_i = \frac{y_i}{x_i} \tag{4-72}$$

As x_i, y_i, and z_i represent mole fractions, three more relations can be stated from the definitions.

$$\sum_{i=1}^{m} x_i = \sum_{i=1}^{m} y_i = \sum_{i=1}^{m} z_i = 1 \tag{4-73}$$

For 1 mole of mixture, $n = 1$,

$$V + L = 1 \tag{4-74}$$

therefore

$$z_i = x_i L + y_i V \tag{4-75}$$

If y_i is replaced by its equivalent $K_i x_i$,

$$z_i = x_i(L + K_i V)$$

If both sides are divided by $(L + K_i V)$, the following equation is obtained:

$$x_i = \frac{z_i}{L + K_i V}$$

If on the right side of the above equation both the numerator and denominator are divided by V, the results are as follows:

$$x_i = \frac{1}{V} \frac{z_i}{(L/V) + K_i} \qquad (4\text{-}76)$$

As x_i represents the mole fraction of the ith constituent in the liquid phase, the sum of the mole fractions of all the constituents is 1. Therefore,

$$\sum_{i=1}^{m} x_i = \frac{1}{V} \sum_{i=1}^{m} \frac{z_i}{(L/V) + K_i} = \sum_{i=1}^{m} \frac{z_i}{L + VK_i} = 1 \qquad (4\text{-}77)$$

or

$$\sum_{i=1}^{m} \frac{z_i}{(L/V) + K_i} = V \qquad (4\text{-}78)$$

A similar equation can be obtained by replacing x_i in Eq. (4-75) by its equivalent y_i/K_i, which, on solving for y_i, yields

$$y_i = \frac{z_i}{(L/K_i) + V}$$

The summation of y_i is equal to 1; thus

$$\sum_{i=1}^{m} y_i = \sum_{i=1}^{m} \frac{z_i}{(L/K_i) + V} = 1 \qquad (4\text{-}79)$$

and

$$\sum_{i=1}^{m} \frac{z_i}{(L/K_i V) + 1} = V \qquad (4\text{-}80)$$

These equations apply within the two-phase region. In equilibrium calculations, the composition of the composite fluid or feed must be known together with the pressure and temperature at which the equilibrium exists. Suitable equilibrium ratios for each component must be available as defined by the conditions of pressure and temperature and the composition of the composite. In general if the system is comprised of m components, the number of unknowns in the calculations are equal to $2m + 2$, m values of x_i, m values of y_i, and the values of L and V. There are available m equations $y_i = K_i x_i$ and m equations $z_i = x_i L + y_i V$. In addition for 1 mole of feed $V + L = 1$ and $\Sigma z_i = \Sigma x_i = \Sigma y_i = 1$. Thus there are available $2m + 2$ equations to solve for $2m + 2$ unknowns. As can be noted from Eqs. (4-74) through (4-77), it is not possible to solve analytically for explicit values of the unknowns. However, in Eqs. (4-74) through (4-77) the number of unknowns have been reduced by substitution to two, L and V. Thus any of Eqs. (4-77) through (4-80) can be used together with Eq.

(4-74) to solve for L and V, the mole fractions of liquid and vapor, respectively. The two equations thus available for the solution for two unknowns must be solved by a trial-and-error procedure. Several alternative forms of Eqs. (4-77) through (4-80) can be derived. The form selected depends on the estimated magnitude of L and the computing method to be used.

At the bubble-point pressure, $L = 1$ and $V \cong 0$, since all the mixture can be considered as a single-phase liquid. Equations (4-76) through (4-78) become meaningless at the bubble point. Likewise at the dew point, $L \cong 0$ and $V = 1$ and Eqs. (4-79) and (4-80) are meaningless. However, at the bubble point

$$z_i \cong x_i$$

$$z_i = \frac{y_i}{K_i} \quad \text{or} \quad y_i = z_i K_i$$

and
$$\sum_{i=1}^{m} y_i = \sum_{i=1}^{m} z_i K_i = 1 \tag{4-81}$$

which is a reduced form of Eq. (4-79).

Thus at the bubble point, the mixture is in equilibrium with an infinitesimal amount of gas having a composition defined by

$$y_i = z_i K_i$$

At the dew point,

$$z_i \cong y_i$$

$$z_i = x_i K_i \quad \text{or} \quad x_i = \frac{z_i}{K_i}$$

and
$$\sum_{i=1}^{m} x_i = \sum_{i=1}^{m} \frac{z_i}{K_i} = 1 \tag{4-82}$$

which is a reduced form of the last segment of Eq. (4-77).

The mixture at the dew point is in equilibrium with an infinitesimal quantity of liquid having a composition defined by

$$x_i = \frac{z_i}{K_i}$$

Experimentally Determined Equilibrium Ratios.   The equilibrium ratio is defined as

$$K_i = \frac{y_i}{x_i}$$

The value of K_i is dependent upon the pressure, temperature, and composition of the hydrocarbon system. Equilibrium ratios for a low-shrinkage oil and a condensate at temperatures of 200°F are shown in Figs. 4-63 and 4-64 as functions of pressure. The equilibrium ratios for all constituents

(K values) for both types of fluids are shown to converge to a value of 1 at about 5,000 psia. This point is termed the convergence pressure. If the temperature at which the equilibrium ratios were presented was the critical temperature of the mixture, then the convergence pressure would be the critical pressure. For all temperatures other than the critical temperature, the convergence of K values is only an "apparent" convergence. The system will have either a dew point or a bubble point at some pressure less than the convergence pressure and exist as a single-phase fluid at the conditions expressed by the point of apparent convergence. As equilibrium ratios are undefined in the single-phase region, it is the extrapolation of the actual values which apparently converges to 1.

The apparent convergence pressure is a function of the composition of the mixture. The effect of composition is correlated by means of convergence pressures in the most extensive set of published equilibrium ratios for hydrocarbons, the NGAA "Equilibrium Ratio Data Book."[43] Equilibrium ratios for hydrocarbons, methane through decane, and nitrogen for a 5,000-psia convergence pressure are reproduced in Figs. 4-65 through 4-77 by permission of the NGAA. Data for carbon dioxide and hydrogen sulfide from the same source are presented in Figs. 4-78 and 4-79, respectively, for a 4,000-psia convergence pressure. The values for 5,000-psia convergence pressures can be used for a large number of hydrocarbon systems.

The effect of composition on K values is shown in Fig. 4-80, where values for 1,000- and 5,000-psia convergence pressures are compared at a temperature of 100°F. The differences in K values for the two convergence pressures at pressures below 100 psia are not significant for the lighter hydrocarbons ethane through pentane. For fluids with convergence pressures of 4,000 psia or greater, the values of the equilibrium ratios are essentially the same to pressures of 1,000 psia. It then becomes apparent that at low pressures and temperatures equilibrium ratios are nearly independent of composition. To define the K values for a mixture at higher pressures it is necessary to define the convergence pressure of the mixture. A discussion of the selection of the proper convergence pressure is presented in Chap. 5.

In conventional analyses of hydrocarbon fluids everything heavier than hexane is grouped together and reported as heptanes plus. This fraction is a mixture of materials of varying volatility. Because the vapor-pressure curves and critical properties of the hydrocarbons heavier than hexane are fairly close together, it is possible to characterize the mixture by an average set of K values. The properties of heptanes plus can be estimated from the properties of heavier hydrocarbons such as nonane or decane. Normally a more satisfactory procedure for characterizing the heptanes plus is to use correlated experimental data for heptanes-plus fractions of fluids with similar properties to those under consideration. For this purpose equilibrium ratios for the heptanes-plus fractions reported by Katz and

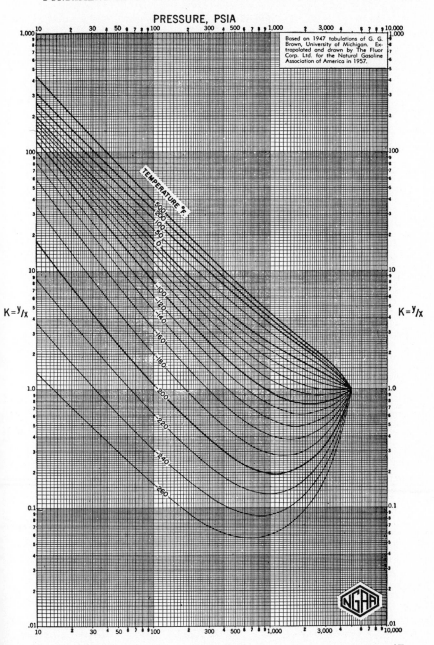

PRESSURE, PSIA

Based on 1947 tabulations of G. G. Brown, University of Michigan. Extrapolated and drawn by The Fluor Corp. Ltd. for the Natural Gasoline Association of America in 1957.

TEMPERATURE °F.

$K = {}^{y}/_{x}$

$K = {}^{y}/_{x}$

Fig. 4-65. Equilibrium ratios for methane, 5,000-psia convergence pressure. (*From NGAA.*[43])

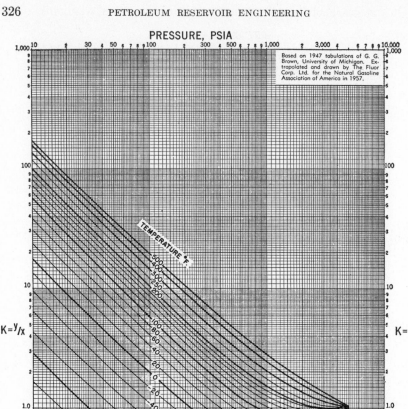
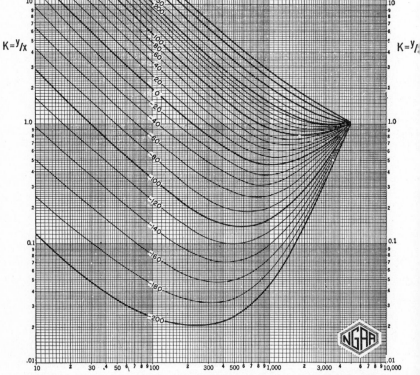

FIG. 4-66. Equilibrium ratios for ethane, 5,000-psia convergence pressure. (*From NGAA.*[43])

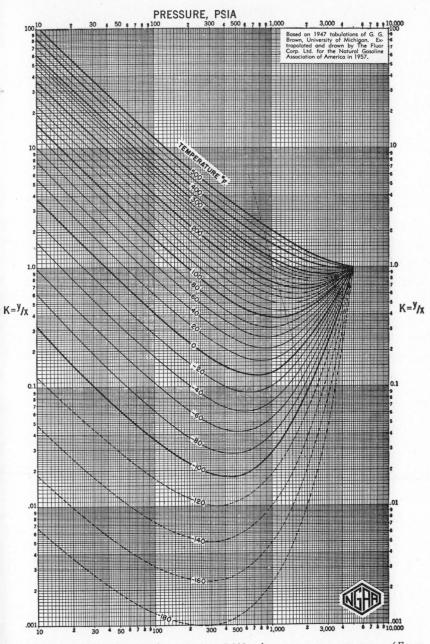

FIG. 4-67. Equilibrium ratios for propane, 5,000-psia convergence pressure. (*From NGAA.*[43])

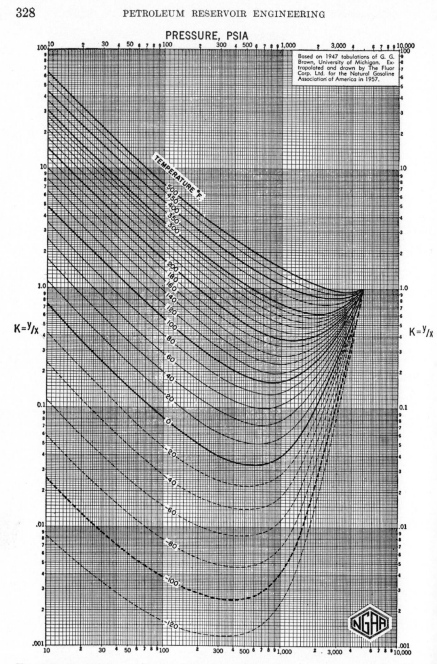

Fig. 4-68. Equilibrium ratios for isobutane, 5,000-psia convergence pressure. (*From NGAA.*[43])

normal
Butane

C4

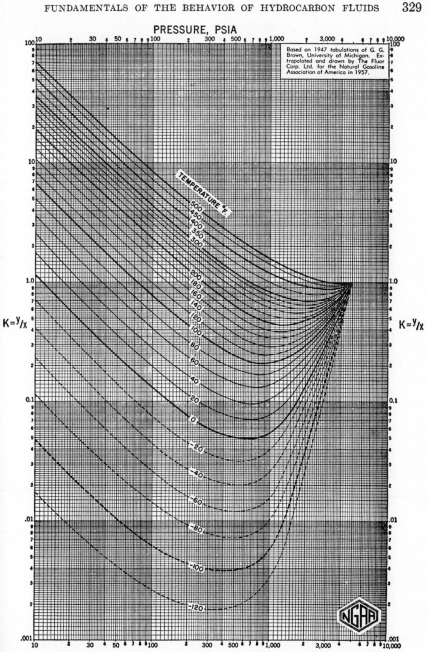

FIG. 4-69. Equilibrium ratios for normal butane, 5,000-psia convergence pressure. (From NGAA.[43])

ISO BUTANE

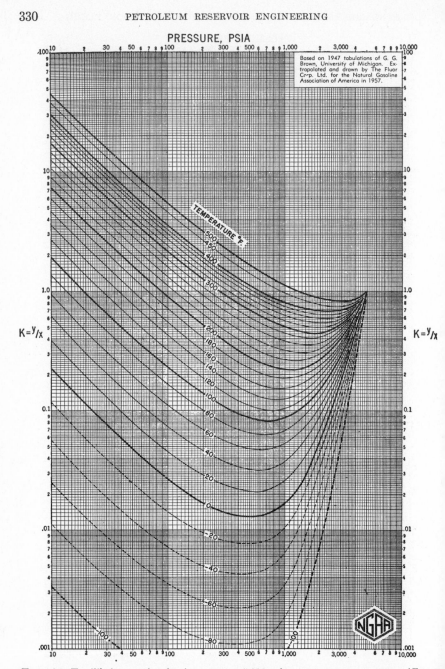

FIG. 4-70. Equilibrium ratios for isopentane, 5,000-psia convergence pressure. (*From NGAA.*[43])

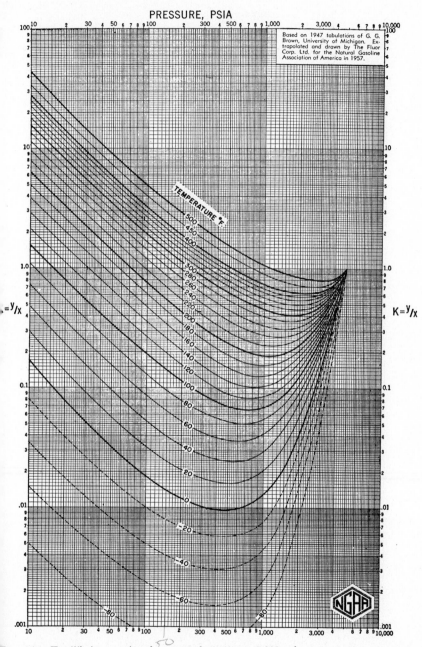

FIG. 4-71. Equilibrium ratios for normal pentane, 5,000-psia convergence pressure. From NGAA.[43])

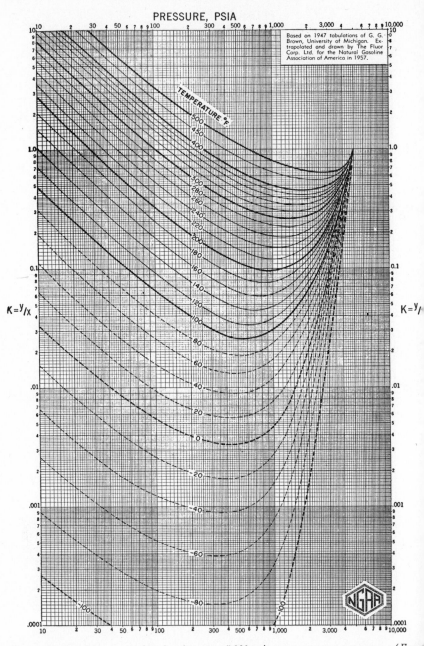

FIG. 4-72. Equilibrium ratios for hexane, 5,000-psia convergence pressure. (*From NGAA.*[13])

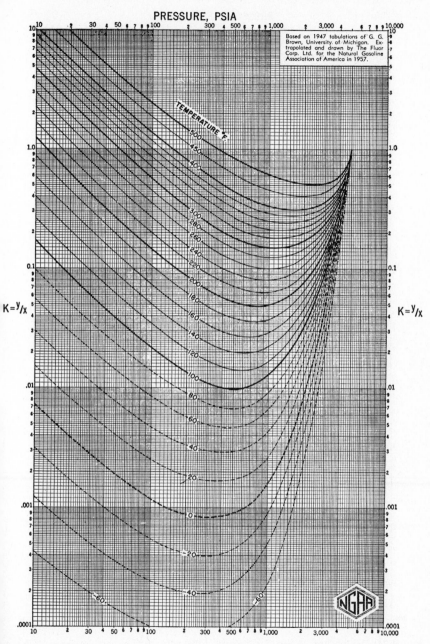

Fig. 4-73. Equilibrium ratios for heptane, 5,000-psia convergence pressure. (*From NGAA.*[43])

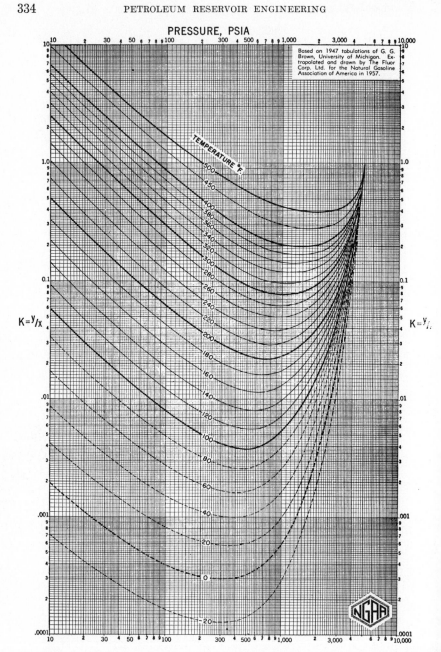

FIG. 4-74. Equilibrium ratios for octane, 5,000-psia convergence pressure. (*From NGAA.*[43])

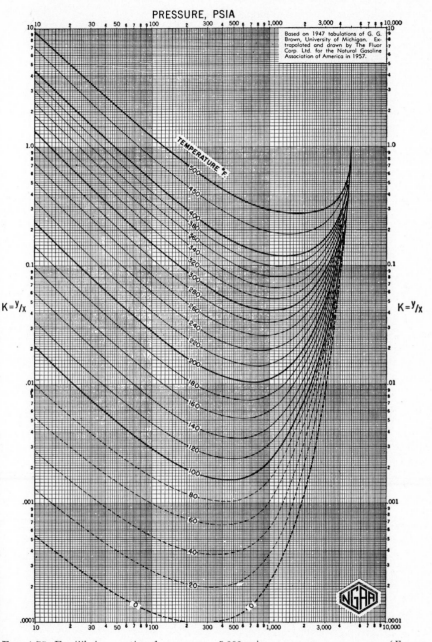

FIG. 4-75. Equilibrium ratios for nonane, 5,000-psia convergence pressure. (*From NGAA.*[43])

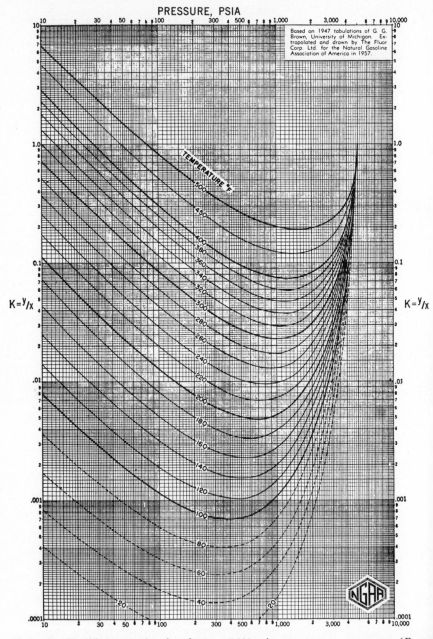

Fig. 4-76. Equilibrium ratios for decane, 5,000-psia convergence pressure. (*From NGAA.*[43])

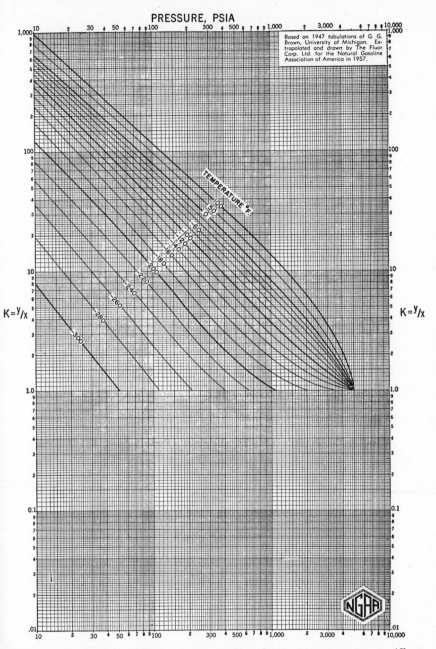

Fig. 4-77. Equilibrium ratios for nitrogen, 5,000-psia convergence pressure. (*From NGAA.*[43])

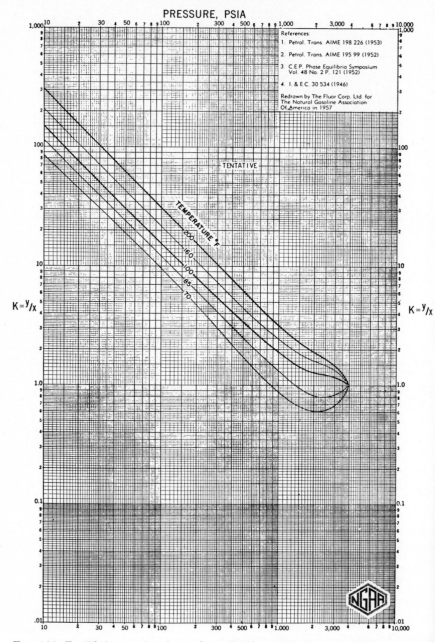

FIG. 4-78. Equilibrium ratios for carbon dioxide, 4,000-psia convergence pressure. (*From NGAA.*[43])

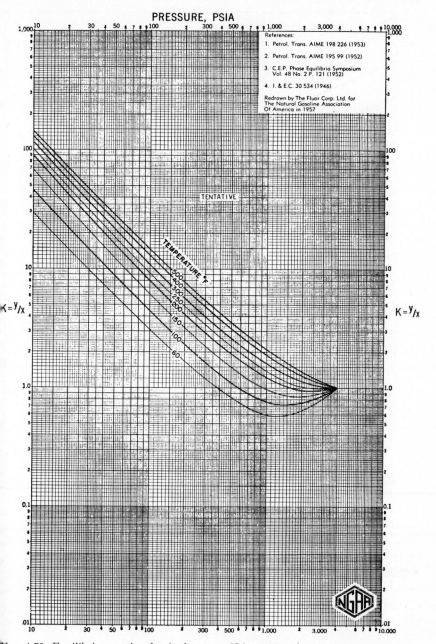

FIG. 4-79. Equilibrium ratios for hydrogen sulfide, 4,000-psia convergence pressure. (*From NGAA.*[43])

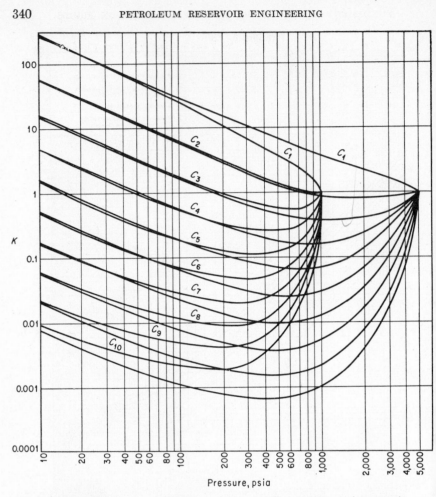

K

Pressure, psia

Fig. 4-80. Comparison of equilibrium ratios at 100°F for 1,000- and 5,000-psia convergence pressure. (*From NGAA.*[43])

Hachmuth[41] and Roland, Smith, and Kaveler[42] are plotted in Fig. 4-81. The data of Katz are preferred for crude-oil systems, and the data of Roland et al. are preferred for condensate fluids.

Calculating Procedures for Solution of Equilibrium Problems. The basic relations developed in Eqs. (4-74) through (4-77) apply to hydrocarbon systems which exist at pressures and temperatures within the two-phase region. Equations (4-78) and (4-79) apply to the special conditions existing at the bubble-point and dew-point pressures.

The solution of the equations to determine the dew-point and bubble-point pressures requires the selection of appropriate equilibrium ratios at the temperature of interest. An estimate of the pressure is made, and the

K values so determined used in Eq. (4-81) or (4-82) depending on whether a bubble point or dew point is sought. If the sum of the calculated values does not equal 1, another pressure is selected and the calculations repeated. Trials are continued until the sum is determined within the desired accuracy.

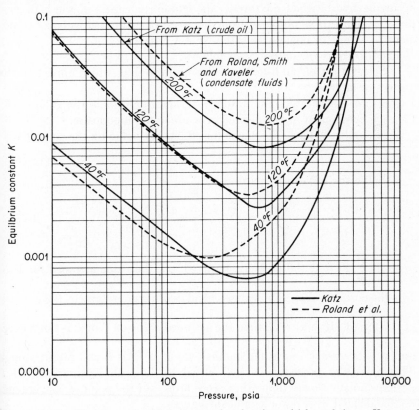

Fig. 4-81. Equilibrium ratios of heptanes-plus fraction. (*Adapted from Katz and Hachmuth*[41] *and Roland et al.*[42])

The primary use of dew-point or bubble-point pressure calculations in petroleum-production engineering work is in connection with adjusting equilibrium ratios to observed bubble-point or dew-point pressures.

The calculation of phase equilibrium within the two-phase region involves the selection of equilibrium constants for the conditions of pressure and temperature under consideration and the solution of the relations expressed in Eqs. (4-77) through (4-80). Essentially two types of problems are of interest: the equilibrium between phases at reservoir conditions and at separator conditions. The calculative methods for these cases are identical except in the final volume calculation of the liquid. However, the

selection of appropriate equilibrium ratios for reservoir conditions is mor
difficult. The calculation of such equilibriums based on adjusted equilib
rium ratios will be discussed in Chap. 5.

For separator calculations at pressures of 500 psi or less, the 5,000-ps
convergence pressure charts of the NGAA can be used for most naturall
occurring hydrocarbon mixtures. The equilibrium ratio of the heptanes
plus fraction can be obtained from Fig. 4-81 or estimated as being equiv
alent to octane or nonane from the NGAA charts.

The procedure of solution is as follows:

√1. Select a set of published equilibrium-ratio data which applies to th
system of interest.

2. From the selected equilibrium-ratio data determine the value of K
for each component at the desired pressure and temperature.

3. Assume a value of V, thus fixing the other value by the equation

$$L + V = 1 \qquad (4\text{-}74)$$

4. Solve the following equations:

$$\sum_{i=1}^{m} x_i = \sum_{i=1}^{m} \frac{z_i}{L + VK_i} = 1 \qquad (4\text{-}77)$$

5. If $\sum_{i=1}^{m} x_i = 1$, the problem is solved and the value assumed for V o
L is the correct value. The values calculated for x_i's represent the analysi
of the liquid. If $\sum_{i=1}^{m} x_i \neq 1$, then the assumed value of V or L is not correc
and it is necessary to assume another value of V or L and repeat step 4
This particular procedure is well suited to slide-rule calculations. It i
perhaps a little slower than other procedures, but the point of convergenc
can be defined with fewer significant figures.

A variation of the calculating procedure which is particularly advan
tageous when using a desk calculator is as follows:

4a. Solve the equation

$$\sum_{i=1}^{m} \frac{z_i}{(L/V_a) + K_i} = (V)_{calc} \qquad (4\text{-}78)$$

5a. If the calculated value $(V)_{calc}$ is equal to the assumed value V_a, th
problem is solved and the composition of the liquid can be determined fro
the

$$x_i = \frac{1}{V} \frac{z_i}{(L/V) + K_i} \qquad (4\text{-}76)$$

and that of the vapor from

$$y_i = K_i x_i \quad \text{or} \quad y_i = \frac{z_i - x_i L}{V}$$

6a. If the calculated value $(V)_{\text{calc}}$ is not equal to the assumed value V_a, a new assumption for V must be made and the calculation repeated.

7a. By several successive trials V can be calculated to an accuracy of about 0.0001.

The procedure outlined is in most convenient form for solution with a desk calculator which has a cumulative multiplication feature. Successive trials can be made with a minimum transcription of numbers from the calculator to a table. A plot of the calculated V as a function of the assumed V is of aid in making estimates for later trials. The number of trials required depends on the experience of the engineer in estimating values of V.

The above relations apply when liquid is expected to be the major mole fraction. Although they are valid even though the vapor is the major fraction, it is desirable to use in that case either Eq. (4-79) or (4-80). This results in determining the composition of the major phase with greater accuracy.

The solution of equilibrium problems on high-speed computors has stimulated greater interest in flash calculations, particularly in solving reservoir equilibrium problems and multistage separator problems.

Rachford[44] has reported on a procedure which utilizes a form of the equilibrium relation developed as follows:

$$z_i = L x_i + V K_i x_i \tag{4-83}$$

and

$$z_i = \frac{L y_i}{K_i} + V y_i$$

Since

$$L = 1 - V$$

$$x_i = \frac{z_i}{(K_i - 1)V + 1}$$

$$y_i = \frac{K_i z_i}{(K_i - 1)V + 1}$$

and

$$\Sigma x_i = \Sigma y_i = 1$$

Therefore,

$$\sum_{i=1}^{m} (y_i - x_i) = \sum_{i=1}^{m} \frac{(K_i - 1)z_i}{(K_i - 1)V + 1} = f(V, K_i, z_i) = 0 \tag{4-84}$$

must be assumed, and trial calculations performed. The correct value of V makes the function $f(V, K_i, z_i) = 0$. For other than the correct value the graph of the function has the form shown in Fig. 4-82. The function has the unique property of yielding negative values if V (assumed) is too great and positive values if V (assumed) is too small. Trials can be made by always assuming $V = 0.5$ for the initial trial, then successively moving to higher or lower values in a systematic fashion by successive halving of the steps in the assumed values. This procedure yields rapid convergence on the correct value.

Another procedure proposed by Holland and Davison[45] uses the following form of the equilibrium relation:

$$g(L,K_i,z_i) = \sum_{i=1}^{m} \frac{z_i}{1 - L(1 - 1/K_i)} - 1 = 0 \qquad (4\text{-}85)$$

In this relation L is assumed and the function is calculated. If the function is not equal to zero with the desired degree of accuracy, a new value

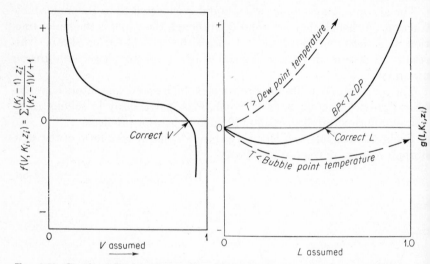

FIG. 4-82. Graph of Rachford's equilibrium function. (*From Rachford and Rice.*[44])

FIG. 4-83. Graph of Holland's equilibrium function. (*From Holland and Davison.*[45])

of L is assumed and the calculation repeated. A graph of the function is shown in Fig. 4-83. If, at the pressure selected, the selected temperature is greater than the dew-point temperature of the mixture, no solution exists except the trivial solution $L = 0$. Also if the selected temperature is less than the bubble-point temperature, the trivial solution $L = 0$ is the only solution obtained. The dashed curves represent solutions of such situations. The solid curve represents the function if the pressure and temperature conditions are truly in the two-phase region. In applying the method, a value of L very close to 1 is first assumed. The function is evaluated. If the function is not equal to zero within the desired accuracy, a new L is assumed as follows:

$$L_j = L_{j-1} - \frac{g(L,K_i,z_i)_{j-1}}{g'(L,K_i,z_i)_{j-1}} \tag{4-86}$$

where $g'(L,K_i,z_i)_{j-1}$ is the first derivative of $g(L,K_i,z_i)_{j-1}$ and is equal to

$$g'(L,K_i,z_i) = \sum_{i=1}^{m} \frac{(1 - 1/K_i)z_i}{[1 - L(1 - 1/K_i)^2]} \tag{4-87}$$

L_j is the new assumed value of L, and L_{j-1} is the previous assumed value of L.

Separator Problems

The application of equilibrium calculations to separator problems requires a brief discussion of the separation process. In a separator, a stream of fluid, referred to as the feed, is brought to equilibrium at the separator temperature and pressure. The pressure of the separator is subject to direct control by means of pressure-regulating devices. The temperature is usually determined by the temperature of the fluid entering the separator and the prevailing atmospheric temperature. Thus, the temperature of an oil-field separator may vary from a low at night to a high during the day. Seasonal variations also occur. In some instances separator temperatures are controlled by heating or by refrigeration.

The vapor and liquid are removed from contact on leaving the separator. Several separators may be operated in series, each receiving the liquid phase from the separator operating at the next higher pressure. Each condition of pressure and temperature at which vapor and liquid are separated is called a stage of separation. Examples of two- and three-stage separation processes are shown in Fig. 4-84. It can be noted that a process using one separator and a stock tank is a two-stage process unless the conditions of pressure and temperature of the two are identical.

Separator calculations are performed to determine the composition of the products, the oil volume factor, and the volume of gas released per barrel of oil and to determine optimum separator conditions for the particular conditions existing in a field.

Example 4-15 presents a sample calculation of two-stage separation applied to the reservoir fluid of Table 4-13. The equilibrium ratios were taken from Figs. 4-65 through 4-72 and 4-81. Two calculating procedures are presented in detail in the example. The calculations are illustrated in suitable form for the application of desk calculators. Graphs for estimating the value of V to be used in additional trials after the first two trials are illustrated in Figs. 4-85 and 4-86 for methods A and B, respectively. In method A, V (assumed) is plotted for each calculated Σx_i. The correct value of V corresponds to $\Sigma x_i = 1$. Therefore, the extrapolation of a curve through the calculated points to the line $\Sigma x_i = 1$ yields a corrected estimate of V. For method B, V (assumed) is plotted for each V (calculated). The correct solution should lie on a 45° line (slope of 1) which passes

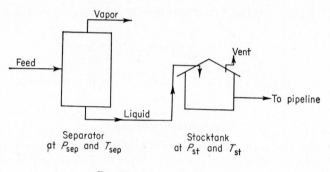

Two-stage separation

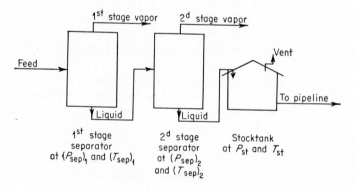

Three-stage separation

Fig. 4-84. Schematic drawing of separation processes.

through the origin. The intersection of the 45° and a curve through the calculated points yields the new estimate for V.

The calculation procedure by method A yields directly the values of x_i, the mole fraction of a component in the liquid. The mole fractions of the components in the vapor are calculated from the definition of the equilibrium ratio.

$$y_i = K_i x_i$$

Frequently, a satisfactory solution for the values of L, V, and x_i will be obtained from which the calculated values of y_i do not sum to 1. This is because of minor discrepancies in the K values, and the error can be distributed uniformly over the composition of the vapor to yield corrected

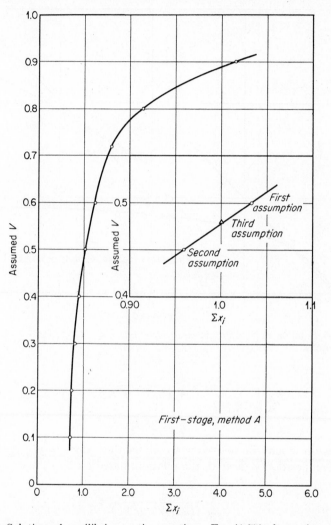

FIG. 4-85. Solution of equilibrium ratio equations, Eq. (4–77), for various assumed gas fractions.

y_i values, or they can be calculated by $y_i = (z_i - x_iL)/V$. For this reason, if it is expected that the vapor phase composition is critical, the alternative relation $\Sigma \dfrac{z_i}{(L/K_i) + V} = 1$, based on Σy_i, should be used.

In method B, after a satisfactory V is obtained, the x_i are calculated from

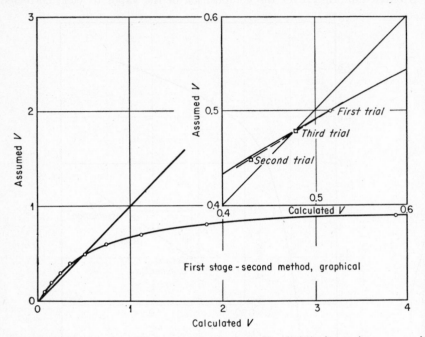

FIG. 4-86. Solution of equilibrium ratio equations, Eq. (4-78), for various assumed gas fractions.

$$x_i = \frac{1}{V} \frac{z_i}{(L/V) + K_i} \qquad (4\text{-}76)$$

and y_i from

$$y_i = K_i x_i$$

Once again it may be necessary to adjust the values of y_i. If the vapor phase composition is desired, the relation, based on Σy_i,

$$\sum_{i=1}^{m} \frac{z_i}{(L/K_i V) + 1} = V \qquad (4\text{-}80)$$

should be used.

The equation $z_i = x_i L + y_i V$ can be used to calculate the composition of the second phase, and no correction will be necessary. This method is longer, but if the problem merits the additional accuracy, it is desirable.

Example 4-15. Calculation of Two-stage Separation Problem.

CALCULATION FOR FIRST-STAGE SEPARATION (35 PSIA, 40°F)
METHOD A [Eq. (4-77)]

Component	Reservoir fluid analysis z_i	K, at 35 psia and 40°F [a]	First assumption: L = 0.5, V = 0.5			Second assumption: L = 0.55, V = 0.45			Third assumption: L = 0.52096, V = 0.47904			
			VK_i	$L+VK_i$	$\frac{z_i}{L+VK_i}=x_i$	VK_i	$L+VK_i$	$\frac{z_i}{L+VK_i}=x_i$	VK_i	$L+VK_i$	$\frac{z_i}{L+VK_i}=x_i$	$Y_i=x_iK_i$
C_1	0.3396	61.0000	30.5000	31.0000	0.0109	27.4500	28.0000	0.0121	29.2214	29.7424	0.0114	0.6965
C_2	0.0646	9.000	4.5000	5.0000	0.0129	4.0500	4.6000	0.0159	4.3113	4.8323	0.0133	0.1203
C_3	0.0987	2.2000	1.1000	1.6000	0.0616	0.9900	1.5400	0.0641	1.0539	1.5748	0.0627	0.1379
C_4	0.0434	0.6100[b]	0.3050	0.8050	0.0539	0.2745	0.8245	0.0526	0.2922	0.8131	0.0533	0.0326
C_5	0.0320	0.1570[b]	0.0785	0.5785	0.0553	0.0706	0.6206	0.0516	0.0752	0.5961	0.0537	0.0084
C_6	0.0300	0.0350	0.0175	0.5175	0.0579	0.0157	0.5657	0.0530	0.0167	0.5377	0.0559	0.0019
C_+	0.3917	0.0032	0.0016	0.5016	0.7809	0.0014	0.5514	0.7103	0.0015	0.5224	0.7497	0.0024
Total	1.0000				1.0336			0.9596			1.0000	1.0000

[a] From Figs. 4-65 through 4-72 and 4-81.
[b] Average of iso and normal.

Equations used:

$$x_i = \frac{z_i}{L+VK_i} \qquad \sum_{i=1}^{m} x_i = 1 \qquad y_i = K_i x_i \qquad y_{ic} = \frac{y_i}{\sum_{i=1}^{m} y_i}$$

Final answer:

0.52096 mole liquid
0.47904 mole vapor

(Handwritten annotations):
- GIVEN CHART
- GUESS (over L and V)
- 135 psia
- COMPOSITION of separator gas
- CALCULATE BUBBLE POINT: GUESS PRESSURE

Component	Z_i	$\frac{P_i}{K_i}$	K_i	$Z_i K_i$	$\frac{x}{y}$
C_1					
C_2					

- $\Sigma = 1.000$ AT BUBBLE POINT

CALCULATION FOR FIRST-STAGE SEPARATION (35 PSIA, 40°F) METHOD B [EQ. (4-78)]

Component	Reservoir fluid analysis z_i	K_i at 35 psia and 40°F[a]	L = 0.5, V = 0.5		L = 0.55, V = 0.45		L = 0.52096, V = 0.47904		x_i	$y_i = K_i x_i$
			$\frac{L}{V}+K_i$	$\frac{z}{L/V+K_i}$	$\frac{L}{V}+K_i$	$\frac{z}{L/V+K_i}$	$\frac{L}{V}+K_i$	$\frac{z}{L/V+K_i}$		
C_1	0.3396	61.0000	62.0000	0.0055	62.2222	0.0054	62.0875	0.0055	0.0114	0.6965
C_2	0.0646	9.0000	10.0000	0.0065	10.2222	0.0063	10.0875	0.0064	0.0133	0.1203
C_3	0.0987	2.2000	3.2000	0.0308	3.4222	0.0288	3.2875	0.0300	0.0627	0.1379
C_4	0.0434	0.6100[b]	1.6100	0.0269	1.8322	0.0237	1.6975	0.0255	0.0533	0.0326
C_5	0.0300	0.1570[b]	1.1570	0.0276	1.3792	0.0232	1.2445	0.0257	0.0537	0.0084
C_6	0.0300	0.0350	1.0350	0.0290	1.2572	0.0239	1.1225	0.0268	0.0559	0.0019
C_{7+}	0.3917	0.0032	1.0032	0.3905	1.2254	0.3196	1.0907	0.3591	0.7497	0.0024
Total				$V_c = 0.5168$		$V = 0.4309$		$V = 0.4790$	1.0000	1.0000

[a] From Figs. 4-65 through 4-72 and 4-81.
[b] Average of iso and normal.

Equations used:

$$\sum_{i=1}^{m} \frac{z_i}{L/V+K_i} = V \qquad x_i = \frac{1}{V}\frac{z_i}{L/V+K_i}$$

$$y_i = K_i x_i \qquad y_{ic} = \frac{y_i}{\displaystyle\sum_{i=1}^{m} y_i}$$

Final answer:
0.52096 mole liquid
0.47904 mole vapor

CALCULATION FOR SECOND STAGE OF SEPARATION (15 PSIA, 40°F)

METHOD A

Component	Reservoir fluid analysis z_i	K_i at 35 psia and 40°F[a]	First assumption: $L = 0.95$, $V = 0.05$			Second assumption: $L = 0.9789$, $V = 0.0211$				
			VK_i	$L + VK_i$	$\dfrac{z}{L+VK_i} = x_i$	VK_i	$L + VK_i$	$\dfrac{z}{L+VK_i} = x_i$	$y_i = K_i x_i$	y_i corrected
C_1	0.0114	145.0000	7.2500	8.2000	0.0013	3.0595	4.0384	0.0028	0.4103	0.4109
C_2	0.0133	20.5000	1.0250	1.9750	0.0067	0.4325	1.4114	0.0095	0.1941	0.1944
C_3	0.0627	5.1000	0.2550	1.2050	0.0520	0.1076	1.0865	0.0577	0.2942	0.2945
C_4	0.0534	1.4000[b]	0.0700	1.0200	0.0523	0.0295	1.0084	0.0529	0.0741	0.0742
C_5	0.0537	0.3750[b]	0.0178	0.9678	0.0554	0.0075	0.9864	0.0544	0.0194	0.0194
C_6	0.0558	0.0750	0.0037	0.9537	0.0584	0.0016	0.9805	0.0569	0.0043	0.0043
C_{7+}	0.7497	0.0030	0.0001	0.9501	0.7890	0.0006	0.9789	0.7658	0.0023	0.0023
Total	1.0000				1.0154			1.0000	0.9987	1.0000

[a] From Figs. 4-65 through 4-72 and 4-81.
[b] Average of iso and normal.

Final answer:

0.9789 mole liquid

0.0211 mole vapor

In addition to the compositions and the mole fractions of the mixture which exist in the liquid and vapor phases, three additional parameters are of interest to the engineer. These are the API gravity of the stock-tank liquid, the gas-oil ratio, and the oil-formation volume factor. The API gravity can be calculated from the density of the stock-tank liquid determined from the composition by the procedure of Standing and Katz[27] discussed earlier in this chapter.

The gas-oil ratio is defined as the volume of gas in standard cubic feet per barrel of stock-tank oil. Gas is removed from each stage of the separation process so that the gas-oil ratio can be calculated for each stage or combination of stages. Total gas-oil ratio refers to the sum of the gas volumes in standard cubic feet from all stages divided by the volume in barrels of stock-tank oil.

In equilibrium (flash) calculations it is customary to solve the equilibrium relations for each stage on the basis of 1 mole of feed to that stage. Thus, if n_1 moles of feed enter the first stage, the moles of liquid entering the second stage

$$n_2 = L_1 n_1$$

and
$$n_3 = L_2 n_2 = L_2 L_1 n_1$$

where n is the moles of feed and the subscripts refer to the stage. If the third stage is the stock tank, then

$$n_{st} = L_3 n_3 = L_3 L_2 L_1 n_1$$

where n_{st} is the moles of liquid remaining in the stock tank for n, moles into the first separator. In general terms,

$$n_{st} = n_1 \prod_{i=1}^{m} L_i \tag{4-88}$$

where m = number of stages
L_i = mole fraction of liquid off ith stage
n_1 = moles of feed to first stage

If $n_1 = 1$, then

$$\bar{n}_{st} = \prod_{i=1}^{m} L_i \tag{4-89}$$

and $\bar{n}_{st}$ is the mole fraction of stock-tank oil in the feed.

In a similar manner the number of moles of gas can be evaluated. Let n_{gi} be the moles of gas off stage i. Then

$$n_{g1} = V_1 n_1$$

The moles of gas off the second stage

$$n_{g2} = V_2 n_2 = V_2 L_1 n_1$$

and from the third stage,

$$n_{g3} = V_3 n_3 = V_3 L_2 L_1 n_1$$

In general terms the total gas off all stages

$$n_{gT} = \sum_{i=1}^{m} n_{gi} = n_1 \sum_{i=1}^{m} V_i \prod_{j=1}^{i-1} L_j = n_1 - n_{st} \qquad (4\text{-}90)$$

If $n_1 = 1$, then

$$\bar{n}_{gT} = \sum_{i=1}^{m} V_i \prod_{j=1}^{i-1} L_j \qquad (4\text{-}91)$$

where $\bar{n}_{gT}$ is the mole fraction of total gas in the feed, such that $n_1 = \bar{n}_{st} + \bar{n}_{gT}$. The number of moles of gas can be readily converted to standard cubic feet by multiplying the number of moles by the molal volume V_m from Table 4-5 for the desired standard conditions. Thus the total gas volume per mole of feed $= \bar{n}_{gT} V_m$.

The volume of stock-tank oil per mole of feed can be calculated from the density and molecular weight of the stock-tank oil as follows:

$$(V_{st})_m = \frac{\bar{n}_{st} M_{st}}{\rho_{st}} \qquad (4\text{-}92)$$

where $(V_{st})_m$ = volume of stock-tank oil per mole of feed, bbl
M_{st} = molecular weight of stock-tank oil
$\bar{n}_{st}$ = moles of stock-tank oil per mole of feed
ρ_{st} = density of stock-tank oil at 60°F and atmospheric pressure, lb/bbl

The total gas-oil ratio is given by

$$R_T = \frac{\bar{n}_{gT} V_m}{(V_{st})_m} = \frac{\bar{n}_{gT} V_m \rho_{st}}{\bar{n}_{st} M_{st}} \qquad (4\text{-}93)$$

where R_T is the total gas-oil ratio.

If it is known from other data that the feed to the first-stage separator exists as a single-phase liquid at its point of entry into the production stream, then an oil-formation volume factor B_o can be calculated from the data obtained. The analysis of the feed can then be treated as the reservoir fluid, and its density calculated by the method of Standing and Katz.

Let ρ_{res} be the density of the feed in pounds per barrel. Then

$$(V_{res})_m = \frac{M_{res}}{\rho_{res}} \qquad (4\text{-}94)$$

where $(V_{res})_m$ is the volume occupied by 1 mole of feed at reservoir conditions and M_{res} is the molecular weight of the feed. By definition, the oil-formation volume factor

$$B_o = \frac{(V_{res})_m}{(V_{st})_m} = \frac{M_{res}\,\rho_{st}}{\rho_{res}\,M_{st}\bar{n}_{st}} \qquad (4\text{-}95)$$

The calculation of API gravity, total gas-oil ratio, and oil formation volume factor is illustrated in Example 4-16.

The application of equilibrium calculations to the determination of the optimum first-stage separator pressure in a two-stage system (one separator and the stock tank) is illustrated by the data of Table 4-17 and the graphs in Fig. 4-87. The optimum pressure is defined as that pressure at which

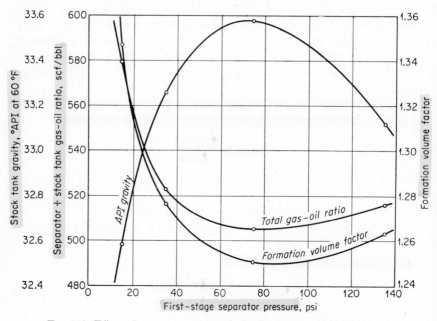

Fig. 4-87. Effect of separator pressure in a two-stage separation process.

the maximum liquid volume is accumulated in the stock tank per volume of reservoir fluid produced. This pressure corresponds to a maximum in the API gravity and a minimum in the gas-oil ratio and oil formation volume factor. The optimum first-stage separator pressure for the system evaluated in Table 4-17 and Fig. 4-87 is 75 psia. The effect of additional stages can be evaluated in a similar manner.

Equilibrium, or "flash," calculations can be used in many other applications. The applications of the methods to calculating the performance of reservoirs containing condensate fluids or volatile oils are becoming increasingly important. Many of these applications will be reviewed later in this text.

Example 4-16.   Calculation of Stock-tank Gravity Separator and Stock-tank Gas-Oil Ratios and Formation Volume Factor.

(1) Component	(2) Mole fraction in liquid phase at 15 psia, 40°F[a]	(3) Mole wt, lb/mole[b]	(4) Relative wt, lb/mole (2) × (3)	(5) Liquid density at 60°F, 14.7 psia[b]	(6) Liquid volume cu ft/mole (4) ÷ (5)
C_1	0.00283	16.042	0.045399		
C_2	0.00947	30.068	0.284744		
C_3	0.05768	44.094	2.543342	31.64	0.0803837
C_4	0.05292	58.120	3.075710	35.71[d]	0.0861302
C_5	0.05441	72.146	3.925464	39.08[d]	0.1004468
C_6	0.05690	86.172	4.903187	41.36	0.1185490
C_{7+}	0.76579	263[c]	201.402770	55.28	3.6433207
Total	1.00000		216.180616		4.0108304

[a] From Example 4-15.   / 2 3 4 9   [b] From Table 4-4.
[c] From Table 4-13.   [d] Average of iso and normal.

$$\text{Density of } C_{3+} = \frac{\text{wt of } C_{3+}}{\text{vol of } C_{3+}} = \frac{215.850}{4.011} = 53.817 \text{ lb/cu ft}$$

$$\text{Wt } \% \ C_2 \text{ in } C_{2+} = \frac{0.285}{216.135217} \times 100 = 0.132$$

$$\text{Wt } \% \ C_1 \text{ in } C_{1+} = \frac{0.046}{216.181} \times 100 = 0.021$$

Density of system including C_1 and C_2 = 53.5 lb/cu ft at 60°F and 14.7 psia (from Fig. 4-50) *P 2 9 7*
Pressure correction from 14.7 to 15 psia is negligible. No temperature correction required.
Density = 53.5 lb/cu ft at 60°F and 15 psia
Specific gravity = 0.8588
API gravity = 33.26°   *141.5 / γ − 131.5 = °API* M_{ST}

$$\text{Separator GOR (gas-oil ratio)} = \frac{(380.69 V_i)\rho_{st}}{M_{st}L_1L_2}$$

$$= \frac{380.69(.47904)(53.5)(5.61)}{216.181(.52096)(.9789)}$$

$$= 496.47 \text{ cu ft/bbl}$$

$$\text{Stock-tank (ST) GOR} = \frac{380.69 V_2 \rho_{st}(5.61)}{\text{mol wt ST liq.} \times L_2}$$

$$= \frac{380.69(0.0211)(53.5)(5.61)}{216.181(.9789)}$$

$$= 11.35 \text{ cu ft/bbl.}$$

Total GOR = separator GOR + stock-tank GOR
$$= 496.47 + 11.35 = 507.82 \text{ cu ft/bbl}$$

Sample calculation of formation volume factor:

$$B_o = \frac{\text{bbl res liq/mole res liq}}{\text{ST bbl/mole res liq}} \tag{4-95}$$

$$\text{Reservoir bbl/mole res liquid} = \frac{M_{\text{res}}}{\rho_{\text{res}}}$$

$$= \frac{122.1748}{46.6(5.61)} = 0.467$$

use charts

$$\text{Stock-tank vol/mole stock-tank oil} = \frac{216.181}{53.5(5.61)} \quad \frac{MW}{(\rho)\, 5.61}$$

$$= 0.72027792 \text{ bbl/mole}$$

$$\text{Stock-tank bbl/mole res fluid} = \text{bbl/mole ST liq} \times L_1 L_2$$
$$= 0.720(0.52096)(0.9789)$$
$$= 0.367$$

$$B_o = \frac{0.467}{0.367} = 1.2723$$

TABLE 4-17. EFFECT OF SEPARATOR PRESSURES ON FLUID PROPERTIES

Separator pressure, psia[a]		Gas-oil ratio, cu ft/bbl			Stock-tank oil gravity, °API	B_o, oil formation volume factor[b]
1st stage	2d stage (stock tank)	Separator	Stock tank	Total		
15	15	587.2		582.2	32.58	1.340
35	15	507.7	8.5	516.2	33.26	1.283
75	15	450.4	39.9	490.3	33.58	1.265
135	15	368.1	135.7	503.8	33.12	1.276

[a] Separator and stock-tank temperature, 40°F.
[b] Reservoir pressure, 3,614 psia; reservoir temperature, 200°F.

REFERENCES

1. Uren, L. C.: "Petroleum Production Engineering: Oil Field Exploitation," 3d ed., McGraw-Hill Book Company, Inc., New York, 1953.

2. Daniels, Farrington: "Outlines of Physical Chemistry," John Wiley & Sons, Inc., New York, 1948.

3. Standing, M. B.: "Volumetric and Phase Behavior of Oil Field Hydrocarbon Systems," Reinhold Publishing Corporation, New York, 1952.

4. Brown, G. G., D. L. Katz, G. G. Oberfell, and R. C. Allen: "Natural Gasoline and the Volatile Hydrocarbons," National Gasoline Association of America, Tulsa, Okla., 1948.

5. Katz, D. L., and Fred Kurata: Retrograde Condensation, *Ind. Eng. Chem.*, vol. 32, no. 6, June, 1940.

6. Clark, Norman: It Pays to Know Your Petroleum, *World Oil,* March and April, 1953.

7. Thornton, O. F.: Gas-condensate Reservoirs—A Review, *Petrol. Engr. Reference Ann.*, 1947.

8. Beattie and Bridgeman: *J. Am. Chem. Soc.*, vol. 49, 1927, and vol. 50, 1928.

9. Benedict, M., G. B. Webb, and L. C. Rubin: An Empirical Equation for Thermodynamic Properties of Light Hydrocarbons and Their Mixtures, *Chem. Eng. Progr.*, vol. 47, August, 1951.

10. Kay, W. B.: Density of Hydrocarbon Gases and Vapors at High Temperatures and Pressure, *Ind. Eng. Chem.*, vol. 28, p. 1014, 1936.

11. Sage, Bruce H., and William N. Lacey: "Volumetric and Phase Behavior of Hydrocarbons," Gulf Publishing Company, Houston, Tex., 1949.

12. Matthews, T. A., C. H. Roland, and D. L. Katz: High Pressure Gas Measurement, *Refiner*, vol. 21, June, 1942.

13. Eilerts, C. K., H. A. Carlson, and N. B. Mullens: Effect of Added Nitrogen on Compressibility of Natural Gas, *World Oil*, June and July, 1948.

14. Olds, R. H., B. H. Sage, and W. N. Lacey: Partial Volumetric Behavior of the Methane-Carbon Dioxide System, "Fundamental Research on Occurrence and Recovery of Petroleum," American Petroleum Institute, 1943.

15. Reamer, H. H., R. H. Olds, B. H. Sage, and W. N. Lacey: Methane-Carbon Dioxide System in the Gaseous Region, "Fundamental Research on Occurrence and Recovery of Petroleum," American Petroleum Institute, 1943.

16. Reamer, H. H., R. H. Olds, B. H. Sage, and W. N. Lacey: Volumetric Behavior of Ethane-Carbon Dioxide System, "Fundamental Research on Occurrence and Recovery of Petroleum," American Petroleum Institute, 1945.

17. Keenan, J. H., and F. G. Keyes: "Thermodynamic Properties of Steam," John Wiley & Sons, Inc., New York, 1947.

18. McCarthy, E. I., W. L. Boyd, and L. S. Reid: The Water Vapor Content of Essentially Nitrogen-free Natural Gas Saturated at Various Conditions of Temperature and Pressure, *Trans. AIME*, 1950, p. 189.

19. Carson, D. B., and D. L. Katz: Natural Gas Hydrates, *Trans. AIME*, vol. 146, 1942.

20. Katz, D. L.: Prediction of Conditions for Hydrate Formation in Natural Gases, *Trans. AIME*, vol. 160, 1945.

21. Vondy, D., N. B. Zaremba, and L. L. Lawrence: "Lease Size Low Temperature Gasoline Plants," Black, Sivalls and Bryson, Inc., Bulletin 3302.

22. Bicher, L. B., and D. L. Katz: Viscosity of Natural Gases, *Trans. AIME*, vol. 155, 1944.

23. Carr, N. L., R. Kobayashi, and D. B. Burrows: Viscosity of Hydrocarbon Gase under Pressure, *Trans. AIME*, 1954, p. 201.

24. Herning, F., and L. Zipperer: Calculation of the Viscosity of Technical Gas Mix tures from the Viscosity of the Individual Gases, *Gas- u. Wasserfach*, vol. 79, 1936

25. Natural Gasoline Supply Men's Association: "Engineering Data Book," 6th ed. Natural Gasoline Association of America, Tulsa, Okla., 1951.

26. Eshbach, Ovid W.: "Handbook of Engineering Fundamentals," John Wiley & Sons, Inc., New York, 1952.

27. Standing, M. B., and D. L. Katz: Density of Crude Oils Saturated with Natura Gas, *Trans. AIME*, vol. 146, 1942.

28. Trube, Albert S.: Compressibility of Undersaturated Hydrocarbon Reservoi Fluids, *Trans. AIME*, vol. 210, 1957.

29. National Standard Petroleum Oil Tables, *Natl. Bur. Standards Circ.* C410, 1936

30. "ASTM Standards on Petroleum Products and Lubricants," American Societ for Testing Materials, 1953.

31. "Tag Manual for Inspectors of Petroleum," 27th ed., C. J. Tagliabue Corp. Newark, N.J.

32. Katz, D. L., and W. Saltman: Surface Tension of Hydrocarbons, *Ind. Eng Chem.*, vol. 31, no. 1, January, 1939.

33. Katz, D. L., R. R. Monroe, and R. R. Trainer: Surface Tension of Crude Oil Containing Dissolved Gases, *Petrol. Technol.*, September, 1943.

34. Sugden: *J. Chem. Soc.*, vol. 125, 1924.

35. Brown, G. G.: Continuous Tables, *Petrol. Engr.*

36. Sage, B. H., J. E. Sherborne, and W. N. Lacey: *API Proc., Bull.* 216, 1935.

37. Continuous Tables, *Petrol. Engr.*

38. Perry, J. H., "Chemical Engineers' Handbook," 3d ed., McGraw-Hill Boo Company, Inc., New York, 1950.

39. Lewis, G. N., and M. Randall: "Thermodynamics and the Free Energy of Chemi cal Substances," McGraw-Hill Book Company, Inc., New York, 1923.

40. Dodge and Newton: *Ind. Eng. Chem.*, vol. 29, 1937.

41. Katz, D. L., and K. H. Hachmuth: Vaporization Equilibrium Constants in Crude Oil–Natural Gas System, *Ind. Eng. Chem.*, vol. 29, p. 1072, 1937.

42. Roland, C. H., D. E. Smith, and H. H. Kaveler: Equilibrium Constants for Gas-distillate System, *Oil Gas J.*, vol. 39, no. 46, p. 128, Mar. 27, 1941.

43. "Equilibrium Ratio Data Book," Natural Gasoline Association of America, Tulsa Okla., 1957.

44. Rachford, H. H., and J. D. Rice: Procedure for Use of Electronic Digital Com putors in Calculating Flash Vaporization Hydrocarbon Equilibrium, *Trans. AIME* vol. 195, 1952.

45. Holland, C. D., and R. R. Davison: Simplify Flash Distillation Calculation *Petrol. Refiner*, March, 1957.

CHAPTER 5

DETERMINATION AND APPLICATION
OF RESERVOIR-FLUID PROPERTIES

INTRODUCTION

The volumetric behavior of reservoir fluids must be determined as a function of pressure and temperature in order that the engineer can evaluate the production performance of a reservoir. In Chap. 4, the fundamentals of fluid behavior were developed. The volumetric behavior was shown to be related to various physical properties of the hydrocarbons composing the fluid. Mixtures of hydrocarbons may be characterized by specific gravity, molecular weight, compositional analyses, and other properties which are readily subject to measurement.

The fluids must be identified by direct measurement of certain properties. These measurements involve securing samples of the fluids from the field. The simplest procedure is to make measurements of the properties of the produced fluids. The API gravity of stock-tank oil and the producing gas-oil ratio are routinely determined. The gas-oil ratio is ordinarily the ratio of the volume, in standard cubic feet, of gas from the separator to the volume of stock-tank oil both measured over the same time interval. If good field practice is followed, the gravity of the separator gas is determined by the Ac-Me balance or similar device. These data, together with estimates or observations of the reservoir pressure and temperature, can be used to estimate the properties of the reservoir fluids. Such estimates may be of limited accuracy.

Improved estimates of the properties of the reservoir fluids can be made by securing samples which are representative of the reservoir fluids and subjecting these samples to various laboratory analyses. The proper sampling of fluids is of greatest importance in securing accurate data.

Samples are usually collected by field or laboratory technicians especially trained in performing the operation. However, field and reservoir engineers must be familiar with sampling techniques and the various laboratory analyses available. The engineer involved must decide when a sample is required, what sampling techniques should be used, and how the well should be prepared for sampling. Direct supervision of the preparation and sampling of the well is essential. The reservoir engineer must understand sampling methods in order to evaluate the accuracy of reported laboratory data.

SAMPLING RESERVOIR FLUIDS

There are essentially three sampling techniques for obtaining reservoir-fluid samples for analysis of the pressure, volume, and temperature (PVT) relations. These three techniques are commonly known as:
1. Bottom-hole sampling
2. Recombination sampling
3. Split-stream sampling

The general manner of preparing a well and obtaining a sample for each of these three testing techniques is presented in the following sections.

Bottom-hole Sampling

A well must be selected and properly prepared before any of the sampling techniques can be applied. The engineer should select a well with a high productivity so as to maintain as high a pressure as possible in the formation surrounding the well. It is suggested that the well be as new as possible so as to minimize free gas saturation. The well selected should not be producing free water. If the only available well does produce water, special care must be exercised in locating the sampling bomb. A series of productivity tests should be conducted on the well to determine the bottom-hole flowing pressures which exist at various rates of flow. These tests aid in selecting the well which will have the highest flowing bottom-hole pressure at the stabilized flow rate. The producing history of the wells should be studied. The well selected for sampling should have been producing with a stabilized gas-oil ratio. If a well exhibits a rapidly increasing gas-oil ratio, the saturation conditions would probably prohibit the collecting of a representative sample.

Once the well to be sampled has been selected and all necessary tests performed, it must be properly prepared. The time required for well preparation is dependent upon the past history and the productivity of the well. The well should be flowed at as low a stabilized rate as possible. The stabilized rate of flow should be continued until such time as a constant gas-oil ratio is indicated on the surface. It is desired, but not always obtained, that the low flow rates will cause such a small pressure gradient around the well bore that the gas saturation in the surrounding area will be reduced, by solution and compression, to an immobile state.

The length of time required to reach a constant producing gas-oil ratio is dependent upon the free gas saturation when the well preparation began. The well is then shut in to permit the pressure to build up in the formation adjacent to the well bore. The duration of the shut-in period will be dependent upon the productivity of the well. In some instances, the time period may be 2 to 3 hr, whereas in others, it may be as high as 72 hr. Fluid entering the well bore during the shut-in period enters under increasing back pressures. The fluid which is opposite the perforations after the maximum well pressure is obtained should be gas-saturated at the pressure and temperature at the bottom of the hole.

A tubing pressure survey is conducted to locate the gas-oil and water-oil interfaces. A special sampling bomb is run on a wire line. This bomb is activated at the surface so as to retrieve a bottom-hole fluid sample under pressure. There are several different types of bottom-hole sampling devices on the market. In Fig. 5-1 is a schematic drawing of a sampler. The valves are locked open at the surface, the bomb is located at the desired sampling position, and the valves are activated by dropping a metal bar or by a preset clock mechanism. Instruction for the operation of these various devices will normally accompany the instrument and hence will not be discussed here. If properly used, all the devices are successful in obtaining appropriate samples.

The foregoing well preparation and sampling procedure can be used with any normal gas–crude-oil system. This procedure is often modified when reservoir fluid and well conditions meet special requirements. When, at a low stabilized flow rate, the flowing bottom-hole pressure of the test well is thought to exceed the

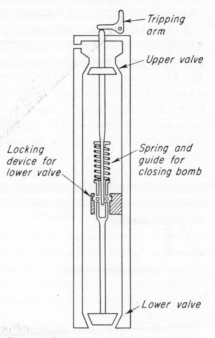

Fig. 5-1. Schematic drawing of bottom-hole fluid-sample bomb. (*From Pirson.*[1])

bubble-point pressure of the reservoir fluid, a bottom-hole sample can be collected with the well flowing. An additional step is added to the well-preparation procedure previously discussed. After the shut-in period, the well is allowed to flow at a low stabilized rate. The sample bomb is run, and the sample collected with the well flowing. The sample collected should be representative of the reservoir fluid. This is the ideal sampling procedure, as the oil entering is fresh oil at all times and is representative of the reservoir fluid.

The bomb and its fluid sample are brought to the surface. The bomb is checked at the surface for possible leaks. The pressure in the bomb at the surface is measured to indicate whether or not the bomb was properly activated in the hole. The pressure in the bomb should be slightly less than the bottom-hole pressure at which the sample was collected. Normally, the saturation pressure of the collected sample at surface temperature is measured by pumping water or mercury into the sample and observing the pressure-volume behavior. Once again, this pressure should be less than the pressure at which the sample was collected. If it exceeds the sampling pressure, then the sampling device either collected free gas or leaked oil. To protect against possible leakage, at least two fluid samples are collected and checked against each other.

If the reservoir were initially at its saturation pressure, the bottom-hole sample obtained would probably indicate a solution-gas–oil ratio and bubble-point pressure different from that of the original reservoir fluid. The saturation pressure obtained from the bottom-hole fluid sample will normally be lower than the bubble-point pressure of the original reservoir fluid. As the pressure around the well declines with production, gas is released from solution. This gas either remains in the reservoir or is produced. The oil that enters the well bore has a lower saturation pressure than that originally existing in the reservoir. Any free gas that enters the well will migrate upward in the tubing owing to gravity. Hence the oil collected by the bottom-hole sampler exhibits a saturation pressure corresponding to the existing reservoir pressure rather than to the initial reservoir pressure.

These inaccuracies in sampling can be corrected either by a mathematical technique, which will be discussed later in this chapter, or by adding additional gas to the samples in the laboratories. Both methods have limitations but yield a sufficiently close approximation to the actual reservoir fluid for use in reservoir calculations.

Recombination Samples

A second technique used in obtaining fluid samples from which to determine PVT relations is known as recombination sampling. For a recombination sample, the fluids are collected at the surface. A sample of separator oil and separator gas are collected, and these samples are recombined in the laboratory in the proper proportions as determined by production characteristics measured at the surface during sampling operations.

The selection of the well from which to collect the sample is based upon the same conditions specified for bottom-hole sampling. The same presampling tests are desired.

Once the well is selected, it is flowed for a period of time sufficient to stabilize the producing gas-oil ratio at the surface. It is desired to have the gas-oil ratio checked over at least three comparable time intervals.

These checks may be over 2-hr, 4-hr, or longer time intervals if necessary to obtain the desired stability of the producing gas-oil ratio. Once the producing gas-oil ratio stabilizes, the well is ready for sampling.

Gas and liquid samples are collected from the same separator at the same flow conditions. A larger quantity of separator gas must be collected because of its high compressibility compared with the liquid. The manner of collecting these fluid samples varies with company and individual preference. Pressure-control devices are attached to the separator to maintain stabilized conditions. The sampling containers can be attached to the separator as indicated in Fig. 5-2. The oil-control valve should be regulated

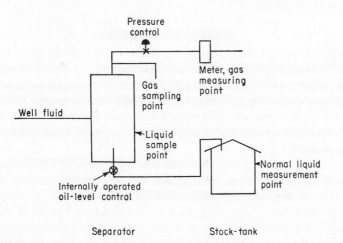

Fig. 5-2. Schematic layout of production facilities with indicated sample points for recombined samples.

so that the oil sample outlet is always submerged. Regardless of the method of collecting the fluid samples, the following data should be recorded:

1. A volume of oil in the separator compared with a volume of oil in the stock tank. This information permits the field calculation of a shrinkage factor for separator oil. The final shrinkage factor for separator oil is determined in the laboratory by flashing to stock-tank conditions.

2. The temperature and pressure of the separator.

3. The temperature and pressure of the stock tank.

4. The specific gravity of the stock-tank oil.

5. The amount of separator gas produced per stock-tank barrel (GOR, gas-oil ratio).

6. The gravity of the separator gas obtained in field or laboratory to correct meter measurements.

7. The flowing bottom-hole pressure and temperature.

8. The shut-in bottom-hole pressure and temperature.

PETROLEUM RESERVOIR ENGINEERING

With these data it is possible to obtain an analysis of the fluid entering the separator by properly recombining the separator liquid and separator gas

The recombination method of sampling is just as good as the bottom hole sampling technique for reservoirs where the flowing pressure exceeds the bubble-point pressure of the reservoir fluid. When the bottom-hole flowing pressure is less than the bubble-point pressure, free gas is produced from the reservoir. The bubble-point pressure for a recombination sample may be in excess of the original bubble-point pressure of the reservoir fluid owing to the excess gas. In most cases, these errors can be found and corrections made by taking into account the other data measured while collecting the sample.

Split-stream Sampling

The third method of sampling is split-stream sampling. This method is primarily used in sampling of gas condensate wells. The same qualifications and procedures are used in selecting and preparing a well for split-stream sampling as were used for obtaining a recombination sample. In addition to the data measured for recombination sample, it is necessary to note the temperature and pressure of the flow stream of the point at which the sample is collected.

In split-stream sampling, a small-diameter tube is inserted into the middle of the flow stream. Part of the flow is diverted through this tube into either an auxiliary separator or sampling bottles. In most cases, this sample is obtained by inserting the tube in tubing to 8 or 10 ft below the surface well head connections or in the flow stream just upstream of the separator. Precaution must be taken to center the tube properly in the main flow stream. The velocity of fluid flow in the sampling tube should be maintained equal to the velocity in the pipe. The quantity of fluid removed by the sampling tube should not exceed the quantity of fluid which is flowing in a comparable area within the main flow stream. This method is fast and, if used in conjunction with a small, temperature-controlled separator, will permit an analysis of the reservoir fluid in the field.

The sampling tube can be connected directly to the sample bottle. The sample thus collected is comparable to a bottom-hole sample taken under flowing conditions. The fluid from the sampling tube can be separated so as to collect separate gas and liquid samples as in the recombination method. Greater accuracy is obtained by separating the sample stream and collecting individual gas and liquid samples. Any variations in instantaneous gas-liquid rates are usually averaged out during the period of separation.

The split-stream method of sampling loses its accuracy with high-liquid content fluids. It is difficult to ensure the proper entry of gas and liquid into the sampling tube for high flowing liquid-gas ratios. Much of the liquids will be concentrated along the wall of the pipe owing to friction. The sample tube located in the middle of the pipe collects a greater proportion of gas than actually exists. More detailed information on sampling techniques is available in the literature.[2,3]

Once the samples have been collected, they are shipped to a laboratory for complete analysis. The type of laboratory analysis is dependent upon the type of reservoir and the information desired.

LABORATORY ANALYSIS OF RESERVOIR-FLUID SAMPLES

Laboratory analyses are the same regardless of the method used in collecting the sample. Before any tests can be performed, it is necessary that a certain amount of preparation be made on the field sample.

In the case of a bottom-hole sample it is necessary to raise the temperature and pressure of the field sample to reservoir conditions. The sample is then transferred to an appropriate test cell for analysis.

The preparation of a recombination sample is more complex. The gas and oil must be recombined in correct proportions to obtain a representative reservoir-fluid sample. The quantities which must be measured in the laboratory prior to recombination are briefly enumerated. A mole analysis of the separator liquid and gas sample are obtained usually by means of a fractional distillation column. A small volume of the separator liquid is raised to field separator test conditions and flashed to field stock-tank test conditions. The volume of separator liquid and the resulting volume of stock-tank liquid are used in calculating a separator shrinkage factor.

$$S_s = \text{separator shrinkage factor} = \frac{\text{volume stock-tank oil}}{\text{volume separator oil}} \qquad (5\text{-}1)$$

This shrinkage factor is used to determine the ratio of separator gas to separator liquid. The separator gas-oil ratio is then calculated, and the number of cubic feet of gas to be recombined with a given volume of separator liquid is determined.

$$R_p S_s = R_{sp} \qquad (5\text{-}2)$$

where R_p = separator GOR expressed with respect to stock-tank liquid
S_s = separator fluid shrinkage factor
R_{sp} = separator GOR expressed with respect to separator liquid

The separator liquid and gas samples are recombined in the proportion dictated by R_{sp}. The mole composition of the produced fluid can be calculated from the analysis of the separator liquid and gas. The procedure for calculating the composition of the produced fluid from separator fluid analyses is presented in Example 5-1.

When the fluids are recombined, the resulting fluid is equivalent to a bottom-hole sample, and it is only necessary to raise the pressure and temperature of the sample to reservoir conditions for further analysis.

The preparation of a split-stream sample is the same as a recombination sample if it was separated and collected as gas and liquid. If the sample was collected without separation, it is treated in the same manner as a bottom-hole sample.

There are many analyses which can be made on a reservoir-fluid sample. The amount of data desired determines the number of tests performed in the laboratory. There are three laboratory tests which are measured on all gas–crude-oil reservoir-fluid samples. These three tests determine the composite or total formation volume factor by flash liberation, the differential liberation formation volume factors and solution-gas–oil ratio, the gas-compressibility factor, and a study of the effect of surface separator conditions on flash volume factors.

Example 5-1. Calculation of Composition of Produced Fluid Analysis from Analyses of Separator Liquid and Gas.

1. Calculation of liquid density:

(1) Component	(2) Mole fraction[a]	(3) Mol wt	(4) Relative wt, lb/mole (2) × (3)	(5) Wt fraction (4)/Σ(4)	(6) Liquid density, lb/cu ft	(7) Liquid volume, cu ft/mole (4)/(6)
C_1	0.0238	16.042	0.38180	0.00239		
C_2	0.0069	30.068	0.20747	0.00130		
C_3	0.0155	44.094	0.68346	0.00429	31.64	0.02160
iC_4	0.0230	58.120	1.33676	0.00839	35.08	0.03810
nC_4	0.0239	58.120	1.38907	0.00872	36.35	0.03821
iC_5	0.0329	72.146	2.37360	0.01489	38.90	0.06102
nC_5	0.0440	72.146	3.17442	0.01992	39.27	0.08084
C_6	0.0610	86.172	5.25649	0.03298	41.36	0.12709
C_{7+}	0.7690	188.00	144.57200	0.90712	52.77	2.73966
			159.37507	1.00000		3.10652

[a] From laboratory and field data.

$$\text{Density of } C_{3+} = \frac{159.37507 - 0.38180 - 0.20747}{3.10652} = 51.11372$$

$$\text{Wt \% } C_1 \text{ in } C_{1+} = 0.239$$

$$\text{Wt \% } C_2 \text{ in } C_{2+} = \frac{0.20747}{159.37507 - 0.38180} = 0.1304$$

Density of separator liquid = 51.0 lb/cu ft

Gravity of stock-tank oil* = 29.2°API at 60°F

GOR* = 338.5 cu ft/bbl

SG = 0.8289   density of stock-tank oil = 51.64 lb/cu ft

Separator shrinkage factor = 0.960

Separator-gas–separator-liquid ratio = 338.5 cu ft/bbl (0.960)

= 325 cu ft/bbl

2. Calculation of composition of produced fluid:

(8) Component	(9) Mole fraction gas, $y_i{}^*$	(10) Component in gas, lb-moles/bbl $\dfrac{325}{380.69} \times$ (9)	(11) Mole fraction liquid, $x_i{}^*$	(12) Component in liquid, lb-moles/bbl $1.79672 \times$ (11)	(13) (10) + (12)	(14) Mole fraction composite, (13)/Σ(13) z_i
N	0.0088	0.00751			0.00751	0.00283
CO_2	0.0260	0.02220			0.02220	0.00838
H_2S	0.0140	0.01195			0.01195	0.00451
C_1	0.6929	0.59154	0.0238	0.04276	0.63430	0.23932
C_2	0.1401	0.11960	0.0069	0.01240	0.13200	0.04980
C_3	0.0731	0.06241	0.0155	0.02785	0.09026	0.03405
iC_4	0.0119	0.01016	0.0230	0.04132	0.05148	0.01942
nC_4	0.0210	0.01793	0.0239	0.04294	0.06087	0.02297
iC_5	0.0049	0.00418	0.0329	0.05911	0.06329	0.02388
nC_5	0.0046	0.00393	0.0440	0.07906	0.08299	0.03131
C_6	0.0010	0.00085	0.0610	0.10960	0.11045	0.04167
C_{7+}	0.0017	0.00145	0.7690	1.38168	1.38313	0.52186
	1.0000				2.65043	1.00000

* From laboratory and field data.

Number of moles of separator liquid/bbl of separator liquid

$$= \frac{\text{density of separator liquid, lb/bbl}}{\text{mol wt of separator liquid}} = \frac{51(5.61)}{159.37507} = 1.79672$$

Relative Total Volume

The relative total volume is measured by an equilibrium, or "flash," iberation process commonly called the pressure-volume, or PV, test. The est is started with a sample of reservoir fluid in a high-pressure cell at eservoir temperature and at a pressure in excess of the reservoir pressure. The volume in the cell under these conditions is known. The pressure in he cell is lowered by increasing the space available in the cell for the fluid. Depending on the cell, the volume is increased by withdrawal of mercury

* From laboratory and field data.

from the cell or the removal of a piston. A schematic representation of the
test is shown in Fig. 5-3. The cell pressure is lowered in small increments,
and the volume change for each pressure increment recorded. This procedure is repeated until a large change in the pressure-volume slope is indicated. This change in slope occurs when gas is liberated from solution.
The pressure at which the large change in the pressure-volume slope occurs
is considered the bubble point (see Fig. 5-4). After gas is liberated, the
procedure is altered and the sample is brought to equilibrium after each
change in volume. To obtain equilibrium, the sample is thoroughly agi-

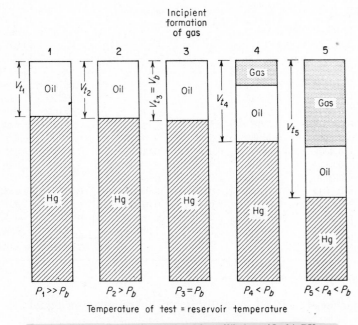

Fɪɢ. 5-3. Schematic representation of equilibrium (flash) PV test.

tated, either with an internal mixing device or by shaking the test cell.
The equilibrium pressure is recorded. This procedure is followed until the
volume capacity of the laboratory cell is reached. The resulting data are
expressed as the relative total volume. A procedure will be discussed later
for calculating the total formation volume factor using relative total volume data.

The pressure-volume test is conducted on a hydrocarbon mixture of
definite composition. As no hydrocarbon material is removed from the
cell during the test, the composition of the total hydrocarbon mixture in
the cell remains fixed at the original composition. The test is equivalent
to determining the volume relations along an isotherm of a phase diagram

such as Fig. 4-6. The gas liberated from solution is the equilibrium vapor phase which forms in the two-phase region.

Differential Oil Formation Volume Factor and Gas in Solution

The differential formation volume factor and gas in solution test begins in the same manner as the relative total volume test. The sample is placed in a high-pressure cell with the pressure above the reservoir bubble-point pressure and the temperature of the cell at reservoir temperature. The pressure is lowered in increments, and the volume change in the cell noted. The pressure is lowered until such time that free gas is liberated in the cell. Then for predetermined pressure or volume increments, mercury is withdrawn from the cell, gas is released from solution, and the cell is agitated until the liberated gas is in equilibrium with the oil. A schematic representation of the test is shown in Fig. 5-5.

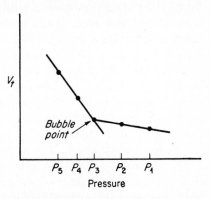

FIG. 5-4. Determination of bubble point from PV relations.

The total volume of the gas and oil is determined by the mercury-volume changes during the test. All the free gas is ejected from the cell at

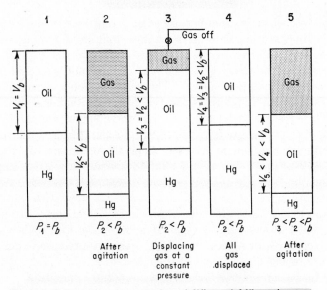

FIG. 5-5. Schematic representation of differential liberation test.

a constant pressure by injecting mercury. The volumes of the free gas displaced and the oil remaining in the cell are thus measured at cell conditions. The free gas is also measured at standard conditions. Depending upon future tests, either the free gas is analyzed at each stage of liberation or all the gas is collected in a sample bottle. This procedure is repeated for all the pressure increments until only oil remains in the cell at reservoir temperature and atmospheric pressure. The gas which is liberated by a differential process, from the bubble-point pressure to atmospheric pressure at reservoir temperature, can be calculated from these data.

A shrinkage factor for the oil, due to temperature change, is determined by ejecting the oil out of the cell into a container so that its volume can be measured at 60°F. With the appropriate calculating procedures, the differential formation volume factors at all the various pressure intervals can be computed. If all the liberated gas is collected in one sample container, compressibility factors for the composite gas can be measured.

The differential liberation process as conducted in the laboratory is a stepwise equilibrium process. At each pressure decrement vapor and liquid phases are brought to equilibrium.

The essential difference between the equilibrium test and the differential test is the removal of a portion of the fluid from the cell during the differential test. As a result the composition of the material remaining in the cell is progressively changed during the test. The materials removed are predominantly lighter hydrocarbons, methane, ethane, and propane.

Flash Separation Test

Another test which is often performed as a routine test is that of determining flash separation data on a bubble-point fluid sample. A cell is charged with a reservoir sample at a pressure above the original bubble-point pressure of the reservoir. Then part of this fluid is ejected from the cell into a stage separation system. The volume ejected is carefully measured and is flashed through the separation process, either one, two, or three stages. The pressure and temperature of these stages are carefully controlled. The volume of gas from each stage of separation and the volume of residue liquid remaining in the last stage of separation are measured. Thus, an oil formation volume factor for flash separation of a bubble-point sample and the flash gas-oil ratio off each stage of separation can be calculated.

This process is repeated for several first- and second-stage pressure combinations for a three-stage system. If a two-stage system is used, the process is repeated for several first-stage separator pressures. The following data are reported as a result of these tests:

The oil formation volume factor for the separator condition

The gravity of the stock-tank oil

The separator gas-oil ratios
The stock-tank gas-oil ratio
The total gas-oil ratio

The flash separation tests are essentially small-scale field separation processes. The results could be calculated from the composition of the reservoir
fluid as described in Chap. 4. However, the laboratory tests are easily
made and are preferred.

Gas Compressibility

The compressibility of the liberated gas or of the separator gas can be
determined in much the same manner as the composite volume factor of
the oil. The gas sample is charged into an evacuated pressure cell, and
the mass of gas in the cell is calculated. The pressure of the gas is increased
by injecting mercury into the cell. The volume of gas in the cell, at the cell
pressure, is determined by the amount of mercury injected. From these
measurements it is possible to calculate the compressibility factor for the
gas by utilizing equations shown in Chap. 4.

The compressibility factor of the gas liberated at each pressure decrement during a differential liberation process can be calculated from the
volumes occupied by the displaced gas at cell pressure and at atmospheric
pressure as measured in a receiver.

The analysis or gas gravity of the gas liberated at each pressure decrement can be determined. Using the analysis or gravity of the gas displaced
at each pressure step, the compressibility factor as a function of pressure
is calculated by the use of pseudo-critical properties and compressibility
curves. The methods of calculation are presented in Chap. 4. A sample
calculation from the measured gas gravity for a sample is presented later
in this chapter.

There are numerous other laboratory tests which are often requested on
reservoir-fluid samples. These analyses are normally considered in addition
to conventional PVT analysis. Some of the more frequently requested
analyses are fluid viscosity, differential-flash formation volume factors,
equilibrium ratios, and fluid composition. Following are brief discussions
of the procedures used in performing some of these laboratory tests.

Fluid Viscosity

If the oil viscosity is desired at reservoir pressure and temperature, it is
necessary to use a high-pressure rolling-ball viscosimeter. This instrument
measures the time required for a precision steel ball to roll a given distance
in a tube filled with oil. The time of travel is converted to viscosity by
means of a calibration curve for the instrument. The clearance between
the ball and tube can be changed by changing the ball diameter. The lower
the fluid viscosity, the smaller the clearance used.

In order to measure the reservoir-oil viscosity, the rolling-ball viscosimeter is charged with a reservoir-fluid sample at a pressure in excess of the original reservoir pressure. The pressure in the viscosimeter is lowered by differential separation of gas and liquid inside the viscosimeter. The viscosity of the liquid remaining from the separation is measured. To reduce the pressure in the rolling-ball viscosimeter, the top valve of the viscosimeter is opened slightly, permitting some liberated gas to escape. The pressure on the system is lowered, and more gas is liberated. The flow of gas is stopped, and the viscosimeter is rotated, permitting the ball to travel up and down the tube. This agitates the system and permits the oil and gas to reach equilibrium. The pressure existing in the cell when equilibrium is reached is the pressure at which the liquid viscosity is measured.

A sample can be differentially liberated in a rolling-ball viscosimeter provided the fluid shrinkage is not greater than approximately 50 per cent. There is excess volume in the top of the viscosimeter for the accumulation of the free gas, thus keeping the tube completely immersed in oil at all times. Should the sample have a shrinkage greater than 50 per cent, it would be necessary to perform the tests in two stages. The first stage would be with a sample above the original bubble point. This sample would be permitted to shrink approximately 50 per cent. The viscosimeter would then be charged with a sample already differentially liberated to some intermediate pressure. The viscosity of the liquid phase over the remaining pressure range would then be measured.

The rolling-ball viscosimeter can also be used for measuring gas viscosities. It is rather difficult, as the clearance between the ball and the tube must be extremely small. Any impurities or any small obstruction causes the ball to hang, yielding inaccurate readings. In most cases, the gas viscosity can be calculated from the gas analysis and the curves presented in Figs. 4-40 through 4-45. Thus, the engineer is able to calculate a gas viscosity at each pressure at which he has the analysis of the liberated gas. This is cheaper, easier, and perhaps more accurate than measuring these data in the laboratory.

Differential-Flash Oil Formation Volume Factors and Gas in Solution

The differential formation volume factor is not the same as the flash formation volume factor. Nor are the differential and flash gas-oil ratios the same. Thus, regardless of the testing procedures—flash or differential—some correction must be made on the resulting data to approximate the fluid behavior in the oil-production process.

Another type of test has been suggested by Dodson[4] which represents a combination of differential and flash liberation processes. Thus, this method is called composite liberation. The test is more difficult to perform and requires collection of larger reservoir-fluid samples. The sample is differentially liberated to a pressure. The oil resulting from this differential liberation is then flashed to stock-tank conditions. The gas liberated by the flash is the gas in solution, and the oil volume discharged from the cell compared with the resulting oil volume is the formation volume factor. The differential process is continued to a lower value of pressure. The resulting liquid at this lower pressure is flashed to stock-tank conditions, yielding data as in the previous flash. The process is repeated over a number of pressure steps to secure the complete analysis. The above laboratory procedure can be utilized to obtain a more accurate representation of the actual separation imposed upon an oil-gas system in the production process. This behavior will be discussed more thoroughly later, where means of approximating combination formation volume factors will be discussed. These tests are more expensive and time-consuming than the conventional tests. The question arises, Does the reservoir warrant such an expenditure, or is it necessary for the type of calculations desired? This question must be answered by the engineer.

Equilibrium Ratios

One of the more expensive laboratory tests is the determination of equilibrium ratios (K values) for a reservoir fluid. This particular type of test is a modification of the flash test previously described. The test is performed in the following fashion: A cell is charged with a reservoir-fluid sample above its saturation pressure and at reservoir temperature. The sample is then flashed by dropping the pressure in the cell until gas is liberated within the cell. The oil and gas are agitated until a state of equilibrium is reached. The oil and gas are both removed individually from the cell at the cell pressure, and each analyzed by means of a Podbielniak fractional distillation column. The equilibrium ratios at this particular value of pressure and temperature can be calculated from these analyses. The cell is recharged with another sample of oil above the original saturation pressure. This sample is flashed to a lower pressure than the previous sample. Once again the gas and liquid are removed individually at this sampling pressure, and the analysis of each phase determined. Equilibrium ratios at another pressure and at reservoir temperature are thus obtained. This procedure is continued over a desired range of pressure. The equilibrium ratios obtained by this procedure apply only to a material having the composition of the bubble-point fluid and do not necessarily apply to the gas-liquid systems which actually exist in the reservoir after fluid has been produced.

Fluid Composition

The composition of hydrocarbon fluids is normally determined by fractionation. A low-temperature fractionating column is used for gases, and a high-temperature column is used for liquid hydrocarbons. These columns are essentially pressure- and temperature-controlled fractionating towers in which the components are removed in decreasing order of their vapor pressures at the temperatures of the column. This type of analysis is accurate and requires small sample volumes. Impurities such as CO_2, etc., must be removed and determined by other means such as an Orsat absorption analysis.

An alternate system, the mass spectrometer, is available for gas analysis. The advantages of the mass spectrometer are that it is extremely fast and requires a very small sample. The mass spectrometer will measure other components such as hydrogen sulfide (H_2S), carbon dioxide (CO_2), and nitrogen (N_2), as well as the hydrocarbons in the system. For a gas sample which is to be analyzed by the mass spectrometer, it is necessary that it be collected in a glass- or ceramic-lined container to prevent a reaction between the sample and the walls of the retaining vessel. Analysis by means of the mass spectrometer is inexpensive. The disadvantage of this method is that a calibration or test sample must be made to permit quantitative as well as qualitative analysis, and as the number of components increase, the calculations required increase, so that for n components n simultaneous equations must be solved. Therefore for many analyses high-speed computing equipment is desirable to reduce costs and calculation time.

Chromatography is another means of determining fluid composition. A special column is used which separates the components on molecular weight and structure. This method is fast, requires a small sample, and is inexpensive. Chromatography is becoming a standard means of gas or liquid analysis.

Presentation of a Fluid Analysis for a Gas–Crude-oil System

The form in which data from a laboratory analysis of a gas–crude-oil system are reported is dependent upon the requirements of the individual company or laboratory which analyzed the fluid. There are two basic means of expressing the reservoir volume: (1) as a ratio of the surface volume and (2) as a ratio of some reference reservoir volume.

There are essentially six methods of referring the oil volume in the reservoir to surface or stock-tank volumes.

The most common of these is the differential oil formation volume factor. This volume factor is obtained by dividing the volume occupied by the liquid, at some reservoir pressure, by the volume that liquid would yield if it were differentially liberated to stock-tank conditions. Expressed in equation form, this would be

$$\text{Differential FVF} = B_{od} = \frac{V}{V_{Rd}} \tag{5-3}$$

where V_{Rd} = volume resulting by a differential process to stock-tank conditions

V = volume of liquid at some given pressure and temperature

B_{od} = differential formation volume factor

The second way of expressing reservoir volume relationships is essentially the same as the first, except that the standard surface volume is obtained by a flash liberation process. This quantity is referred to as a flash formation volume factor (FVF); it represents the ratio of the liquid volume at some pressure and temperature in the reservoir compared with the volume resulting from a flash liberation of that reservoir volume to some standard condition. This relation is expressed in Eq. (5-4).

$$\text{Flash FVF} = B_o = \frac{V}{V_{Rf}} \tag{5-4}$$

were V_{Rf} is the residual volume resulting from a flash liberation process and B_o is the formation volume factor by flash liberation.

Another means of expressing volume changes for reservoir fluids is commonly referred to as the shrinkage factor. The shrinkage factor is the ratio of the volume of residual fluid resulting from a liberation process to the volume at reservoir pressure and temperature required to yield that residual volume. There are two shrinkage factors, one resulting from a differential liberation process and another resulting from a flash liberation process. These two shrinkage factors are expressed in equation form as

$$\text{Differential shrinkage factor} = \frac{V_{Rd}}{V} = \frac{1}{B_{od}} \tag{5-5}$$

$$\text{Flash shrinkage factor} = \frac{V_{Rf}}{V} = \frac{1}{B_o} \tag{5-6}$$

The last means of expressing reservoir volumes with respect to surface volumes is the percentage shrinkage of the reservoir crude. As previously indicated for the other two expressions, there are two percentage shrinkage values dependent upon the type of liberation process involved. The percentage shrinkage is an expression of the change in volume from reservoir pressure and temperature to some standard pressure and temperature. The change in volume is compared with the resulting volume at the standard pressure and temperature. Expressed in equation form they are

$$\% \text{ shrinkage differential} = \frac{V - V_{Rd}}{V_{Rd}} \times 100 = (B_{od} - 1)100 \tag{5-7}$$

$$\% \text{ shrinkage flash} = \frac{V - V_{Rf}}{V_{Rf}} \times 100 = (B_o - 1)100 \tag{5-8}$$

The change in reservoir liquid volumes is often referred to some volume base other than surface volumes. The base to which it is most frequently referred is to the volume occupied by the liquid phase at the bubble-point pressure and reservoir temperature. This relationship is normally referred to as the relative oil volume and expressed in equation form is

$$\frac{V}{V_b} = \text{relative oil volume} \tag{5-9}$$

where V_b is the liquid volume at bubble-point pressure and reservoir temperature.

As would be expected, two numerical values exist for any one fluid, dependent upon the means of liberation used in obtaining the resulting volumes. The volume V in the above equation can take on two values, dependent upon the liberation process, resulting in relative oil volume by differential liberation and a relative oil volume due to flash liberation.

At times it is convenient to express reservoir volume relationships in terms of total volume, that is, the reservoir oil volume and its original complement of dissolved gas. These volumes, as in the case of the liquid volume, are referred to either stock-tank volumes or bubble-point volumes. The total volume ratios are expressed as

$$\text{Total vol factor} = B_t = \frac{V_t}{V_{Rf}}$$

$$= \frac{\text{vol of oil and its complement of liberated gas at } P}{\text{vol of stock-tank oil resulting from oil vol at } P}$$

and

$$\text{Relative total vol} = \frac{V_t}{V_b}$$

$$= \frac{\text{vol of oil and its complement of gas liberated at } P}{\text{vol of bubble-point oil required to yield vol } P}$$

As in the case of the other expressions of fluid volume, different values are obtained for different liberation processes. Most laboratory fluid analyses report only the relative total volume by the flash process. As the other total volume relationships must be calculated, their discussion is deferred to the section on correcting laboratory fluid-analysis data.

Other than expressing comparative volumes of reservoir fluids with respect to some standard measure, as either bubble-point liquid or stock-tank liquid, there also must be a means of expressing the gas in solution and the gas liberated from the oil owing to pressure changes. Liberated gas is that gas which is formed as the pressure is dropped below original bubble-point pressure. The gas in solution normally refers to the total amount of gas that has been liberated in bringing the liquid existing at the elevated temperature and pressure to stock-tank conditions and is reported in standard cubic feet per stock-tank barrel. The gas liberated at bubble-point pressure is zero, whereas the gas in solution is a maximum at the bubble-point pressure.

The numerical values obtained for the gas in solution and gas liberated will depend upon the process of liberation, flash or differential. In most cases the gas liberated by a differential liberation process is reported as the pressure is progressively decreased. Most engineers consider that the liberation process in the reservoir more closely approaches a differential process than a flash process. The fluid produced from the reservoir to the surface is considered to undergo a flash process, as it is felt that the liberation in the tubing and in the surface equipment closely approaches a true flash liberation system.

Idealized flash and differential formation volume factors B_o and

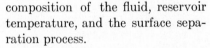

Fig. 5-6. Idealized comparison of flash and differential formation volume factors.

solution-gas–oil ratios R_s are presented in Figs. 5-6 and 5-7. It is noted that the flash liberation values are less than those of the differential process. This relationship between the two processes may occur as shown or in reverse. The exact relation of the two processes depends on the composition of the fluid, reservoir temperature, and the surface separation process.

An illustration of one form in which the results of a laboratory analysis of a crude oil-gas sample are reported is shown in Tables 5-1 through 5-3 and Figs. 5-8 through 5-11. The pressures at which the laboratory measurements were made are listed in Table 5-1, column 1. The relative total volumes resulting from a flash liberation process are listed in Table 5-1, column 2, and are presented graphically in Fig. 5-8. It is noted

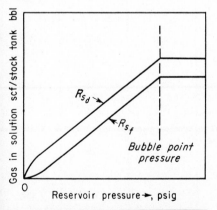

Fig. 5-7. Idealized comparison of flash and differential solution-gas–oil ratios.

that a marked change in slope occurs in the relative volume curve at the bubble-point pressure. The viscosities of the reservoir liquid resulting from a differential liberation are reported in Table 5-1, column 3. It will be noted

TABLE 5-1. RESERVOIR-FLUID SAMPLE TABULAR DATA[6]

(1) Pressure, psi	(2) Pressure-volume relation at 220°F, relative volume of oil and gas $\frac{V_t}{V_b}$	(3) Viscosity of oil at 220°F, cp	(4) Gas-oil ratio liberated per bbl of residual oil	(5) Gas-oil ratio in solution per bbl of residual oil	(6) Differential liberation 220°F $\frac{V}{V_{Rd}}$
5,000	0.9739				1.355
4,700	0.9768				1.359
4,465		1.004			
4,400	0.9799				1.363
4,100	0.9829				1.367
3,970		0.968			
3,800	0.9862				1.372
3,600	0.9886				1.375
3,530		0.931			
3,400	0.9909				1.378
3,200	0.9934				1.382
3,130		0.908			
3,000	0.9960				1.385
2,900	0.9972				1.387
2,820		0.889			
2,800	0.9985				1.389
2,695	1.0000	0.880	0	638	1.391
2,663	1.0038				
2,607	1.0101				
2,560		0.890			
2,512			42	596	1.373
2,503	1.0233				
2,358	1.0447				
2,300			89	549	1.351
2,197	1.0727				
2,008			150	488	1.323
2,000	1.1160				
1,960		0.997			
1,773	1.1814				
1,702			213	425	1.295
1,550	1.2691				
1,470		1.124			
1,351	1.3792				
1,315			290	348	1.260
1,180	1.5117				
1,010			351	287	1.232
992	1.7108				
940		1.300			
711	2.2404				
705			412	226	1.205
540	2.8606				
450		1.570			
410	3.7149				
405			474	164	1.175
289	5.1788				
150			539	99	1.141
0		2.872	638	0	1.066[a]

[a] At 60°F = 1.000
V = volume at given pressure
V_b = volume at saturation pressure at specified temperature

V_{Rd} = residual oil volume at 14.7 psia and 60°F
Gravity of residual oil = 28.8°API at 60°F
Specific gravity of liberated gas = 1.0626

378

that the viscosity decreases with pressure until the bubble point is reached and that above the bubble point the viscosity increases with pressure. The volumes reported in columns 4, 5, and 6 of Table 5-1 were all measured during one laboratory test, a differential liberation of the reservoir sample from pressures above the bubble point to stock-tank conditions of 14.7 psia and 60°F. The gas liberated from solution expressed as standard cubic feet per stock-tank barrel of oil resulting from the differential liberation is

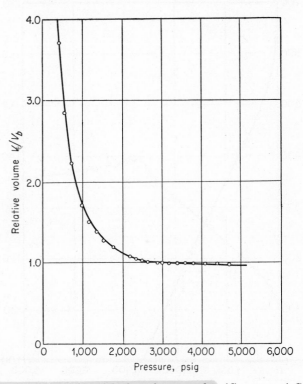

Fig. 5-8. Volumetric behavior of hydrocarbon sample. (*Courtesy of Core Laboratories, Inc.*)

reported in column 4. It is noted that as the pressure is progressively decreased, the amount of gas liberated progressively increases. The gas differentially liberated is shown as a function of reservoir pressure in Fig. 5-9. The gas in solution in the reservoir liquid at reservoir pressure and temperature is tabulated in Table 5-1, column 5, and illustrated graphically in Fig. 5-10. The differential formation volume factors are presented in Table 5-1, column 6, and Fig. 5-9.

The results of flash liberation of a bubble-point oil sample through various combinations of stage separation are indicated in Table 5-2 and Fig. 5-11. These data were obtained by flashing bubble-point oil through a separator system where the first-stage separator operated at 0, 50, 100, and 200 psig and the second stage of separation was always at 0 psig. The separator temperature (column 2) remained fairly stable in the neighborhood of 74 to 77°F. The first-stage separator gas-oil ratio (column 3) progressively decreased as the first-stage separator pressure increased. The second-stage separator gas-oil ratio (column 4) progressively increased.

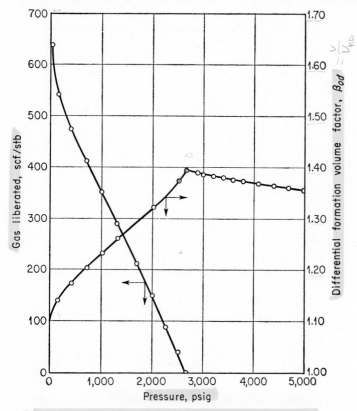

Fig. 5-9. Analysis of hydrocarbon sample; (○) measured data.

The total gas-oil ratio, which is the sum of the gas liberated in the separator and stock tank (sum of columns 3 and 4), reached a minimum value for a separation pressure of approximately 100 psig, after which it began to increase. The gravity of the stock-tank oil reached a maximum value for a first-stage separator pressure of approximately 100 psig. The shrinkage factor, which is the reciprocal of the flash formation volume factor, also reached a maximum at a first-stage separator pressure of 100 psig. The specific gravity of the flashed gas is reported as 0.9725 for the single-stage separator system. The flash separation values reported in Table 5-2 are used in conjunction with the values in Table 5-1 to calculate the fluid properties needed for reservoir calculations. The procedure for calculating the required fluid characteristics is discussed later in this chapter.

The last major part of a standard fluid analysis is the composition of the bubble-point liquid expressed as both weight and mole per cent. The sample analysis is indicated in Table 5-3. In all such analyses the liquid

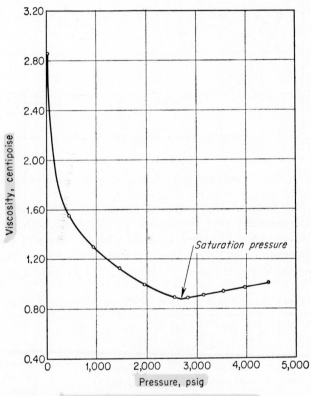

Fig. 5-10. Viscosity of liquid hydrocarbon.

density, the molecular weight, and the API gravity of the heptanes-plus fraction are reported. The reporting of these properties is necessary in choosing satisfactory pseudo-critical properties and equilibrium ratios.

A fluid analysis where the values are reported in terms of a unit volume of oil at the bubble-point pressure and reservoir temperature is presented in Table 5-4.

It is important that the engineer analyze the form in which the fluid-analysis data are reported. The engineer must convert the reported fluid

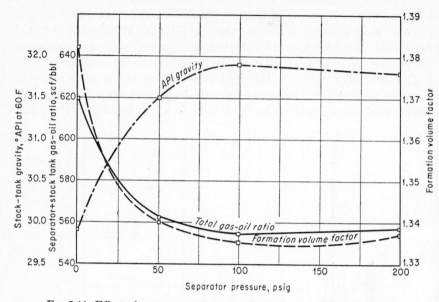

analysis to the basis of field operating conditions. For example, the gas production is normally that liberated at the separator with no record of the gas liberated from the stock-tank oil. In the standard analysis, no data are reported which relate liberated and solution-gas–oil ratios with respect to separator conditions. If differential data were used, an error

TABLE 5-2. SEPARATOR TESTS OF RESERVOIR-FLUID SAMPLE[6]

(1) Separator pressure, psi	(2) Separator temp, °F	(3) Separator gas-oil ratio[a]	(4) Stock-tank gas-oil ratio[a]	(5) Stock-tank gravity, °API at 60°F	(6) Shrinkage factor,[b] V_{Rf}/V_b	(7) Flash formation volume factor[c]	(8) Specific gravity of flashed gas
0	74	620		29.9	0.7236	1.382	0.9725
50	75	539	23	31.5	0.7463	1.340	
100	76	505	49	31.9	0.7491	1.335	
200	77	459	98	31.8	0.7479	1.337	

[a] Separator and stock-tank gas-oil ratio in cubic feet of gas at 60°F and 14.7 psia per barrel of stock-tank oil at 60°F.

[b] Shrinkage factor: V_{Rf}/V_b is barrels of stock tank oil at 60°F per barrel of saturated oil at 2,695 psig and 220°F.

[c] Formation volume factor: V_b/V_{Rf} is barrels of saturated oil at 2,695 psig and 220°F per barrel of stock-tank oil at 60°F.

TABLE 5-3. HYDROCARBON ANALYSIS OF RESERVOIR-FLUID SAMPLE[6]

Component	Weight %	Mole %	Density at 60°F, gm/cc	°API at 60°F	Mol wt
Methane	4.45	33.78			
Ethane	1.59	6.42			
Propane	3.56	9.82			
Isobutane	0.63	1.33			
n-Butane	1.43	2.99			
Isopentane	0.74	1.25			
n-Pentane	1.14	1.93			
Hexanes	2.12	2.99			
Heavier	84.20	38.97	0.8859	28.1	263
Hydrogen sulfide	0.14	0.52			
	100.00	100.00			

approaching 20 per cent for high separator pressures may be made in certain calculations.

In the next section the correction of the reported laboratory data for fallacies in technique in obtaining samples, for technique of measuring properties in the laboratory, and for field measurement practices are presented.

PREPARATION OF FLUID-ANALYSIS DATA FOR USE IN RESERVOIR CALCULATIONS

The manner in which fluid samples are collected, analyzed, and reported has been discussed. It was mentioned that certain corrections of reported data were required before application to a field problem. The type of corrections required will be dependent upon the state of depletion at which the fluid sample was collected and the sampling method used. It is desirable to obtain a fluid sample as early in the life of a field as possible so that the sample will closely approximate the original reservoir fluid. Collection of a fluid sample early in the life of a field reduces the chances of free gas existing in the oil zone of the reservoir.

There are three series of calculations which must be made on laboratory fluid-analysis data so that they can be used in reservoir calculations. First, the laboratory data as reported must be smoothed. This smoothing is to reduce any errors which might have been introduced in laboratory measurements. The smoothing is applied to the relative total volume and differential oil volume data.

The second series of calculations involves the computation of combination (often called flash) formation volume factors and gas-oil ratios. These parameters are calculated assuming that differential liberation occurs in the reservoir and that flash liberation occurs between the reservoir and the stock tank.

The other corrections which are necessary depend upon the sample and when it was obtained. If, from field data, it is apparent that the bubble point of the laboratory sample is in error, it becomes necessary to alter all the values reported in the fluid analysis to fit observed field conditions. There are several field conditions which might be used to indicate the accuracy of a fluid analysis. These conditions will be discussed in more detail when the correcting procedures are amplified.

Smoothing Laboratory Data

The smoothing of laboratory data is accomplished by two means. The total relative volume is fitted to a dimensionless compressibility curve which is referred to as the Y function. The Y function usually is linear with pressure when plotted on rectangular coordinate paper. The relative oil volume factor is fitted to a dimensionless volume change function. This function is referred to as the ΔV or Hurst[5] function. The logarithm of ΔV is usually linear with the logarithm of the difference in pressure and the bubble-point pressure.

In determining the best line which will fit the laboratory data points, two methods are recommended: the mean least-squares method and the method of averages.

Relative Total Volume Data. The pressure-volume relationship of a crude-oil–dissolved-gas system is a flash liberation process. A given mass of the reservoir fluid is expanded in a cell maintained at reservoir temperature, and the equilibrium pressure and volume observed. The laboratory data are usually expressed as relative total volume V_t/V_b. These data frequently require smoothing to correct for laboratory inaccuracies in measuring small volume changes. A dimensionless compressibility function is used to smooth the values reported by the laboratory. This function is defined as

$$Y = \frac{P_b - P}{P[(V_t/V_b) - 1]} \qquad (5\text{-}10)$$

where P_b = bubble-point or saturation pressure
  P = reservoir pressure for which Y is being calculated
  V_t/V_b = relative total volume at the pressure P

The Y function either is a straight-line function of pressure or has only small curvature. To smooth the relative total volume data, the Y function is computed and plotted as a function of the pressure P. The Y data will be erratic near the bubble-point pressure owing largely to difficulties associated with measuring small changes in volume in the laboratory pressure cell.

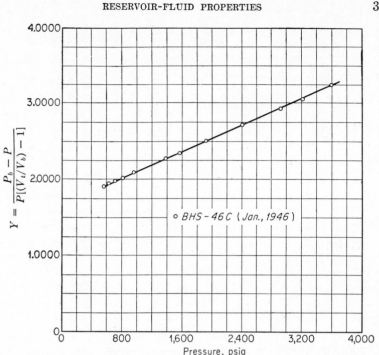

FIG. 5-12. Smoothing of relative total volume data. Bottom-hole sample BHS-46C.

Calculation of Best Y Curve. Straight-line relationships are conveniently expressed mathematically in the form

$$y = a + bx \qquad\qquad (5\text{-}11)$$

where b is the slope and a the intercept.

Several methods are available for fitting an expression such as Eq. (5-11) to an array of data point. Two of these methods will be considered here.

Fitting an equation to a given set of data implies determining the constants a and b such that the resulting straight-line equation will closely express the relationship throughout the range of the data. The method of least squares and method of averages can be used to obtain the equation of the best curve which can be fitted to the measured data points.

The least-squares fitting method can be applied to the Y function regardless of its curvature. If the data points approximate a straight line, the equation to be fitted is $Y = a + bP$. A value of Y corresponding to each pressure is calculated and then plotted as a function of pressure. The curve which best fits these points can then be calculated using the method of least squares. The laboratory relative volume data in Table 5-4 are smoothed by means of the least-squares method in Example 5-2. The resulting least-squares fit is shown in Fig. 5-12.

Note (handwritten, top margin): $(R_o)_B$ at Bubble Point. To convert To $(R_o)_{ST} = (R_o)_B \dfrac{u_B}{u_{F1D}}$

Table 5-4. Sample Analysis of BHS-46C at 224°F

(1) Pressure, psia	(2) $P_b - P$, psi	Flash liberation (3) Relative total volume $\frac{V_t}{V_b}$	(4) $\frac{V_t}{V_b} - 1$	(5) $Y = \frac{P_b - P}{P[(V_t/V_b) - 1]}$	Differential liberation (6) Relative oil volume[a] $\frac{V}{V_b}$	(7) $\Delta V = 1 - \frac{V}{V_b}$	(8) Gas expansion factor v, cu ft at S.C. per cu ft at P and 227°F	(9) Gas compressibility factor Z	(10) Gas gravity	(11) Relative gas volume[b]	(12) Gas liberated, cu ft at S.C. per bbl saturated oil
5,010		0.9859									
4,910		0.9875									
4,810		0.9893									
4,710		0.9909									
4,610		0.9928									
4,510		0.9946									
4,410		0.9966									
4,310		0.9985									
4,260		0.9995									
4,228	0	1.0000			1.0000						
4,210	18	1.0014	0.0014	3.0539							
4,185	43	1.0036	0.0036	2.8541							
4,120	108	1.0077	0.0077	3.4043							
4,060	168	1.0124	0.0124	3.3373							
3,998	230	1.0172	0.0172	3.3446							
3,885	343	1.0268	0.0268	3.2913							
3,810	418				0.9474	0.0526	212.2	0.9286	0.9245	0.0863	102.7
3,780	448	1.0363	0.0363	3.2649							
3,590	638	1.0555	0.0555	3.2021							
3,410	818				0.9050	0.0950	196.0	0.8991	0.9070	0.0782	188.5
3,215	1,013	1.1038	0.1038	3.0355							
3,010	1,218				0.8709	0.1291	175.8	0.8855	0.9080	0.0770	264.3
2,930	1,298	1.1524	0.1524	2.9068							
2,610	1,618				0.8386	0.1614	152.9	0.8832	0.9046	0.0779	331.1
2,415	1,813	1.2794	0.2794	2.6869							
2,210	2,018				0.8091	0.1909	129.1	0.8852	0.8956	0.0858	393.4
2,018	2,210	1.4757	0.4757	2.4839							
1,938	2,290	1.7220	0.7220	2.3259							
1,810	2,418				0.7812	0.2188	105.4	0.8885	0.8972	0.1002	452.9
1,578	2,650	1.9195	0.9195	2.2444							
1,410	2,818				0.7552	0.2448	80.7	0.9029	0.9064	0.1231	508.4
1,380	2,848	2.1663	1.1663	2.1635							
1,200	3,028	2.6111	1.6111	2.0709							
1,010	3,218				0.7307	0.2693	54.8	0.9532	0.9333	0.1709	561.2
975	3,253	3.0561	2.0561	2.0062							
825	3,403	3.5012	2.5012	1.9644							
715	3,513	3.9464	2.9464	1.9383							
610	3,618				0.7039	0.2961	32.8	0.9604	1.0052	0.2945	615.6
570	3,658	4.3918	3.3918	1.8921							
245	3,983				0.6735	0.3265	12.6	1.0061	1.2272	0.7965	671.7

TABLE 5-4 (Continued)

	Separator tests						Fluid analysis		
(13) Separator pressure, psig	(14) Separator gas-oil ratio, cu ft/bbl[c]	(15) Shrinkage factor, bbl STO/bbl saturated oil	(16) Separator gas gravity	(17) STO gravity, °API		(18) Components	(19) Separator gas	(20) Separator oil, mole %	(21) Reservoir fluid
100	1,017	0.6207	0.7447	38.75		N_2	0.20	0.0	0.13
75	1,054	0.6160	0.7570	38.65		CO_2	1.73	0.0	1.11
50	1,083	0.6130	0.7776	38.33		C_1	76.39	2.01	49.65
30	1,109	0.6090	0.8015	37.96		C_2	10.22	1.71	7.16
15	1,154	0.5984	0.8260	37.42		C_3	7.63	4.40	6.47
0	1,202	0.5920	0.8643	36.81		iC_4	1.04	1.98	1.38
						nC_4	1.77	4.01	2.57
						iC_5	0.27	2.17	0.95
						nC_5	0.38	2.49	1.14
						C_6	0.37	5.62	2.26
						C_{7+}	—	75.61	27.18

[a] Oil volume at pressure P per volume of saturated oil.
[b] Gas volume liberated at pressure P_i in dropping the pressure from P_{i-1} to P_i per volume of saturated oil.
[c] Standard cubic feet of gas per barrel of stock-tank oil.

Compressibility of oil = $18.0 \times 10^{-6} \frac{1}{psi}$
Reservoir oil density at P_b = 0.6313 gm/cc
Mol wt C_{7+} = 218
Density C_{7+} = 0.8472 gm/cc at 60°F

Field test conditions:
    Separator gas-oil ratio = 1,085:1
    Separator pressure = 78.5 psig
    Separator temperature = 76°F
    Tank gas-oil ratio = 40 cu ft/bbl
    Tank shrinkage = 2.26 %

Method of Least Squares.   The method of least squares for a straight line can be expressed for practical purposes as follows:

For each data point, write an equation of the form $y_i = a + bx_i$, where a and b are unknowns to be determined. For example, with four data points (x_1,y_1), (x_2,y_2), (x_3,y_3), (x_4,y_4) the array of Eqs. (5-11) becomes

$$
\begin{aligned}
y_1 &= a + bx_1 \\
y_2 &= a + bx_2 \\
y_3 &= a + bx_3 \\
y_4 &= a + bx_4
\end{aligned}
\tag{5-12}
$$

Summing the above equations, obtain the first normal equation:

$$
\sum_{i=1}^{4} y_i = 4a + b \sum_{i=1}^{4} x_i
\tag{5-13}
$$

To obtain the second normal equation multiply each equation by the coefficient of b, so that

$$
\begin{aligned}
y_1 x_1 &= ax_1 + bx_1^2 \\
y_2 x_2 &= ax_2 + bx_2^2 \\
y_3 x_3 &= ax_3 + bx_3^2 \\
y_4 x_4 &= ax_4 + bx_4^2
\end{aligned}
\tag{5-14}
$$

Summing Eqs. (5-14) yields the second normal equation

$$
\sum_{i=1}^{4} y_i x_i = a \sum_{i=1}^{4} x_i + b \sum_{i=1}^{4} x_i^2
\tag{5-15}
$$

Generalizing Eqs. (5-13) and (5-15) so that they apply to n points

$$
\sum_{i=1}^{n} y_i = na + b \sum_{i=1}^{n} x_i
\tag{5-16}
$$

and

$$
\sum_{i=1}^{n} x_i y_i = a \sum_{i=1}^{n} x_i + b \sum_{i=1}^{n} x_i^2
\tag{5-17}
$$

Equations (5-16) and (5-17) are the generalized normal equations for two dimensions. The constants a and b are evaluated by solving Eqs. (5-16) and (5-17) simultaneously, so that

$$
b = \frac{-\sum_{i=1}^{n} x_i \sum_{i=1}^{n} y_i + n \sum_{i=1}^{n} (x_i y_i)}{n \sum_{i=1}^{n} x_i^2 - \left(\sum_{i=1}^{n} x_i \right)^2}
\tag{5-18}
$$

$$a = \frac{\sum\limits_{i=1}^{n} y_i \sum\limits_{i=1}^{n} x_i^2 - \sum\limits_{i=1}^{n} x_i y_i \sum\limits_{i=1}^{n} x_i}{n \sum\limits_{i=1}^{n} x_i^2 - \left(\sum\limits_{i=1}^{n} x_i\right)^2} \qquad (5\text{-}19)$$

To verify the goodness of fit of the calculated line, the standard deviation is computed (see Example 5-2). The standard deviation is calculated by the following equation:

$$\text{Standard deviation} = S = \left[\frac{1}{n} \sum\limits_{i=1}^{n} (y_{0i} - y_{ci})^2\right]^{1/2} \qquad (5\text{-}20)$$

where n = number of data points

y_{0i} = value of y from the data

y_{ci} = value of y calculated from the equation of the curve

The standard deviation has the same units as the data. From Fig. 5-12 it is noted that several points near the bubble point were omitted from the calculations. These were eliminated on the basis of possible errors as previously mentioned.

Example 5-2. Smoothing of Relative Volume Data. A calculation of the best-fitting line by the method of least squares follows (Y function, BHS-46C):

$Y^* = \dfrac{P_b - P}{P[(V_t/V_b) - 1]}$ (1)	(2) Pressure P, psia	(3) YP	(4) P^2
3.4043	4,120	14,025.72	16,974,400
3.3373	4,060	13,549.44	16,483,600
3.3446	3,998	13,371.71	15,984,004
3.2943	3,885	12,798.35	15,093,225
3.2649	3,780	12,341.32	14,288,400
3.2021	3,590	11,495.54	12,888,100
3.0355	3,215	9,759.13	10,336,225
2.9068	2,930	8,516.92	8,584,900
2.6869	2,415	6,488.86	5,832,225
2.4839	1,938	4,813.79	3,755,844
2.3259	1,578	3,670.27	2,490,084
2.2444	1,380	3,097.27	1,904,400
2.2207	1,200	2,664.84	1,440,000
2.0709	975	2,019.13	950,625
2.0062	825	1,655.12	680,625
1.9644	715	1,404.54	511,225
1.9383	630	1,221.13	396,900
1.8921	570	1,078.49	324,900
47.6235	41,804	123,971.57	128,919,682

* Table 5-4, column 5.

The normal equations (1) and (2) are

(1) $\Sigma Y = na + b\Sigma P$

$$47.6235 = 18a + 41,804b$$

$$a = \frac{47.6235 - 41,804b}{18}$$

(2) $\Sigma YP = a\Sigma P + b\Sigma P^2$

$$123,971.57 = 41,804a + 128,919,682b$$

Substituting for a in (2)

$$123,971.57 = 41,804 \frac{47.6235 - 41,804b}{18} + 128,919,682b$$

$$123,971.57 = 110,602.933 - 97,087,467b + 128,919,682b$$

$$31,832,215b = 13,368.64$$

$$b = 0.00041997$$

then

$$a = \frac{47.6235 - (41,804)(0.00042)}{18} = \frac{30.0658}{18} = 1.6703$$

$$Y = a + bP = 1.6703 + 0.000420P$$

Method of Averages. The method of averages is the second means of calculating the equation of a straight line which describes a set of data points. The data are divided into two groups with approximately equal numbers of data points. The arithmetic average coordinates for each group are obtained. The equation which describes the line passing through these two average coordinate points would represent a fitted curve to the over-all group of data points. Mathematically these can be expressed as

$$P_1 = \frac{1}{k} \sum_{i=1}^{k} P_i$$

$$Y_1 = \frac{1}{k} \sum_{i=1}^{k} Y_i$$

(5-21)

$$P_2 = \frac{1}{n - k} \sum_{i=k+1}^{n} P_i$$

$$Y_2 = \frac{1}{n - k} \sum_{i=k+1}^{n} Y_i$$

where (P_1, Y_1) = first average point
(P_2, Y_2) = second average point
k = number of data points used in evaluating first point
n = total number of data points

From the foregoing equation which defines the coordinate points 1 and 2, it is possible to calculate the slope of the straight-line curve by the following equation:

$$b = \frac{Y_2 - Y_1}{P_2 - P_1} \qquad (5\text{-}22)$$

The value of a can readily be calculated once the value of b is obtained by inserting into the basic equation $Y = a + bP$ the value of b and the value of Y and P for one of the average points calculated. The result of the value of a will then be given by Eq. (5-23).

$$a = Y_2 - bP_2 = Y_1 - bP_1 \qquad (5\text{-}23)$$

The method of averages is more appropriately applied when there is more than one fluid sample; hence an illustration of the procedure is presented in Chap. 7 where several fluid analyses are involved.

Smoothing Differential Liberation Oil Volume Data. The flash total volume data were fitted to a straight line by means of a dimensionless compressibility term expressed as a function of reservoir pressure. Hurst[5] found that the differential oil volume data could be smoothed by the use of a dimensionless volume difference term as a function of the pressure difference. The dimensionless volume difference term is denoted as ΔV and is expressed as

$$\Delta V = \frac{V_b - V}{V_b} = 1 - \frac{V}{V_b} \qquad (5\text{-}24)$$

where V/V_b is the relative oil volume, volume of oil at P per volume of oil at the bubble-point pressure. The pressure difference term is denoted as ΔP and is expressed as

$$\Delta P = P_b - P \qquad (5\text{-}25)$$

where P_b = bubble-point pressure
P = reservoir pressure at which V/V_b is determined

Hurst found that the logarithm of ΔV and logarithm of ΔP should plot as a straight line. The equation of the resulting straight line is

$$\log \Delta V = B + C \log \Delta P$$
or $$\Delta V = D \, \Delta P^C$$
where $$D = 10^B \qquad (5\text{-}26)$$
$$C = \text{slope of resulting curve}$$
$$B = \text{value of } \log \Delta V \text{ when } \Delta P = 1, \log \Delta P = 0$$

The previously discussed least-squares or averaging methods can be used to determine the best values of the constants B and C. Once the equation of the best straight line has been determined, the engineer can calculate the relative oil volumes for the smoothed data. The relative oil volumes are expressed as

$$\frac{V}{V_b} = 1 - D\,\Delta P^C \qquad (5\text{-}27)$$

The smoothed differential formation volume factors can be computed from the relative oil volume by the following equation:

$$B_{od} = \frac{V}{V_b}\frac{V_b}{V_{Rd}} = B_{odb}\,(1 - D\,\Delta P^C) \qquad (5\text{-}28)$$

An illustrative example of this type of calculation for bottom-hole sample 46C, Table 5-4, is shown in Examples 5-3 and 5-4, and the resulting graphical plot is shown in Fig. 5-13.

Correction of Laboratory Sample Data for Separator Conditions

As indicated earlier, laboratory data are reported for differential liberation from the bubble-point pressure to standard conditions and for flash

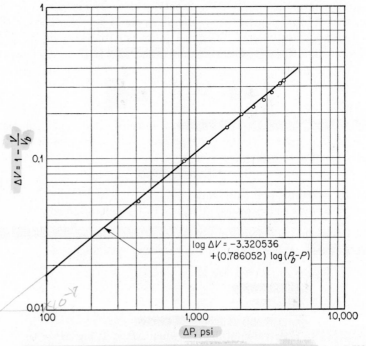

$$\log \Delta V = -3.320536$$
$$\qquad +(0.786052)\,\log(P_b{-}P)$$

FIG. 5-13. Smoothing laboratory-determined differential volume data; (○) BHS-46C.

liberation of a bubble-point sample through various separator systems. In the standard fluid-analysis report the effect of producing differentially liberated oil through a separator system is not considered.

————•◦•————

Example 5-3. Calculation of Best-fitting Line by Method of Least Squares (relative oil volume difference curve, BHS-46C).

Normal equations

(1) $$\Sigma \log \Delta V = an + b\Sigma \log (P_b - P)$$
(2) $$\Sigma[\log \Delta V \log (P_b - P)] = a\Sigma \log (P_b - P) + b\Sigma [\log (P_b - P)]^2$$

$P_b - P$	$\log (P_b - P)$	$[\log (P_b - P)]^2$	ΔV	$\log \Delta V$	$\log \Delta V \log(P_b - P)$
410	2.612784	6.826640	0.0526	−1.279014	−3.341787
818	2.912753	8.484130	0.0950	−1.022276	−2.977637
1,218	3.085647	9.521217	0.1291	−0.892451	−2.753789
1,618	3.208979	10.297546	0.1614	−0.792096	−2.541819
2,018	3.304921	10.922503	0.1909	−0.719194	−2.376879
2,418	3.383456	11.447774	0.2188	−0.659953	−2.232922
2,818	3.449941	11.902093	0.2448	−0.611189	−2.108566
3,218	3.507586	12.303160	0.2693	−0.569764	−1.998496
3,618	3.558469	12.662702	0.2961	−0.528562	−1.880871
3,983	3.600210	12.961512	0.3265	−0.486117	−1.750123
	32.624746	107.329277		−7.560616	−23.962889

(1) $$-7.560616 = 10a + 32.624746b$$
(2) $$-23.962889 = 32.624746a + 107.329277b$$

Solving for a in (1),

$$a = \frac{-32.624746b - 7.560616}{10}$$

Substituting for a in (2),

$$-23.962889 = 32.624746\ \frac{-32.624746b - 7.560616}{10} + 107.329277b$$

$$-23.962889 = -24.666317 - 106.437405b + 107.329277b$$
$$0.891872b = 0.703428$$
$$b = 0.788709$$

$$a = \frac{-32.624746(0.788709) - 7.560616}{10}$$
$$= -3.329301$$

$$\log \Delta V = \overset{b}{a} + \overset{m}{b} \log (P_b - P)$$
$$= -3.329201 + 0.788709 \log (P_b - P)$$

Example 5-4. Calculation of Smoothed Relative Oil Volume Data for BHS-46C.

(1)	(2)	(3)	(4)	(5)	(6)	(7)	(8)
P	$P_b - P$	$\log (P_b - P)$	$\dfrac{0.788709}{\log (P_b - P)}$	$\log \Delta V$	Positive mantissa of $\log \Delta V$	ΔV	Smoothed relative oil volume V/V_b
4,228	0						1.00000
3,690	538	2.73078	2.15379	−1.17541	0.82459−2	0.06677	0.93323
3,410	818	2.91275	2.29731	−1.03170	0.96830−2	0.09296	0.90704
3,010	1,218	3.08565	2.43367	−0.89553	0.10447−1	0.1272	0.8728
2,610	1,618	3.20898	2.53095	−0.79825	0.20175−1	0.1591	0.8409
2,210	2,018	3.30492	2.60662	−0.72258	0.27742−1	0.1894	0.8106
1,810	2,418	3.38346	2.66856	−0.66064	0.33936−1	0.2185	0.7815
1,410	2,818	3.44994	2.72099	−0.60821	0.39179−1	0.2465	0.7535
1,010	3,218	3.50759	2.76646	−0.56274	0.43726−1	0.2737	0.7263
610	3,618	3.55847	2.80659	−0.52261	0.47739−1	0.3002	0.6998
245	3,983	3.60021	2.83951	−0.48969	0.51031−1	0.3238	0.6762

$$\Delta V = 1 - \frac{V}{V_b}$$

$$\log \Delta V = \log D + C \log (P_b - P)*$$
$$C = 0.788709$$
$$\log D = -3.32920 = 7.67946 - 10$$

The normal practice is to consider that differential liberation occurs in the reservoir. It is necessary, then, that any expression of liberated gas should yield the same values as are obtained on the differential liberation test. These values should be expressed as standard cubic feet liberated per barrel of bubble-point oil or per barrel of flash stock-tank oil.

Flash liberation is considered to occur between the reservoir and the separator. The oil that leaves the reservoir is flashed to the separator, necessitating that the solution-gas–oil ratio and formation volume factor be determined by a flash process. To compensate for the simultaneous operation of both liberation processes, combination solution gas, liberated gas, and formation volume factor values are required.

The required data can be measured in the laboratory by the technique indicated by Dodson.[4] The purpose here is to show how a combination liberation system can be approximated by use of the differential and flash

* C and log D obtained from curve-fitting calculation in Example 5-3.

liberation data contained in a conventional fluid-analysis report. In order to calculate the combination fluid-analysis properties from standard analysis data, certain assumptions are required. These assumptions are:

1. Standard cubic feet of gas in solution per barrel of bubble-point oil is defined by the flash liberation test to separator pressure and temperature.

2. The standard cubic feet of gas liberated per barrel of bubble-point liquid is defined by a differential liberation process at reservoir conditions.

3. The standard cubic feet of gas remaining in solution at reservoir conditions which will be liberated upon producing that liquid to the separator by a flash liberation process is the difference between the original gas in solution and the differentially liberated gas corrected for the reservoir shrinkage of the fluid.

4. The relationship between the formation volume factors of flash and differential separated samples remains constant over the entire pressure range of interest.

5. The formation volume factor of the bubble-point liquid is determined by the flash liberation process to separator conditions and then to the stock tank.

The preceding assumptions limit the range of application of the calculating procedure to pressures above 500 psia. Assumptions 3 and 4 above are thought to be the more limiting. It is known that excess produced gas will affect the separator gas-oil ratio, composition of produced liquid and gas, and hence the formation volume factor and gas in solution values. The effect of excess gas production is not normally considered even when measuring combination fluid properties in the laboratory.

Correction of Solution-gas-Oil Ratios. First, consider the calculation of the gas in solution. As stipulated in the first assumption above, the total gas in solution is determined by flash-liberation of a bubble-point fluid sample. The engineer will have to select the separator conditions which most closely approximate field operating conditions. This means that the engineer must convert the gas-oil ratio reported in the analysis to the basis of 1 bbl of bubble-point oil to apply the following procedure.

The conversion from a stock-tank base to a bubble-point base requires that the gas-oil ratio be multiplied by the shrinkage factor (the reciprocal of the formation volume factor). In equation form this can be stated as

$$(R_{sf})_b = R_{sp}\frac{1}{B_{ofb}} \qquad (5\text{-}29)$$

where $(R_{sf})_b$ = gas in solution per barrel of bubble-point oil, scf
R_{sp} = gas liberated at the separator per stock-tank barrel of oil by flashing bubble-point oil, scf
B_{ofb} = bubble-point oil required to yield 1 bbl of stock-tank oil when flashed through the separator system, bbl

The standard cubic feet of gas liberated by a differential process can be reported with respect to stock-tank or bubble-point conditions. If the values are reported with respect to stock-tank oil, it is necessary to refer them to bubble-point oil. This conversion can be expressed as

$$(R_L)_b = (R_L)_{std} \frac{1}{B_{odb}} \tag{5-30}$$

where $(R_L)_b$ is the standard cubic feet liberated by differentially lowering the pressure from the bubble-point pressure P_b to some other reservoir pressure P referred to a barrel of bubble-point oil, $(R_L)_{std}$ is the standard cubic feet of gas liberated by differentially lowering the pressure from the bubble-point pressure to some other reservoir pressure referred to a barrel of liquid at standard conditions, and B_{odb} is the barrels of bubble-point oil required to yield 1 bbl of differentially liberated stock-tank oil.

The gas in solution at any reservoir pressure P with respect to a barrel of bubble-point liquid is the difference in the gas originally in solution and the gas differentially liberated. This can be expressed as

$$(R_s)_b = (R_{sf})_b - (R_L)_b \tag{5-31}$$

where $(R_s)_b$ is the standard cubic feet of gas in solution at P per barrel of bubble-point oil.

Multiplying $(R_s)_b$ by the flash bubble-point formation volume factor converts the gas in solution per bubble-point barrel to gas in solution per stock-tank barrel of oil, so that

$$R_s = (R_s)_b B_{ofb}$$
$$= [(R_{sf})_b - (R_L)_b] B_{ofb} \tag{5-32}$$
$$= R_{sp} - [(R_L)_{st}] \frac{B_{ofb}}{B_{odb}}$$

Calculations illustrating the conversion of the differential liberation gas data to field operating separation conditions is illustrated using fluid sample BHS-46C (Table 5-4) in Example 5-5.

Correction of Oil-volume Relations. It was assumed previously that the relationship between the flash and differential liberation processes would be constant at any reservoir pressure. In this case, the combination flash-differential formation volume factor can be calculated from the differential formation volume factor data and the flash formation volume factor for bubble-point oil.

The combination formation volume factor can be expressed as

$$B_o = \frac{B_{ofb}}{B_{odb}} B_{od} \tag{5-33}$$

where B_{ofb} = bubble-point oil required to yield 1 bbl of stock-tank oil when flashed through the separator to stock-tank conditions, bbl

B_{odb} = bubble-point oil required to yield 1 bbl of stock-tank oil when differentially liberated to stock-tank conditions, bbl

B_{od} = oil at reservoir pressure P required to yield 1 bbl of stock-tank oil when differentially liberated to stock-tank conditions, bbl

B_o = oil at reservoir pressure P required to yield 1 bbl of stock-tank oil when flashed through the separator, bbl. This term is often referred to as simply the flash formation volume factor

————•◆•————

Example 5-5. Gas in Solution Corrected for Field Separation Conditions of 50 Psig and 76°F.

correct FOR Difference ÷flash & Differential

$$R_s = R_{sp} - (R_L)_b B_{ofb}$$

$$B_{ofb}^* = \frac{1}{0.6130} = 1.63132$$

$$R_{sp}^* = 1{,}083 \text{ scf/STB}$$

Pressure	$(R_L)_b$*	$(R_L)_b B_{ofb}$	R_s
4,228	0	0	1083.0
3,810	102.7	167.54	915.46
3,410	188.5	307.50	775.50
3,010	264.3	431.16	651.84
2,610	331.1	540.13	542.87
2,210	393.4	641.76	441.24
1,810	452.9	738.82	344.18
1,410	508.4	829.36	253.64
1,010	561.2	915.50	167.50
610	615.6	1,004.24	78.76
245	671.7	1,095.76	-12.76

* Table 5-4.

Example 5-6. Determination of Combination Formation Volume Factors for 50 Psig and 74°F Separator Conditions (BHS-46C).

$$B_o = B_{od} \frac{B_{ofb}}{B_{odb}} = \frac{V}{V_b} B_{ofb}^*$$

* $B_{ofb} = 1/0.6130 = 1.63132$ (from Table 5-4, column 15).

Pressure	V/V_b†	B_o
4,228	1.0000	1.63132
3,810	0.9474	1.54551
3,410	0.9050	1.47634
3,010	0.8709	1.42072
2,610	0.8386	1.36802
2,210	0.8091	1.31990
1,810	0.7812	1.27439
1,410	0.7552	1.23197
1,010	0.7307	1.19201
610	0.7039	1.14829
245	0.6735	1.09869

† Table 5-4, column 6.

Calculations on the fluid sample of Table 5-4 illustrating the computation of combination formation volume factors are shown in Example 5-6 on page 394.

The calculation of combination volume factors and solution-gas–oil ratios is in close agreement with composite data measured in the laboratory.

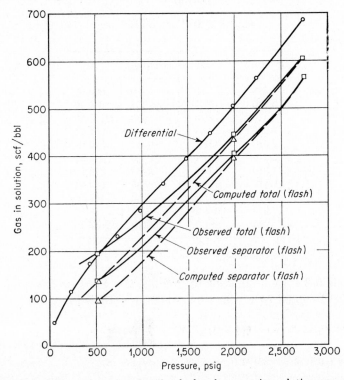

FIG. 5-14. Comparison of measured and calculated composite solution-gas–oil ratios.

Table 5-5 and Figs. 5-14 and 5-15 present a comparison of laboratory-measured differential and composite volume factors and solution ratios and values calculated by the methods described previously. The calculated values closely agree with the measured values to approximately 500 psia.

TABLE 5-5. COMPARISON OF MEASURED AND CALCULATED COMPOSITE VOLUME FACTORS AND SOLUTION RATIOS

Composite liberation data[4]

Reservoir data		Gas-oil ratios, cu ft/bbl			Formation volume factor B_o
Pressure, psig	Temp, °F	Separator	Stock tank	Total	
2,730	140	566.3	36.7	603	1.2552
1,986	140	403.3	40.6	443.9	1.1752
511	140	136.8	58.5	195.3	1.0776

Differential data[4]

Pressure, psig	Gas liberated, cu ft/bbl	Formation volume factor B_{od}
2,730	0	1.3442
1,986	180.86	1.2733
511	501.0[a]	1.1430[a]

Calculated composite

Pressure, psig	Gas-oil ratio, cu ft/bbl		Formation volume factor B_{of}
	Separator	Total	
2,730	566.3	603	1.2552
1,986	397.4	434.1	1.1890
511	98.5	135.2	1.0722

[a] Read from curve of Dodson.

Total Volume Factors

As mentioned previously, the total volume factor B_t is the reservoir volume of liquid and gas required to yield 1 bbl of stock-tank oil. Frequently the total volume factor is referred to as the reservoir volume occupied by one stock-tank barrel of oil and its complement of gas. The total volume factor is a function of the fluid shrinkage and volume of gas liberated. For this reason there are three distinct ways in which the total volume factor can be computed. The three total volume factors result from the different gas liberation processes: flash, differential, and a combination of the two processes.

Flash Total Volume Factor. The total volume factor for flash liberation is computed from data in the fluid-analysis report. The total relative volume (pressure-volume relation) is an expression of the total volume occupied by the liquid and gas originally comprising 1 bbl of bubble-point liquid. The volume of bubble-point oil required to yield a barrel of stock-tank oil (flash formation volume factor) is used to convert the total relative

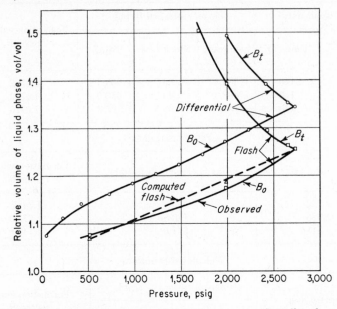

Fig. 5-15. Comparison of measured and calculated composite oil volume factors. (*Adapted from Dodson.*[4])

volume to the total volume factor. The flash total volume factor is given by the following relationships:

$$B_{tf} = \frac{V_t}{V_b}\frac{V_b}{V_{Rf}} = \frac{V_t}{V_b}B_{ofb} = \frac{V_t}{V_b}\frac{1}{(SH)_b} \qquad (5\text{-}34)$$

where V_t = reservoir volume occupied by gas and liquid
  V_b = volume occupied by bubble-point liquid
  V_{Rf} = stock-tank volume resulting in flashing bubble-point oil through separator to stock-tank condition
  B_{ofb} = bubble-point oil required to yield one stock-tank barrel of oil when flashed to separator conditions, bbl
  $(SH)_b$ = shrinkage factor, bbl of stock-tank oil per bbl of bubble-point oil

Differential Total Volume Factor. The second means of computing the total volume factor is to assume a complete differential process of liberation. This volume factor can also be calculated from data in the fluid-analysis report. The differential total volume factor is expressed by the following relation:

$$B_{td} = B_{od} + (R_L)_{st} \frac{B_g}{5.61}$$

$$= B_{odb} \left[\frac{V}{V_b} + (R_L)_b \frac{B_g}{5.61} \right] \qquad (5\text{-}35)$$

where B_{td} = total volume factor at pressure P

B_{od} = differential formation volume factor at pressure P, bbl reservoir oil per bbl of stock-tank oil

$(R_L)_{st}$ = gas liberated by differential liberation from bubble point to pressure P per stock-tank barrel of oil, scf

B_g = gas formation volume factor, reservoir cu ft/scf

5.61 = conversion factor, cu ft/bbl

B_{odb} = bubble-point oil required to yield 1 bbl of oil at stock-tank conditions by differential process, bbl

V/V_b = relative oil volume, volume of oil at P per volume of oil at bubble-point pressure

$(R_L)_b$ = gas liberated from P_b to P by differential process with respect to 1 bbl of oil at bubble-point pressure, scf

Combination Total Volume Factor. Generally it is considered that differential liberation occurs in the reservoir and flash liberation from the reservoir to the surface. For the computation of a more realistic total volume factor, the two liberation processes should be combined. The combination total volume factor is dependent upon the same assumptions used in calculating the combination fluid relationships previously discussed. The total volume factor can be expressed as

$$B_t = B_o + (R_L)_{st} \frac{B_{ofb}}{B_{odb}} \frac{B_g}{5.61}$$

$$= B_{ofb} \left[\frac{V}{V_b} + (R_L)_b \frac{B_g}{5.61} 5.61 \right] \qquad (5\text{-}36)$$

where B_o is the reservoir liquid volume at P required to yield 1 bbl of stock-tank oil by a flash process from P to separator and stock-tank conditions and $(R_L)_{st}$, B_g, B_{ofb}, and B_{odb} are as previously defined.

It is noted that the conversion from surface to reservoir volumes is dependent upon a flash process only. But as indicated in prior discussion of the combination analysis, the relationship between the oil and gas volume at P and the bubble-point oil volume is essentially a differential process.

The total volume factors for the fluid analysis presented in Table 5-4 are calculated by all three methods in Example 5-7. The results of Example 5-7 are graphically compared in Fig. 5-16.

Example 5-7. Calculation of Flash, Differential, and Combination Total Formation Volume Factors for BHS-46C. Separator conditions, 50 psig and 74°F. (See table, page 403.)

$$B_{ofb} = \frac{1}{0.6130} = 1.63132 \text{ from Table 5-4, column 15}$$

$$B_{odb} = \frac{1}{0.655} = 1.52671$$

(1) $B_{tf} = \dfrac{V_t}{V_b}\dfrac{V_b}{V_{Rf}} = \dfrac{V_t}{V_b} B_{odb}$

(2) $B_{td} = B_{odb}\dfrac{V}{V_b} + (R_L)_b \dfrac{B_g}{5.61}$

(3) $B_t = B_{ofb}\dfrac{V}{V_b} + (R_L)_b \dfrac{B_g}{5.61}$

Correcting Fluid-analysis Data to Reservoir Bubble-point Pressure

Sampling procedures sometimes are in error, so that the samples obtained have an erroneous bubble-point pressure. In partially depleted fields or in fields which originally existed at the bubble-point pressure, it is difficult to obtain a fluid sample which actually represents the original oil in the reservoir at the time of discovery. In these cases it is necessary to utilize other field data to determine the actual bubble point of the oil in the reservoir and correct the bubble-point pressure of the sample to this value.

In correlating laboratory bubble-point pressures with field data there are generalities in behavior which will aid in the determination of the actual bubble-point pressure. If the reservoir contained an initial gas cap, it is ordinarily assumed that the reservoir liquid was saturated at the original gas-cap pressure. Exceptions to this rule occur when there is great structural relief. Often, the oil in the lower segments of these reservoirs is undersaturated.

If the reservoir fluid was a single-phase system, it can be assumed that either the reservoir fluid was undersaturated or the bubble-point pressure was the original reservoir pressure. In most fields the production data and pressure behavior can be used to determine whether the fluid is undersaturated or not. If the producing gas-oil ratio remains stable and the reservoir pressure declines very rapidly for small changes in cumulative production, it would be suspected that the reservoir fluid was undersatu-

Pressure	V_t/V_b*	$B_{tf}^{(1)}$	V/V_b†	$(R_L)_b$‡	$1/B_0$§	$(R_L)_b(B_0/5.61)$	$(V/V_b)+(R_L)_b(B_0/5.61)$	$B_{td}^{(2)}$	$B_t^{(3)}$
4,260	0.9995	1.63049							
4,228	1.0000	1.63132							
4,210	1.0014	1.63360							
3,885	1.0268	1.67504							
3,810			1.0000	0		0	1.0000	1.52671	1.63132
3,780	1.0363	1.69054							
3,590	1.0555	1.72186							
3,410	1.1038	1.80065							
3,215			0.9474	102.7	212.2	0.08626	1.03366	1.57810	1.68623
3,010			0.9050	188.5	196.0	0.17143	1.07643	1.64340	1.75600
2,930	1.1524	1.87993							
2,610	1.2794	2.08711							
2,415			0.8709	264.3	175.8	0.26798	1.13888	1.73874	1.85788
2,210	1.4757	2.40734							
1,938			0.8386	331.1	152.9	0.38600	1.22460	1.86961	1.99771
1,810	1.7220	2.80913	0.8091	393.4	129.1	0.54318	1.35228	2.06454	2.20600
1,578			0.7812	452.9	105.4	0.76594	1.54714	2.36203	2.52388
1,410			0.7552	508.4	80.7	1.12297	1.87817	2.86742	3.06390
1,380	1.9195	3.13132							
1,200	2.1663	3.53393							
1,010	2.6111	4.25954	0.7307	561.2	54.8	1.82546	2.55616	3.90251	4.16991
975	3.9464	6.43784							
630			0.7039	615.2	0.8	3.34333	4.04723	6.17895	6.60233
610	4.3918	7.16443							
570									
245			0.6735	671.7	12.6	9.50258	10.17608	15.53592	16.6004

*Table 5-4, column 3. †Table 5-4, column 6. ‡Table 5-4, column 12. §Table 5-4, column 8.

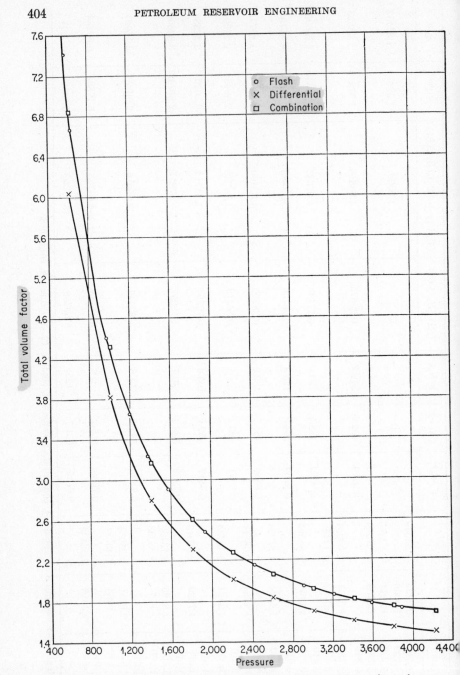

Fɪɢ. 5-16. Comparison of methods of calculating total formation volume factors.

rated. If the gas-oil ratios tend to increase early in the life of the field and the reservoir pressure does not decline at a very rapid rate, it could be concluded that the original reservoir pressure was the bubble-point pressure of the fluid. If there is no evidence from field data that the oil is under-saturated, the normal procedure is to consider the original reservoir pressure to be the bubble-point pressure of the oil in the reservoir.

It was mentioned previously that the sample analysis could be corrected to agree with field-observed data by changing the volume of gas recombined with the separator liquid sample. A calculation procedure can be used which assumes that the error in the laboratory-measured values is due to an inaccurate solution-gas volume.

The calculating procedure assumes that the gas required to correct the analysis can be added or subtracted according to the laboratory gas liberation data. All the other fluid factors required in reservoir computations are corrected using the above assumption.

Correction of Liberated Gas-Oil Ratio. The laboratory reports the gas liberated with respect to a barrel of fluid at the bubble-point or stock-tank conditions resulting from differential liberation. A set of total liberated gas-oil ratio values is reported with respect to a stock-tank barrel of oil resulting from flash liberation of a bubble-point sample through various separator conditions.

Consider first the necessity of correcting the differential liberation data to the field-observed bubble-point pressure. Assume that only differential liberation of the solution gas occurs in the reservoir. Also, assume that the reason the bubble-point pressure of the reservoir is in error is that too much or not enough of the liberated gas was collected in the sample. If this is the case, then the sample data can be corrected by removing or adding the quantity of gas required to satisfy the true bubble-point conditions. Before the required quantity can be calculated, it is necessary to assume that the solution behavior exhibited by the reservoir sample will be valid for correcting the liberated gas data.

The simplest way to make the desired correction is on a differential solution-gas–oil ratio curve such as Fig. 5-17. If the field-determined bubble-point pressure is greater than the sample bubble-point pressure, the new quantity of gas in solution is obtained by extrapolating the curve to the field-determined bubble-point pressure. When the field-determined bubble-point pressure is less than the sample bubble-point pressure, just stop the solution-gas–oil ratio curve at the field-determined value.

The differential liberated gas-oil ratio is corrected by adding or subtracting the difference observed between the total solution-gas–oil ratio at the sample and the field bubble points. When the field-determined bubble point is greater than the sample bubble point, add the observed difference to the sample liberated gas data. If the field bubble-point value is less than the sample value, subtract the observed difference from the sample values. Altering the gas liberation data when the bubble point is changed from 4,228 to 3,690 psia is illustrated in Example 5-8. Liberated gas is used in

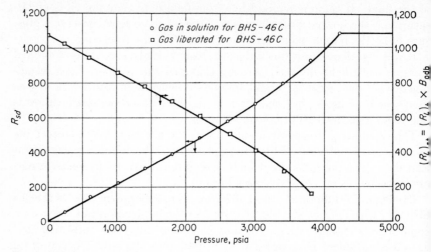

Fig. 5-17. Solution and liberated gas-oil ratios as determined by a differential liberation process.

this example, as differential solution-gas-oil ratios are not available to stock-tank conditions.

Example 5-8. Correction of Differentially Liberated Gas Data for Change in Bubble Point (BHS-46C).

Pressure P, psia	$(R_L)_{bs}$*	$(R_L)_{bs} - (R_L)_{sb}$†	$(R_L)_b$‡
4,228—sample BP		−130	
3,690—field BP	130	0	
3,410	188.5	58.5	62.7
3,010	264.3	134.3	143.9
2,610	331.1	201.1	215.5
2,210	393.4	263.4	282.3
1,810	452.9	322.9	346.1
1,410	508.4	378.4	405.6
1,010	561.2	431.2	462.2
610	615.6	485.6	520.5
245	671.7	541.7	580.6

* $(R_L)_{bs}$ is the gas liberated per barrel of bubble-point oil for the laboratory sample.

† $(R_L)_{sb}$ is the gas liberated from the sample bubble-point pressure to the observed field bubble-point pressure per barrel of sample bubble-point oil.

‡ $(R_L)_b$ is the gas liberated per barrel of field bubble-point oil. $(R_L)_b = [(R_L)_{bs} - (R_L)_{sb}] V_b/V_{sb}$, where V_b/V_{sb} is the relative oil volume of the field bubble-point oil to the sample bubble-point oil.

Correction of Differential Relative Oil Volume. To correct the differential relative oil volume to bubble-point conditions other than those measured in the laboratory it is necessary to assume that the dimensionless volume change quantity ΔV used in smoothing the laboratory data will define the new fluid system. In smoothing the differential liberation data it was shown that

$$\log \frac{V_b - V}{V_b} = \log \Delta V = \log D + C \log (P_b - P) \qquad (5\text{-}26)$$

terms as defined previously.

The above equation will reduce to

$$\Delta V = 1 - \frac{V}{V_b} = D(P_b - P)^C$$

Rewriting in terms of the relative oil volume factor, the following is obtained:

$$\frac{V}{V_b} = 1 - D(P_b - P)^C \qquad (5\text{-}27)$$

The values of C and D are determined from the smoothed sample data. The field-determined bubble-point pressure is now used as P_b. The new value of the relative oil factor at any reservoir pressure P can be calculated with this relationship.

The preceding equations must be used to calculate at least one relative oil volume if the field bubble-point pressure is greater than the sample bubble-point pressure. If the field value is less than the sample value of the bubble-point pressure, the relative oil volumes can be calculated from the existing sample data.

Equation (5-37) is required to calculate the differential oil formation volume factor at the field-determined bubble-point pressure when that pressure is above the bubble-point pressure reported for the laboratory sample. The formation volume factors reported for the laboratory sample are still valid for the reservoir fluid sample at all pressures below its reported bubble point.

The differential oil formation volume factor B_{odb} for field determined bubble-point conditions can be calculated using corrected oil volume data and the sample differential oil volume factor as follows:

$$\Delta V \frac{1}{B_{ods}} = \frac{V_b - V}{V_b} \frac{V_{Rd}}{V} = \frac{V_{Rd}}{V} - \frac{V_{Rd}}{V_b}$$

$$= \frac{1}{B_{ods}} - \frac{1}{B_{odb}}$$

$$B_{odb} = \frac{B_{ods}}{1 - \Delta V}$$

$$= \frac{B_{ods}}{1 - D(P_b - P)^C}$$

$$= \frac{B_{ods}}{V/V_b} \tag{5-37}$$

where B_{ods} is the laboratory-reported differential formation volume factor at pressure P and V/V_b is the relative oil volume factor at pressure P corrected for the change in bubble-point pressure.

The corrected relative oil volume can be calculated by dividing the sample relative oil volume at each pressure by the sample relative oil volume at the corrected bubble-point pressure. This relationship is stated as

$$\frac{V}{V_b} = \frac{V/V_{bs}}{V_b/V_{bs}} \tag{5-38}$$

This relationship is valid because in a differential liberation process the gas is removed at each pressure step. Thus the liquid volumes resulting from differential liberation actually evolve from the liquid volume existing at each preceding pressure. This method of correcting the relative oil volume is demonstrated in Example 5-9 for the bottom-hole sample in Table 5-4.

— · • • · —

Example 5-9. Adjustment of Relative Oil Volume and Differential Formation Volume Factors for BHS-46C.

Pressure, psia	Sample relative oil volume V^*/V_{bs}	Adjusted relative oil volume V/V_b	Adjusted differential formation volume factor B_{od}^*
4,228	1.0000		
3,810	0.93301	1.0000	1.42444
3,410	0.90687	0.9720	1.38453
3,010	0.8727	0.9354	1.33236
2,610	0.8408	0.9012	1.28366
2,210	0.8106	0.8688	1.23755
1,810	0.7817	0.8378	1.19343
1,410	0.7538	0.8079	1.15083
1,010	0.7267	0.7789	1.10946
610	0.7004	0.7507	1.06931
245	0.6768	0.7254	1.03328

* B_{od} remains unchanged from the sample data.

$$\frac{V}{V_b} = \frac{V/V_{bs}}{V_b/V_{bs}} = \frac{V/V_{bs}}{0.93301}$$

$$B_{od} = \frac{V}{V_{bs}} B_{odb} = \frac{V}{V_{bs}} \quad (1.52671)$$

— · ● · —

Correction of Flash Formation Volume Factors for Changes in Bubble-point Pressure. The assumptions made in calculating the combination differential-flash system are used in calculating the new flash formation volume factor and solution-gas–oil ratio. In the case of the combination system it was shown that

$$B_o = B_{od} \frac{B_{ofb}}{B_{odb}} \qquad (5\text{-}33)$$

Rewriting the above equation to solve for B_{ofb} results in

$$B_{ofb} = \frac{B_o}{B_{od}} B_{odb} = (B_{ofb})_s \frac{V_b}{V_{bs}} \qquad (5\text{-}39)$$

where B_o and B_{od} = flash and differential formation volume factors at a given reservoir pressure, usually sample bubble-point pressure

B_{odb} = corrected differential formation volume factor at new bubble-point pressure

V_b/V_{bs} = relative oil volume reported in sample for fluid at field bubble-point pressure

The new value of gas in solution is calculated by the same equations used in calculating the gas in solution for the combination system. It is

$$R_s = (R_{sp})_s \pm (R_L)_{st} \frac{B_{ofb}}{B_{odb}} \qquad (5\text{-}32)$$

where R_{sp} is the gas in solution from the sample analysis and $(R_L)_{st}$ is the standard cubic feet of gas added to or subtracted from the sample liberated gas values to correct for the change in bubble-point pressure. The flash formation volume factor and solution-gas–oil ratio for the fluid sample of Table 5-4 are corrected in Example 5-10.

— · ● · —

Example 5-10. Correction of Adjusted Sample for Surface Separator Conditions of 50 Psig and 74°F.

Pressure, psia	$(R_L)_b(B_{ofb})_s$*	R_s†	Adjusted relative oil volume V/V_b	B_o‡
4,228				
3,690	212.1	870.9	1.0000	1.52207
3,410	307.5	775.5	0.9720	1.47945
3,010	431.2	651.8	0.9354	1.42374
2,610	540.1	542.9	0.9012	1.37169
2,210	641.8	441.2	0.8688	1.32237
1,810	738.8	344.2	0.8378	1.27519
1,410	829.4	253.6	0.8079	1.22968
1,010	915.5	167.5	0.7789	1.18554
610	1,004.2	78.8	0.7507	1.14262
245	1,095.8	−12.8	0.7254	1.10411

* $B_{ofb} = (B_{ofb})_s(V_b/V_{bs}) = (1/0.6130)0.93301 = 1.52207$
† $R_s = R_{si} - (R_L)_b(B_{ofb})_s = 1,083 - (R_L)_b(B_{ofb})_s$
‡ $B_o = (V/V_b)B_{ofb} = (V/V_b)1.52207$

Correction of Total Volume Factors for Different Bubble-point Pressure.
The only total formation volume factor which requires additional corrections is the flash total formation volume factor. This factor is calculated from the relative volume factor as reported in the fluid analysis. In order to correct the total volume factor, it is necessary to correct the total relative volume.

The original total relative volume data were smoothed by means of the Y function. If it is assumed that the slope and the intercept of the fitted Y function curve are valid, the new relative total volume can be calculated from the *equation of the Y* function.

The Y function is defined as

$$Y = a + bP = \frac{P_b - P}{P[(V_t/V_b) - 1]}$$

Rewriting the above equation for the relative total volume results in

$$\frac{V_t}{V_b} = 1 + \frac{P_b - P}{aP + bP^2} = 1 + \frac{1 - (P_b/P)}{a + bP} \qquad (5\text{-}40)$$

where a and b are defined from the Y curve fitted to laboratory data, P_b is the new bubble-point pressure, and P is the reservoir pressure at which V_t/V_b is desired.

The corrected flash total volume factor can now be calculated using the definition of the flash total volume factor,

$$B_{tf} = \frac{V_t}{V_b} B_{ofb}$$

provided both the relative total volume V_t/V_b and the formation volume factor B_{ofb} refer to the new bubble-point conditions. The above correcting procedure was applied to the fluid sample in Table 5-4. The calculations are presented in Examples 5-10 and 5-11.

FLUID-ANALYSIS DATA ON GAS-CONDENSATE SYSTEMS

Gas-condensate systems are analyzed by a technique different from that discussed for a gas–crude-oil system. The methods used in analyzing such a sample will be discussed in the succeeding section.

As was mentioned earlier, a bottom-hole sampling technique cannot be used on a gas-condensate well because of the accumulation of liquid in the bottom of the hole. Thus, all reservoir fluid samples used for the analysis of gas-condensate fluids are of either the recombination or split-stream variety. In most cases, gas and liquid are collected from a high-pressure separator. The same field measurements are made as previously discussed in the collection of a recombination separator sample. The quantities collected are brought to the laboratory and carefully analyzed and recombined to represent the reservoir fluid. The same precautions apply for a gas-condensate fluid that applied for a gas–crude-oil fluid with respect to recombination sample.

Laboratory Measurements

In the laboratory a standard analysis consists of measuring the pressure-volume relationship, a pressure depletion history, the analysis of the well stream effluent at various stages of the pressure depletion, a volume-pressure depletion relationship, and compressibility factors for the produced gas. There are other analyses and special calculations which an engineer may desire. These extra analyses must be requested in addition to the standard fluid analysis. As these extra analyses are special, they will not be discussed in great detail in the following text.

Relative Volume. In measuring the relative volume relationship the same procedure is used as was used in the gas–crude-oil system. The difference in the two fluid-measuring systems is that for a gas-condensate system, the pressure cell has a glass window covering its entire length. This glass window permits visual observation of sample changes resulting from changing the pressures. The relative volume relationship does not behave as does the crude-oil–natural-gas system. A sharp change normally does not exist in the shape of the pressure-volume curve at the dew point as occurred at the bubble point of a gas–crude-oil system. Relative volume data for a condensate fluid are illustrated in Table 5-6 and Fig. 5-18. It is noted that the dew point had to be determined by visual observation and not by the change in the slope of the relative volume curve.

Example 5-11. Adjusting Relative Total Volume to New Bubble Point and Calculation of Adjusted Total Flash Formation Volume Factor.

Pressure, psia	$\dfrac{P_b}{P}$	$\dfrac{P_b}{P}-1$	$0.00042P$	$1.6703 + 0.00042P$	$\dfrac{(P_b/P)-1}{1.6703 + 0.00042P}$	Adjusted relative total volume V_t/V_b	B_{tf}
3,690	1.0000	0				1.0000	1.5221
3,410	1.0821	0.0821	1.4322	3.1025	0.0265	1.0265	1.5624
3,010	1.2259	0.2259	1.2642	2.9345	0.0770	1.0770	1.6393
2,610	1.4138	0.4138	1.0962	2.7665	0.1496	1.1496	1.7498
2,210	1.6697	0.6697	0.9282	2.5985	0.2577	1.2577	1.9143
1,810	2.0387	1.0387	0.7602	2.4305	0.4274	1.4274	2.1726
1,410	2.6170	1.6170	0.5922	2.2625	0.7147	1.7147	2.6099
1,010	3.6535	2.6535	0.4242	2.0945	1.2669	2.2669	3.4504
610	6.0492	5.0492	0.2562	1.9265	2.6209	3.6209	5.5113
245	15.0612	14.0612	0.1029	1.7732	7.9298	8.9298	13.5918

TABLE 5-6. PRESSURE-VOLUME RELATIONSHIP FOR CONDENSATE FLUID SAMPLE[6]

Pressure, psi	Relative volume
5,000	0.8293
4,800	0.8509
4,600	0.8753
4,400	0.9034
4,300	0.9186
4,200	0.9340
4,100	0.9513
3,900	0.9880
3,830 (dew point)	1.0000
3,797 (reservoir pressure)	1.0094
3,600	1.0552
3,200	1.1662
2,803	1.3185
2,400	1.5325
2,000	1.8424
1,622	2.2886
1,200	3.1160
932	4.0870
756	5.0572

Pressure Depletion Study. Another part of a standard laboratory analysis is a pressure depletion relationship for the fluid system. During this study the volume produced is measured and the well stream effluent is sampled at various stages of depletion. These samples are used to determine the composition and compressibility of the well stream effluent. The depletion test is performed in the completely windowed cell. When the laboratory test is made, the cell is charged with a known volume of reservoir fluid. The pressure in the cell is lowered by bleeding gas from the top of the cell, simulating a well producing only gas with the liquid remaining in the reservoir. The volume of gas produced is measured and expressed in standard units. At predetermined pressures, the produced well stream is sampled so that an analysis of the producing stream can be obtained. Also, compressibility factors are determined on these well stream samples. From these data, volume per cent produced at any stage of pressure depletion is calculated by dividing the volume produced by the volume originally in the cell, both at standard conditions. The results of a pressure depletion study for a gas-condensate fluid are illustrated in Tables 5-7 and 5-8 and Figs. 5-19 to 5-21.

From a study such as reported in Tables 5-7 and 5-8, it is possible to design the proper gas facilities for the well stream effluent. It is also possible to determine the advisability and economics of a cycling project to recover fluids that would condense during pressure depletion.

TABLE 5-7. DEPLETION STUDY OF CONDENSATE FLUID SAMPLE[6]

	Reservoir pressure, psig						
	3,830	3,797	3,500	2,800	2,100	1,400	600
Component:							
Methane	84.80	84.81	84.95	85.61	86.07	85.94	85.41
Ethane	5.95	5.95	5.97	5.97	5.95	6.08	5.99
Propane	2.55	2.55	2.60	2.57	2.55	2.57	2.68
Isobutane	0.47	0.47	0.44	0.47	0.48	0.48	0.49
n-Butane	0.75	0.75	0.74	0.70	0.72	0.73	0.75
Isopentane	0.30	0.30	0.29	0.28	0.27	0.29	0.31
n-Pentane	0.21	0.21	0.21	0.21	0.21	0.24	0.25
Hexanes	0.37	0.37	0.35	0.34	0.35	0.41	0.48
Heptanes plus	2.24	2.23	2.09	1.48	1.03	0.88	1.20
Carbon dioxide	2.36	2.36	2.36	2.37	2.37	2.38	2.44
	100.00	100.00	100.00	100.00	100.00	100.00	100.00
Heptanes plus, mol wt	128	128	126	124	121	119	119
gpM:							
Propane plus		2.554	2.470	2.144	1.929	1.883	2.124
Butanes plus		1.855	1.771	1.442	1.227	1.178	1.403
Pentanes plus		1.465	1.384	1.061	0.843	0.791	1.007

TABLE 5-8. DEPLETION STUDY OF CONDENSATE FLUID SAMPLE ILLUSTRATING DEVIATION FACTOR AND VOLUME PRODUCED[6]

	Reservoir pressure, psig						
	3,797	3,500	2,800	2,100	1,400	600	0
Deviation factor Z	0.9430	0.9351	0.9245	0.9210	0.9397	0.9771	
Volume % produced	0	6.490	23.571	42.462	61.730	83.535	99.060[a]
	Ideal expansion						
Deviation factor Z	0.943	0.931	0.907	0.905	0.923	0.961	1.000
Volume % produced	0	6.602	23.216	42.174	62.096	84.182	100.000

[a] Residual liquid:
   Gallons at 60°F per MMscf of original fluid = 426.6
   Mol wt = 159
   Density at 60°F = 0.8344

Gas Compressibility. The compressibility factor for the produced gas is measured in the same fashion as was reported for the gas of a gas–crude-oil system. The compressibility factors for the sample used as an illustra-

tion are reported in Fig. 5-21 and Table 5-8. A calculated compressibility factor where no condensation is considered is also reported in Table 5-8 for comparison. From this comparison it is noted that the compressibility factors where no liquid condensation is considered are smaller than the compressibility factors for the produced gas. This is as would be expected, as the condensed liquid occupies less space than the same number of moles of gas would occupy if it were under like pressure conditions.

Use of Condensate Analysis. A careful study of the condensate fluid analysis that is reported in Tables 5-6 through 5-8 enables the engineer to evaluate better the behavior of a gas-condensate system. It also enables him to gain an understanding of fluid behavior as affected by composition and changes in pressure and temperature. As an example, observe the change in well stream composition as affected by retrograde condensation. It is noted that the fraction of the well stream comprised of methane and ethane changes very little regard-

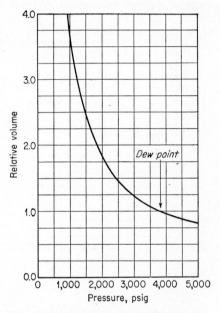

Fig. 5-18. Pressure-volume relationship for a gas-condensate fluid. (*Courtesy of Core Laboratories, Inc.*[6])

less of the pressure of the reservoir. It becomes quite apparent that only minute volumes of methane and ethane are retained in the condensate in the reservoir. The primary component which changes with changing pressures is the heptanes plus (C_{7+}). It is noted that as the pressure continues to decline, the heptanes-plus (C_{7+}) fraction in the well stream effluent goes through a minimum. If the pressure depletion study were carried to an ultimate abandonment pressure of 14.7 psia, it would be found that all the heptanes plus would not be vaporized. All the heptanes plus do not revaporize because most of the lighter components, methane and ethane, have been produced. Thus, insufficient quantities of volatile constituents remain to cause the heptanes-plus fraction to revaporize.

APPLICATIONS OF FLUID-ANALYSIS DATA

The data reported in conventional fluid-analysis studies have many applications in reservoir engineering. Perhaps the application with which most engineers are familiar is the use of fluid-analysis data in reservoir material-balance studies. The application of these data to material-balance studies are discussed in Chap. 8 and a companion volume. Of course, in the case of a gas-condensate system they are also used as an aid in the design of surface separation systems as well as the evaluation of reservoir

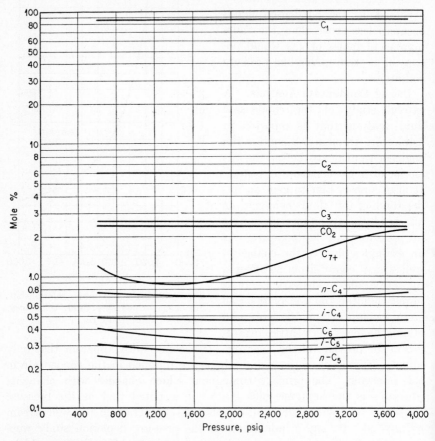

FIG. 5-19. Variation in well stream composition with pressure decline for a gas-condensate sample. (*Courtesy of Core Laboratories, Inc.*[6])

performance. Evaluation of surface separator conditions, for condensate and gas–crude-oil systems was illustrated in Chap. 4.

One other important application of fluid-analysis data is the calculation of equilibrium-ratio data which can be applied at reservoir pressures and temperatures. The data resulting from these equilibrium ratios can be used to calculate combination volume factors and to make mass material-balance studies. Equilibrium ratios can also be used to calculate the entire analysis of the depletion study as was reported in the standard gas-condensate fluid analysis. The method of making the necessary calculations for the determination of the above-mentioned quantities was discussed in Chap. 4.

FITTING PUBLISHED EQUILIBRIUM-RATIO DATA TO LABORATORY FLUID-ANALYSIS DATA

There are essentially two methods for using fluid-analysis data to calculate or determine the appropriate equilibrium ratios for use in reservoir calculations. One method uses the laboratory bubble-point or dew-point

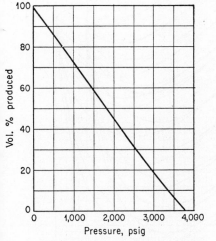

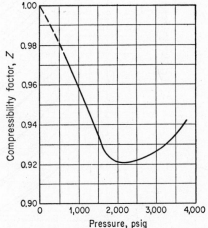

FIG. 5-20. Cumulative produced well stream volume as a function of pressure decline for a gas-condensate fluid. (*Courtesy of Core Laboratories, Inc.*[6])

FIG. 5-21. Variation in compressibility factor of well stream with pressure decline for a gas-condensate fluid. (*Courtesy of Core Laboratories, Inc.*[6])

analysis to select a set of published equilibrium curves which satisfy the dew-point or bubble-point condition at reservoir temperature and pressure. The other method calculates four equilibrium-ratio points for each component in the reservoir fluid and then smooths in curves to these four points using published equilibrium-ratio data as a guide.

Consider first the case where the fluid analysis is used in selecting an appropriate set of published equilibrium ratios. In order to select the proper set of curves it is necessary to evaluate the flash equations discussed in Chap. 4, using published equilibrium-ratio data at bubble-point or dew-point conditions. The equilibrium-ratio curves which satisfy the flash equations at bubble- or dew-point pressure and reservoir temperature are the ones selected. Curves similar to those shown in Figs. 4-63 through 4-81

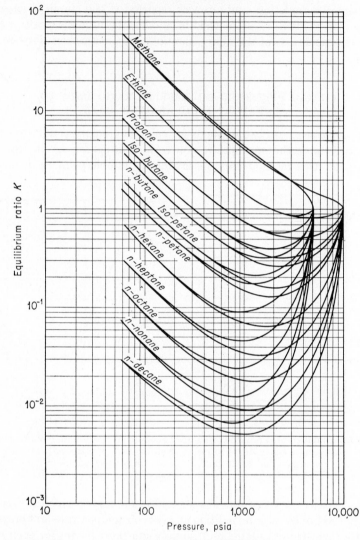

Fig. 5-22. Equilibrium vaporization ratio K at 220°F for convergence pressures of 5,000 and 10,000 psi. (*From school taught by M. J. Rzasa at University of Tulsa, 1948.*)

or Fig. 5-22 can be used as a basic group from which to select the equilibrium-ratio curves to be tested.

An apparent convergence pressure must be estimated for the fluid. This convergence pressure must be at least 10 per cent greater than the dew-point or bubble-point pressure reported in the fluid analysis. The NGAA "Equilibrium Ratio Data Book"[7] presents a chart for estimating apparent convergence pressures. Crude-oil and gas-condensate fluids usually have apparent convergence pressures in the 5,000-psia range. Select a convergence pressure, and then read the equilibrium ratios corresponding to the bubble-point or dew-point pressure and reservoir temperature. Using the appropriate set of flash equations, Chap. 4, determine the equilibrium ratio to be used for the C_{7+} fraction.

It is usually necessary that an arbitrary curve be drawn to represent the heptanes-plus fraction. The actual location of this curve will be determined by the behaviors of the heavier components in the system.

Consider the case of a condensate sample whose composition is that given in Table 5-7. When this material is tested to select proper equilibrium ratios, the apparent convergence pressure of 5,000 lb per in. is chosen. Tests were made wherein heptane plus was represented by normal nonane and by normal decane. With data for normal nonane, the calculations yielded results which indicated that the heptanes-plus fraction had the characteristics of a heavier constituent. Normal decane was then tried for the heptanes-plus fraction, and the results indicated that a lighter fraction should be used. A fit was tried using an apparent convergence pressure of 6,000 psia. The results indicated that a convergence pressure of 6,000 psia was too high. The values to be used to represent the system are equilibrium ratios for a 5,000-psia convergence pressure with the heptanes-plus fraction fitted to a curve between the curves of n-nonane and n-decane. The value of the equilibrium ratio required for the heptanes plus to balance the system is calculated. Using the calculated heptanes-plus equilibrium ratio at the dew point or bubble point, a curve can be constructed lying between the n-nonane and n-decane curve so as to represent the heptane-plus fraction in the system. This type of curve fitting can be applied both to crude-oil–gas systems and gas-condensate systems.

Example 5-12 illustrates the selection of a set of published data to describe a particular fluid which has a dew point of 3,810 psia at a temperature of 220°F.

The second method of determining equilibrium ratios from reservoir fluid-analysis data is by the use of published equilibrium ratios and empirical correlation charts. The method is a little more laborious but can yield very satisfactory results.

Four equilibrium ratios are determined for each component in the fluid. Each component has two pressure points at which the equilibrium ratios are equal to 1. The equilibrium ratios when plotted at a constant temperature apparently converge at an equilibrium ratio of 1. The pressure at which this convergence occurs is dependent upon the analysis of the fluid under consideration. As in the preceding method the apparent convergence pressure must exceed the bubble-point or dew-point pressure by at least 10 per cent. The equilibrium ratio of each component is one at the

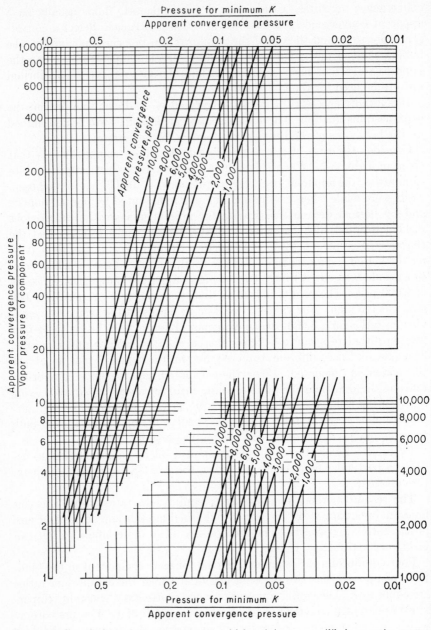

Fig. 5-23. Correlation of the pressure at which minimum equilibrium ratio occurs with the ratio of apparent convergence pressure to the vapor pressure of the component. (*From Standing.*[10])

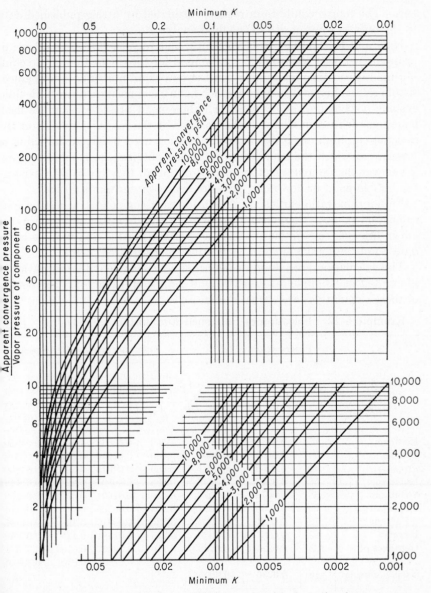

Fig. 5-24. Correlation of minimum equilibrium ratio with the ratio of apparent convergence pressure to the vapor pressure of the component. (*From Standing.*[10])

vapor pressure of that component, thus yielding the second set of equilibrium-ratio points. A third equilibrium ratio, the minimum value for each component, is determined empirically from Figs. 5-23 and 5-24 using the vapor pressure of the components and the apparent convergence pressure. The fourth equilibrium ratio is calculated by fitting published data to the bubble-point or dew-point fluid analysis.

Katz and Hachmuth;[8] Roland, Smith, and Kaveler;[9] or some other appropriate set of published data can be used for the calculation of the equilibrium ratio at the fluid bubble-point or dew-point pressure. The engineer should select the published data which have a composition nearest his fluid system. Calculate the dew-point or bubble-point pressure using the flash equations defined in Chap. 4, the analysis of the reservoir fluid, and the selected equilibrium ratios. No consideration need be given the value of pressure at which the equilibrium ratios are obtained from the literature. Whenever the desired set of values are obtained, they are said to exist at the bubble-point or dew-point pressure and reservoir temperature.

Equilibrium-ratio curves are drawn for each component using the four calculated points. Equilibrium ratios form the literature are used as an aid in drawing smooth curves. These data are plotted and used as extra data points and guide lines in connecting the calculated points.

The calculation procedure for determining a set of equilibrium ratios is illustrated in Example 5-13 for the fluid sample of Table 5-1. The results of these calculations are shown in Fig. 5-25.

Example 5-12.   Selection of Published Equilibrium-ratio Data.

Component	Mol wt	Reservoir temp 220°F		Reservoir pressure 3,810 psia			
		Analysis	Corrected analysis z_i	Apparent convergence, 5,000 psia		Apparent convergence, 6,000 psia	
				K_i^*	z_i/K_i	K_i	z_i/K_i
Methane	16	0.8481	0.8686	1.550	0.56038	1.60	0.54280
Ethane	30	0.0595	0.0609	0.840	0.07250	0.81	0.07518
Propane	44	0.0255	0.0261	0.650	0.04015	0.56	0.04660
Isobutane	58	0.0047	0.0048	0.535	0.00897	0.42	0.01142
n-Butane	58	0.0075	0.0077	0.480	0.01604	0.35	0.02200
Isopentane	72	0.0030	0.0031	0.405	0.00765	0.21	0.01476
n-Pentane	72	0.0021	0.0022	0.355	0.00619	0.19	0.01157
Hexanes	86	0.0037	0.0038	0.275	0.01381	0.125	0.03040
Heptanes plus	128	0.0223	0.0228	0.110†	0.20727	0.070†	0.32571
				0.082‡	0.27804	0.059‡	0.38644
Carbon dioxide	46	0.0236	0.0000				
		1.0000	1.0000	$\Sigma = 0.93296$†		$\Sigma = 1.08044$†	
				$\Sigma = 1.00373$‡		$\Sigma = 1.14117$‡	

* From Fig. 5-22. K_i required for C_{7+} for apparent convergence pressure of 5,000 = 0.08603.

† Using n-nonane for heptane plus.

‡ Using n-decane for heptane plus.

Example 5-13. Determination of Correct K-value Data at Reservoir Temperature. It has been shown that when a hydrocarbon mixture has a large weight per cent of heptanes plus and a small weight per cent of methane (Fig. 4-4), the critical point would be to the right of the cricondenbar and the system would essentially be a gas–crude-oil system; hence, this example system will be classified as a gas–crude-oil system. Katz and Hachmuth[8] published one of the better sets of equilibrium-ratio data for

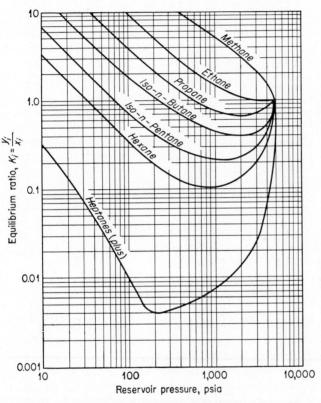

FIG. 5-25. Adjusted equilibrium ratios for a crude-oil sample.

this type of system. Therefore, their data will be used here except for methane, where Brown's[11] data will be used.

Step 1. Roughly sketch the 200°F equilibrium-ratio data presented by Katz (Fig. 4-63) on a log plot of K_i against reservoir pressure.

Step 2. From Katz's data, determine a set of K values which will satisfy the bubble-point conditions that

$$y_i = z_i K_i \qquad \Sigma y_i = 1 \qquad L \to 1 \qquad V \to 0$$

Regardless of the pressure indicated by Katz's data, it will be assumed that these data apply at the bubble-point pressure of 2,695 psig. The calculations are presented in tabular form below.

Component	Wt %	Mole %	Mole % hydrocarbons only	K_i at 200°F, 3,000 psia	z_iK_i	K_i at 200°F, 2,000 psia	z_iK_i	K_i at 200°F, 2,450 psia	z_iK_i
Methane	4.45	33.78	33.96	2.10	0.71316	2.75	0.93390	2.46	0.83542
Ethane	1.59	6.42	6.46	0.95	0.06137	1.10	0.07106	0.995	0.06428
Propane	3.56	9.82	9.87	0.66	0.06514	0.64	0.06317	0.650	0.06416
Isobutane	0.63	1.33							
n-Butane	1.43	2.99	4.34	0.43	0.01866	0.38	0.01649	0.395	0.01714
Isopentane	0.74	1.25							
n-Pentane	1.14	1.93	3.20	0.30	0.00960	0.225	0.00720	0.246	0.00787
Hexanes	2.12	2.99	3.00	0.22	0.00660	0.15	0.00450	0.162	0.00486
Heavier (C_{7+})	84.20	38.97	39.17	0.024	0.00940	0.015	0.00588	0.017	0.00666
Hydrogen sulfide	0.14	0.52	0.00						
	100.00	100.00	100.00		0.88393		1.10220		1.00039

Density of C_{7+} = 0.8859 gm/cc at 60°F
°API gravity of C_{7+} = 28.1 at 60°F
Mol wt of C_{7+} = 263 lb/mole
Reservoir bubble-point pressure = 2,695 psig
Reservoir bubble-point temperature = 200°F

The values at 2,500 psia on Katz's curve are sufficiently close to satisfy present conditions and therefore will be used to represent the K values of this sample at 2,695 psig and 200°F.

The vapor pressure of the ith constituent at 200°F is obtained from the literature. The heptanes-plus vapor pressure must be estimated either by selecting a fluid of comparable molecular weight or by letting nonane or decane represent the heptanes plus.

Katz's data at 200°F has an apparent convergence pressure of 5,000 psia. As this pressure is sufficiently greater than the observed bubble point, it will be tried as the first apparent convergence pressure, thus yielding a third set of equilibrium ratios.

The fourth K value for each component is the minimum value of K_i at 200°F. These values are obtained by using Figs. 5-23 and 5-24. Determine the ratio of the apparent convergence pressure to the vapor pressure of each constituent, and read the pressure at which the minimum K value exists. With the use of the same ratio, the magnitude of the minimum K value is obtained from Fig. 5-24.

When the four K values determined by the previously discussed methods for each component and the K values determined by Katz at pressures less than 1,000 psia were used, equilibrium-ratio curves were constructed for the fluid at reservoir temperature. The curves constructed for this fluid are shown in Fig. 5-25.

A sample calculation for determination of minimum K value appears on page 426.

———•••———

OTHER METHODS OF DETERMINING HYDROCARBON FLUID PROPERTIES

The calculation of reservoir volume factors by means of fluid analysis and the assumption of modified ideal solutions was discussed in Chap. 4. Earlier in this chapter the laboratory measurement of fluid properties was presented. As laboratory analysis or all the information required by the calculation method is not always available, other methods for approximating reservoir fluid properties have been developed.

Modified Ideal Solutions

In the ideal-solutions method presented in Chap. 4, the information required was the stock-tank fluid analysis, total produced gas analysis, producing gas-oil ratio, reservoir temperature, and reservoir pressure. Modifications of this method have been devised requiring less information. These modified methods are presented in order of decreasing data requirements. *not in handout*

No Analysis of Stock-tank Liquid. If it is assumed that the stock-tank liquid is comprised of nothing lighter than propane, then the apparent density of the propane-plus fraction of the total produced fluid can be calculated from the gas analysis, producing gas-oil ratio, and stock-tank-oil gravity. The calculating procedure is the same as in the case of the ideal-solution method presented in Chap. 4 except that the apparent density of propanes plus is defined by Eq. (5-41).

$$\rho_{C3+} = \frac{350\gamma_{st} + (R_p/380) \sum_{i=3}^{m} y_i M_i}{5.61 + (R_p/380) \sum_{i=3}^{m} [y_i M_i/(\rho_i)_L]} \tag{5-41}$$

where y_i = mole fraction of ith component in total produced gas
        M_i = molecular weight of ith component, lb

CALCULATION OF MINIMUM K VALUES

(1) Component	(2) Apparent convergence pressure, psi	(3) Vapor pressure of pure component at 200°F[a]	(4) 2 ÷ 3	(5) $\dfrac{\text{Pressure for min } K[d]}{\text{Apparent convergence pressure}}$	(6) Pressure for min K	(7) Minimum K'
Methane	5,000	1930.[b]	2.591	0.6	3,000	0.95
Ethane	5,000	580	8.62	0.42	2,100	0.67
Propane	5,000	225	22.20	0.32	1,600	0.41
Butanes	5,000	77.5	64.52	0.231	1,155	0.205
Pentanes	5,000	30.0	166.66	0.175	875	0.10
Hexanes	5,000					
Heptanes plus	5,000	0.007732[c]	646,000	0.048[e]	240	0.0039[•]

[a] From vapor-pressure curve, Fig. 4-60, unless otherwise indicated.

[b] Obtained by straight-line extrapolation of vapor-pressure curve (Fig. 4-60).

[c] Value for hexadecane, mol wt 226.4, density of liquid 0.7758 gm/cc at 60°F. Hexadecane is not so dense as the heptanes plus of the sample but is the closest data available.

[d] From Fig. 5-23.

[e] This represents the minimum value shown in Fig. 5-23. The values of the ratio of apparent convergence pressure to vapor pressure goes only to 10,000. Therefore, it is known that the minimum pressure will be somewhat lower than that calculated.

[f] From Fig. 5-24.

[•] Taken as the minimum value of K_i at 5,000-psia convergence pressure. Same reason as in e above.

γ_{st} = specific gravity of stock-tank liquid at 14.7 psia and 60°F
R_p = total produced gas-oil ratio, scf per stock-tank barrel
$(\rho_i)_L$ = liquid density of ith component at 14.7 psia and 60°F, lb/cu ft
ρ_{C3+} = apparent density of propanes plus at 14.7 psia and 60°F, lb/cu ft

The weight per cent of ethane in the ethane plus and weight per cent of methane in the system are defined by Eq. (5.42).

$$\text{Wt \% } C_2 \text{ in } C_{2+} = \frac{(R_p/380)y_2 M_2}{350\gamma_{st} + (R_p/380)\sum_{i=2}^{m} y_i M_i} \quad (5\text{-}42)$$

$$\text{Wt \% } C_1 \text{ in } C_{1+} = \frac{(R_p/380)y_i M_i}{350\gamma_{st} + (R_p/380)\sum_{i=1}^{m} y_i M_i}$$

The apparent density of the total mixture at 14.7 psia and 60°F is determined from Fig. 4-50. Corrections for reservoir pressure and temperature are made with the use of Figs. 4-51 and 4-53. Oil formation volume factors are calculated by Eq. 5-44.

No Surface Liquid or Gas Analysis. Katz[12] further simplified the ideal-solution method by the use of an empirical correlation. These correlations eliminate the necessity of knowing the gas analysis. The engineer need know only the total gas-oil ratio, the stock-tank liquid gravity, and the total produced gas gravity. Using the gas gravity and stock-tank gravity, the apparent gas density is determined from Fig. 5-26. Knowing the apparent gas density, the total produced gas-oil ratio, and the gas gravity, it is possible to calculate the apparent liquid density of the produced gas. Thus the apparent total density at surface conditions can be obtained by use of the following equations:

$$\text{Lb of gas} = \frac{R_p}{380}\gamma_g \times 28.96$$

$$(\rho_o)_{app} = \frac{350\gamma_{st} + (R_p/380)\gamma_g \times 28.96}{5.61 + (R_p/380)\gamma_g[28.96/(\rho_g)_{app}]} \quad (5\text{-}43)$$

where γ_g = specific gravity of gas (air = 1)
γ_{st} = specific gravity of stock-tank oil at 60°F and 14.7 psia
$(\rho_g)_{app}$ = apparent liquid density of produced gas as obtained from Fig. 5-26
$(\rho_o)_{app}$ = apparent density of stock-tank oil and its dissolved gas expressed at 14.7 psia and 60°F

The apparent density at surface conditions having been obtained, the procedure is the same as in the previous method, where the density at reservoir conditions is calculated by the use of Figs. 4-51 and 4-53. The density
at reservoir conditions having been obtained, the formation volume factor
corresponding to this produced gas-oil ratio and the given reservoir temperature and pressure can be calculated with the following equation:

$$B_o = \frac{(350)\gamma_{st} + (R_p/380)\gamma_g(28.96)}{5.61(\rho_o)_{res}} \tag{5-44}$$

where $(\rho_o)_{res}$ is the density of the reservoir liquid in pounds per cubic foot
and B_o is the formation volume factor. In all the preceding calculations

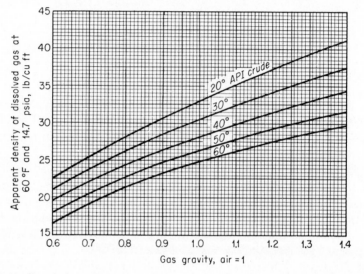

Fig. 5-26. Correlation of apparent density of dissolved gas and gas gravity. (*From
Katz.*[12])

it should be pointed out that the gas-oil ratio represents the total gas produced per stock-tank barrel and would be the sum of the gas from each
stage of separation. The gas gravity is the gravity of all the gas produced
and if not measured as such would have to be calculated from the gravity
of gas off each individual stage of separation.

All the preceding means of calculating formation volume factors are
accurate within approximately 3 per cent provided the producing gas-oil
ratio actually represents the total solution gas. Of course, if the wells are
producing at excess gas-oil ratios, then the volume factors as calculated
cannot be expected to correlate with the formation volume factor resulting
from a laboratory analysis. These calculating and correlating procedures
are invaluable in obtaining a first estimate of formation volume factors to

be used in making preliminary economic studies prior to the time that complete fluid-analysis data are either warranted or available.

Empirical Methods

Empirical methods differ from the modified ideal-solution methods in that they depend on curves or equations derived from correlations of laboratory analysis. The empirical methods usually depend on gas gravity, stock-tank gravity, reservoir pressure, and temperature.

Katz's Curves. Katz[12] prepared a correlation from data on Mid-Continent crudes for calculation of the reservoir formation volume factor. The use of these charts requires the reservoir temperature, reservoir pressure, gas in solution, and API gravity of the crude. These two curves are

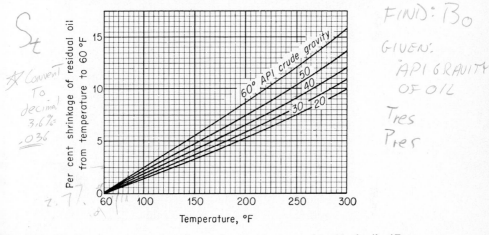

FIG. 5-27. Fluid-volume correction factor for temperature of residual oil. (*From Katz.*[12])

presented in Figs. 5-27 and 5-28. If the gas-oil ratio required in Fig. 5-28 is considered to be the producing gas-oil ratio, then these curves can be used to get an estimate of the formation volume factor at that point. These two curves can be used in conjunction with a correlating curve presented by Beal[13] (Fig. 5-29), in which the gas in solution is correlated with stock-tank gravity and saturation pressure. If the gas in solution is determined from Fig. 5-29 as a function of stock-tank gravity and reservoir pressure, Figs. 5-27 and 5-28 can be used to calculate the formation volume factor for the reservoir oil. These three curves will then permit the calculation of a complete formation volume factor and solution-gas–oil ratio curves by using various pressures and assuming the stock-tank gravity to remain constant.

The procedure for calculating formation volume factors using Beal's and Katz's empirical curves is illustrated in Example 5-14.

Example 5-14. Calculation of Formation Volume Factor and Solution-gas–Oil Ratios by Beal and Katz Charts.

1. Knowing the API stock-tank oil gravity assume a saturation pressure and determine the solution-gas–oil ratios from Fig. 5-29.

2. From Fig. 5-27 determine the fluid shrinkage due to the change from reservoir temperature to 60°F. Denote this shrinkage value by the term S_t.

3. From Fig. 5-28 determine the fluid shrinkage due to the liberation of the solution gas as the pressure decreased from saturation pressure to atmospheric. Denote this shrinkage value by S_p.

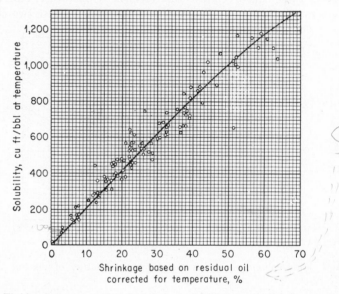

Fig. 5-28. Fluid-volume correction factor for solubility (corrected for temperature of residual oil). (*From Katz.*[12])

4. The oil formation volume factor at the saturation pressure of Step 1 is then defined by the following equation:

$$B = (1 + S_t)(1 + S_p)$$

where S_t and S_p are expressed as fractions.

A calculation of formation volume factor from data compiled by Beal and Katz follows:

$R_{si} = 1{,}202$	Reservoir temperature = 224°F
Gas gravity = 0.8643	$S_t = 0.075$ (from Fig. 5-27)
Oil gravity = 36.81°API	$B_o = (1 + S_t)(1 + S_p)$
Separator pressure = 0 psi	

P_b	R_s*	S_p†	$1 + S_p$	B_o
4,228	1,202‡	0.620	1.620	1.7415
3,810	1,150	0.585	1.585	1.7039
3,410	1,035	0.515	1.515	1.6286
3,010	910	0.445	1.445	1.5534
2,610	790	0.385	1.385	1.4889
2,210	670	0.325	1.325	1.4244
1,810	550	0.265	1.265	1.3599
1,410	425	0.205	1.205	1.2954
1,010	310	0.150	1.150	1.2363
610	210	0.100	1.100	1.1825
245	120	0.057	1.057	1.1363

* From Fig. 5-29.
† From Fig. 5-28.
‡ Not from Fig. 5-29. Value is the initial solution ratio from field data.

Standing's Correlation. Another empirical correlation has been presented which requires the total gas-oil ratio, the gravity of the stock-tank oil and produced gas, and the reservoir temperature. This correlation was presented by Standing[14] for California fluids. The formation volume factor is expressed by the following equation:

$$B_o = \text{GOR}\, \frac{\gamma_g^{0.5}}{\gamma_o} + 1.25t \qquad (5\text{-}45)$$

where all the symbols are as defined previously except t, which is defined as the reservoir temperature expressed in degrees Fahrenheit.

Standing further amplified the correlation to permit the calculation of bubble-point pressures and total volume factors. It is expected that the results obtained from these correlations would be more accurate than those obtained from Beal's and Katz's data, as these charts account for the gas gravity. These charts are presented in Figs. 5-30 and through 5-32. Figure 5-31 represents a graphical solution of Eq. (5-45). If it is assumed that the producing gas-oil ratio represents the solution ratio, the formation volume factor can be determined from Fig. 5-31. Working with Fig. 5-30, it is possible to take these same parameters, gas-oil ratio, gas gravity, tank-oil gravity, and reservoir temperature, and determine the pressure at which a given amount of gas would be in solution. This curve essentially accomplishes the same results as the data presented by Beal but includes more variables; therefore, it is felt to be more accurate. By combining Figs. 5-30 and 5-31 it is possible to determine the formation volume factor and solution-gas–oil ratio pressure relationships. If various gas-oil ratios are assumed and either a constant gas gravity and constant stock-tank oil gravity or some predetermined variation with pressure, the pressure and formation volume factor corresponding to each value of gas-oil ratio can be determined. Thus, formation volume factors as a function of pressure and gas in solution as a function of pressure can be approximated from these two

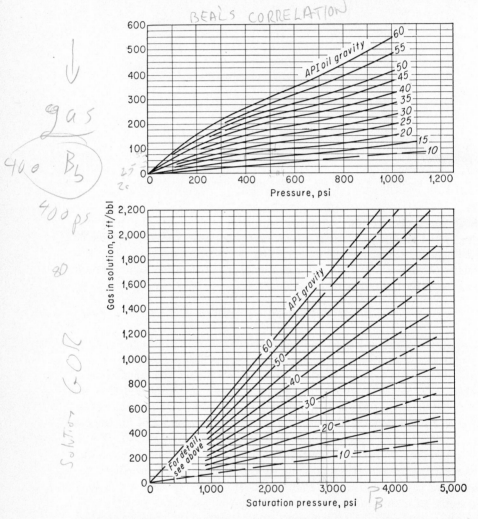

FIG. 5-29. Prediction of solubility from saturation pressure and gravity of crude oil Represents average conditions for 508 observations from 164 samples taken from 151 oil fields. Average deviation, 22.0 per cent. (*From Beal.*[13])

curves. The total formation volume factor can be estimated from Fig. 5-32 This volume factor represents the reservoir volume occupied by 1 bbl o oil and its complement of liberated gas. It is actually a combination o Figs. 5-30 and 5-31 which permits the calculations of the expansion of the

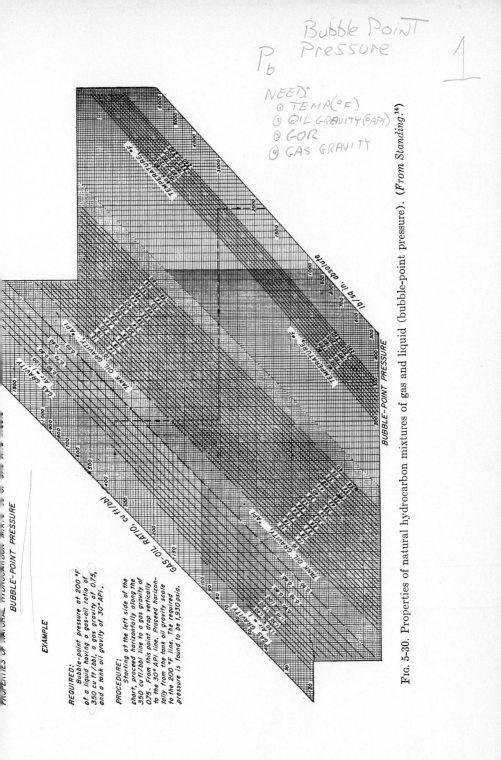

Fig. 5-30. Properties of natural hydrocarbon mixtures of gas and liquid (bubble-point pressure). (*From Standing.*[14])

PROPERTIES OF NATURAL HYDROCARBON MIXTURES OF GAS AND LIQUID

BUBBLE-POINT PRESSURE

EXAMPLE

REQUIRED:

Bubble-point pressure at 200 °F of a liquid having a gas-oil ratio of 350 cu ft/bbl, a gas gravity of 0.75, and a tank oil gravity of 30° API.

PROCEDURE:

Starting at the left side of the chart, proceed horizontally along the 350 cu ft/bbl line to a gas gravity of 0.75. From this point drop vertically to the 30° API line. Proceed horizontally from the tank oil gravity scale to the 200 °F line. The required pressure is found to be 1,930 psia.

Bubble Point Pressure

P_b

1

NEED:
① TEMP (°F)
② OIL GRAVITY (°API)
③ GOR
④ GAS GRAVITY

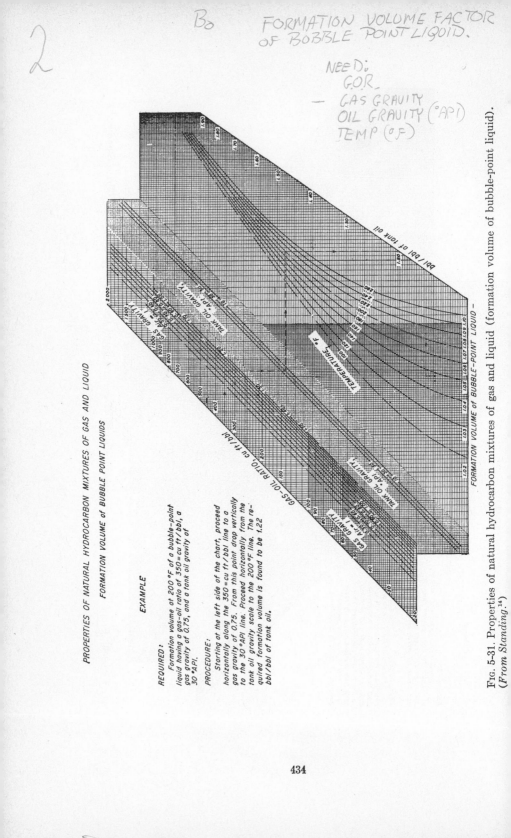

2

B_o — FORMATION VOLUME FACTOR OF BOBBLE POINT LIQOID.

NEED:
G.O.R.
— GAS GRAVITY
OIL GRAVITY (°API)
TEMP (°F)

PROPERTIES OF NATURAL HYDROCARBON MIXTURES OF GAS AND LIQUID

FORMATION VOLUME OF BUBBLE POINT LIQUIDS

EXAMPLE

REQUIRED:
Formation volume at 200°F of a bubble-point liquid having a gas-oil ratio of 350 = cu ft/bbl, a gas gravity of 0.75, and a tank oil gravity of 30 °API.

PROCEDURE:
Starting at the left side of the chart, proceed horizontally along the 350 = cu ft/bbl line to a gas gravity of 0.75. From this point drop vertically to the 30 °API line. Proceed horizontally from the tank oil gravity scale to the 200 °F line. The required formation volume is found to be 1.22 bbl/bbl of tank oil.

FIG. 5-31. Properties of natural hydrocarbon mixtures of gas and liquid (formation volume of bubble-point liquid). (From Standing.[14])

434

liberated gas. An example problem is not included, as the charts presented by Standing incorporate example problems.

A comparison between laboratory data and empirical calculated data is presented in Table 5-9 and Fig. 5-33. In this particular case the fluid sample used is tending toward a high-shrinkage crude, so that the empirical

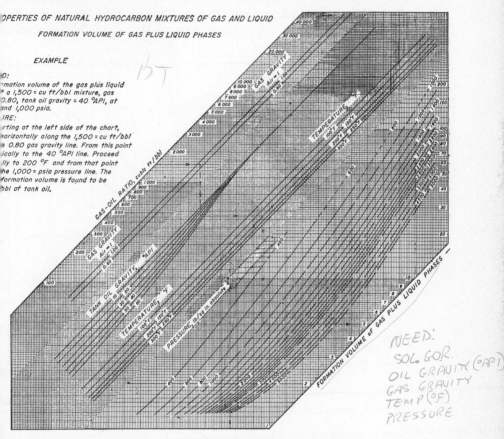

FIG. 5-32. Properties of natural hydrocarbon mixtures of gas and liquid; formation volume of gas plus liquid phases. (*From Standing.*[14])

correlations do not give so good agreement throughout the pressure range as one might expect. In most cases, these empirical correlations yield values which are comparable within 3 or 4 per cent to measured laboratory data.

Calculating Reservoir Volumes for Gas-condensate Systems

Gas-condensate systems are most frequently treated as gas systems. Correlations are employed to recombine the condensed liquid resulting from surface separation with the separator gas. These correlation procedures convert the produced gas gravity to a reservoir gas gravity. In all the calculations which follow, it is assumed that the produced material is a gas in the reservoir and that no liquid was produced from the reservoir. This does not mean that there cannot be liquid existing in the reservoir, simply that none of that liquid is produced and included in the calculations.

Recombination with Equation of State. The first method for converting surface volumes to reservoir volumes requires that the following data be

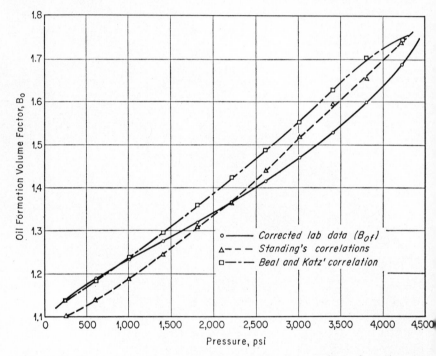

FIG. 5-33. Comparison of measured and calculated oil formation volume factors.

known: analysis of the produced gas, analysis of the condensate, the total gas-oil ratio, and the reservoir temperature and pressure. Using these data it is possible to calculate the composition of the reservoir gas by the methods indicated with respect to recombined samples in an earlier section of this chapter. By means of the recombined composition, the compressibility factor can be determined through the use of reduced temperatures and pressures. When the value of the compressibility factor is known, the actual volume in the reservoir can be calculated by use of the equation of state $PV = ZnRT$. This would give the volume in the reservoir per mole of fluid. The molecular weight of the surface gas and liquid can be calcu-

TABLE 5-9. COMPARISON OF MEASURED AND CALCULATED FORMATION VOLUME
FACTORS AND SOLUTION-GAS–OIL RATIOS*

Bubble-point pressure, psi	Calculated from data compiled by Beal and Katz		Calculated using Standing's correlations		Corrected laboratory data	
	B_o	R_s	B_o†	R_s‡	B_{of}§	R_s¶
4,228	1.7415	1,202	1.740	1,202	1.6892	1,202
3,810	1.7039	1,150	1.655	1,090	1.6003	1,028.5
3,410	1.6286	1,035	1.595	950	1.5287	883.6
3,010	1.5534	910	1.520	830	1.4711	755.6
2,610	1.4889	790	1.442	700	1.4165	642.7
2,210	1.4244	670	1.366	570	1.3667	537.5
1,810	1.3599	550	1.310	450	1.3196	437.0
1,410	1.2954	425	1.242	335	1.2757	343.2
1,010	1.2363	310	1.189	225	1.2343	254.0
610	1.1825	210	1.139	127	1.1890	162.1
245	1.1363	120	1.100	41	1.1376	67.4

* R_{si} = 1,202, gas gravity = 0.9643, oil gravity = 36.81°API, separator pressure =
0 psi, and reservoir temp = 224°F.
† From Fig. 5-31.
‡ From Fig. 5-30.
§ $B_{of} = B_{od}(B_{ofb}/B_{odb})$.
¶ $R_s = 1,202 - (R_L)_d(B_{ofb}/B_{odb})$.

lated from their respective analyses. The moles of fluid produced per day
can be calculated from the gas-oil ratio and daily gas production. The
reservoir voidage per day is then given by

Reservoir voidage = (mole produced/day)(cu ft/mole reservoir gas)   (5-46)

Correlation Charts and Equation of State. Another method of converting surface volumes to reservoir volumes is similar to the one previously
discussed except that correlation charts are used for obtaining the gravity
of the reservoir gas. In order to use the correlation charts presented by
Standing,[10] it is necessary that the following data be known: the produced
gas gravity, the barrels of condensate per million cubic feet, and the condensate gravity. From the correlating chart shown in Fig. 5-34, it is possible to calculate the gravity of a single-phase fluid which would result
from the recombination of the surface gas and liquid. The correlating
chart refers to this single-phase fluid as the well fluid. When the well fluid
gravity is used, pseudo-critical properties and compressibility factors for
the fluid are obtained from the fluid property charts in Chap. 4. When the
compressibility factor is known, the solution is the same as in the preceding
example, where the equation of state is used to calculate the reservoir voidage per day or per standard cubic foot produced. Example 5-15 illustrates
the calculating procedure.

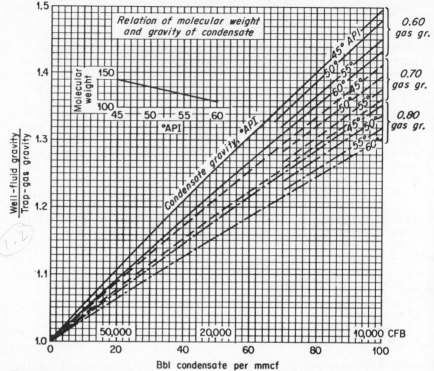

FIG. 5-34. Effect of condensate volume on the ratio of surface gas gravity to well fluid gravity. (*From Standing.*[10])

Sage's and Olds's Correlation. An empirical method for calculating reservoir volumes for gas-condensate systems was presented by Sage and Olds,[15] where the volume relations depend on the volume of condensate rather than the volume of produced gas. This relation is expressed as

$$B = \frac{AR_pT}{P} \qquad (5\text{-}47)$$

where B = formation volume factor reservoir, bbl/stock-tank bbl
A = empirical correlating constant, Table 5-10 and Fig. 5-35
R_p = producing gas-oil ratio, scf/stock-tank bbl
T = reservoir temperature, °R
P = reservoir pressure, psia

To apply this calculating procedure it is necessary to know the producing gas-oil ratio and the reservoir pressure and temperature and to have access to the table of correlating factors. The calculating procedure is illustrated in Example 5-15.

TABLE 5-10. VALUES OF COEFFICIENT A USED TO CALCULATE FORMATION
VOLUME OF GAS-CONDENSATE SYSTEMS BY METHOD OF SAGE AND OLDS[15]

$$(B = AR_pT/P)$$

Pressure, psia	$A(10)^3$					
	100°F	130°F	160°F	190°F	220°F	250°F
600	4.58	4.67	4.75	4.83	4.89	4.93
800	4.46	4.57	4.67	4.76	4.83	4.87
1,000	4.35	4.47	4.59	4.69	4.77	4.82
1,250	4.21	4.35	4.49	4.62	4.71	4.77
1,500	4.09	4.25	4.41	4.55	4.66	4.73
1,750	3.99	4.17	4.34	4.50	4.62	4.71
2,000	3.93	4.11	4.29	4.46	4.59	4.69
2,250	3.89	4.08	4.26	4.43	4.57	4.68
2,500	3.88	4.06	4.25	4.42	4.57	4.68
2,750	3.89	4.07	4.26	4.44	4.58	4.69
3,000	3.92	4.10	4.29	4.47	4.61	4.71

Example 5-15. Calculation of Daily Reservoir Voidage.

Liquid gravity, 53.3°API
Separator gas-oil ratio, 40.795 Mcf/bbl
Stock-tank gas-oil ratio, 2.780 Mcf/bbl
Separator gas gravity, 0.6174
Stock-tank gas gravity, 1.0900
Reservoir temperature, 190°F
Reservoir pressure, 2,900 psia
Separator gas rate, 3.130 MMscf/day
Stock-tank gas rate, 0.213 MMscf/day
Condensate rate, 76.725 bbl/day

1. *Using Katz-Standing correlations*

$$\text{Av gas gravity} = \frac{3.130(0.6174) + 0.213(1.0900)}{3.130 + 0.213} = 0.6475$$

$$\text{Bbl cond/MMscf} = \frac{1,000}{43.575} = 22.949 \text{ bbl/MMscf}$$

From Fig. 5-34,

$$\frac{\gamma_w}{\gamma_s} = 1.125$$

Well fluid gravity = 1.125(0.6475) = 0.7284

From Fig. 4-30,

$$_pP_c = 664$$

$$_pT_c = 391$$

$$P_r = \frac{2,900}{664} = 4.367$$

$$T_r = \frac{650}{391} = 1.662$$

From Fig. 4-25,

$$Z = 0.847$$

$$\text{Res vol/mole of composite} = \frac{ZRT}{P} = \frac{0.847(10.71)(650)}{2,900}$$

$$= 2.03316 \text{ cu ft/mole}$$

$$\text{Moles gas produced/bbl condensate} = \frac{\text{total GOR}}{380} = \frac{43,575}{380}$$

$$= 114.6711 \text{ moles gas/bbl}$$

$$\text{Sp gr of liquid} = \frac{141.5}{°\text{API} + 131.5} = \frac{141.5}{53.3 + 131.5} = 0.7657$$

From Fig. 5-34,

Mol wt of liquid = 121

$$\frac{\text{Moles liquid produced}}{\text{Bbl of condensate}} = \frac{350 \times (\text{sp gr})}{\text{mol wt}} = \frac{350(0.7657)}{121}$$

$$= 2.2148 \text{ moles liquid/bbl}$$

$$B = \frac{\text{bbl res space}}{\text{bbl of condensate}}$$

$$= \frac{(\text{moles liquid/bbl} + \text{moles gas/bbl})(\text{res cu ft/mole of composite})}{5.61}$$

$$= \frac{2.03316(114.6711 + 2.2148)}{5.61} = 42.36145$$

Daily res voidage = $76.725B$ = 3250.182 bbl/day

2. *Using Sage's and Olds's correlation*

$$B = \frac{AR_pT}{P}$$

From Table 5-10,

$$A = 4.458$$

$$B = \frac{4.458(43.575)(650)}{2,900} = 43.54036$$

Daily res voidage = 43.54036(76.725)

= 3340.634 bbl/day

3. *Using Standing's correlation chart*

From Fig. 5-32,

$B = 44.5$

Daily res voidage = 44.5(76.725)

= 3414.263 bbl/day

Standing's Correlation. A method of converting surface volumes to reservoir volumes was presented by Standing for condensate systems. This is the same correlation (Figs. 5-30 to 5-32) used in the estimation of the PVT relationship for a gas–crude-oil system.

By correlating field data Standing developed correlation charts which are dependent upon the gas-oil ratio, the gas gravity, reservoir temperature and pressure, and stock-tank oil gravity. These curves can be used for gas-condensate systems because at high gas-oil ratios, the tank oil gravity becomes insignificant. Figure 5-32 is the empirical chart to be used with a condensate. This chart permits a rapid calculation of the formation volume factor for a condensate system. The formation volume factor is expressed as barrels of reservoir fluid per barrel of stock-tank liquid. An example calculation using this chart for a gas–crude-oil system is incorporated with the figure. For the crude and condensates systems from which this chart was prepared, the accuracy was approximately 3 per cent. It is expected that the accuracy for other systems should still be within 3 to 5 per cent.

Review of Correlations. It should be pointed out that these empirical procedures are not meant to supplant or replace laboratory analysis of reservoir fluids. Empirical relationships are presented as an aid to the engineer so that he can estimate the physical properties of the reservoir fluids in order to determine the best means of obtaining a reservoir sample and whether a reservoir sample is warranted and to obtain estimates of fluid performance prior to collecting a sample. These correlating devices can also be used as a check against fluid analysis. If the results obtained by using these empirical relations and the laboratory analysis results are extremely far removed, then in all probability some error was made in collecting the sample, and the engineer should feel justified in requesting a new sample.

LAST PAGE of HANDOUT

Correlations for Fluid Viscosities

Viscosity of Oils. Beal[13] presented empirical correlations for the determination of reservoir fluid viscosities dependent upon the stock-tank gravity, reservoir pressure, and temperature. The viscosity of a gas-free crude oil is presented as a function of API gravity and temperature in Fig. 5-35. The gas in solution, dependent on reservoir pressure and stock-tank gravity, can be obtained from Fig. 5-29. The viscosity of the reservoir liquid is then read from Fig. 5-36 or 5-37, depending on whether the

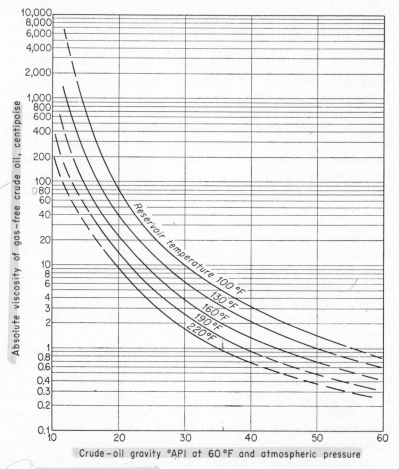

Fig. 5-35. Gas-free crude viscosity as a function or reservoir temperature and stock-tank crude gravity. (*From Beal.*[13])

oil is saturated or undersaturated. Take, for example, the case when the crude-oil gravity is 32½°API at stock-tank conditions and reservoir temperature is 175°F. Then from Fig. 5-35 it is found that the viscosity of the gas-free crude oil at reservoir temperature is 2.3 centipoises. From Fig. 5-36 entering with the gas in solution, 500 scf at 2,000 psia from Fig. 5-29 and reading to the gas-free crude-oil viscosity from Fig. 5-35, it is found that the reservoir oil viscosity is 0.8 centipoise. Figure 5-37 permits the calculation of the oil viscosity for undersaturated crude oils.

These three curves were used to calculate the viscosity of the reservoir fluid presented in Table 5-1. A comparison of the empirical and measured viscosity values are shown in Fig. 5-38. It is noted that the viscosities calculated by Beal's correlation are slightly greater than the viscosities actually measured in the laboratory. Exact agreement cannot be expected, as

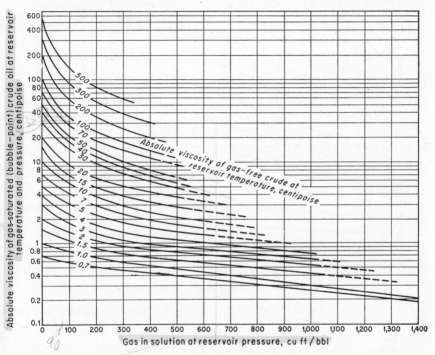

FIG. 5-36. Reservoir crude-oil viscosity from gas-free crude-oil viscosity and gas in solution. Correlation based on 351 viscosity observations from 41 crude-oil samples representing average conditions for 29 oil fields. Average deviation, 13.4 per cent. (*From Beal.*[13])

Beal's correlations have an accuracy of approximately 80 per cent. These correlations are useful in flow calculations when laboratory fluid data are not available.

It is noted in Fig. 5-35 that for a constant API gravity the viscosity of a gas-free crude oil decreases with increasing temperature. From Fig. 5-36 it is seen that for a constant gas-free viscosity at a fixed reservoir temperature the reservoir viscosity decreases with increasing solution gas (increasing pressure). In Chap. 4 it was stated that increasing the pressure on a liquid increases the viscosity of that liquid. The effect of gas entering solu-

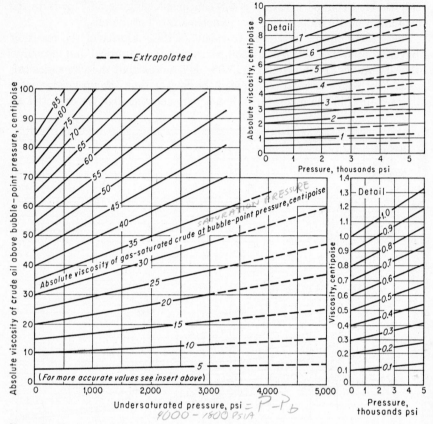

FIG. 5-37. Viscosity of crude oil above the bubble-point pressure. Average deviation, 2.7 per cent. (*From Beal.*[13])

tion so far outweighs the effect of pressure that the viscosity of the reservoir fluid decreases with increasing pressure.

Viscosity of Gases. The gas viscosity can be calculated by the procedures of Carr[16] outlined in Chap. 4. The application of the procedure to calculation of the viscosity of the liberated gas from sample analysis BHS-46C is presented in Example 5-16.

⎯⎯⎯•◆•⎯⎯⎯

Example 5-16. Calculation of, Gas Viscosity (BHS-46C).

Reservoir temp = 224°F or 684°R

$$\mu = \left(\frac{\mu}{\mu_1}\right)\mu_1$$

(1)	(2)	(3)	(4)	(5)	(6)	(7)	(9)	(8)	(10)
P	Gravity*	Mol wt	$_pT_c$†	$_pP_c$†	T_r	P_r	μ/μ_1‡	μ_1§	μ
3,810	0.9245	26.81	460	656	1.487	5.808	2.3	0.0118	0.0271
3,410	0.9070	26.30	454	657	1.507	5.190	2.1	0.0119	0.0250
3,010	0.9080	26.33	454	657	1.507	4.581	1.92	0.0119	0.0228
2,610	0.9046	26.23	453	657	1.510	3.973	1.75	0.0119	0.0208
2,210	0.8956	25.97	451	658	1.517	3.359	1.6	0.0120	0.0192
1,810	0.8972	26.02	451	658	1.517	2.751	1.4	0.0120	0.0168
1,410	0.9064	26.29	454	657	1.507	2.146	1.29	0.0119	0.0154
1,010	0.9333	27.07	462	656	1.481	1.540	1.19	0.0118	0.0140
610	1.0052	29.15	483	653	1.416	0.934	1.08	0.0116	0.0125
245	1.2272	35.59	555	638	1.232	0.384	1.02	0.0110	0.0112

* From Table 5-4, column 10.
† From Fig. 4-30.
‡ From Fig. 4-45.
§ From Fig. 4-43.

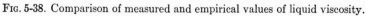

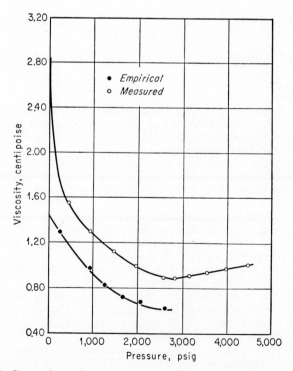

FIG. 5-38. Comparison of measured and empirical values of liquid viscosity.

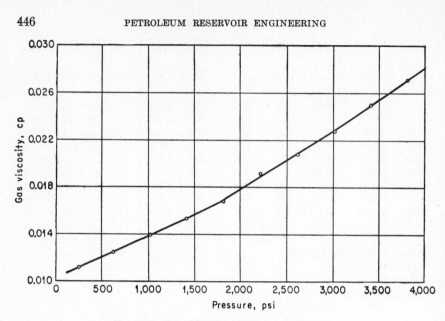

FIG. 5-39. Calculated gas viscosity, BHS-46C.

REFERENCES

1. Pirson, S. J.: "Oil Reservoir Engineering," 1st ed., McGraw-Hill Book Company, Inc., 1950.

2. Reudelhuber, F. O.: Petroleum Technology, *AIME Conf. Proc. of* 1956, Texas A and M College.

3. Frank Purdum Co.: "Laboratory Procedures for Testing Samples."

4. Dodson, C. R., D. Goodwill, and E. H. Mayer: Application of Laboratory PVT Data to Reservoir Engineering Problems, *AIME Petrol. Trans.*, vol. 198, 1953.

5. Hurst, W., Personal Communication.

6. Core Laboratory Example Reservoir Crude Oil Analysis Trade Literature, Core Laboratories, Inc.

7. "Equilibrium Ratio Data Book," Natural Gasoline Association of America, Tulsa, Okla., 1957.

8. Katz, D. L., and K. H. Hachmuth: Vaporization Equilibrium Constants in a Crude Oil-Natural Gas System, *Ind. Eng. Chem.*, vol. 29, 1937.

9. Roland, C. H., D. E. Smith, and H. H. Kaveler: Equilibrium Constants for a Gas-distillate System, *Oil Gas J.*, vol. 39, no. 46, Mar. 7, 1941.

10. Standing, M. B.: "Volumetric and Phase Behavior of Oil Field Hydrocarbon Systems," Reinhold Publishing Corporation, New York, 1952.

11. Brown, G. G.: Charts presented in "Natural Gasoline Supply Men's Association Technical Manual," 5th ed., Natural Gasoline Supply Men's Association, 1946.

12. Katz, D. L., Prediction of the Shrinkage of Crude Oils, *Drilling and Production Practice*, 137, American Petroleum Institute, 1942.

13. Beal, C.: The Viscosity of Air, Water, Natural Gas, Crude Oil and Its Associated Gases at Oil Field Temperatures and Pressures, *AIME Petrol. Trans.*, vol. 165, 1946.

14. Standing, M. B.: A Pressure-Volume-Temperature Correlation for Mixtures of

California Oils and Gases, *Drilling and Production Practice*, 275, American Petroleum Institute, 1947.

15. Sage, B. H., and R. H. Olds: Volumetric Behavior of Oil and Gas from Several San Joaquin Valley Fields, *AIME Petrol. Trans.*, vol. 170, 1947.

16. Carr, N. L., R. Kobayaski, and D. B. Burrows: Viscosity of Hydrocarbon Gases under Pressure, *AIME Petrol. Trans.*, vol. 201, 1954.

CHAPTER 6

PROPERTIES OF WATER

INTRODUCTION

The petroleum engineer is concerned with and must have a knowledge of the physical and chemical properties of water because petroleum accumulations are found associated with water and rarely is petroleum production obtained without accompanying water production. In fact, in many cases the volume of water associated with petroleum reservoirs exceeds that of the petroleum accumulation and the total volume of water production far exceeds that of petroleum.

The petroleum engineer is directly concerned with water because of the necessity for observing and predicting its location, direction of movement, rate of movement, and association with other fluids both at the surface and in the reservoir. More specifically, the petroleum-reservoir engineer is charged with the study, evaluation, and prediction of the volume of water in the reservoir, the rate of movement of the water through the reservoir, the water influx into the reservoir, and the accompanying problems which develop in petroleum production, such as water coning. He may also use water data as an exploratory tool to find petroleum through chemical composition and electrical resistivity correlations between fields or producing zones within a field. Also, water data are useful in determining the entrance of extraneous fluids into the reservoir and for determining the effectiveness of any completion operation or water shutoff procedure. In water flooding practices, water data are used for the reasons cited before and also to ascertain the possibility of formation plugging due to reaction of injected waters with the reservoir water and to predict fluid injectivity rates.

In order to make a complete and comprehensive petroleum-reservoir engineering study, it is necessary to have a complete water analysis, including both physical and chemical property data. Perhaps the most frequently used physical properties are compressibility and viscosity. However, it is quite often desirable, if not necessary, to include gas solubility, density, volume factor, and salinity data.

A chemical property analysis should be available on the water in every petroleum reservoir. The analysis should be of such scope and completeness as to permit calculations to predict and solve future problems arising

448

from the characteristics of the water. The analysis should show the total solids and the parts per million of each positive and negative ion and/or radical. From this information it will be possible to represent the analysis graphically and to calculate reacting values, products, and properties of reaction.

Whenever possible it is recommended that representative samples of the particular reservoir water be obtained and their physical and chemical properties determined through the services of a reputable laboratory. Quite often this procedure is not feasible owing to timing, economics, or other reasons. If circumstances are such, the petroleum-reservoir engineer may then find it expedient to resort to empirical data or correlation charts. The majority of this chapter is devoted to the consideration of solution of water problems through use of these data and correlations. Prior to the discussion of the use of these correlations it is appropriate to consider the scope of the research on which the correlations were developed and the limitations of their utility.

Historically much confusion exists in the early petroleum-reservoir engineering literature regarding identification and classification of reservoir waters. In an effort to alleviate this difficulty, the American Petroleum Institute, in 1941, acting through its Subcommittee on Core Analysis, conducted a study on reservoir waters. Questionnaires submitted to petroleum technologists requesting reservoir water classifications yielded some 300 different types and terminology, the majority of which have appeared in the literature. Although there was considerable difference of opinion, the subcommittee was successful in establishing widespread and common usage of such water terms as connate, interstitial, residual, and free, to name only a few. The discussions which follow in this chapter are applicable to all types and classifications of reservoir waters.

Since reservoir pressures as high as 15,000 psi and temperatures as high as 350°F have been encountered in vastly different geological environments, reservoir waters exhibit widely varying physical and chemical properties. As a result, salinities in the hundreds of thousands of parts per million, gas solubilities of 50 cu ft per bbl, water-formation volume factors exceeding 1.20 bbl per bbl, compressibilities of 4×10^{-6} bbl per bbl per psi, and viscosities of less than 0.10 centipoise have been observed. In this chapter all the gas volumes are expressed at 14.73 psia and 60°F and the water volumes are expressed at 60°F unless designated otherwise. Furthermore, reservoir waters are assumed to be saturated with natural gas at the reservoir conditions unless specified otherwise.

The physical properties of water are dependent upon its chemical composition, temperature, and pressure. Water may be pure or may contain dissolved, entrained, or suspended salts; inert materials; or gases.

There is much information in the technical literature on the properties of pure water at or near atmospheric conditions. The calculations which the petroleum-reservoir engineer will be called upon to make will be expedited through his knowledge of these properties of pure water, since the empirical data usually employed in the solution of problems concerning reservoir water are referred to pure water properties.

There are fairly complete data in the literature showing the effect of temperature and pressure on the properties of pure water over a temperature range from 32 to 250°F at pressures ranging from 0 to 6,000 psia. Literature relative to the effect of composition on the properties is meager and is limited to gas-solubility data within the aforementioned temperature and pressure ranges. There are very few data in the literature on the effect of pressure, temperature, and composition on the physical properties of pure or reservoir waters at pressures exceeding 5,000 psia, temperatures exceeding 250°F, and salinities exceeding 30,000 ppm.

PHYSICAL PROPERTIES OF WATER

Solubility of Natural Gas in Water

The solubility of natural gas in pure water has been studied[1] and has been shown to be dependent upon the temperature and pressure of the water as illustrated in Fig. 6-1a. The solubility is expressed in cubic feet of gas at 14.7 psia and 60°F per barrel of water at 60°F. It should be noted that at 5,000 psia and 260°F, the solubility of natural gas in pure water may be greater than 20 cu ft per bbl. At even higher pressures and temperatures, it would be expected that higher gas solubilities would be obtained, probably owing to the effect of pressure on gas solubility.

––––––––––

Example 6-1. Determination of Solubility of Natural Gas in Pure Water. A relatively shallow petroleum reservoir is known to have a pressure of 5,000 psia and a temperature of 200°F, and the connate water produced from the reservoir is known to be relatively pure. Estimate the probable gas solubility in the water.

From Fig. 6-1a read the gas solubility in pure water as 20 cu ft per bbl.

The solubility of natural gas in reservoir water has been found to be dependent upon the pressure, temperature, and salinity of the water. Saline reservoir water has a lower gas solubility than does pure water at the same temperature and pressure. Dodson prepared the graph shown in Fig. 6-1b for the purpose of correcting gas-solubility values of pure water obtained from Fig. 6-1a for the effect of salinity. Using Dodson's data, Jones[2] proposed the following empirical relationship for the same purpose:

$$R_{sw} = R_{swp}\left(1 - \frac{XY}{10,000}\right)$$ (6-1)

where R_{swp} = solubility of natural gas in pure water, cu ft/bbl
    R_{sw} = solubility of natural gas in reservoir water, cu ft/bbl
    Y = salinity of water, ppm
    X = salinity correction factor

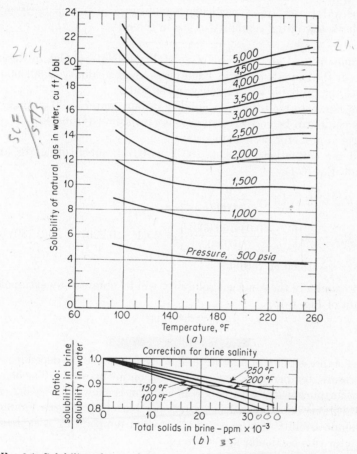

Fig. 6-1. Solubility of natural gas in water. (*From Dodson and Standing.*[1])

The correction factor X is primarily dependent upon the temperature of the water, and the following values have been suggested by Jones:[2]

Temperature, °F	Correction factor X
100	0.074
150	0.050
200	0.044
250	0.033

From these data it can be shown that a reservoir water with a salinity of 50,000 ppm will contain about 75 per cent as much dissolved natural gas at 5,000 psia and 200°F as pure water at the same conditions of temperature and pressure.

Gas-solubility calculations are important in estimating reservoir volumes of water and are of particular significance in petroleum reservoirs which have low initial solubility of gas in the oil.

Example 6-2. Determination of Solubility of Natural Gas in Reservoir Water. If the connate water in Example 6-1 had a salinity of 50,000 ppm, estimate the probable gas solubility in the water.

From Fig. 6-1a read the gas solubility in pure water as 20 cu ft per bbl. This value can be corrected to account for salinity using Eq. (6-1) and the correction-factor data as follows:

Read the correction factor at 200°F as 0.044, substituting the known values in Eq. (6-1).

$$R_{sw} = R_{swp}\left(1 - \frac{XY}{10,000}\right)$$

$$R_{sw} = 20\left[1 - \frac{0.044(50,000)}{10,000}\right] = 20(1 - 0.220) = 20(0.780)$$

$$= 15.60 \text{ cu ft/bbl}$$

Approximately the same gas solubility will be obtained by extrapolating the data of Fig. 6-1b.

Compressibility of Water

The compressibility of pure water has been shown[1] to be dependent upon the pressure, temperature, and gas in solution in the water. The compressibility for pure water with no gas in solution is shown in Fig. 6-2, where compressibility is expressed in barrels per barrel per degree Fahrenheit. The compressibility of pure water at constant temperature is expressed as follows for this particular application:

$$c_{wp} = -\frac{1}{V}\left(\frac{\Delta V}{\Delta P}\right)_T \tag{6-2}$$

where c_{wp} = compressibility of pure water, 1/psi
  V = volume of pure water, bbl
  ΔV = change in volume of pure water, bbl
  ΔP = change in pressure, psi

It should be noted that there is a wide range of compressibilities. Increasing pressures have the effect of reducing the value, whereas increasing temperatures have the effect of producing an increase. The compressibility of pure water at 6,000 psia and 200°F is approximately 2.9×10^{-6} bbl per bbl per psi. Since with increasing depth higher pressures and temperatures are encountered, it is expected that the compressibility will increase but the magnitude will be dependent upon the relative increases in pressure and temperature.

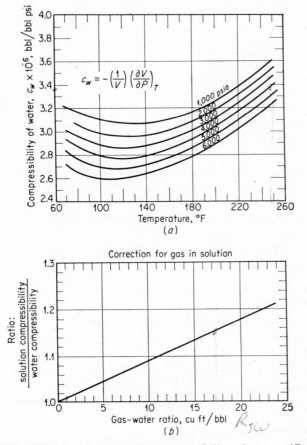

Fig. 6-2. Effect of dissolved gas upon the compressibility of water. (*From Dodson and Standing.*[1])

At a given pressure and temperature, the effect of gas in solution in pure water is to increase the compressibility over that of pure water at the same pressure and temperature. Dodson[1] prepared the graphical method of correction for gas solubility shown in Fig. 6-2b. Using Dodson's[1] data, Jones[2] proposed the following empirical method of solution:

$$c_w = c_{wp}(1 + 0.0088R_{sw}) \tag{6-3}$$

454   PETROLEUM RESERVOIR ENGINEERING

R_{sw} = solubility of gas in reservoir water, cu ft/bbl
c_{wp} = compressibility of pure water, 1/psi
c_w = compressibility of reservoir water, 1/psi

The effect of gas solubility on the compressibility of water is considerable, as a reservoir water containing 20 cu ft of natural gas per barrel will have a compressibility approximately 18 per cent greater than that of pure water at the same pressure and temperature.

Since reservoir waters contain salts and the salinity affects the gas solubility, it is evident that this correction must be applied to the gas solubility prior to its use in Eq. (6-3). The procedure for making this correction is described under the section on Solubility of Natural Gas in Water in an earlier part of this chapter.

The compressibility of a reservoir water is useful in estimating reservoir fluid volumes and in predicting the mobility of invasion of water into the oil-producing zone.

Example 6-3. Determination of Compressibility of Pure Water. A petroleum reservoir is known to have a reservoir pressure and temperature of 4,000 psia and 140°F, respectively, and the connate water in the reservoir is believed to be relatively pure. Estimate the probable compressibility factor for the water.

From Fig. 6-2a read the compressibility of pure water as 2.8 × 10⁻⁶ bbl per bbl per psi.

Example 6-4. Determination of Compressibility of Reservoir Water.

If the connate water in Example 6-3 is known to have a salinity of 30,000 ppm, compute the compressibility factor for the water.

From Fig. 6-2a read the compressibility of pure water as 2.9 × 10⁻⁶ bbl per bbl per psi.

This value must be corrected for salinity. Hence, the gas solubility for pure water is read from Fig. 6-1a as 17.6. This value can be corrected to connate-water salinity through use of Fig. 6-1b or Eq. (6-1) and the X factor. Using Eq. (6-1),

$$R_{sw} = R_{swp}\left(1 - \frac{XY}{10,000}\right) = 17.6\left(1 - \frac{0.055 \times 30,000}{10,000}\right) = 14.7$$

$$= 14.7 \text{ cu ft/bbl}$$

Then using Eq. (6-3),

$$c_w = c_{wp}[1 + 0.0088(R_{sw})]$$
$$= 2.9 \times 10^{-6}[1 + 0.0088(14.7)]$$
$$= 3.27 \times 10^{-6} \text{ bbl/(bbl)(psi)}$$

A similar answer can be obtained through use of Fig. 6-2b.

Knowing the gas solubility in the connate water R_{sw} to be 14.7 cu ft per bbl, use Fig. 6-2b and read a correction factor of 1.13.

To obtain the compressibility of the connate water, multiply the compressibility of pure water by the correction factor.

$$c_w = (2.9 \times 10^{-6})(1.13) = 3.27 \times 10^{-6} \text{ bbl/(bbl)(psi)}$$

Thermal Expansion of Water

NOT RELATED TO B_w (Volume Factor) COMPRESSIBILITY

The thermal expansion of pure water can be illustrated in a number of different ways, but it is believed that the method shown in Figs. 6-3 and

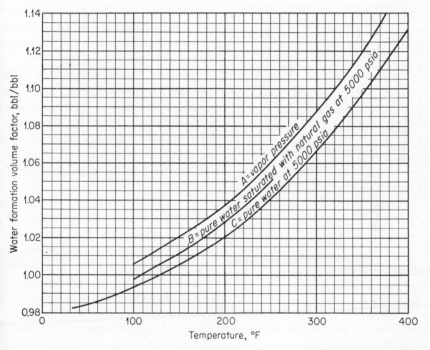

FIG. 6-3. Water-formation volume factor. (*From Keenan and Keyes*[3]; *and Dodson and Standing.*)

6-4, in which the volume factor is plotted versus temperature, is the most convenient. The thermal expansion of pure water is the slope of the curve at any given set of conditions. The thermal expansion is expressed in barrels per barrel per degree Fahrenheit temperature. The thermal expansion of pure water at constant pressure can be expressed as follows:

$$\beta = \frac{1}{V}\left(\frac{\Delta V}{\Delta T}\right)_P \tag{6-4}$$

456   PETROLEUM RESERVOIR ENGINEERING

where β = thermal expansion coefficient of pure water, $1/°F$
        V = volume of water, bbl
        ΔV = Change in volume of water, bbl
        ΔT = Change in temperature of water, °F

The curve in Fig. 6-3 for pure water at its vapor pressure indicates that with an increase in temperature from 60 to 250°F an increase of approximately 6 per cent in water volume results. The other curves indicate the relative importance of pressure and gas solubility on thermal expansion. In general, over most of the range of pressures and temperatures found in petroleum reservoirs, the pressure and gas solubility have a negligible

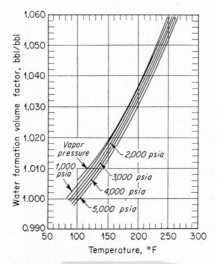

FIG. 6-4. Formation volume factor of water saturated with natural gas. (*From Dodson and Standing.*[1])

effect upon the thermal expansion of water. From practical considerations, it is obvious that the pressure would be important only in so far as it influences the gas solubility.

However, at low temperatures (32 to 125°F), consideration must be given to the effect of pressure and gas solubility on thermal expansion. Consequently, the salinity of the water must be considered, since it has an effect on the gas solubility of the water.

Example 6-5. Determination of Thermal Expansion of Pure Water. A relatively pure connate water is known to exist in a reservoir at 5,000 psia and 200°F. Estimate the thermal-expansion coefficient for this water.

Using curve C in Fig. 6-3 locate the point corresponding to the reservoir conditions. Construct a line tangent to the curve at this point, and determine the slope as follows:

$$\beta = \frac{\Delta B_w}{B_w \, \Delta T} = \frac{1}{1.02}\left(\frac{1.055 - 0.986}{300 - 100}\right) = 0.00469$$

Example 6-6. Determination of Thermal Expansion of Reservoir Water. If the connate water in Example 6-5 is known to be saturated with natural gas at the reservoir conditions, compute the probable thermal-expansion coefficient for the water.

Using curve B (Fig. 6-3), locate the point corresponding to the reservoir conditions. Construct a line tangent to the curve at this point, and note that for practical purposes the slope is identical with that obtained in Example 6-5 and hence the thermal-expansion coefficients are approximately equal.

Water-formation Volume Factor

The volume factor for pure water is dependent upon its pressure and temperature, and the relationship is illustrated in Fig. 6-5 and in Tables 6-1 and 6-2. It is obvious that in accordance with the compressibility and

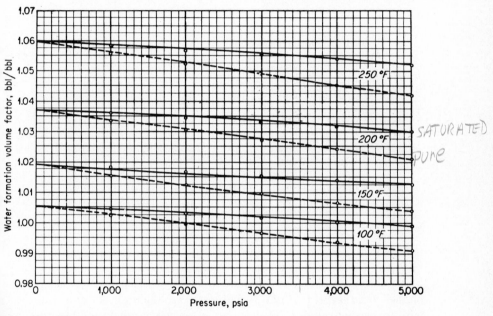

FIG. 6-5. Water-formation volume factor for pure water (dashed lines) and pure water saturated with natural gas (solid lines) as a function of pressure and temperature. (*From Dodson and Standing.*[1])

thermal-expansion characteristics discussed previously, an increase in pressure produces a decrease in the volume factor whereas, at constant pressure, an increase in temperature produces an increase in the volume factor. At a pressure of 5,000 psia and 250°F the volume factor for water is 1.042 relative to a volume factor of 1.000 at 14.73 psia and 60°F. From Fig. 6-5 it is seen that a change in temperature from 100 to 250°F produces a much greater effect on the water-formation volume factor than a pressure change from 0 to 5,000 psia.

Pure water at elevated pressures and temperatures often contains dissolved gas. In Fig. 6-5 data are presented for pure water saturated with

TABLE 6-1. WATER-FORMATION VOLUME FACTOR FOR PURE WATER SATURATED
WITH NATURAL GAS[1]

Saturation pressure, psia	Water-formation volume factor, bbl/bbl, at °F			
	100	150	200	250
1,000	1.0045	1.0183	1.0361	1.0584
2,000	1.0031	1.0168	1.0345	1.0568
3,000	1.0017	1.0154	1.0330	1.0552
4,000	1.0003	1.0140	1.0316	1.0537
5,000	0.9989	1.0126	1.0301	1.0522

TABLE 6-2. WATER-FORMATION VOLUME FACTOR FOR PURE WATER[1,3]

Pressure, psia	Water-formation volume factor, bbl/bbl, at °F						
	32	100	150	200	250	300	400
200	0.9982	1.0050	1.0207	1.0363	1.0617	1.0872	
600	0.9967	1.0037	1.0193	1.0349	1.0599	1.0852	1.1592
1,000	0.9953	1.0025	1.0153	1.0335	1.0560	1.0835	1.1566
2,000	0.9920	0.9995	1.0125	1.0304	1.0523	1.0792	1.1498
3,000	0.9887	6.9966	1.0095	1.0271	1.0487	1.0749	1.1433
4,000	0.9855	0.9938	1.0067	1.0240	1.0452	1.0707	1.1371
5,000	0.9822	0.9910	1.0039	1.0210	1.0418	1.0666	1.1311
6,000	0.9791	0.9884	1.0031	1.0178	1.0402	1.0626	1.1254

natural gas. Naturally, at a given pressure and temperature, gas-saturated pure water has a higher volume factor than pure water.

In addition to dissolved gas, most reservoir waters contain salts. As explained in an earlier part of this chapter, gas solubility in water is decreased with increasing salinity. Hence, at a given pressure and temperature, an increase in the salinity of water produces a decrease in the volume factor.

The volume factor for a reservoir water can be computed with aid of graphs as follows: (1) At the given pressure and temperature the volume factors for pure water and pure water saturated with natural gas are read from Fig. 6-5, (2) the gas solubility for pure water is read from Fig. 6-1a and corrected for salinity using Fig. 6-1b or Eq. (6-1), (3) assuming the effect of gas solubility on the volume factor to be linear, the volume factor at the desired pressure and temperature is computed by interpolation using the gas solubility of gas-saturated pure water and the gas solubility of the reservoir water as the basis for the interpolation.

In the case that the reservoir temperature does not correspond to those for which curves have been illustrated on Fig. 6-5, it is recommended that at the reservoir pressure, water-formation volume factors be read at the four temperatures illustrated and these values plotted versus temperature. Interpolation of this graph will yield the desired water-formation volume factor at the reservoir temperature.

Water-formation volume factors are used in estimating reservoir volumes of water and find particular applicability in material-balance calculations.

Example 6-7. Determination of Formation Volume Factor for Pure Water. Estimate the water-formation volume factor for a relatively pure connate water at a reservoir pressure of 5,000 psia and a temperature of 250°F.

From Fig. 6-5 read the formation volume factor for pure water as 1.044 bbl per bbl.

Example 6-8. Determination of Formation Volume Factor for Reservoir Water. Estimate the water-formation volume factor for a connate water, salinity 50,000 ppm, at a reservoir pressure of 5,000 psia and a temperature of 250°F.

From Fig. 6-5 read the following: formation volume factor of pure water saturated with natural gas = 1.054 bbl/bbl, formation volume factor of pure water = 1.044 bbl/bbl.

From Fig. 6-1a read the gas solubility in pure water as 21 cu ft per bbl. Correct this value for salinity using Fig. 6-1b or Eq. (6-1), factor X. In this case, use Fig. 6-1b and extrapolate the 250°F curve to obtain a correction of 0.84. Hence, the gas solubility in 50,000 ppm connate water is estimated as

$$21 \times 0.84 = 17.64$$

Since the pure water saturated with 21 cu ft of natural gas per barrel of water had a water-formation volume factor of 1.054, the formation volume factor for the 50,000 ppm connate water having 17.64 cu ft of gas dissolved per barrel can be estimated as follows:

$$1.044 + (1.054 - 1.044) \frac{17.64}{21} = 1.052 \text{ bbl/bbl}$$

Example 6-9. Determination of Formation Volume Factor for Reservoir Water by Various Methods. The following production data are available for a field in which the oil-producing zone is at 5,750 ft. The reservoir pressure is 2,675 psia, and the reservoir temperature is 193°F.

Method 1. Correct the total quantity of water production to its equivalent volume at reservoir conditions, assuming that the reservoir pressure remains constant and that the water has a specific gravity of 1.10 at 60°F/60°F. Assume that the average annual surface temperature is 60°F and the pressure 14.73 psia.

From curve A (Fig. 6-7) read a salinity of 143,000 ppm corresponding to a specific gravity of 1.10.

(1)	(2)	(3)	(4)
Year	Average oil-production rate, bbl/day	Water-oil ratio, bbl/bbl	Gas-oil ratio, cu ft/bbl
1936	240	0.25	
1937	130	0.28	1,903
1938	84	0.38	2,822
1939	54	0.60	3,252
1940	43	0.66	4,579
1941	36	0.81	5,952

From Fig. 6-4 (as estimated from Table 6-1) read a volume factor of 1.0365 bbl per bbl for pure water at its vapor pressure.

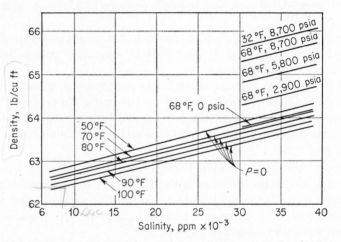

FIG. 6-6. Effect of salt concentration and temperature on water density. (*From Rowe.*[5])

From Fig. 6-1a read 14 cu ft per bbl as the solubility of natural gas in pure water at reservoir conditions.

Using Eq. (6-1) and the correction factor compute the solubility of natural gas in 143,000 ppm reservoir water as follows:

$$R_{sw} = R_{swp}\left(1 - \frac{XY}{10,000}\right)$$

$$= 14\left(1 - \frac{0.045 \times 143,000}{10,000}\right)$$

$$= 4.98 \text{ cu ft/bbl}$$

A similar result can be obtained using Fig. 6-1b as follows. Assuming the relationship to be linear, the extrapolated value of the ratio would be 0.350. This value can be obtained by reading the value of the ratio at 28,600 ppm (143,000/5) and 193°F as 0.870, multiplying the difference between unity (1) and this value (0.870) by 5, subtracting this result from unity, and thereby obtaining $0.350[1 - (1 - 0.87)5]$. Then

$$14 \times 0.350 = 4.90 \text{ cu ft/bbl}$$

Although either result is sufficient for use in practice, the value obtained

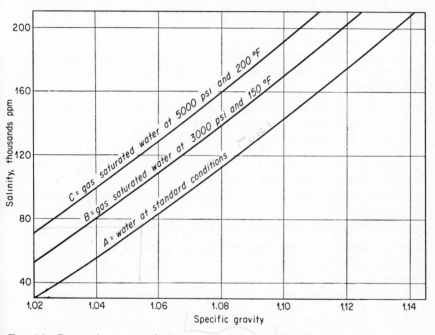

FIG. 6-7. Reservoir water salinity represented as a function of specific gravity. (*From Jones.*[2])

through use of the equation is the more accurate. Hence, it will be used in succeeding calculations.

From Fig. 6-5 the change in volume factor at 193°F per cubic foot of gas solubility can be estimated as follows:

$$\frac{1.0345 - 1.0305}{14} = \frac{0.0040}{14} = 0.000286 \text{ bbl/(bbl)(cu ft)}$$

From Fig. 6-2a read a compressibility factor of 3.10×10^{-6} bbl per bbl per psi for pure water. This value can be corrected for the effect of solubility and salinity effects through use of Eq. (6-3).

$$c_w = c_{wp}(1 + 0.0088R_{sw})$$
$$= 3.10 \times 10^{-6}(1 + 0.0088 \times 4.98)$$
$$= 3.24 \times 10^{-6} \text{ bbl/(bbl)(psi)}$$

A similar result can be obtained using Fig. 6-2b. At a gas-water ratio of 4.98 cu ft per bbl read 1.04. Hence the compressibility for the reservoir water is

$$3.10 \times 10^{-6}(1.04) = 3.22 \times 10^{-6} \text{ bbl/(bbl)(psi)}$$

Here again either result can be used in practice, but the result obtained through use of the equation is the more accurate and will be used in succeeding calculations.

Hence, since the formation volume factor for pure water at 193°F was found to be 1.0365 bbl per bbl and the correction for compressibility was 3.22×10^{-6} bbl per bbl per psi, the formation volume factor for the reservoir water is

$$B_w = 1.0365 - 3.22 \times 10^{-6} \times 2,675 + 0.000286 \times 4.98$$
$$= 1.0365 - 0.0086 + 0.0014$$
$$= 1.0293 \text{ bbl/bbl}$$

Method 2. From the data of Fig. 6-5 obtain the following information at 2,675 psia and 193°F by interpolation.

Volume factor for pure water saturated with natural gas = 1.032 bbl/bbl

Volume factor for pure water = 1.028 bbl/bbl

From Method 1 the gas solubility in pure water at the reservoir conditions was found to be 14 cu ft per bbl and the gas solubility for the saline water at the reservoir conditions was found to be 4.98 cu ft per bbl.

Hence, the water-formation volume factor for the reservoir water is

$$B_w = 1.028 + \frac{4.98}{14}(1.032 - 1.028) = 1.0294 \text{ bbl/bbl}$$

This value compares favorably with that obtained by Method 1.

Hence the production data can be corrected as follows:

Year	Annual oil production, bbl	Water-oil ratio, bbl/bbl	Annual water production, bbl
1936	87,700	0.25	21,900
1937	45,500	0.28	13,300
1938	30,700	0.38	11,650
1939	19,750	0.60	11,850
1940	15,700	0.66	10,350
1941	13,200	0.81	10,650
			79,700

Therefore, the equivalent reservoir volume of this quantity of water is

$$79{,}700 \times 1.0294 = 82{,}000 \text{ bbl}$$

Density, Specific Volume, and Specific Gravity

The density of water is expressed in mass per unit volume; the specific volume, in volume per unit of mass and specific gravity, relates the observed density to the density at some set of conditions ordinarily referred to as base conditions. The base conditions to which all values are referred in this chapter are 14.73 psia and 60°F.

The magnitude and units of the more common methods of expression of these properties for pure water at 14.73 psia and 60°F are as follows:

0.999010 gm/cc
8.334 lb/gal
62.34 lb/cu ft
350 lb/bbl (U.S.)
0.01604 cu ft/lb

The aforementioned quantities can then be related as follows:

$$\gamma = \frac{\rho_w}{62.34} = \frac{1}{62.34 v_w} = 0.01604 \rho_w = \frac{0.01604}{v_w} \tag{6-5}$$

where γ = specific gravity
ρ_w = density, lb/cu ft
v_w = specific volume, cu ft/lb

The relative density and relative volume of pure water over a range of temperature from 0 to 400°F are given in Table 6-3.

TABLE 6-3. RELATIVE DENSITY AND VOLUME OF PURE WATER

Property	Temperature, °F							
	0	100	150	200	250	300	350	400
Relative density, gm/ml	0.99987	0.99306	0.98026	0.96301	0.9426	0.9184	0.8900	0.8571
Relative volume, ml/gm	1.00013	1.00699	1.02014	1.03715	1.0610	1.0890	1.1243	1.1669

The petroleum-reservoir engineer often needs to determine the density of reservoir water. This value can be obtained readily by observing that the density of the reservoir water is related to the density of pure water at base conditions in the following manner:

$$\frac{v_w}{v_{wb}} = \frac{\rho_{wb}}{\rho_w} \cong B_w \tag{6-6}$$

where v_{wb} = specific volume of water at base conditions, lb/cu ft

ρ_{wb} = density of water at base conditions, lb/cu ft

B_w = formation volume factor for water, reservoir volumes per unit volume at base conditions

Therefore, if the density of water at base conditions and the volume factor for the water are available from either direct measurement or use of empirical correlations, the density of the water at reservoir conditions can be calculated.

Quite often it is necessary for the reservoir engineer to make calculations in which the density is required. In the absence of appropriate laboratory data Figs. 6-6 and 6-7 can be used to estimate the magnitude of this property.

Example 6-10. Determination of Density of Pure Water. A connate water is known to be relatively pure and exists in the reservoir at 5,000 psia and 200°F. If it is assumed that the water is saturated with natural gas, what is the density of the water at reservoir conditions?

Read B_w from Table 6-1 or Fig. 6-5 (1.0301 bbl/bbl).

Hence, the density is

$$\rho_w = \frac{\rho_{wb}}{B_w} = \frac{62.34}{1.0301} = 60.51 \text{ lb/cu ft}$$

Example 6-11. Determination of the Salinity of Reservoir Water. A connate water is found to have a specific gravity of 1.04 at standard conditions. What is the probable salinity of the water?

Read salinity from curve A of Fig. 6-7 (55,000 ppm).

Viscosity of Water

Few data have been published on the viscosity of either pure or reservoir waters. The paper of Beal[6] in which the work of Bridgman[7] on pure water and others[4] was compiled is perhaps the most comprehensive. These results are presented in Tables 6-4 and 6-5 and in Fig. 6-8.

The viscosity of pure water at its vapor pressure decreases from 1.79 centipoises at 0.0886 psia to 0.174 centipoise at 89.6 psia. The viscosity of pure water decreases from 1.4 to 0.3 centipoise with an increase in temperature from 50 to 200°F. Bridgman's results show very little change in viscosity of pure water over a range of 14.2 to 7,100 psia.

The other curve in Fig. 6-8 is that which was presented by Jones[2] for a reservoir water containing 60,000 ppm of salt. Unpublished data indicate that brine viscosity increases with salinity over a temperature range of 32 to 300°F at pressures not exceeding 5,000 psia.

There is a real need for information relative to the effect of salinity and

TABLE 6-4. VISCOSITY OF WATER AT VARIOUS TEMPERATURES AND AT VAPOR PRESSURE[6]

Temp, °F	Viscosity, cp	Vapor pressure, psia[a]
32	1.79	0.0886
50	1.31	0.180
68	1.00	0.339
86	0.801	0.616
104	0.656	1.07
122	0.549	1.79
140	0.469	2.89
158	0.406	4.52
176	0.357	6.87
194	0.316	10.18
212	0.284	14.7
230	0.256	20.8
284	0.196	52.4
321	0.174	89.6

[a] Pressure is that of the saturated vapor at the indicated temperature.

TABLE 6-5. VISCOSITY OF WATER AT HIGH PRESSURES AND TEMPERATURES[6,7]

Pressure, psia	Absolute viscosity, cp, at °F			
	32	50.5	86	166.6
14.2	1.792	1.40	0.871	0.396
7,100	1.680	1.35	0.895	0.411
14,200	1.65	1.33	0.921	0.428
21,300	1.67	1.33	0.950	0.443
28,400	1.71	1.35	0.986	0.461

gas solubility on the viscosity of reservoir waters at elevated pressures and temperatures.

Example 6-12. Determination of the Viscosity of Pure Water. A connate water is relatively pure and is found in a reservoir having a pressure of 1,000 psia and a temperature of 150°F. Estimate the viscosity of the water.

Read the viscosity from Fig. 6-8 (approximately 0.5 centipoise).

Example 6-13. Determination of the Viscosity of Reservoir Water. If the connate water in Example 6-12 had a salinity of 50,000 ppm and was found at a reservoir pressure of 7,000 psia and a temperature of 150°F, estimate the probable viscosity of the water.

From Fig. 6-8 it is seen that salinities up to 60,000 ppm and pressures up to 7,000 psia have very little effect on the viscosity. Hence, the viscosity would be 0.5 centipoise.

Solubility of Water in Natural Gas

The solubility of water in natural gas is an important physical property, since it influences the treating, processing, and transporting of natural gas.

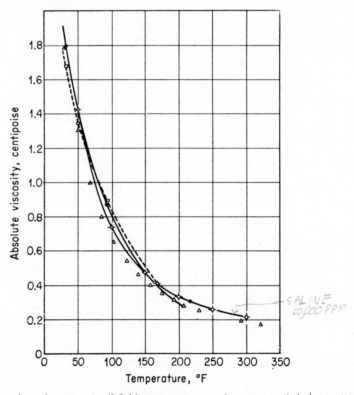

FIG. 6-8. Viscosity of water at oil-field temperature and pressure; (-⧄-⧄-) saline water (60,000 ppm) at 14.7 psia pressure, (•——•) at 14.2 psia pressure, (☐ - - - ☐) at 7,100 psia pressure, (△) at vapor pressure. (*From Van Wingen.*[8])

Hence, in many operations it is necessary to exert strict control of this property. The theoretical principles and laws upon which this property are dependent are presented in Chap. 4. Among other considerations, a detailed discussion of the factors influencing the formation of hydrates is included.

The solubility of water in natural gas is dependent upon the pressure, temperature, and composition of both the water and natural gas. The relationship between the solubility of pure water in natural gas and the pressure and temperature developed by Dodson[1] is illustrated in Fig. 6-9a. A method for correcting the pure-water–solubility data for salinity is presented in Fig. 6-9b. Water-solubility data are limited to a maximum pres-

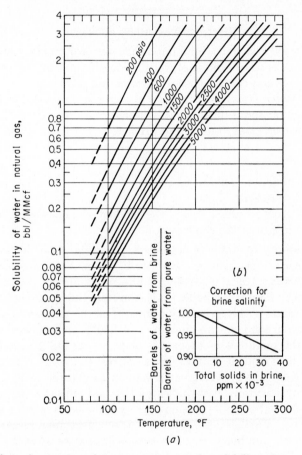

FIG. 6-9. Effect of pressure and temperature upon the solubility of water in natural gas. (*From Dodson and Standing.*[1])

sure of 5,000 psia and a temperature of 300°F, indicating the desirability of additional data.

Example 6-14. Determination of the Solubility of Water in Natural Gas. What is the solubility of a brine having a salinity of 20,000 ppm in natural gas at 3,000 psia and 250°F?

From Fig. 6-9 read the solubility of pure water in natural gas at the specified conditions as 2 bbl per 1,000 Mcf of gas.

From Fig. 6-9 the correction factor for the salinity of the water is found to be 0.95, and hence, the solubility of the 20,000 ppm brine in the natural gas at the reservoir conditions is

$$2 \times 0.95 = 1.90 \text{ bbl/Mcf of gas}$$

Electrical Resistivity of Water

The electrical resistivity is an important physical property of water and is utilized in electrical logging of wells to identify and correlate formations as well as to locate contacts between water and oil.

The resistivity (specific resistance) of water is a measure of its electrolytic conduction and is directly proportional to its cross-sectional area and inversely proportional to its length. Hence,

$$R = r \frac{A}{L} \tag{6-7}$$

where R = resistivity, ohm-meters
r = resistance, ohms
A = cross-sectional area of the conductor, meters squared
L = length of the conductor, meters

The resistivity of water is dependent primarily upon the temperature and chemical composition of the water in the manner illustrated in Fig. 6-10. Pure water has a relatively infinite resistivity as compared with a water having a very low salinity. It is seen from the figure that for a water of a given salinity, the resistivity decreases as the temperature increases. Since reservoir temperatures in excess of 350°F and reservoir water salinities in the hundreds of thousands of parts per million have been encountered, it is evident that the range of resistivities of waters found in petroleum reservoirs is even greater than that indicated in Fig. 6-10.

Recent investigations have indicated that not only the quantity of salt present in the water but its chemical composition has a marked effect upon the resistivity. Furthermore, pressure has an effect upon resistivity, since it influences gas solubility, which in turn is dependent upon the salinity of the water. The effect of gas solubility on water resistivity at elevated pressures and temperatures has not been defined.

In view of these uncertainties it is recommended that at high pressures, temperatures, and salinities the correlation of Fig. 6-10 be used only if laboratory data for the particular reservoir water sample are unavailable.

Example 6-15. Determination of the Resistivity of Reservoir Water. A connate water, salinity of 50,000 ppm, is found in a reservoir having a pressure of 5,000 psia and a temperature 100°F. Determine the resistivity of this water at reservoir conditions.

The resistivity, estimated from Fig. 6-9, is 0.1 ohm-meter.

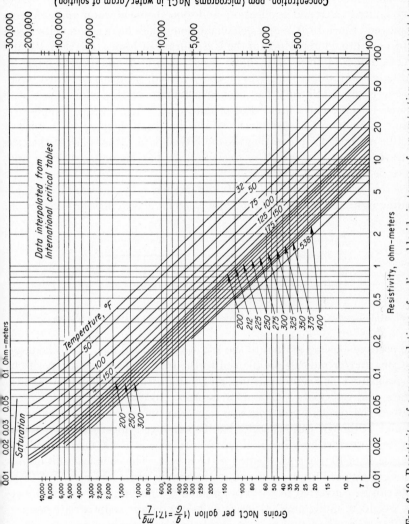

Fig. 6-10. Resistivity of aqueous solutions of sodium chloride in terms of concentrations at constant temperatures. (*From Schlumberger.*[12])

469

CHEMICAL PROPERTIES OF WATER

The early analyses[9,10] of oil-field waters reported only specific gravity and total solids concentrations. However, such analyses were known to have limited value and application, and soon chemical water analyses were employed. Since 1920, major petroleum companies have accumulated such data, and through the efforts of the U.S. Bureau of Mines, AIME, API, and AAPG, much of these data have been assembled and efforts made to correlate and interpret the data on geographical, regional, and areal bases. However, great difficulty has been experienced in this effort, since it is desired to represent a great deal of data concerning a water sample in a simple manner. The most recent effort[11] has been directed to the use of graphic methods for presenting the analytical data. It is believed that the method proposed by Stiff[11] is the simplest, has maximum utility, and, hence, is the most popular.

In this graphical method, illustrated in Fig. 6-11, horizontal lines extending right and left from a centrally located vertical line form the graph.

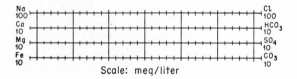

Fig. 6-11. Essential feature of the water pattern analysis system. (*From Stiff.*[11])

The positive ions are plotted to the left of the vertical line, while the negative ions and/or radicals are plotted to the right. Characteristic positions are designated for sodium, calcium, magnesium, iron, chloride, bicarbonate, sulfate, and carbonate ions and/or radicals. In case there is a difference between the positive and negative ions, the difference is represented as sodium. Although various scales can be employed, most reservoir waters may be plotted with sodium and chloride on a scale of 100 milliequivalents and a scale of 10 milliequivalents for the others. The chemical unit of milliequivalents per liter is employed in the graphs presented here, but these units can be converted to parts per million by multiplying by the equivalent weight in milligrams. If other units are desired, appropriate conversion factors can be found in a standard chemical handbook.[4]

When the water-analysis data are plotted on the graph and the adjacent points are connected by straight lines, a closed "pattern" is formed as illustrated in Fig. 6-12. The resulting "patterns" have many different sizes and shapes but, to the experienced eye of the specialist in this field, characterize the water. Characteristic shapes are observed for fresh water, sea water, oil-field brines, etc. Stiff observed that one of the distinctive features of the method is that the pattern maintains its characteristic shape upon dilution of the sample, thereby permitting a qualitative determination of the total salt concentration. Another advantage of this method is that the choice of scale can be made to exaggerate or minimize a particular chemical

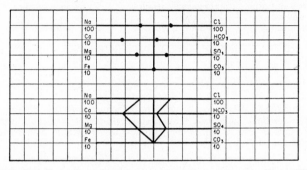

FIG. 6-12. Method of constructing water analysis pattern. (*From Stiff.*[11])

characteristic of the water, thereby facilitating identification of such a characteristic in future samples.

This method has been employed in many practical applications for correlating producing formations as illustrated in Fig. 6-13. The characteristic

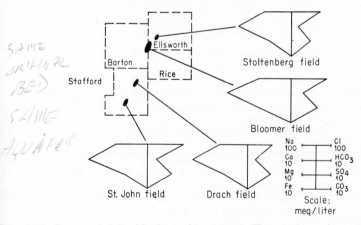

FIG. 6-13. Course of Arbuckle formation through Kansas shown by water patterns. (*From Stiff.*[11])

pattern of the Arbuckle is evident. The method has also been used in tracer studies in water flooding and in conjunction with drill-stem testing to identify the source of water. In Fig. 6-14 an application for detecting foreign water encroachment and locating its source is illustrated. Through this study the water leak was eliminated expeditiously and at a minimum of expense.

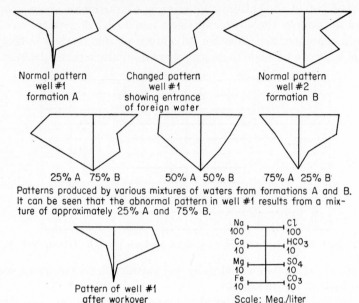

Normal pattern
well #1
formation A

Changed pattern
well #1
showing entrance
of foreign water

Normal pattern
well #2
formation B

25% A 75% B   50% A 50% B   75% A 25% B

Patterns produced by various mixtures of waters from formations A and B. It can be seen that the abnormal pattern in well #1 results from a mixture of approximately 25% A and 75% B.

Pattern of well #1
after workover

	Na 100		Cl 100	
	Ca 10		HCO₃ 10	
	Mg 10		SO₄ 10	
	Fe 10		CO₃ 10	

Scale: Meq/liter

FIG. 6-14. Detection of foreign water and determination of its source. (*From Stiff.*[11])

REFERENCES

1. Dodson, C. R., and M. B. Standing: Pressure-Volume-Temperature and Solubility Relations for Natural Gas-Water Mixtures, *Drilling and Production Practice*, American Petroleum Institute, 1944.

2. Jones, Park J.: "Petroleum Production," Reinhold Publishing Corporation, 1946.

3. Keenan, J. H., and F. G. Keyes: "Thermodynamic Properties of Steam," John Wiley & Sons, Inc., New York, 1936.

4. Hodgman, C. D.: "Handbook of Chemistry and Physics," Chemical Rubber Publishing Company, 1951.

5. Rowe, W. E.: Effect of Salinity on Physical Properties of Water, "Secondary Recovery of Oil in the United States," American Petroleum Institute, 1950.

6. Beal, Carlton: The Viscosity of Air, Water, Natural Gas, Crude Oil and Its Associated Gases at Oil Field Temperatures and Pressures, *Trans. AIME*, vol. 165, 1946.

7. Bridgman, D. W.: "The Physics of High Pressure," The Macmillan Company, New York, 1931.

8. Van Wingen, N.: Viscosity of Air, Water, Natural Gas, and Crude Oil at Varying Pressures and Temperatures, "Secondary Recovery of Oil in the United States," American Petroleum Institute, 1950.

9. Tickell, E. G.: "Report of the California State Oil and Gas Supervisor," 1921.

10. Reistle, C. E.: *U.S. Bur. Mines Tech. Paper* 404, 1927.

11. Stiff, H. A., Jr.: The Interpretation of Chemical Water Analysis by Means of Patterns, *Trans. AIME*, vol. 192, 1951.

12. Schlumberger Well Surveying Corporation: Document 4.

CHAPTER 7

DATA EVALUATION
FOR RESERVOIR CALCULATIONS

INTRODUCTION

The fundamental concepts of reservoir rock and fluid properties were presented in the preceding chapters. The reservoir engineer utilizes these concepts, together with field and laboratory data, to describe petroleum reservoirs and reservoir processes. It is the purpose of this chapter to review methods of formation and data evaluation to provide the engineer with the average parameters describing the physical characteristics of reservoirs which enable him to make volumetric estimates of the quantity of hydrocarbon originally in place.

Guthrie[1] presented an outline of a data-processing procedure for petroleum-engineering data. This outline (Fig. 7-1) summarizes the sources of data and the general evaluation process required to reduce the data to descriptive parameters to be used in reservoir calculations. Basic data are obtained from both the laboratory and field and can be grouped in two broad categories: (1) reservoir-fluid and production data and (2) formation evaluation data.

In the top row of Fig. 7-1 are listed the primary types of data which normally are obtained from a hydrocarbon reservoir. The remaining block titles represent common means of consolidating, reducing, and presenting the data from the various sources.

Sources and taking of data are discussed in this chapter only to the extent necessary to identify the characteristics of the data. Emphasis is placed on the reduction and preparation of the data for engineering calculations.

Hydrocarbon reservoirs are tapped by wells, and the wells are basically the source of all information concerning the reservoir. Formation evaluation data are obtained during the drilling and completion of the well. Data of this type must be obtained during particular phases of the drilling and completion operation. If not obtained at the appropriate time, certain types of data (i.e., core samples) may be lost to the records.

Reservoir-fluid and production data are obtained largely after the wells

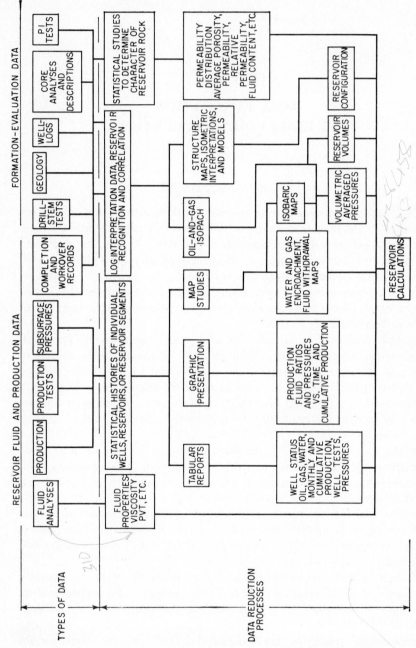

FIG. 7-1. Processing procedure for petroleum-engineering data. (*After Guthrie.*[1])

are completed, and consequently the operator of the wells has greater latitude in taking such data.

FIELD RECORDS

Completion and Workover Records

Completion and workover records are primarily inventories of the physical equipment (casing, tubing, etc.) in a well and of the condition of the well bore. The completion record is also a check list of the tests and other observations taken during the progress of initially drilling and completing the well. On occasion after the original completion, mechanical equipment must be replaced, repaired, or altered or the well must be drilled deeper. An operation to effect these changes is termed a "workover." A supplementary workover record is made and included in the well file.

Completion and workover records are invaluable sources of information for both engineering and operating personnel. These records should be kept as accurately as possible and should be consulted in planning additional tests or evaluating data. Example 7-1 is an example of a completion record on a well. Although this record includes the major components of any completion record, it is not intended as a suggested form. The record includes casing and tubing setting depths; perforation (completion) interval; intervals cored, tested, or logged; and other pertinent data. Some of the observations mentioned will be discussed in succeeding sections of this chapter.

Production Records

The engineer is interested in the oil, gas, and water production from the reservoir and the gas and water injection into a reservoir. The manner of fluid accounting may vary among companies and among fields, but the records are all based on the same measurements.

Oil Production. The volume of oil produced from a reservoir is measured by volumetric vessels or positive displacement meters. The volumetric methods may be a large storage tank or a measuring tank used in a lease automatic custody transfer unit. In either case, the liquid volume is measured at the pressure and temperature of the vessel according to a calibrated volume scale for the vessel. Positive displacement meters measure the volume of fluid flowing at flowing temperature and pressure. Like volumetric vessels, positive displacement meters must be calibrated against some standard.

Usually oil production from the wells on a lease is collected and measured at one or more central locations at which lease storage tanks are erected. A group of tanks is referred to as a "tank battery." Several wells

Example 7-1. Individual Well Record.

Company _____Amcot_____ Lease _____ State _____ Well No. ___3___

Completion data

Elev. __498 ft__ K.B. __496 ft__ D.F. __486 ft__ Gr.   Total depth __5,415 ft__

Comm. __2-26-54__ Comp. __3-14-54__ Reworked _____

Prod. from zone_____ Perfs. __5391–5396 & 5398–5404__ Subsea __−4893–4898 &__

                                                                         −4900–4906

I.P. __106 BOPD__ CK __⅛ in.__ GOR __885:1__ Grav. __42.6__ TP __690 lb__

Prod. from zone _____ Perfs. _____ Subsea _____

I.P. _____ CK _____ GOR _____ Grav. _____ CP _____

Logs run _____

Casing record	Mechanical equipment in well
__163 ft__ of __9⅝ in.__ set at ___ ft with __82__ SX.	
__5,415 ft__ of __5½ in.__ set at ___ ft with ___ SX.	Bonner Packer (N. 1) at 5,375 ft
___ ft of ___ in. set at ___ ft with ___ SX.	
___ ft of ___ in. set at ___ ft with ___ SX.	
___ ft of ___ in. set at ___ ft with ___ SX.	
__5,369 ft__ of tubing run with pkr. at __5,369 (2 in.)__	

Casing perforation record

Date	Perforations Actual depth From	To	Subsea depth From	To	Shots Number	Kind	Zone name	Production tests and squeeze data
3-14-54	5,391	5,396	−4,893	−4,898	40			Pkr at 5,375 ft. Leak developed.
	5,398	5,404	−4,900	−4,906	48			Set Pkr No. 2 at 5,365 ft. Swbd well. Prod. 30.8 bbl in 24 hr. 80 lb sur press. ¼ in. ck. (1.28 bbl/hr)
3-16-54								Flowed into tank making 35 BPD 75 lb pressure.
3-23-54								Gelfraced w/1,000 gal. Broke formation at 2,500 lb. Well brought back in Pot. 106 BOP ⅛ in. ck 1,075 lb pressure.

Well history

Date	Remarks
2-26-54	Moved on location. Rigging up
2-27-54	Ran 163.26 ft of 9⅝ in. csg. cmtd w/82 sx. (3 jts., 36 lb on bottom w/Baker guide shoe,
	3 jts. 32.30 lb H-40 on top)
2-28-54	Drilled to 1,055 ft
3-1-54	Cored 1,055 ft to 1,094 ft. Drld 1,094 to 1,410 ft. Cored 1,410 to 1460½ ft
	(See Core Record)
3-2-54	Drld 1,460½ to 1,818½ ft to 1,864 ft 30 in. DST 1,859½–1,864 ft (See Core Record)
3-3-54	Cored 1,864 to 1,892 ft 30 in. DST 1,882–1,892 ft (See Core Record)
3-4-54	Drlg ahead at 3,717 ft
3-8-54	Drld to 5,354 ft
3-9-54	Cored 5,354 to 5,415 ft 30 in. DST 5,390–5,415 ft (See Core Recrod)
3-10-54	Ran Halliburton Log to TD 5,415 ft. Took 14 side-wall cores. (See Core Record)
3-12-54	Ran 5,415 ft of 5½-ft casing. 2 cement jobs—from 5,415 ft up 150 sx. From DV Tool 2440
	up 313 sx common cmt w/78 sx stratacrete
3-13-54	Ran tbg. Hit bottom plug at 5,381 ft. Drld. cmt to 5,415 ft. Continued mud measured
	out of hole. Ran gamma-ray neutron
3-14-54	Perf: 5391–96 & 5398–5404. Set Bonner Pkr. at 5,375 ft. Ran 5,369 ft 2 in. tbg
3-15-54	Swbd. well until it made pipeline oil. Tubing or Pkr developed leak. Casing showed
	vacuum. Made mud & died. Prep. to reset pkr
3-16-54	Set Bonner Packer No. 2 at 5,365 ft. Swabbing well at 5,000 ft 100% oil. Released rig.
	Well flowing into tank at 10 PM making 35 bbl/day. 75 lb pressure
3-23-54	Gelfraced w/1,000 gal. (1 lb sand per gal.) Broke formation at 2,500 lb. Well brought
	back into potentialing 106 BOPD ⅛ lb ck. 1,075 lb pressure

normally produce into a single tank battery, and frequently the oil is commingled before measurement. This is particularly true if oil production is gauged volumetrically. Even when positive displacement meters are used, the oil is frequently collected in central storage and a volumetric gauge taken. The records from the displacement meters are used to allocate the volumetrically gauged production to the wells. Thus, the basic oil-production measurement is for the group of wells connected to the tank battery. If positive displacement meters are not used, individual well oil production is allocated from battery records based on periodic well tests.

Regardless of how the oil volume is measured, there are essentially three production values reported. The first value reported is the volume actually produced, not corrected for temperature and B.S. and W. (nonsalable content). The second set of production records are those which give the actual volume of salable oil produced at the standard temperature of 60°F. The third set of records indicates the amount of oil sold. The corrected amount of oil produced minus the amount in storage should equal the amount sold.

The question arises as to which of these three sets of records to use in making an engineering study of a reservoir. The one most frequently used is the actual salable oil production corrected to 60°F. Although this oil volume does not account for all the hydrocarbon liquids and solids removed from the reservoir, it is perhaps the most accurate value available. Some of the nonsalable products (B.S. and W.) actually were produced from the reservoir. These unsalable products are primarily water, dirt, and solidified hydrocarbons (paraffin). It is impossible using standard field procedures to determine the exact volume of dirt and paraffin comprising the B.S. and W. fraction. Therefore, it is not possible to evaluate the hydrocarbon volume produced in the solid state. In most cases, the solids volume is so small as to be negligible in any engineering calculation.

Gas Production. In the process of stabilizing a reservoir fluid into storageable or salable quantities, it is necessary to remove the more volatile constituents. These volatile constituents combine to form the gas production. In order to maximize the volume of stable liquid, the gas can be removed in several stages at different pressures and temperatures. The number of stages in the separation and the pressure and temperature of these stages have a great effect on the economics of producing an oil reservoir and on engineering calculations pertaining to the reservoir.

The amount of gas obtained during the process of separation is usually expressed at standard conditions with respect to the volume of oil produced. The actual volume of gas produced can be determined from two sets of records. If gas is being sold to a gasoline plant or gas-transmission line, records will be available on the gas sales volume. The engineer must determine in each field how much of the produced gas is represented by plant records. Seldom is the gas sold that is liberated from the liquid in the stock tank. In some cases, the gas from the low-pressure separators is not sold. If plant records are used to determine the gas production, then the reservoir-fluid properties (see Chap. 5) should be corrected to the prevailing conditions.

As in the case of oil production, the gas is frequently collected and measured at a tank battery after the production of several wells has been commingled. If individual well-production records are required, the gas production must be allocated to the wells on the basis of well tests.

The second method of obtaining gas-production records is through periodic well-production tests. These tests should be conducted under normal operating conditions, measuring the gas production from all separators but not the stock tank. The oil volume produced during the period should be corrected for tank temperature and B.S. and W. The total gas produced is divided by the corrected total stock-tank oil production to give the producing gas-oil ratio for this particular well. It is usually assumed that this gas-oil ratio applies to half of the time period between the last production test and the present test and to half of the period between the present test and the next production test.

The total gas production is calculated using the gas-oil ratio for a prescribed time period and the oil volume produced during that period.

Water Production. The water produced from oil reservoirs is usually unpalatable and unsuitable for irrigation and hence has no economic value. For this reason records of water production are usually not sufficient for most engineering purposes.

Water-production data can usually be compiled from monthly well-production tests required by most companies and some state regulatory bodies. The accuracy of the water production reported on these tests depends on the individual performing the test. The water production obtained on the monthly well-production test is reported as either a water-oil ratio or a "water cut." The water-oil ratio is the barrels of water produced per corrected barrel of stock-tank oil. The water cut is the fraction or percentage of water in the total liquid production.

The cumulative water production is calculated in the same manner as was the cumulative gas production from gas-oil ratio test data. The cumulative water production must be calculated for each well, using the oil-production records of the well.

In some fields, excellent water-production records are available. The produced water is measured, using positive displacement meters, weirs, or some other metering device. These data are usually found on fields under pressure maintenance or secondary recovery operations or with severe water-disposal problems. In this case, the water-production records are maintained in the same fashion as the oil-production records. It is to the engineer's advantage to have these complete records available.

An example calculation of individual well-production records based on production test data and tank-battery production records is presented in the section on well tests.

Well Tests

There are many types of well tests. Some are performed before the well is completed or even before a formation is known to be productive; others are performed during the life of a producing well. From these well tests many important pieces of information are gathered. Some of the most important information is whether the formation is or is not productive of oil or gas, the capacity or capability of the well to produce, the permeability of the rock adjacent to the hole, and the average permeability of that part of the formation which constitutes the drainage area. The approximate drainage characteristics of the formation and an estimate of the rate of decline of the individual well are obtained from periodic tests. Some of these tests aid in the determination of the gas-oil and oil-water contacts.

Drill-stem Tests. In drilling and completing wells with rotary tools, the mud-laden fluid normally exerts a hydrostatic pressure in excess of the formation pressure. The formation fluids are thus sealed off from the well bore. To determine the producible fluid content of the formations and estimate the productivity of the formation, it is necessary to relieve the formation of the hydrostatic pressure of the mud column. This can be done by completing the well and displacing the mud fluid. Completing the well is expensive. Therefore, it is desirable to determine the producible fluid content by some other means. The drill-stem test provides such a means of evaluation, as it is in essence a temporary completion.

A packer and valve assembly are lowered on the drill pipe to a position opposite the formation to be tested. The valve assembly is so constructed as to prevent fluid entry into the drill pipe during placement. The packer is set above the formation to be tested and expanded to seat against the wall of the well bore, thus excluding the mud fluid in the annular space from the test interval. The valve assembly is then opened so that the formation is subjected to the reduced pressure existing in the drill pipe. The formation fluids, if mobile, can then flow into the drill pipe and subsequently be recovered by closing the valve assembly and withdrawing the drill pipe.

Drill-stem tests can be run in either open (uncased) hole or cased hole and are normally referred to as open-hole and perforation tests, respectively. There are two general open-hole testing programs or methods in field use:

1. Test possible productive zones as the zones are penetrated by the drill. This type of program is usually conducted in conjunction with mud logging and/or coring programs. The hydrocarbon shows are determined from examination of the mud, cuttings, and the cores.

2. Test possible productive zones after drilling through to greater depths or to total depth. In this method, side-wall cores, mud logs, and other well logs aid in selecting test intervals. To test in this fashion, it is necessary to use straddle packers or to set successive cement plugs to isolate the intervals.

Perforation tests are conducted in cased holes on intervals defined by perforations in the casing. Casing must be set and cemented prior to testing by this method. While the perforation test yields valuable information with respect to positive evidence of oil or gas, production of salt water or no production must be viewed in general as nondefinitive. The possibilities of leaks behind the pipe, nonpenetration of bullets, and other similar problems must be carefully evaluated in interpreting the data. The mechanical procedure of drill-stem testing is more satisfactory in cased than in open hole.

A schematic drawing of conventional drill-stem test tools is presented in Fig. 7-2.

The results of a drill-stem test are interpreted from pressure data observed at the surface chokes during the test, the recovery obtained on withdrawing the drill pipe or circulating out the entrapped fluid, and the pressure data recorded by a recording pressure gauge located in the tool. The pressure may be recorded either at the surface or on a chart within the instrument in the tool.

The detailed interpretation of a drill-stem test is rather involved. However, the following general rules apply:

1. Formation productive of gas if high surface pressures are observed and little or no liquid is recovered in drill pipe

2. Formation productive of oil if moderate surface pressures are observed and liquid recovery is oil free of water

3. Formation productive of water if low surface pressures are observed and water is recovered in drill pipe

Many tests cannot be interpreted by these elementary rules. The more advanced techniques of interpretation are beyond the scope of this chapter.

Drill-stem tests yield valuable information on gas-oil and oil-water contacts as well as on the potential productivity of the formation. In many instances a contact will occur within a test interval. A gas-oil contact is evidenced by a high surface pressure and a small to moderate recovery of oil. A water-oil contact is evidenced in beds having permeabilities in excess of about 100 millidarcys by recovery of both oil and water. In beds of lower permeability, capillary phenomena create transition zones of appreciable extent from which both water and oil may be produced, thus obscuring the location of the contact by a drill-stem test. Further discussion of the use of drill-stem tests for determining fluid contacts for a field is presented in a later section of this chapter.

A subsurface pressure record and other pertinent data on a drill-stem test are presented in Fig. 7-3. The pressure record reflects the operation of the tool.

In some test tools, an initial "shut-in" or "closed-in" formation pressure is obtained prior to allowing the formation to produce into the drill pipe. This initial closed-in pressure is valuable in interpreting test data and is an aid in estimating original formation fluid pressures.

Production Tests. Routine production tests are made periodically on oil wells. The tests may be of a few hours' or several days' duration, depending on the well characteristics and the desired results. Most frequently

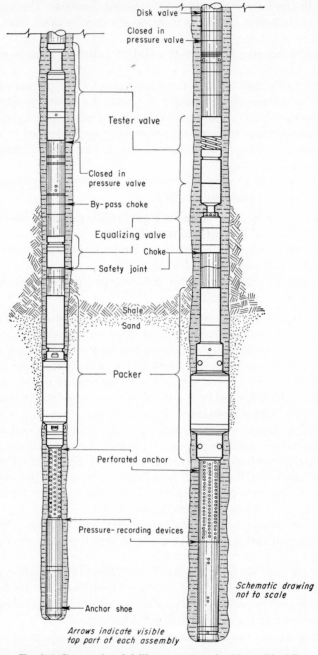

Disk valve

Closed in
pressure valve

Tester valve

Closed in
pressure valve

By-pass choke

Equalizing valve

Choke

Safety joint

Shale

Sand

Packer

Perforated anchor

Pressure-recording devices

*Schematic drawing
not to scale*

Anchor shoe

*Arrows indicate visible
top part of each assembly*

FIG. 7-2. Conventional drill-stem test tools. (*From Black.*[2])

the well is tested for 24 hr or less and the data corrected to a 24-hr basis. The tests are conducted by means of a portable test separator or a test separator and test tank provided at the tank battery. Oil, water, and gas production are gauged over the test period. The gas and water production are used to calculate gas-oil ratios and water-oil ratio or cuts.

Production tests are required by state regulatory bodies as well as for operational purposes. In Table 7-1 is presented a modified GO-2 form as required by the Texas Railroad Commission's Oil and Gas Division. The last three columns are added to provide for recording additional pertinent data.

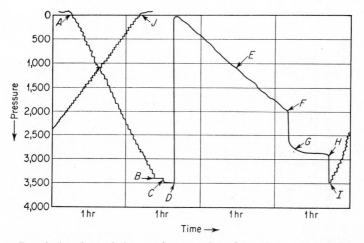

FIG. 7-3. Descriptive data of time tool open, where located, recovery, and pressure. *A*, test started; *B*, reached packer seat; *C*, packer seated; *D*, test opened; *E*, flowing pressure; *F*, test closed; *G*, build-up pressure; *H*, packer unseated; *I*, started out of hole; *J*, reached surface. Interval tested: 6,356 to 6,380 ft. Time open: 1 hr 20 min. Time shut: 30 min. Recovery: 62 joints oil, 2 joints oil-cut mud, no water. (*Halliburton Oil Well Cementing Company.*)

The allowable is the permissible rate of production per producing day. The allowable is assigned to a particular well and must be produced from that well unless transfer to another well or wells is permitted by state rules. A total monthly allowable for a field or well is set by state regulatory bodies after hearing nominations of purchases of crude oil by major oil buyers. Allowables are based on a number of factors including the depth of the well, the ability of the well to produce, gas-oil ratio of the well, and field conditions. Most states have tables for determining the allowable for producing wells. Table 7-2 presents the discovery allowable and 1947 amended schedule for producing wells in the state of Texas. Wells producing at gas-oil ratios in excess of 2,000 scf per bbl are normally penalized.

TABLE 7-1. PRODUCTION TEST RECORD

Lease and battery	Well No.	Date of test	Producing method	Choke size	Surface pressures, psig Tubing	Casing Begin	Casing End	Duration of test, hr	Allowable, bbl	Production during test Water, bbl	Oil, bbl	Gas, Mcf	Gas-oil ratio, cu ft/bbl	Water-oil ratio, bbl/bbl[a]	Separator Conditions Temp, °F[a]	Pressure, psig[a]
Amcot no. 1	1	12-2	Flow	¼				24	100	5	105	60.30	600	0.048		
	3	12-3	Flow	¼				24	100	2	101	65.65	650	0.020		
	4	12-4	Flow	¼				24	100	4	96	57.76	610	0.042		
	5	12-5	Flow	¼				24	100	Trace	98	66.64	680			
	6	12-7	Flow	¼				24	80	30	81	48.60	600	0.370		
Amcot no. 2	2	12-3	Flow	⅛				24	60	Trace	60	200.00	3333			
	7	12-4	Flow	⅛				24	40	Trace	42	210.00	5000			
	8	12-5	Flow	⅛				24	80	0	78	195.00	2500	0		
	9	12-7	Flow	⅛				24	100	0	102	70.38	690	0		
	10	12-8	Flow	⅛				24	100	0	104	73.84	710	0		

[a] These columns are not on the standard GO-2 form.

TABLE 7-2

Discovery allowable schedule (effective Mar. 20, 1950)		Amended 1947 yardstick (effective Apr. 1, 1950)			
Interval of depth	Daily well allowable, bbl	Depth	10 acres	20 acres	40 acres
0– 1,000	20	0– 1,000	18	28	
1,000– 2,000	40	1,000– 1,500	27	37	57
2,000– 3,000	60	1,500– 2,000	36	46	66
3,000– 4,000	80	2,000– 3,000	45	55	75
4,000– 5,000	100	3,000– 4,000	54	64	84
5,000– 6,000	120	4,000– 5,000	63	73	93
6,000– 7,000	140	5,000– 6,000	72	82	102
7,000– 8,000	160	6,000– 7,000	81	91	111
8,000– 9,000	180	7,000– 8,000	91	101	121
9,000–10,000	200	8,000– 8,500	103	113	133
10,000–10,500	210	8,500– 9,000	112	122	142
10,500–11,000	225	9,000– 9,500	127	137	157
11,000–11,500	225	9,500–10,000	152	162	182
11,500–12,000	290	10,000–10,500	190	210	230
12,000–12,500	330	10,500–11,000		225	245
12,500–13,000	375	11,000–11,500		255	275
13,000–13,500	425	11,500–12,000		290	310
13,500–14,000	480	12,000–12,500		330	350
14,000–14,500	540	12,500–13,000		375	395
		13,000–13,500		425	445
		13,500–14,000		480	500
		14,000–14,500		540	560

SOURCE: Railroad Commission of Texas.

(The allowable is reduced in proportion to gas production in excess of 2,000 cu ft per bbl.)

In Table 7-3 are presented production data for a lease having two tank batteries. The entries enclosed on the tabulation are from gauged volumes at the battery corrected for temperature. The remaining quantities are calculated from well test data and the battery production data as shown in Example 7-2.

The utilization of test data in computing gas production and the allocation of observed battery oil and water production data to individual wells are illustrated in Example 7-2.

Individual well oil production was allocated in the example both on the basis of test rates of production and on the basis of allowable. Water and gas productions were computed from average water-oil and gas-oil ratio data from well tests. The computed water production was used to allocate

TABLE 7-3. PRODUCTION DATA

Lease and battery	Well No.	Previous cumulative production			Current month production			Current cumulative production		
		Oil, bbl	Water, bbl	Gas, Mcf	Oil, bbl	Water, bbl	Gas, Mcf	Oil, bbl	Water, bbl	Gas, Mcf
Amcot no. 1		154,225	12,338	92,535	9,264	800[a]	5,730	163,489	13,138	98,265
	1	36,110	1,503	21,810	1,929	83	1,161	38,039	1,586	22,971
	3	34,153	671	20,003	1,871	44	1,188	36,024	715	21,191
	4	29,418	1,214	17,650	1,947	87	1,178	31,365	1,301	18,828
	5	28,727		18,098	1,966		1,288	30,693		19,386
	6	25,817	8,950	14,974	1,551	586	915	27,368	9,536	15,889
Amcot no. 2		91,164	0	151,495	7,022	0[a]	13,393	98,186	0	164,888
	2	16,410	0	44,307	1,116	0	3,534	17,526	0	47,841
	7	11,851	0	36,738	786	0	3,670	12,637	0	40,408
	8	19,144	0	42,117	1,499	0	3,718	20,643	0	45,835
	9	23,703	0	14,933	1,792	0	1,183	25,495	0	16,116
	10	20,056	0	13,400	1,829	0	1,288	21,885	0	14,688

[a] Gross volumes measured at battery.

Example 7-2. Calculation of Gas Production and Allocation of Oil, Water, and Gas Production to Individual Wells.

Lease and battery	Well No.	Previous test Gas-oil ratio, cu ft/bbl	Previous test Water-oil ratio, bbl/bbl	Current test Gas-oil ratio, cu ft/bbl	Current test Water-oil ratio, bbl/bbl	Average test data Gas-oil ratio, cu ft/bbl	Average test data Water-oil ratio, bbl/bbl	Average test data Test rate, bbl/day	Fractional rate	Allowable, bbl/day	Fractional allowable	Prod. based on average test rate Oil, bbl	Water, bbl	Gas, Mcf	Prod. based on allowable Oil, bbl	Water, bbl	Gas, Mcf	Based on average test rate Fractional water	Corrected water, bbl	Based on allowable Fractional water	Corrected water, bbl
Amcot no. 1	1	604	0.032	600	0.048	602	0.040	102	0.2082	100	0.2083	1,929	77	1,161	1,930	77	1,162	0.1039	83	0.1043	83
	3	620	0.024	650	0.020	635	0.022	99	0.2020	100	0.2083	1,871	41	1,188	1,930	42	1,226	0.0553	44	0.0569	45
	4	600	0.040	610	0.042	605	0.041	103	0.2102	100	0.2083	1,947	80	1,178	1,930	79	1,168	0.1080	87	0.1071	86
	5	630		680		655		104	0.2122	100	0.2084	1,966		1,288	1,930		1,264				
	6	580	0.330	600	0.370	590	0.350	82	0.1674	80	0.1667	1,551	543	915	1,544	540	911	0.7328	586	0.7317	586
								490	1.0000	480	1.0000	9,264	741	5,730	9,264	738	5,731	1.0000	800	1.0000	800
Amcot no. 2	2	3,000		3,333		3,167		61	0.1589	60	0.1580	1,116		3,534	1,109		3,512				
	7	4,340		5,000		4,670		43	0.1120	40	0.1053	786		3,670	739		3,451				
	8	2,460	0	2,500	0	2,480	0	82	0.2135	80	0.2105	1,499	0	3,718	1,478	0	3,665				
	9	630	0	690	0	660	0	98	0.2552	100	0.2631	1,792	0	1,183	1,848	0	1,220				
	10	698	0	710	0	704	0	100	0.2604	100	0.2631	1,829	0	1,288	1,848	0	1,301				
								384	1.0000	380	1.0000	7,022		13,393	7,022		13,149				

487

the gauged water production to the wells. The computed gas production was summed to yield the battery gas production.

The computed water production differed from the gauged production by about 7 per cent. The computed gas production based on an oil allocation from allowables differed by about 1.5 per cent from that based on an oil allocation from test rates. Frequently greater discrepancies in computed gas production may occur. Gas production metered at the battery has been observed to differ from that computed from production test gas-oil ratios by as much as 15 per cent. Well test and production records must be carefully taken and recorded to provide data of sufficient accuracy for engineering purposes.

Pressure Tests. Pressures are measured in wells for three primary purposes. The average reservoir pressure is calculated using bottom-hole shut-in pressures. Well performance is determined by measuring flowing and shut-in bottom-hole pressures as functions of time. The position of the fluid level in a well, needed for equipment design, is determined from well pressure surveys.

The pressures are measured by inserting a pressure element and recording mechanism in the well. As the pressure bomb descends, the pressure inside the bomb increases because of the fluid head. The bomb is stopped at predetermined depths for a short period of time. The pressure bomb is removed after reaching final depth, and the recording of the extensions of the pressure element is converted to units of pressure. The results of such a well survey, flowing and static, are shown in Fig. 7-4.

It is not possible for the operator to determine a water level in the well from measurements made at the surface, whereas with some instruments he can determine the approximate oil level in the well. Actually, it is not necessary for the operator to locate the oil and water levels mechanically, as they can be calculated from the data obtained while lowering the pressure bomb in the hole. As gas usually has a gradient of 0.1 psi per ft or less, oil has a gradient ranging between 0.38 and 0.28 psi per ft, and water has a gradient ranging between 0.465 and 0.43 psi per ft; it is possible to select the intervals in which the fluid column changes from gas to oil or oil to water. Once the interval is selected, the location of the interface can be calculated by Eq. (7-1).

$$H_c = H_t + \frac{(P_b - P_t) - G_{dh} \Delta H}{G_{dl} - G_{dh}} \tag{7-1}$$

where H_c = depth to interface, ft
    H_t = depth to top of interval in which interface occurs, ft
    P_t = pressure at top of interval, psi
    P_b = pressure at bottom of interval, psi

G_{dh} = gradient of heavier fluid, determined from next lower interval, psi/ft

G_{dl} = gradient of lighter fluid determined from upper interval, psi/ft

ΔH = distance between points of measurement of P_t and P_b, ft

Because of restrictions in the tubing such as crossover valves and chokes, it may be impossible to measure the pressure opposite the well perforations.

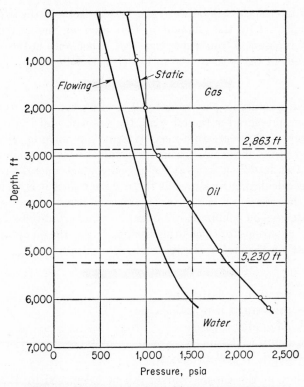

Fig. 7-4. Static and flowing tubing pressure survey.

When the pressure cannot be measured at the perforations, it is necessary to calculate the pressure from the other measurements. In this calculation it is assumed that the fluid existing at the last pressure point in the tubing exists between that point and the perforations. This assumption can easily be in error in the cases where the last pressure point must be located high in the tubing string because of obstructions. The pressure at the perforation is calculated by Eq. (7-2) (see Example 7-3).

$$P_p = P_b + \frac{P_b - P_{b-1}}{H_b - H_{b-1}} (H_p - H_b) \tag{7-2}$$

where P_p = pressure opposite top of perforations, psi
$\quad\quad P_b$ = pressure at lowest depth of survey, psi
$\quad\quad P_{b-1}$ = pressure at next to lowest depth of survey, psi
$\quad\quad H_b$ = depth of lowest measured pressure P_b, ft
$\quad\quad H_{b-1}$ = depth of P_{b-1}, ft
$\quad\quad H_p$ = depth of perforations, ft

If a fluid interface exists between the next to last and the last pressure point as indicated by the gradient in that interval, the gradient of the denser fluid is assumed from experience with other wells in the area. This assumed gradient G_{dh} is then used in Eq. (7-2) as follows:

$$P_p = P_b + G_{dh}(H_p - H_b) \tag{7-2a}$$

In analyzing reservoir performance, it is often necessary to determine some average pressure for the oil zone, the gas cap, and the water zone. As the majority of the pressure measurements are made in oil wells, it is necessary to adjust these measurements to values which would exist at the pressure datum in the oil zone, at the gas-oil contact, and at the water-oil contact. In calculating pressures at these three points it is assumed that the fluid in the reservoir is in a state of static equilibrium. It is also assumed that a continuous oil column exists from the well perforations to all three points. The equations for calculating the pressure at the oil datum, usually the volumetric mid-point of the reservoir, is

$$P_d = P_p + G_{ro}(H_d - H_p) \tag{7-3}$$

where P_d = pressure at datum, psi
$\quad\quad P_p$ = pressure at perforations, psi
$\quad\quad G_{ro}$ = oil gradient, psi/ft
$\quad\quad H_d$ = depth to datum, ft
$\quad\quad H_p$ = depth to perforations, ft

The pressure at the gas-oil contact can be calculated by Eq. (7-3) by replacing the depth of the datum by the depth of the gas-oil contact. The same substitution is made to calculate the pressure at the water-oil contact.

Once production begins, a reservoir is never in static equilibrium. For this reason only the wells which penetrate the gas cap or water zone or are in close proximity thereto should be used to calculate the pressure at the gas-oil and water-oil contacts. If wells far removed from the gas cap are used for calculating the pressure at the gas cap, transient pressure variations will be introduced which will cause the average gas-cap value to be in error.

No correction procedure has been described for adjustment of pressure measurements made in gas wells. The same procedure and equations apply to gas wells as to oil wells.

Example 7-3. Pressure Adjustments Made from a Well Pressure Survey.

Measured data		Calculated data	
Depth	Pressure	Pressure difference	Pressure gradient
0	800		
1,000	900	100	0.1
2,000	1,000	100	0.1
3,000	1,140	140	0.14
4,000	1,470	330	0.33
5,000	1,800	330	0.33
6,000	2,220	420	0.42
6,200	2,310	450	0.45

Top of perforations at 6,300 ft
Surface elevation, 200 ft
Gas-oil contact at 6,050 ft subsea or 6,250 ft from surface
Water-oil contact at 6,150 ft subsea or 6,350 ft from surface
Oil-zone datum at 6,125 ft subsea or 6,325 ft from surface

1. Depth of gas-oil interface:
From Eq. (7-1)

$$H_c = H_t + \frac{(P_b - P_t) - G_{dh}\,\Delta H}{G_{dl} - G_{dh}}$$

$$= 2,000 + \frac{(1,140 - 1,000) - 0.33(1,000)}{0.1 - 0.33}$$

$$= 2,000 + \frac{140 - 330}{-0.23}$$

$$= 2,863.6 \text{ ft}$$

2. Depth of oil-water interface:

$$H_c = 5,000 + \frac{(2,220 - 1,800) - 0.450(1,000)}{0.33 - 0.45}$$

$$= 5,000 + \frac{420 - 450}{0.12}$$

$$= 5,230.7 \text{ ft}$$

3. Calculation of pressure at perforation:
From Eq. (7-2)

$$P_p = P_b + \frac{P_b - P_{b-1}}{H_b - H_{b-1}} (H_p - H_b)$$

$$= 2,310 + \frac{2,310 - 2,220}{6,200 - 6,000} (6,300 - 6,200)$$

$$= 2,310 + \frac{90}{200} (100)$$

$$= 2,355 \text{ psi}$$

4. Calculation of pressure at gas-oil contact:
From Eq. (7-3)

$$P_{GOC} = P_p + G_{ro} (H_{GOC} - H_p) = 2355 + 0.33 (6,250 - 6,300)$$

$$= 2,338.5 \text{ psi}$$

5. Calculation of pressure at oil datum:
Substituting in Eq. (7-3)

$$P_d = 2,355 + 0.33 (6,325 - 6,300) = 2,355 + 0.33(25)$$

$$= 2,363.2 \text{ psi}$$

6. Calculation of pressure at water-oil contact:
Substituting in Eq. (7-3)

$$P_{WOC} = 2,355 + 0.33 (6,350 - 6,300) = 2,355 + 0.33(50)$$

$$= 2,371.5 \text{ psi}$$

Productivity Tests on Oil Wells. The productivity of an oil well is determined by a series of flow and pressure tests. The static or shut-in bottom-hole pressure is measured, and the flowing bottom-hole pressure is measured for various rates of oil production. The pressure difference is plotted as a function of the flow rate in stock-tank barrels per day. The slope of the resulting curve is the productivity index with units of barrels per day per pound per square inch.

In theory this quantity can also be equated to the permeability and thickness of the producing formation. To eliminate as many of the variables as possible it is suggested that the equation be evaluated in the limit as ΔP and Q both approach zero, so that the slope of the curve as it approaches the ordinate would be used for the evaluation of the permeability and thickness of the formation. From Eq. (2-34):

$$Q_o = \frac{7.082 k_o h (P_e - P_w)}{B_o \mu_o \ln r_e / r_w} \tag{7-4}$$

where
Q_o = rate of production, stock-tank bbl/day
k_o = effective permeability to oil, darcys
h = sand thickness, ft
P_e and P_w = pressures at effective radius of drainage r_e and the well radius r_w, respectively, psi
μ_o = oil viscosity, centipoises
B_o = oil formation volume factor at P_e

From the definition of productivity index (PI)

$$PI = \frac{Q_o}{P_{si} - P_f} = \frac{Q_o}{P_e - P_w} = \frac{7.082 k_o h}{B_o \mu_o \ln (r_e/r_w)} \tag{7-5}$$

where shut-in pressure P_{si} is assumed to be equal to P_e and the well pressure flowing P_f and P_w are equal by definition.

If $\ln (r_e/r_w)$ is assumed to be equal to 7.082, for an r_w of 4 in. r_e is 386 ft and for an r_w of 6 in. r_e is 540 ft. For 20-acre well spacing r_e is frequently taken as about 417 ft (half the distance between wells), and for 40-acre well spacing r_e is about 660 ft. Thus for wells of between 20- and 40-acre spacing, evaluated as ΔP and Q approach zero:

$$PI = \frac{k_o h}{B_o \mu_o} \tag{7-6}$$

where B_o and μ_o are defined at P_e.

The bottom-hole pressure after the well has been shut in for some extended period of time, 24 to 72 hr, is usually the first value determined in performing a productivity test on an oil well. The actual time of shut in will be dependent upon the characteristics of the wells. The well is then opened to some small choke size and permitted to flow with the rate of production being recorded as a function of time. When the rate is stabilized, then the bottom-hole pressure at that time is recorded. In order to obtain these bottom-hole pressures, a pressure bomb is usually run into the hole and left there during all the production tests. Once the stabilized rate has been obtained and the time recorded, so that a correlation can be made with the bottom-hole pressure recorded by the bomb, the choke is opened to increase the flow rate. When the flow rate once again has become stabilized, the pressure and time are again noted.

This procedure is followed for three or four different production rates. The materials are then reduced to a graphical form similar to that shown in Fig. 7-5. Here it is noted that the bottom-hole pressure continues to decline until it approximately stabilizes for some given rate, and then when the production rate is changed, it rapidly declines again and begins to stabilize at the new rate. The values read are at the same time. Actually many values can be calculated from these decline curves where both the rate and pressure are recorded as functions of time.

The productivity test data are correlated and the selected rates and their corresponding ΔP's are plotted as shown in Fig. 7-5. This curve is extrapolated to the zero ordinate, where the slope of the curve is the reciprocal of the productivity index. This is the value used to calculate the permeability and net thickness of the formation. It must be pointed out here that the calculated permeabilities apply to the immediate volume around the well bore. The calculated values are sensitive to well completion and

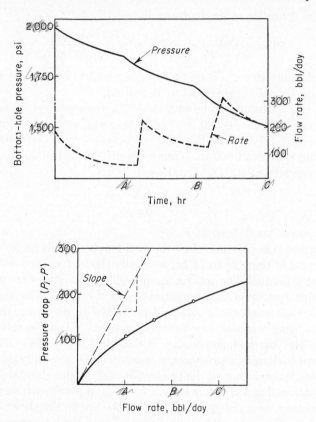

FIG. 7-5. Productivity measurement of an oil well. (*After Evinger and Muskat.*[3])

damage or improvement brought about at the well bore during drilling and completion operations. If the formation is plugged, then low values of permeability will be indicated. If the formation has been fractured or acidized, then high values of permeability will be calculated.

Productivity tests can be run at various states of depletion of the reservoir, and the productivity at some given shut-in bottom-hole pressure will be obtained. If these productivities are plotted as a function of shut-in reservoir pressure, the decline in productive capacity will be noted (Fig. 7-6).

This decline, unless remedial operations have affected the immediate vicinity of the well bore, indicates the declining production capabilities of the individual well. The decline is caused by several factors, one of which is the increased viscosity of the oil with decreased reservoir pressure. Another factor causing a decline is the decreased oil saturation in the vicinity of the well bore which in turn decreases the oil productivity of the formation. A PI decline curve can be extrapolated to indicate the productive life of a well and the reservoir pressure at which it will no longer be economical to operate (Fig. 7-6).

These curves can also be used in conjunction with other data to indicate the probable reservoir pressure at which this well will no longer be capable of flowing its prorated allowable; hence, at this time artificial lifting equipment will have to be installed.

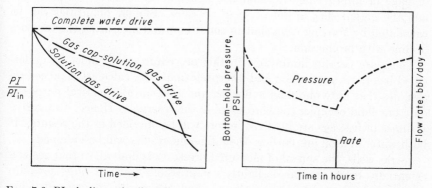

FIG. 7-6. PI decline of oil well under various drive mechanisms.

FIG. 7-7. Theoretical pressure drawdown and build-up performance.

Build-up and Draw-down Tests on Oil Wells. In the field of reservoir engineering certain equations have been adapted from electrical and heat flow dealing with unsteady-state flow systems. Through the years various authors have adapted these equations to special applications in reservoir engineering. One of these applications is in the evaluation of the drainage volume of a well by what is normally referred to as a build-up or draw-down test. A build-up curve is nothing more than shutting a well in and measuring the bottom-hole pressure as a function of time (Fig. 7-7).

Through the use of transient pressure equations, the build-up data, and the performance history of the well, it is possible to calculate the permeability of the formation and obtain some idea as to the formation damage by completion practices. Such analyses are beyond the scope of this chapter.

A build-up test can be run prior to a productivity test simply by running the bomb in the well before it is shut in to allow for the 24- to 72-hr pressure build-up required for establishing the shut-in pressure for the productivity test. From these data, the shut-in pressure can be calculated as if the well were left shut in for infinite time, and under certain conditions the drainage area of a formation can also be determined.

The procedure for running a build-up test is relatively simple. First, the production of a well prior to starting the test must be determined. Second, it is best to flow the well at some given rate for a period of 5 to 10 days prior to shutting in so as to establish a fairly stable pressure distribution within the drainage area of the well. The well is shut in, and the bottom-hole pressure recorded as a function of time. Knowing the past production, the elapsed time, the average flow rate during the time prior to shut-in, the time of shut-in, and the pressure history during the shut-in period, it is possible to calculate the permeability of the formation, the damage or improvement around the well bore denoted as "skin effect," and the actual shut-in bottom-hole pressure. Similar quantities can be calculated by starting from shut-in conditions and measuring the pressure decline with production.

There are certain limitations to this procedure which must be understood before its application and interpretation. The calculating procedure assumes that only one fluid is flowing and that for all practical purposes, only one fluid occupies the pore space in the reservoir. Hence, if excessive volumes of free gas are involved, errors will be obtained in the results. If both water and oil are flowing, then different results will be obtained.

In gas wells, the spread of measured pressures is limited, so that gas can essentially be treated as a slightly compressible fluid. If large pressure drops occur, the method cannot be applied. Primarily, its greatest value is in the initial testing of wells and their classification. This type of data can be collected on a drill-stem test and interpreted to give the relative magnitudes of formation permeability and possibly the drainage area of the well.

Back-pressure Tests on Gas Wells. A back-pressure test on a gas well measures the bottom-hole pressures of the well at shut-in conditions and for three different stabilized flow rates. The recommended procedure is to use four flow rates, but a minimum of three is required.

Figure 7-8 illustrates the results of plotting the difference in the squares of the pressures against the measured flow rates. Normally this set of data should form a straight line when plotted on logarithmic paper. The intercept of this straight line is a measure of the productivity or producing capacity of the formation. The equation which defines this line is given as Eq. (7-7) where

$$Q = C(P_s^2 - P_f^2)^n \qquad (7-7)$$

the quantity P_f refers to the bottom-hole pressure corresponding to flow rate Q, P_s refers to the shut-in bottom-hole pressure, the quantity n is the reciprocal of the slope of the curve as plotted in Fig. 7-8, and C is the intercept of that curve when the difference in the squares of the pressures is equal to 1. The quantity C can be related to the permeability and net thickness of the productive section. These tests are used to classify gas wells. They can be extrapolated and used to estimate the ability of the well to produce against any given surface pressure. They can also be used to estimate the rate of decline of the well with declining reservoir pressure. A value determined from these curves known as the absolute open flow (the flow capacity when atmospheric pressure is imposed at the face of the formation) is used in allocating allowables among wells within a field by most state regulatory bodies. As an example in a particular field, the rules

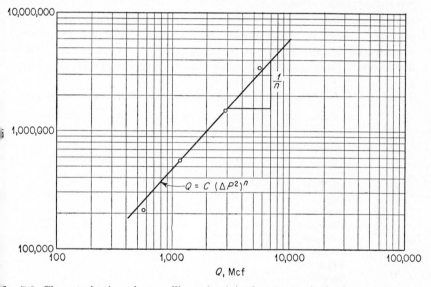

FIG. 7-8. Characterization of gas well's productivity by means of a back-pressure test.

may specify that no well can produce at a rate greater than one-fourth of its absolute open flow. Thus a limiting rate is applied to all wells within a field. If its prorated share of the production of the field exceeds its ability to produce, then it is limited to the minimum quantity.

AVERAGE FLUID PROPERTIES

In Chaps. 4 and 5 of this volume properties of hydrocarbon fluids were discussed. Fluid samples and analyses are relatively expensive; therefore, the data are taken sparingly as compared with other tests such as production and pressure tests. However, multiple fluid samples and analyses are desirable if only to confirm the accuracy of sampling and analysis. The problem then arises as to how to utilize the additional data obtained by multiple sampling and analysis. It is the purpose of this section to present methods of developing average properties for use in calculations.

Reservoir Fluids Existing in the Gaseous State

Gas reservoirs and the gas caps of associated oil and gas reservoirs contain mixtures of hydrocarbons which exist in the gaseous state at reservoir conditions. The most common test performed on such reservoir fluids is a fractional analysis of a recombined sample. The analysis is usually reported in mole per cent or mole fraction by components through hexanes or heptanes plus and including impurities such as carbon dioxide, hydrogen sulfide, and nitrogen. The hydrocarbon components heavier than ethane are considered to be condensable; therefore, the liquid content of the gas is calculated in terms of gallons of condensable liquid per thousand standard cubic feet of the gas (see Example 7-4). Physical properties of the reservoir fluids are usually calculated from the fractional analyses.

Example 7-4.   Calculation of Liquid Content of a Natural Gas.

Component	Mole %*	Equivalent liquid† volume, gal/Mcf	Liquid content, gal/Mcf
Methane	94.35		
Ethane	3.80		
Propane	0.29	27.38	0.0794
i-Butane	0.26	32.57	0.0847
n-Butane	0.34	31.41	0.1068
i-Pentane	0.23	36.41	0.0837
n-Pentane	0.09	36.07	0.0325
Hexane	0.18	40.94	0.0737
Heptanes plus	0.46	47.60‡	0.2190

* Analysis of sand 2, Table 7-4.
† From Table 4-4.

$$‡ \; 0.3155 \frac{M}{S} = \frac{0.3155(120)}{0.7954} = 47.60$$

where M = molecular weight = 120
        S = specific gravity = 0.7954

In Table 7-4 are reported gas analyses obtained from a field containing a number of separate gas sands. Eleven sands are identified in numerical order of increasing depth. Single gas samples were obtained from eight of the sands. Multiple samples were obtained from three of the sands. For each sand having more than one analysis an arithmetic average analysis

TABLE 7-4. HIGH-PRESSURE GAS SAMPLE ANALYSES[a]

Field A

Sand sequence	1	2	3	4	5	5	5	5
Component, mole %:								
Methane	93.57	94.35	91.43	92.38	91.66	90.86	90.96	91.00
Ethane	3.36	3.80	4.38	3.29	3.84	4.48	4.53	4.35
Propane	1.08	0.29	1.66	1.69	1.71	1.74	1.74	1.76
i-Butane	0.38	0.26	0.50	0.50	0.54	0.46	0.34	0.43
n-Butane	0.32	0.34	0.54	0.44	0.40	0.53	0.58	0.54
i-Pentane	{0.38	0.23	0.20	0.50	0.45	0.29	0.30	0.31
n-Pentane		0.09	0.16			0.20	0.21	0.21
Hexanes	0.47	0.18	0.33	0.50	0.60	{1.44	0.40	0.35
Heptanes plus	0.44	0.46	0.80	0.70	0.80		0.94	1.05
Total	100.00	100.00	100.00	100.00	100.00	100.00	100.00	100.00
Gpm:								
Propane	0.296	0.080	0.4554	0.464	0.469	0.4774	0.477	0.4829
i-Butane	0.124	0.085	0.1631	0.163	0.176	0.1498	0.110	0.1401
n-Butane	0.101	0.107	0.1699	0.138	0.126	0.1667	0.182	0.1698
i-Pentane	{0.138	0.084	0.0730	0.181	0.163	0.1058	0.109	0.1131
n-Pentane		0.032	0.0578			0.0722	0.076	0.0758
Hexanes	0.193	0.074	0.1354	0.205	0.246	{0.6764	0.163	0.1434
Heptanes plus	0.200	0.219	0.3906	0.330	0.377		0.505	0.5119
Total:								
Propane plus	1.052	0.681	1.4452	1.481	1.557	1.6483	1.622	1.6370
i-Butane plus	0.756	0.601	0.9898	1.017	1.088	1.1709	1.145	1.1541
i-Pentane plus	0.531	0.409	0.6568	0.716	0.786	0.8544	0.853	0.8442
Well number	5	9	2	3	4	1	7	6
Perforations	4,535–4,545	4,758–4,764	4,824–4,842	5,050–5,058	5,236–5,246	5,382–5,395	5,382–5,395	5,343–5,355
Date sampled	8-15-40	6-2-45	4-9-49	8-7-40	7-27-40	7-30-48	7-30-48	10-2-48

Sand sequence	5	6	6	6	6	6	7	8
Component, mole %:								
Methane	91.12	91.06	90.87	90.60	90.78	90.83	90.53	90.97
Ethane	4.30	3.94	4.50	4.37	4.35	4.29	4.58	4.15
Propane	1.74	1.72	1.82	2.01	1.80	1.84	1.94	1.72
i-Butane	0.44	0.42	0.40	0.52	0.47	0.45	0.43	0.34
n-Butane	0.51	0.58	0.65	0.74	0.54	0.63	0.60	0.57
i-Pentane	{0.49	0.52	0.23	0.23	0.28	0.48	0.28	{0.45
n-Pentane			0.17	0.24	0.23		0.23	
Hexanes	{1.40	0.72	1.37	0.25	0.37	1.49	0.26	0.60
Heptanes plus		1.04		1.04	1.18		1.15	1.20
Total	100.00	100.00	100.00	100.00	100.00	100.00	100.00	100.00
Gpm:								
Propane	0.4766	0.472	0.499	0.551	0.4938	0.5040	0.532	0.472
i-Butane	0.1440	0.137	0.130	0.169	0.1531	0.1473	0.140	0.111
n-Butane	0.1611	0.182	0.204	0.233	0.1698	0.1972	0.189	0.180
i-Pentane	{0.1787	0.188	0.084	0.084	0.1022	0.1723	0.102	0.163
n-Pentane			0.061	0.087	0.0830		0.084	
Hexanes	{0.6557	0 295	0.656	0.102	0.1518	0.7003	0.107	0.246
Heptanes plus		0.491		0.510	0.5954		0.590	0.566
Total:								
Propane plus	1.6161	1.765	1.634	1.736	1.7491	1.7211	1.744	1.738
i-Butane plus	1.1395	1.293	1.135	1.185	1.2553	1.2171	1.212	1.266
i-Pentane plus	0.8344	0.974	0.801	0.783	0.9324	0.8726	0.883	0.975
Well number	11	13	14	16	23	19	8	8
Perforations	5,550–5,560	5,626–5,632	5,760–5,772	5,654–5,661	5,690–5,700	5,690–5,700	5,780–5,790	5,980–5,988
Date sampled	7-15-40	5-18-45	7-19-45	10-13-48	6-8-45	12-3-48	7-2-40	6-20-40

TABLE 7-4. HIGH-PRESSURE GAS SAMPLE ANALYSES[a]

Field *A* (Continued)

Sand sequence	9	10	11	11	11	11	11	11
Component, mole %:								
Carbon dioxide							0.49	0.10
Methane	92.92	92.35	91.71	90.69	90.58	90.27	90.61	90.77
Ethane	2.73	3.03	3.29	4.35	4.40	4.44	3.85	4.07
Propane	1.44	1.58	1.60	1.74	1.74	1.87	1.72	1.73
i-Butane	0.36	0.39	0.48	0.41	0.40	0.36	0.38	0.41
n-Butane	0.46	0.50	0.44	0.49	0.56	0.54	0.52	0.51
i-Pentane	{0.38	0.40	0.47	0.29	0.21	0.35	0.22	0.46
n-Pentane				0.23	0.16	0.20	0.18	
Hexanes	0.37	0.31	0.45	0.42	0.24	0.46	0.15	0.34
Heptanes plus	1.32	1.44	1.56	1.38	1.71	1.51	1.88	1.61
Total	100.00	100.00	100.00	100.00	100.00	100.00	100.00	100.00
Gpm:								
Propane	0.395	0.433	0.439	0.4774	0.477	0.5130	0.472	0.4757
i-Butane	0.117	0.127	0.157	0.1336	0.130	0.1173	0.124	0.1324
n-Butane	0.145	0.157	0.139	0.1541	0.176	0.1698	0.163	0.1604
i-Pentane	{0.138	0.145	0.170	0.1058	0.077	0.1277	0.080	0.1677
n-Pentane				0.0830	0.058	0.0722	0.065	
Hexanes	0.152	0.127	0.185	0.1721	0.098	0.1885	0.061	0.1409
Heptanes plus	0.608	0.664	0.780	0.7624	0.945	0.8375	1.022	0.8694
Total:								
Propane plus	1.555	1.653	1.870	1.8884	1.961	2.0260	1.987	1.9465
i-Butane plus	1.160	1.220	1.431	1.4110	1.484	1.5130	1.515	1.4708
i-Pentane plus	0.898	0.936	1.135	1.1233	1.178	1.2259	1.228	1.1780
Well number	8	8	21	28	32	35		
Perforations	6,110–6,120	6,270–6,275	6,500–6,524	6,593–6,598	6,860–6,870	6,918–6,923		
Date sampled	6-15-40	5-29-40	3-25-49	6-7-45	10-20-48	7-16-45		

[a] Where required, data in computation were taken from Natural Gasoline Association of America, Standard Table o Physical Constants for the Paraffin Hydrocarbons, *NGAA Standard* 2145, adopted 1942, revised 1945.

by components was calculated. The average analyses were used to develop the physical properties, such as volume factors, for the reservoir fluids in these reservoirs.

From inspection of Table 7-4 it may be noted that the percentage of heavier components increase in the deeper sands. This is most evident in the gpM data, where the propanes-plus content increases from 1.052 in sand 1 to 1.9465 in sand 11. This is a common phenomenon in multisand fields where the sands are of the same geologic age. The variation is usually of sufficient uniformity that plots of composition as a function of depth can be used to verify the accuracy of single samples from a sand. Also, the composition of gas in a sand from which a sample is not available can be estimated from correlations of composition with depth.

More elaborate laboratory tests of gas-phase reservoir fluids can be partially verified by comparing the composition of the fluid tested with other samples on which only fractional analyses were obtained. Depletion and pressure-volume tests are frequently performed on condensate gases. As these tests are usually performed on recombined samples, the compositions can be closely controlled to reflect the average values obtained from other amples. Seldom is it necessary to average the results from such tests, as more than one such analysis on a fluid is rare. Arithmetic averages of corresponding results are usually satisfactory if multiple tests are available.

Reservoir Fluids Existing in the Liquid State

Physical properties of hydrocarbon mixtures which exist in the liquid state at initial reservoir conditions are determined from PVT tests of samples of the reservoir fluids. As discussed in Chap. 5 these PVT tests consist of a group of tests including pressure-volume relations at reservoir temperature, a differential liberation test at reservoir temperature, separator tests from the bubble-point conditions to various surface separator conditions, oil-viscosity measurements at reservoir temperature, and other related measurements. Although fractional analysis of the bubble-point fluid is frequently reported in connection with such tests, the analysis is seldom used to compute physical property data. Confirmation of sampling and analysis requires consideration of the various physical property measurements from two or more samples. In many instances the samples will differ substantially in bubble-point pressure and in composition, but the physical properties may check when compared on a proper basis.

Data from a PVT analysis of BHS-46C were discussed in Chap. 5. In Tables 7-5 and 7-6 are presented data from PVT tests of BHS-47 and BHT 1-155a, respectively. The three tests were conducted on samples taken from the same oil reservoir at different times during the early producing life of the reservoir. BHS-46C and BHS-47 were recombined samples, while BHT 1-155a was a bottom-hole sample. The bubble-point pressures of the samples ranged from 3,599 to 4,451 psia. The reservoir-fluid compositions were reported on BHS-46C and BHS-47. The concentration of methane for BHS-47 was 52.5 mole %, while that for BHS-46C was 49.65. BHS-47 also had the higher bubble-point pressure, 4,451 psia, compared with 4,228 psia for BHS-46C. It is evident that further comparisons must be made to define the physical properties.

Volume measurements in PVT tests are commonly reported as relative volume referred to either the volume at the bubble point or the volume at some specified residual condition. The residual condition can be chosen from differential liberation or separator tests. In the samples of Tables 5-4, 7-5, and 7-6, the relative total volume, relative oil volume, and the gas liberated are referred to a barrel of saturated oil at the sample bubble point. The data are presented graphically in Figs. 7-9 and 7-10. The curves do not coincide but do exhibit parallel trends, indicating that the data may become coincident on adjustment of the reference volume.

The relative total volume can be expressed in terms of the Y function, a dimensionless compressibility function. The data from the three samples

TABLE 7-5. FLUID-ANALYSIS REPORT ON RECOMBINED SAMPLE BHS-47[4]
Reservoir temperature 227°F

		Flash liberation			Differential liberation							
(1)	(2)	(3)	(4)	(5)	(6)	(7)	(8)	(9)	(10)	(11)	(12)	(13)
Pressure, psia	$P_b - P$, psi	Relative total volume V_t/V_b	$\dfrac{V_1}{V_b}$	$Y = \dfrac{P_b - P}{P[(V_t/V_b)-1]}$	Relative oil volume[a] V/V_b	$V = 1 - \dfrac{V}{V_b}$	Oil density, gm/cc	Gas-expansion factor r, scf /cu ft at P and 227°F	Gas compressibility factor Z	Gas gravity	Relative gas[b] volume	Gas liberated, scf/bbl saturated oil
4,813		0.9934					0.619					
4,713		0.9953					0.618					
4,613		0.9972					0.617					
4,513		0.9992					0.616					
4,451	0	1.0000			1.0000		0.615					
4,413	38	1.0027	0.0027	3.19059								
4,343	108	1.0075	0.0075	3.31593								
4,273	178	1.0124	0.0124	3.35912								
4,198	253	1.0183	0.0183	3.29341								
4,148	303	1.0221	0.0221	3.30533								
4,075	376	1.0270	0.0270	3.41818								
3,988	463	1.0378	0.0378	3.25768	0.9382	0.06180	0.634	224.1	0.920	0.922	0.0972	121.8
3,963	488	1.0476	0.0476	3.19774								
3,863	588	1.0575	0.0575	3.15724								
3,768	683	1.0673	0.0673	3.14737								
3,673	778	1.1167	0.1167	3.01334	0.8915	0.10850	0.650	204.4	0.894	0.917	0.0854	219.8
3,538	913	1.1663	0.1663	2.88467								
3,293	1,158	1.2164	0.2164	2.79634	0.8502	0.14980	0.666	187.2	0.882	0.901	0.079	303.6
3,088	1,363	1.3166	0.3166	2.64372								
3,008	1,443	1.4170	0.4170	2.53667								
2,773	1,678	1.5631	0.5631	2.40857	0.8147	0.18530	0.680	154.9	0.880	0.896	0.0876	379.8
2,638	1,813	1.7144	0.7144	2.32208								
2,423	2,028	1.9669	0.9669	2.22102								
2,193	2,258	2.2703	1.2703	2.12305								
2,163	2,288	2.6245	1.6245	2.04483	0.7834	0.21660	0.693	131.3	0.862	0.890	0.0936	448.8
1,889	2,562	3.0802	2.0802	1.9787								
1,674	2,777	3.5359	2.5359	1.93042	0.7506	0.24940	0.708	97.2	0.899	0.894	0.1324	521.1
1,414	3,037	3.9919	2.9919	1.86619								
1,234	3,217	4.4478	3.4778	1.86157	0.7224	0.27760	0.721	69.0	0.921	0.911	0.1599	583.0
1,204	3,247											
1,030	3,421											
870	3,581											
815	3,636				0.6974	0.30260	0.734	44.8	0.934	0.955	0.2151	637.1
755	3,696											
670	3,781											
600	3,851											
415	4,036				0.6698	0.33020	0.749	22.0	0.965	1.067	0.4446	692.0
215	4,236				0.6516	0.34840	0.758	10.9	1.000	1.279	0.5377	725.0

TABLE 7-5 (Continued)

	Separator tests					Fluid composition, mole %			
(14) Separator pressure, psig	(15) Separator gas-oil ratio, cu ft/bbl[c]	(16) Shrinkage factor, bbl STO/bbl saturated oil	(17) Separator gas gravity	(18) Tank oil gravity, °API	(19) Component	(20) Separator gas	(21) Separator oil	(22) Reservoir fluid	
0	1394	0.561	0.844	36.85	N_2	0.28		0.19	
15	1331	0.577	0.819	37.68	CO_2	1.82	2.21	1.20	
30	1269	0.590	0.790	38.14	C_1	78.33	1.85	52.50	
50	1253	0.590	0.773	38.53	C_2	8.86	4.66	6.48	
75	1211	0.597	0.756	38.72	C_3	6.86	1.38	6.11	
100	1172	0.600	0.745	38.96	$i C_4$	1.14	4.59	1.22	
					$n C_4$	1.65	2.31	2.65	
					$i C_5$	0.42	2.27	1.06	
					$n C_5$	0.33	6.11	.99	
					C_6	0.31		2.28	
					C_7^+		74.62	25.32	

[a] Oil volume at pressure P per volume of saturated oil.

[b] Gas volume liberated at pressure P_i in dropping the reservoir pressure from $P_i - 1$ to P_i per volume of saturated oil.

[c] Standard cubic feet of gas per barrel of stock tank oil.

Field test conditions:
Separator gas-oil ratio = 1,198:1
Separator pressure = 88 psig
Separator temperature = 78°F
Tank gas-oil ratio = 41 cu ft/bbl
Tank shrinkage % = 2.56

Compressibility of reservoir fluid = 18.02×10^{-6} 1/psi at 4,613 psi, 224°F
Mol wt C_7^+ = 218
Density C_7^+ = 0.8487 gm/cc at 60°F

503

TABLE 7-6. FLUID ANALYSIS OF BOTTOM-HOLE SAMPLE BHT 1-155a[4]

Reservoir Temperature 226°F

(1) Pressure, psig	(2) $P_b - P$, psi	Flash liberation			Differential liberation				
		(3) Relative total volume V_t/V_b	(4) $\dfrac{V_t}{V_b} - 1$	(5) $Y = \dfrac{P_b - P}{P[(V_t/V_b) - 1]}$	(6) Relative oil volume V/V_b^{a}	(7) Oil viscosity, cp	(8) $\Delta V = 1 - \dfrac{V}{V_b}$	(9) Gas in solution, cu ft/bbl saturated oil	Gas liberated, scf/bbl saturated oil
4,500		0.9848							
4,300		0.9881							
4,100		0.9914							
3,900		0.9950				0.31			
3,700		0.9980							
3,584	0	1.0000	0	0	1.0000	0.30		590	0
3,566	18	1.0019	0.0019	2.64554					
3,539	45	1.0046	0.0046	2.75256					
3,511	73	1.0074	0.0074	2.79774					
3,481	103	1.0101	0.0101	2.91705					
3,450	134	1.0127	0.0127	3.04507					
3,424	160	1.0155	0.0155	3.00162					
3,397	187	1.0182	0.0182	3.01134					
3,100	484	1.0510	0.0510	3.04661					
3,000	584								
2,983	601								
2,901	683	1.0787	0.0787	2.97617	0.945	0.36	0.0550	476	114
2,683	901	1.1136	0.1136	2.93971	0.908		0.0920	397	193
2,485	1,099	1.1561	0.1561	2.81614					
2,400	1,184								
2,280	1,304	1.2058	0.2058	2.76089	0.871	0.44	0.1290	313	277
2,121	1,463	1.2559	0.2559	2.67653					
2,008	1,576								
1,911	1,673	1.3350	0.3350	2.59295					
1,700	1,884				0.841	0.55	0.1590	245	345
1,682	1,902	1.4507	0.4507	2.48680					
1,517	2,067	1.5668	0.5668	2.40764					
1,507	2,077								
1,333	2,251	1.7123	0.7123	2.34435					
1,135	2,399	1.9457	0.9457	2.20586	0.810	0.71	0.1900	175	415
1,002	2,582								
1,000	2,584								
830	2,754	2.5308	1.5308	2.12906					
650	2,934	3.1051	2.1051	2.09587	0.776		0.2240	103	487
497	3,087								
428	3,156	4.5720	3.5720	1.99444					
296	3,288	6.6257	5.6257	1.87929	0.759		0.2410	64	526
248	3,336				0.688		0.3120	0	590
0	3,584				at 60°F = 0.631				

504

TABLE 7-6 (Continued)

(10) Separator pressure, psig	(11) Separator gas-oil ratio, cu ft/bbl[b]	(12) Stock-tank gas-oil ratio, cu ft/bbl	(13) Stock-tank gravity, °API	(14) Shrinkage factor bbl STO/bbl saturated oil	(15) Formation volume factor	(16) Gas gravity
Separator tests						
0	1,093	0	36.3	0.545	1.835	0.853
10	963	7	37.1	0.619	1.616	
30	899	21	37.6	0.635	1.575	

Analyses of separator gas O, psi

Components	Mole %
C_1	83.31
C_2	5.45
C_3	4.86
iC_4	1.16
nC_4	2.04
iC_5	1.05
nC_5	0.81
C_6	0.64
C_{7+}	0.68

[a] Oil volume at pressure P per volume of saturated oil.
[b] Standard cubic feet of gas per barrel of stock-tank oil.

Reservoir oil compressibility = 16.59×10^{-6} 1/psi
Oil density at P_b = 0.580 gm/cc
Bubble point = 3,584 psig at 226°F

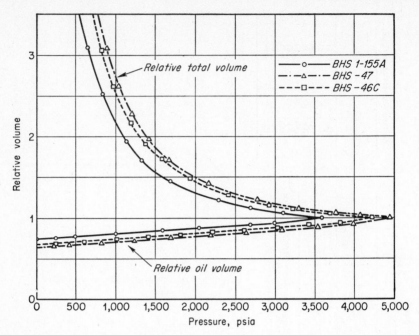

Fig. 7-9. Pressure-volume relations of crude oils, field *B*. (*Courtesy of Shell Oil Company.*[4])

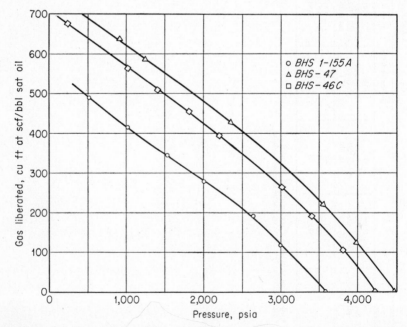

Fig. 7-10. Gas liberated by differential liberation, field *B*. (*Courtesy of Shell Oil Company.*)

are plotted in Fig. 7-11. The data once again are essentially parallel but not coincident. The data of BHS-46C and BHS-47 are considered to be sufficiently in agreement that the data can be averaged.

As the Y function is essentially linear with pressure, the data can be fitted by a straight line. The points indicated as omitted were in the proximity of the sample bubble points—a region in which the data are frequently unreliable. Two procedures can be used to determine the average line passing through the data points of both BHS-46C and BHS-47. The first procedure consists of fitting straight lines to each set of data and then averaging the coefficients. The second procedure is to fit a straight line to the data points of the samples considered together.

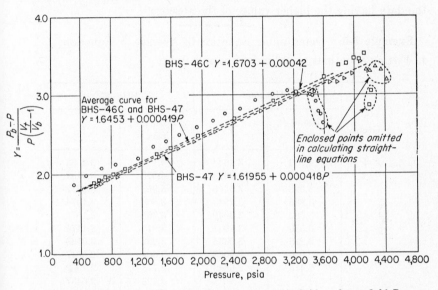

FIG. 7-11. Correlation of Y function for three reservoir-fluid analyses, field B.

The method of averages is used in Example 7-5 to fit lines to both the individual sample data and the combined data. The method of averages consists of determining the average coordinates of two groups of the data and using these average coordinates to determine the slope and the intercept of the line. The groups should be essentially of equal size.

In general terms to fit an equation of the form

$$Y = a + bP$$

let

$$\overline{Y}_1 = \frac{\sum\limits_{j=1}^{k} Y_j}{k} \qquad \overline{P}_1 = \frac{\sum\limits_{j=1}^{k} P_j}{k} \qquad (7\text{-}8)$$

$$\overline{Y}_2 = \frac{\sum\limits_{j=k+1}^{n} Y_j}{n-k} \qquad \overline{P}_2 = \frac{\sum\limits_{j=k+1}^{n} P_j}{n-k} \qquad (7\text{-}8)$$

where $\overline{Y}_1$ and $\overline{P}_1$ are the average coordinates of group I, which contains k pairs of data points Y_j and P_j, and $\overline{Y}_2$ and $\overline{P}_2$ are the average coordinates of group II, which contains $n-k$ pairs of data points Y_j and P_j. Then

$$b = \frac{\overline{Y}_1 - \overline{Y}_2}{\overline{P}_1 - \overline{P}_2}$$

and

$$a = \overline{Y}_1 - b\overline{P}_1 \qquad (7\text{-}9)$$

The conventions on signs are properly considered if group I is assigned to the data having the higher values of the abscissa.

Example 7-5. Determining Equation of Average Y Function.

1. Fitting equations to individual sample data:

Group	BHS-46C		BHS-47	
	Pressure P, psia	Y	Pressure P, psia	Y
I	4,120	3.4043	4,148	3.3053
	4,060	3.3373	4,075	3.4182
	3,998	3.3446	3,963	3.2577
	3,885	3.2943	3,863	3.1974
	3,780	3.2649	3,768	3.1572
	3,590	3.2021	3,673	3.1474
	3,215	3.0355	3,293	3.0133
	2,930	2.9068	3,008	2.8847
	2,415	2.6869	2,773	2.7963
			2,423	2.6437
	31,993	28.4767	34,987	30.8213
	$\overline{P}_1 = 3555$	$\overline{Y}_1 = 3.1641$	$\overline{P}_1 = 3499$	$\overline{Y}_1 = 3.0821$
II	1,938	2.4839	2,163	2.5367
	1,578	2.3259	1,899	2.4086
	1,380	2.2444	1,674	2.3221
	1,200	2.1635	1,414	2.2210
	975	2.0709	1,204	2.1231
	825	2.0062	1,030	2.0448
	715	1.9644	870	1.9787
	630	1.9383	755	1.9304
	570	1.8921	670	1.8862
			600	1.8616
	9,811	19.0896	12,279	21.3131
	$\overline{P}_2 = 1,090$	$\overline{Y}_2 = 2.1210$	$\overline{P}_2 = 1,228$	$\overline{Y}_2 = 2.1313$

$$Y = a + bP$$

where
$$b = \frac{\overline{Y}_1 - \overline{Y}_2}{\overline{P}_1 - \overline{P}_2} \quad \text{and} \quad a = \overline{Y}_1 - b\overline{P}_1$$

For BHS-46C

Slope $b = \dfrac{3.1641 - 2.1210}{3,555 - 1,090}$

$= 0.00042$

$a = 3.1641$

$\quad - 0.00042\,(3,555)$

$= 1.6710$

$Y = 1.6710 + 0.00042P$

For BHS-47

Slope $b = \dfrac{3.0821 - 2.1313}{3,499 - 1,228}$

$= 0.000418$

$a = 3.0821$

$\quad - 0.000418\,(3,499)$

$= 1.6196$

$Y = 1.6196 + 0.000418P$

2. Average coefficients from coefficients of equations for the individual samples:

$$b = \frac{0.00042 + 0.000418}{2} = 0.000419$$

$$a = \frac{1.6710 + 1.6196}{2} = 1.6453$$

Therefore the equation for average properties is

$$Y = 1.6453 + 0.000419P$$

3. Average coefficients from combined data:

Group I (combined)

$$\overline{P}_1 = \frac{31,993 + 34,987}{9 + 10} = \frac{66,980}{19} = 3,525$$

$$\overline{Y}_1 = \frac{28.4767 + 30.8213}{9 + 10} = \frac{59.2980}{19} = 3.1209$$

Group II (combined)

$$\overline{P}_2 = \frac{9,811 + 12,279}{9 + 10} = \frac{22,090}{19} = 1,163$$

$$\overline{Y}_2 = \frac{19.0896 + 21.3131}{9 + 10} = \frac{40.4027}{19} = 2.1264$$

Thus

$$b = \frac{\overline{Y}_1 - \overline{Y}_2}{\overline{P}_1 - \overline{P}_2} = \frac{3.1209 - 2.1264}{3,525 - 1,163} = \frac{9,945}{2,362} = 0.000421$$

$$a = \overline{Y}_1 - b\overline{P}_1 = 3.1209 - 0.000421(3,525) = 1.6369$$

therefore

$$Y = 1.6369 + 0.000421P$$

The equations developed by either procedure used in Example 7-5 are satisfactory representations of the data. The choice of procedure depends on whether it is desired to place equal weight on each data point or on each sample. The procedure of averaging coefficients places equal weight on the sample.

The method of least squares or a simple balancing of points visually in drawing an average line on the graph paper can also be used in fitting average lines to an array of data.

The relative oil-volume difference, or ΔV function, can be used to aver-

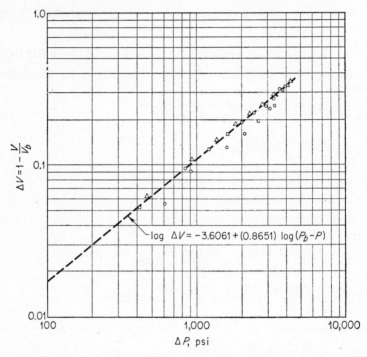

FIG. 7-12. Relative oil-volume difference by differential vaporization, field B.

age relative oil-volume data. This function was also discussed in Chap. 5. In Fig. 7-12 are presented the data for the three samples BHS-46C, BHS-47, and BHT 1-155a. The data for BHS-46C and BHS-47 once again coincide satisfactorily. The fitting of an average line to the data is presented in Example 7-6.

As pointed out earlier in this section, the bubble-point pressures from the three samples differ substantially. The correlations of the Y and ΔV functions were made without adjusting the bubble-point pressure. The equations of the average data can be adjusted to any selected bubble point by the methods presented in Chap. 5.

Example 7-6.   Determining Equation of Average ΔV Function.

Group	BHS-46C		BHS-47	
	Pressure log ΔP, psia	log ΔV	Pressure log ΔP, psia	log ΔV
I	2.6128	8.7210–10	2.6656	8.2089–10
	2.9128	8.9777–10	2.9605	9.0355–10
	3.0858	9.1109–10	3.1345	9.1756–10
	3.2089	9.2079–10	3.2584	9.2679–10
	3.3049	9.2808–10	3.3532	9.3357–10
	15.1252	45.2983–50	15.3722	45.0236–50

Group I (combined):

$$\log \Delta P_1 = \frac{15.1252 + 15.3722}{10} = 3.0447$$

$$\log \Delta V_1 = \frac{(45.2983 - 50) + (45.0236 - 50)}{10} = -0.9678$$

II	3.3834	9.3401–10	3.4404	9.3969–10
	3.4499	9.3888–10	3.5072	9.4434–10
	3.5076	9.4303–10	3.5606	9.4809–10
	3.5585	9.4714–10	3.6070	9.5188–10
	3.6002	9.5139–10	3.6269	9.5421–10
	17.4996	47.0445–50	17.7421	47.3821–50

Group II (combined):

$$\log \Delta P_2 = \frac{17.4996 + 17.7421}{10} = 3.5242$$

$$\log \Delta V_2 = \frac{(47.0445 - 50) + (47.3821 - 50)}{10} = -0.5573$$

Therefore the coefficients for the combined data are

$$b = \frac{-0.9678 - (-0.5573)}{3.0497 - 3.5242} = \frac{-0.4105}{-0.4745} = 0.8651$$

$$a = \log \Delta \overline{V} - b \log \Delta \overline{P}_1 = -0.9678 - 0.8651(3.0497)$$

$$a = -3.6061$$

and the equation is

$$\log \Delta V = -3.6061 + 0.08651 \log \Delta P$$

The correlation of the gas liberated from several samples requires the adjustment of the data to the basis of common bubble-point pressure or to the basis of residual oil. To adjust the gas liberated to a new bubble-point pressure requires the correction of the data for the excess gas liberated between the sample bubble point P_{bs} and the field bubble point P_{bf} and for the shrinkage due to the evolution of the excess gas. The equations for this adjustment are as follows:

$$(R_L)_b = [(R_L)_{bs} - (R_L)_{sb}] \frac{V_{bs}}{V_b} \tag{7-10}$$

where $(R_L)_b$ is the standard cubic feet of gas liberated per barrel of saturated oil at the field bubble point, $(R_L)_{bs}$ is the standard cubic feet of gas liberated per barrel of saturated oil at the sample bubble point, $(R_L)_{sb}$ is the excess gas liberated between the sample bubble point and the field bubble point in standard cubic feet per barrel of saturated oil at the sample bubble point, and V_{bs}/V_b is the volume of sample bubble-point oil per volume of field bubble-point oil.

If the field bubble point is greater than the sample bubble point, the curves of gas liberated (Fig. 7-10) must be extrapolated to the field bubble-point pressure. $(R_L)_{sb}$ is then negative, so that the excess gas is, in effect, added. The quantity V_{bs}/V_b is the relative oil volume calculated using the field bubble-point pressure as P_b in the ΔV equation. If the field bubble-point pressure is less than the sample bubble-point pressure, the quantity $(R_L)_{sb}$ is positive and the excess gas is subtracted. The quantity V_{bs}/V_b is the reciprocal of the relative oil volume calculated using the same bubble-point pressure as P_b in the ΔV equation.

The correction of the gas liberated from a higher sample bubble-point pressure to a lower field bubble-point pressure is shown in Example 5-8.

With the data converted to the same bubble-point conditions, an average curve can be fitted to the resulting data.

Rather than correlating the liberated gas data from several samples, the most expedient approach is to correlate the gas in solution based on residual oil. Many laboratories report gas in solution and gas liberated data based on residual oil obtained from the differential liberation. If the data are expressed on the basis of residual oil, the data from the several samples can be plotted on the same graph and an average curve fitted to the data.

If the data are reported as in Tables 5-4, 7-5, and 7-6, the data can be converted to gas in solution based on a selected separator pressure. An example of this correction is shown in Example 5-5. The data of the three samples were converted to gas in solution based on a 30-psi separator pressure. The converted data are presented in Fig. 7-13. The data for BHS-47 and BHT 1-155a check very closely, while the data of BHS-46C are approximately 50 cu ft per bbl low. The shaded points are particularly significant, as they represent the gas in solution from the 30-psi separator tests on each sample. The dashed line drawn through those points and visually balanced between the trend of data points is an acceptable average curve.

The oil formation volume factor based on a selected separator condition can be calculated for each sample as in Example 5-6. A plot of data calculated for BHS-46C, BHS-47, and BHT 1-155a is presented in Fig. 7-14. The data are based on a 30-psi separator. Values for BHS-46C and BHS-47 check very closely, and a curve fitted visually to these data is a satisfactory average curve. The data from BHT 1-155a do not check with sufficient accuracy to be included in the average.

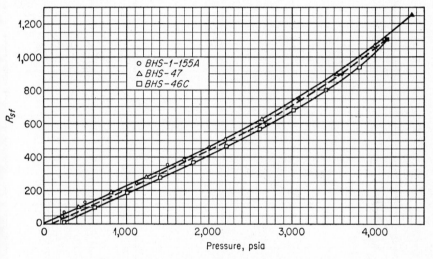

FIG. 7-13. Separator gas in solution based on 30-psi separator, field B.

Separator test data can be correlated with separator pressure, bubble-point pressure, and other parameters. In Fig. 7-15 are presented correlations of separator data with separator pressure. The values of separator gas-oil ratio and shrinkage factor differ because of the different bubble-point pressures of the samples. The data for BHS-46C and BHS-47 show parallel trends. The sample having the higher bubble point has the higher gas-oil ratio and lower shrinkage factor as would be expected.

Correlations of gas gravity, compressibility factor, and gas-expansion factor $1/B_g$ are shown in Fig. 7-16. The data correspond closely and can be fitted by balancing a curve through the data points.

In the foregoing discussions, the reservoir fluid was assumed to have a bubble-point or saturation pressure which was invariant with position in the reservoir. The variation in bubble-point pressure among the samples

BHS-46C, BHS-47, and BHT 1-155a was assumed to be a result of sampling rather than a variation of fluid properties in the reservoir. In fact, the samples BHS-46C and BHS-47 were obtained from the same well having the same completion interval at the time of sampling.

Variations in fluid properties with depth usually occur in fields having separate accumulations in several different formations. The variations are similar to those discussed with respect to the gas sands of Table 7-4. The bubble-point pressure and solution-gas–oil ratio normally increase with depth in formations of similar geologic age on the same structure.

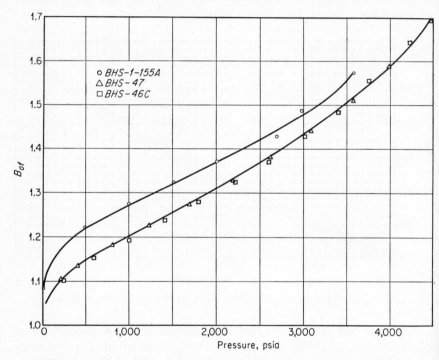

FIG. 7-14. Oil formation volume factor based on 30-psi separator, field B.

In reservoirs having low relief within the oil column, the oil usually exhibits a uniform bubble-point pressure. However, in reservoirs of high relief, variations in fluid properties may occur within the oil column of a single reservoir.

An example of such variations in bubble-point pressure is shown in Fig. 7-17. The data are from the Weber sandstone reservoir of the Rangely oil field of Colorado.[5] The reservoir contained an original gas cap with a gas-oil contact at 320 ft below sea level. The original water-oil contact was found at 1,160 ft below sea level. The oil column was continuous over an

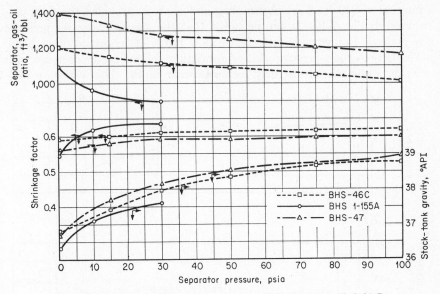

FIG. 7-15. Laboratory flash separator tests of bubble-point oil, field *B*.

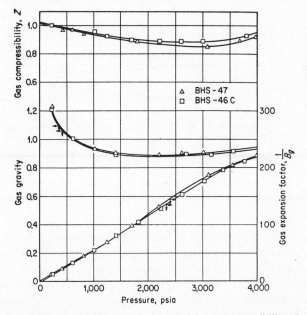

FIG. 7-16. Gas gravity, gas-expansion factor, and gas-compressibility factor by differential liberation, field *B*.

interval of 840 ft. Careful sampling and analysis provided the data presented in Fig. 7-17. It can be noted that the original reservoir pressure is normal; it increases with depth as a result of increasing fluid head. The bubble-point pressure and oil formation volume factor decrease with depth, while the oil viscosity increases with depth. This variation indi-

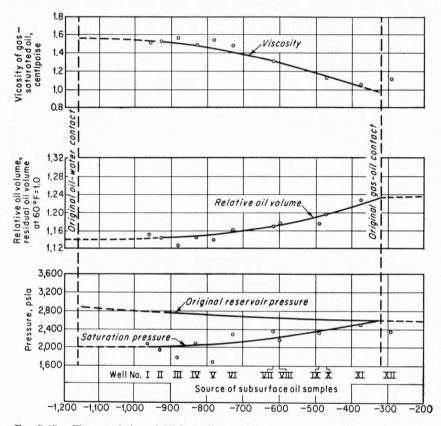

Fig. 7-17. Characteristics of Weber oil at original reservoir conditions correlated with the elevation of the top of Weber sandstone, Rangely Field, Colo. (*From Cupps et al.*[5])

cates an increasing proportion of heavy components with depth. The data suggest gravitational segregation of components within the oil column. Sage and Lacey[6] investigated such phenomena in the laboratory and suggested a procedure for calculating the segregation of components. Cupps et al.[5] applied the procedure of Sage and Lacey to the Weber data and found that the observed segregation was much greater than that calculated.

Espach and Fry[7] presented similar data on the Tensleep sandstone reservoir of the Elk Basin oil field. Their hypothesis for the variation in physical properties was that an oil and gas accumulation existed in the region prior to the final deformation of the rock beds in the area. The deformation caused increased hydrostatic pressures in the trap, thus compressing the gas cap and confining the oil in a smaller area with a greater thickness of oil column. The compressed gas tended to diffuse into the oil column, thus increasing the bubble-point pressure of the oil. It was presumed that the reservoir was discovered before equilibrium could be obtained; thus the lower portions of the reservoir contained oil which was little affected by the diffusion of gas downward from the gas cap. The variation in bubble-point pressure with depth can be attributed to the extent to which the oil achieved equilibrium. The oil nearest the top of the structure would be in equilibrium or nearly so, while that lower in the column would be the furthest removed from equilibrium.

In some fields of great areal extent, variations in physical properties may be found in oil columns of moderate relief. The Scurry County area of West Texas exhibits variations of bubble-point pressure with depth. Cook et al.[8] investigated the variations and proposed a procedure for evaluating such variations with a limited number of PVT tests.

The Scurry County Canyon Reef reservoir contained oil which was initially undersaturated throughout the reservoir. Thus, producing gas-oil ratios, accurately determined, reflect the gas in solution in the oil at the depth at which the well was completed. A number of wells were tested under carefully controlled separator conditions, and the resulting gas-oil ratios plotted as a function of the depth of the mid-point of the completion interval. The gas-oil ratio data and completion intervals are shown in Fig. 7-18.

Bottom-hole samples were obtained from a limited number of the wells tested. The bubble-point pressure and separator gas-oil ratios of the samples were determined in the laboratory. The correlation of bubble-point pressure and gas-oil ratio from the laboratory data is presented in Fig. 7-19. The data of Figs. 7-18 and 7-19 were combined to yield a correlation of bubble point pressure with depth, which is presented in Fig. 7-20.

EVALUATION OF ROCK VOLUME

Once the drill has found an oil-productive formation, the accumulation must be evaluated as to areal extent and hydrocarbon content. The evaluation of any hydrocarbon storage represents the accumulation and analysis of data gathered during the drilling of the wells. It will be assumed here that the reader is familiar with the data gathered during the drilling of oil wells, so that only illustrative examples need be presented.

Areal Extent

The projected surface area of a hydrocarbon deposit can be completely defined only by the drill. Correlation of the information on the producing formation from all wells, producing and nonproducing, must be used. From drillers logs, mud logs, electric and radioactive logs, such as Figs. 7-21 and 7-22, the lithologic top and bottom of the formation can be determined

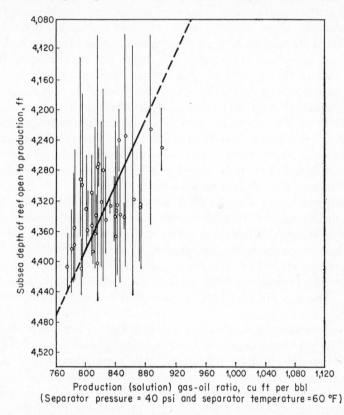

FIG. 7-18. Variation in solution-gas content of reef reservoir oil with depth (showing thickness of exposed zone), Scurry County, Texas. (o) data from Kelly-Snyder area; vertical lines show reef rock exposed in well bore. (*From Cook et al.*[8])

for each well. The values obtained from the logs are converted to subsea measurements and plotted on an areal map. Lines representing points of equal depth are drawn (Fig. 7-23) to represent contour maps of the top and bottom of the formation. These maps alone do not define the areal extent of the reservoir. They define only the structure of the formation. Other information must be considered to determine the limits of the hydrocarbon portion of the reservoir.

In Chap. 1, several types of reservoir traps were discussed. It was stated that hydrocarbon reservoirs are essentially bounded in four ways: (1) by "pinchout" or disappearance of the formation, (2) by facies change or loss of permeability, (3) by faulting, and (4) by water. The limits can be defined by any one or a combination of all these four factors.

Areal limits defined by a disappearance of the formation would become apparent during the construction of the structural contour map. The cause for disappearance of the formation is determined by a detailed study of the formations above and below. Such detailed studies usually indicate unconformities or lenticular formations.

Limits resulting from facies change or loss of permeability are determined from a study of well logs, core analysis, and drill-stem tests. Usually

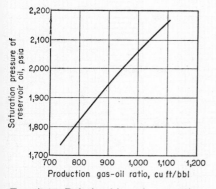

Fig. 7-19. Relationship of saturation pressure of reef reservoir oil with production (solution) gas-oil ratios as determined from fluid sample analyses, Kelly-Snyder area, Scurry County, Texas. (*From Cook et al.*[8])

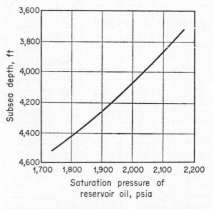

Fig. 7-20. Relationship of saturation pressure of reef reservoir oil with depth, Kelly-Snyder area, Scurry County, Texas. (*From Cook et al.*[8])

these changes are also noted during the construction of the structural map.

A fault boundary is often hard to define. To define the boundary requires a detailed study of the formations above and below. If the displacement of the fault is small, it is difficult at times to determine if the fault plane is a closed boundary without the aid of pressure surveys and production history. Once the fault has been located, its position is indicated on the structural map of the top and base of the formation.

The final means by which the hydrocarbon-bearing portion can be limited is by the absence of hydrocarbons. Because of the conditions under which most formations were deposited, if they do not contain a hydrocarbon, they contain water. Hence, the limiting boundary is the point at which water completely fills the pore space in the formation.

The depth at which any given well ceases to produce oil and produces

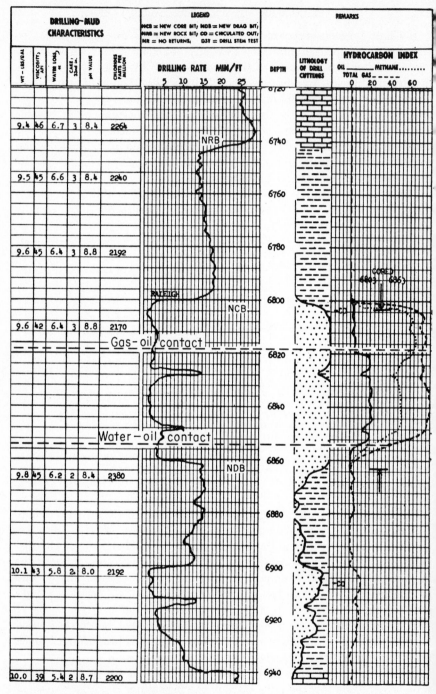

Fig. 7-21. Mud log and drilling time log. (*Adapted from Core Laboratories, Inc.*[9])

only water can be determined from drill-stem tests or core analysis (Fig. 7-24). The depth at which the formation becomes saturated with water can be determined from the resistivity curve of the electric log. As was mentioned in Chap. 3, there are two water-oil contacts, both of interest to the engineer.

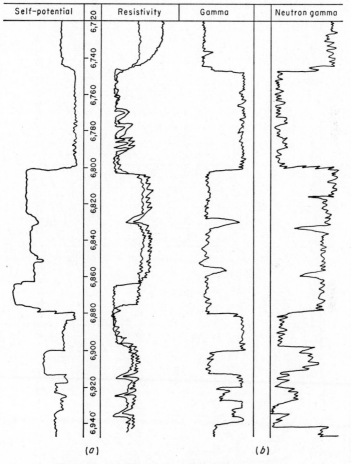

Fig. 7-22. (a) Electric log; (b) radioactive log.

The oil-water contact is selected on the core log as the point at which the oil saturations of the samples decrease and the water saturations increase. This is the water-oil contact defined as *the level below which the fluid production is 100 per cent water.*

Unfortunately not all wells drilled penetrate the water-bearing portion of the formation. It thus becomes necessary to determine the limits of the

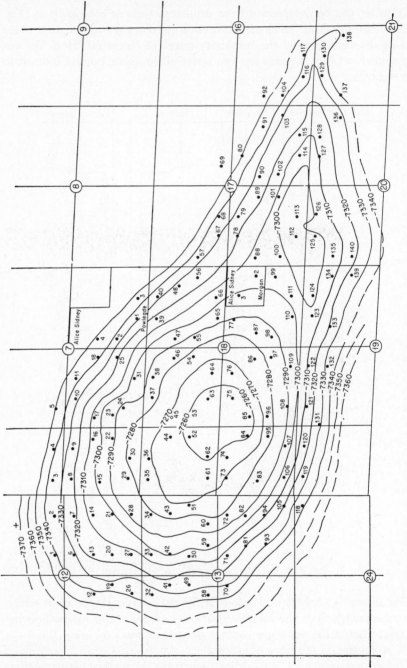

Fig. 7-23. Structure contours on top of sand, Jones Sand, Shuler Field, Ark. (*From Kaveler.*[10])

CORE ANALYSIS AND INTERPRETATION COMPLETION COREGRAPH

SAMPLE NUMBER	DEPTH FEET	PERMEABILITY MILLIDARCYS	POROSITY %	RESIDUAL LIQUID SATURATION % PORE SPACE OIL	TOTAL WATER	PROBABLE PROD.
43	6803.5	201	20.3	2.0	51.5	COND.
44	04.5	354	23.6	2.0	57.2	COND.
45	05.5	106	22.1	1.9	53.4	COND.
46	06.5	25	22.2	2.0	59.0	COND.
47	07.5	498	23.1	2.0	50.0	COND.
48	08.5	906	22.0	2.1	58.4	COND.
49	09.5	146	23.3	2.1	48.5	COND.
50	10.5	747	23.9	2.0	60.0	COND.
51	11.5	410	21.9	2.1	46.5	COND.
52	12.5	536	23.7	2.0	44.1	COND.
53	6816.5	693	23.8	2.0	50.3	COND.
54	17.5	560	23.2	6.8	48.6	COND.
55	18.5	448	22.0	8.0	50.7	COND.
56	19.5	612	23.1	14.1	46.3	OIL
57	20.5	448	22.9	18.3	48.9	OIL
58	21.5	327	22.6	16.2	52.3	OIL
59	22.5	25	20.1	16.5	68.2	OIL
60	23.5	418	21.7	16.1	44.2	OIL
61	24.5	203	22.0	17.8	51.0	OIL
62	25.5	146	21.2	16.0	48.9	OIL
63	6828.5	347	21.7	17.6	56.3	OIL
64	29.5	572	24.0	16.4	49.4	OIL
65	30.5	628	24.9	18.9	43.0	OIL
66	31.5	981	26.6	20.0	46.1	OIL
67	32.5	862	25.6	20.3	42.5	OIL
68	33.5	1040	26.0	20.1	45.4	OIL
69	34.5	2200	28.1	18.6	47.6	OIL
70	35.5	3200	27.3	17.9	47.7	OIL
71	36.5	1162	25.8	19.0	44.2	OIL
72	37.5	1320	27.8	20.0	40.9	OIL
73	38.5	1282	27.2	18.6	49.2	OIL
74	39.5	1141	26.2	18.0	46.4	OIL
75	40.5	257	25.6	18.3	47.2	OIL
76	41.5	482	24.7	16.3	48.1	OIL
77	42.5	844	24.8	20.0	52.1	OIL
78	43.5	665	24.1	17.8	47.6	OIL
79	44.5	221	24.8	19.3	44.3	OIL
80	45.5	961	23.9	19.8	42.7	OIL
81	46.5	0.0	9.8	7.7	68.5	
82	47.5	0.0	6.9	0.0	76.0	
83	49.5	793	25.0	14.0	47.2	OIL
84	50.5	137	26.1	11.9	48.7	OIL
85	51.5	323	25.2	12.6	56.1	OIL
86	52.5	127	25.1	7.6	60.6	WATER
87	53.5	97	23.9	7.3	65.2	WATER
88	54.5	196	25.8	4.0	67.8	WATER
89	55.5	202	24.9	6.3	69.2	WATER
90	56.5	301	24.0	1.9	68.1	WATER
91	57.5	396	23.9	4.1	65.5	WATER
92	58.5	340	24.7	2.0	63.9	WATER
93	59.5	336	23.9	6.9	67.8	WATER

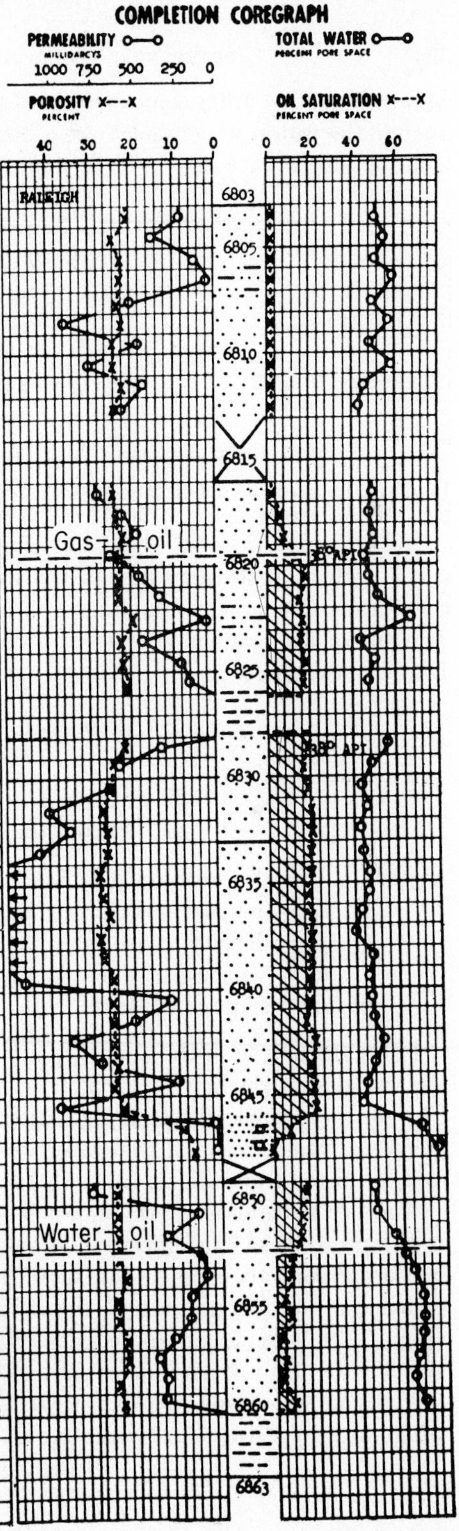

COMPLETION COREGRAPH

PERMEABILITY ○—○ MILLIDARCYS 1000 750 500 250 0

TOTAL WATER ○—○ PERCENT PORE SPACE

POROSITY X--X PERCENT 40 30 20 10 0

OIL SATURATION X--X PERCENT PORE SPACE 0 20 40 60

Fig. 7-24. Core log of an oil-producing formation illustrating selection of gas-oil and water-oil contacts. (*From Core Laboratories, Inc.*[9])

523

water-oil contact from a meager set of data. The normal procedure is to assume the water-oil contact to be a horizontal plane. Thus, if the subsea level of the contact can be determined in several wells, it is possible to establish the contact for the reservoir.

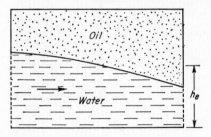

FIG. 7-25. Idealized model of tilted contact. (*From Yuster.*[11])

There are cases when the water-oil contact in a reservoir cannot be assumed to be a horizontal plane. The water table may be tilted if the water zone is in a dynamic state, if the formation is highly variable in permeability, or it may be two horizontal planes if an impermeable barrier exists in the oil zone.

A water table is changed from its static horizontal state by flow of the water. The change in the height of the water-oil contact from its static level h_e is illustrated in Fig. 7-25. The tilt of the contact is related to the pressure drop in the water by Darcy's law where

$$v_w = 0.00624 \frac{k}{\mu} \left(\frac{\rho_w - \rho_o}{144} \right) \frac{dh}{dx} \tag{7-11}$$

or

$$\frac{dh}{dx} = \frac{144 v_w \mu}{0.00624 k} (\rho_w - \rho_o)$$

where
x = distance along horizontal axis, ft
ρ_w = density of water, lb/cu ft
ρ_o = density of oil, lb/cu ft
v_w = velocity of water, ft/day
dh/dx = slope of contact

From an analysis of the above equation, it may be noted that the greater the velocity of the water, the greater the slope of the water-oil contact. The effect of increased velocity is illustrated in Fig. 7-26, where the velocity increases from a to c. If the water velocity is great enough, the oil zone can be shifted from the top of the structure to the flank.

Another cause of a water table not being a continuous horizontal plane is the presence of an impermeable barrier in the oil zone. For a barrier in the oil zone to be effective, there must be a gas cap and water zone which are continuous around the barrier. An idealized structural map (Fig. 7-27) with superimposed gas-oil and water-oil contacts illustrates the effect which a large impermeable barrier may have.

Mention was made in Chap. 3 of a horizontal permeability variation which could cause a tilted water table. The tilt in the level of the zone of complete water saturation is caused by differences in capillary-pressure behavior of the rock for the variation in permeability.

At least three possible reasons for a water table not being horizontal have been presented in previous discussions. The engineer must analyze

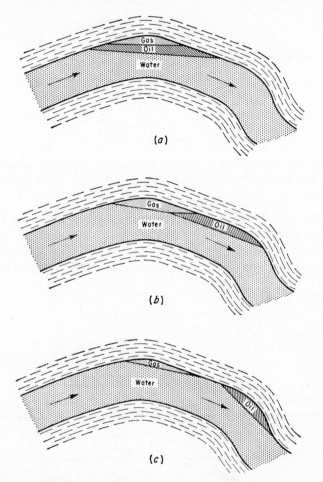

Fig. 7-26. Types of hydrocarbon oil and gas accumulations in gently folded thick sand. (a) Gas entirely underlain by oil; (b) gas partly underlain by oil; (c) gas and oil traps separated. (*From Hubbert.*[12])

all available data to determine the nature of the water table in any given reservoir. To aid in determining the position of the water-oil contact it is suggested that all open-hole drill-stem tests, production and completion tests, core analysis, and log data be plotted on the structural map of the top of the formation. A portion of the required data is plotted on Fig.

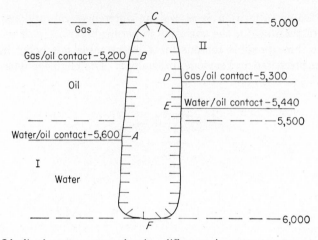

FIG. 7-27. Idealized contour map showing difference in contacts across a shaled-out area. Gas gradient = 0.06 psi/ft; oil gradient = 0.30 psi/ft; water gradient = 0.45 psi/ft.

7-28 illustrating the placement of the contour line representing the intercept of the gas-oil and water-oil contacts with the top of the formation. From a study of the data presented in Fig. 7-28, it is seen that the water-oil contact is defined by only two wells, one a drill-stem test and one a

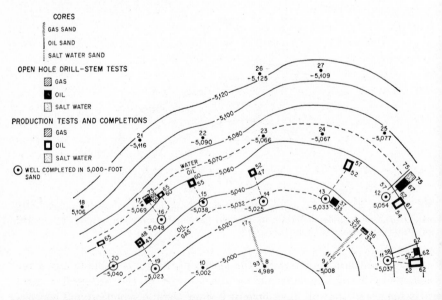

FIG. 7-28. Graphical representation of water-oil contact data on a structure map. Imaginary field. Structure contours on top of "5,000-ft" sand. Contour interval, 20 ft. (*From Alexander.*[18])

production test. The drill-stem test indicates the contact to be between 5,069 and 5,073 ft subsea while the production test indicates it to be between 5,067 and 5,075 ft subsea. A horizontal contact was assumed to exist at 5,070 ft subsea, and its intercept with the top of the sand drawn as the 5,070-ft subsea contour. Figure 7-28 does not have the log data plotted which could also be used as an aid in selecting the water-oil contact, especially in those wells which penetrated but did not test the sand.

Isopach Maps

To evaluate the amount of hydrocarbon volumetrically in an underground reservoir, it is usually necessary to construct an isopach map. Any isopach map indicates the rock volume between limiting boundaries. There are four basic isopach maps which can be constructed: (1) gross isopach map, (2) total net thickness map, (3) net oil isopach map, and (4) net gas isopach map.

Gross Isopach Maps. Gross isopach maps are constructed using the top of the formation, the bottom of the formation, the water-oil contact, and any pinchout, fault, or facies change as the limiting boundaries. In a case involving only the top and bottom of the formation and the water-oil contact, it is possible to construct the isopach from the two structural maps. The water-oil-contact intercept should be plotted on both of the structural maps.

The effect of gas-oil and water-oil contacts on the location of the isopach lines for two different types of structures is shown in Figs. 7-29 and 7-30. Figure 7-29 shows the placement of isopach contour lines for a domal-type trap. The maximum areal extent (zero isopach line) of the accumulation is defined by the contact of the water table with the top of the formation. This figure demonstrates the existence of the oil-water wedge in the formation caused by the horizontal attitude of the water-oil contact and the dip of the bed. It is noted that the isopach lines are equal to the total thickness of the formation when the hydrocarbon-bearing portion is not underlain by water. Figure 7-30 shows the location of the isopach lines in a formation which is limited by a pinchout updip and water underneath. In this case the areal extent is limited on one side by the pinchout and on the other side by the intercept of the water table with the top of the sand. Wedging of the hydrocarbon zone is caused by both the pinchout and the water table and formation dip.

Net Isopach. The gross section must be corrected for two factors in order to obtain the net thickness values to use in the construction of a net isopach map. The first correction is to deduct from the gross thickness that part which is of a lithology not normally considered hydrocarbon-bearing. In sands this may be shale lenses which must be deducted. In limestones there may be chert or calcite lenses which would not contain hydrocarbons. The presence and thickness of these zones of different lithology can be determined from logs or core analysis. Figure 7-31 illustrates the correction of gross sand thickness for the existence of a shale stringer. The thickness value obtained after correcting for shale lenses, etc., is sometimes referred to as net sand thickness.

The second correction made on the gross section to determine the net

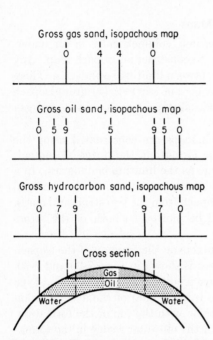

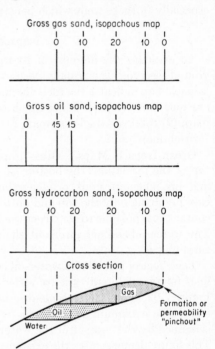

FIG. 7-29. Sketch showing relationship between the subsurface cross section and isopach lines in the vicinity of the cross section for a reservoir containing oil and free gas. Extent of the reservoir limited by the oil-water contact and the bottom of the sand. (*From Vance.*[14])

FIG. 7-30. Sketch showing relationship between the subsurface cross section and isopach lines in the vicinity of the cross section for a reservoir containing oil and free gas. Extent of the reservoir limited in one direction by the oil-water contact and in the other direction by a formation or permeability "pinchout." (*From Vance.*[14])

thickness values depends on the productivity of the formation. Dense sections of essentially zero permeability would not be considered as "pay," since any hydrocarbons which they may contain cannot be produced at economic rates. In limestones, zones of extremely low porosity can be considered as not connected and hence be treated as "nonpay." The existence of dense stringers can be determined from a study of logs and core analysis. The core-analysis log in Fig. 7-32 has a 3-ft section of sand at the top of the

formation which would not be considered net pay because of its low permeability. This core log also indicates the presence of shale stringers the
thicknesses of which are deducted from the gross section. A detailed analysis of the core log of Fig. 7-32 is presented in Example 7-7. A more detailed illustration of the selection
of productive pay is included in
the discussion of averaging permeability and porosity.

A procedure for constructing a
net isopach map is first to construct
a total net thickness map which includes the water-saturated zone
(Fig. 7-33a).

The projections of the intercepts
of the water table with the top and
bottom of the formation are superimposed on this map. The total
net values within the line defined
by the intercept of the water table

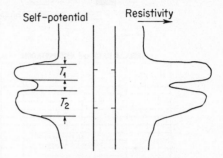

Fig. 7-31. Total thickness as indicated
on an electrical log for a sandstone containing a shale break. (*From Vance.*[14])

and the bottom of the sand represent the net values for the hydrocarbon
isopach. The net hydrocarbon thickness lines between the two water-oil
contact contours are then constructed. Therefore the contours from the
total net thickness map can be traced in this region. The final two steps
of this procedure are illustrated in Fig. 7-33b and c. A complete net
hydrocarbon isopach map for the Jones Sand, Shuler Field, Arkansas, is
shown in Fig. 7-34. This reservoir has a water-oil contact only at the
extreme limits. The location of the zero-thickness contour had to be
estimated from extensions of the top of formation map, as no wells were
available on the edges for complete definition.

Example 7-7. Typical Core Analysis of the Dakota *J* Sand from Core
Log Shown in Fig. 7-32, Specimen Core Analysis* (Smith No. 1 Well File
UAP-1).

This core analysis has been selected to demonstrate a number of interesting and
important points. It represents the *J* Sand of the Dakota series in the northeastern
portion of the Denver-Julesburg Basin.

Coring was commenced in the shale section overlying the *J* Sand, the coring
point probably having been picked from stratigraphic markers present in other
wells in the general area. The actual top of the *J* Sand is at 4,805.0 ft, but from
that depth to 4,808.5 ft, the formation is shaly and tight as shown by the permeability values measured.

Development of permeability at 4,808.5 ft indicates the top of the pay section.

* Courtesy of Core Laboratories, Inc.

COMPANY **GOOD OIL COMPANY**   DATE ON **9-1-52**   FILE NO. **UAP-1(PC)**
WELL **SMITH NO. 1**   DATE OFF **9-16-52**   ENGRS. **EFM-TBG**
FIELD **WILDCAT**   FORMATION **DAKOTA "J" SAND**   ELEV. **4277' KB**
COUNTY **CHEYENNE**   STATE **NEBRASKA**   DRLG. FLD. **WATER BASE MUD**   CORES **DIAMOND**
LOCATION   REMARKS **SERVICE NO. 5**

SAND   LIMESTONE   CONGLOMERATE   CHERT
SHALE   DOLOMITE

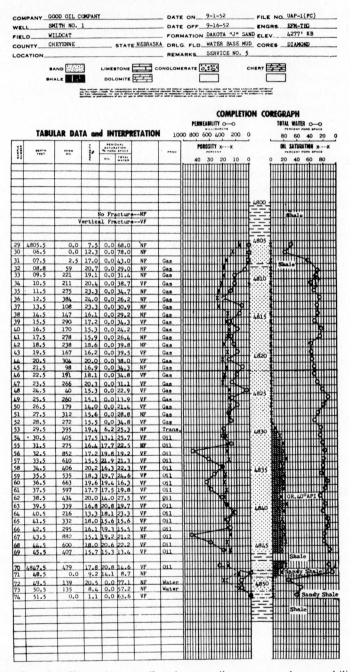

COMPLETION COREGRAPH

TABULAR DATA and INTERPRETATION

PERMEABILITY o—o MILLIDARCYS 1000 800 600 400 200 0
TOTAL WATER o—o PERCENT PORE SPACE 80 60 40 20 0
POROSITY x---x PERCENT 40 30 20 10 0
OIL SATURATION x---x PERCENT PORE SPACE 0 20 40 60 80

No Fracture--NF
Vertical Fracture--VF

SAMPLE NO.	DEPTH FEET	PERM MD	POROSITY %	RESIDUAL SATURATION % PORE SPACE OIL	RESIDUAL SATURATION % PORE SPACE TOTAL WATER	FRACTURE	PROD
29	4805.5	0.0	7.5	0.0	68.0	NF	
30	06.5	0.0	12.3	0.0	78.0	NF	
31	07.5	2.5	17.0	0.0	43.0	NF	Gas
32	08.8	59	20.7	0.0	29.0	NF	Gas
33	09.5	221	19.1	0.0	31.4	NF	Gas
34	10.5	211	20.4	0.0	38.7	VF	Gas
35	11.5	275	23.3	0.0	34.7	NF	Gas
36	12.5	384	24.0	0.0	26.2	NF	Gas
37	13.5	108	23.3	0.0	30.9	NF	Gas
38	14.5	147	16.1	0.0	29.2	NF	Gas
39	15.5	290	17.2	0.0	34.3	VF	Gas
40	16.5	170	15.3	0.0	24.2	NF	Gas
41	17.5	278	15.9	0.0	26.4	NF	Gas
42	18.5	238	18.6	0.0	39.8	NF	Gas
43	19.5	167	16.2	0.0	39.5	VF	Gas
44	20.5	304	20.0	0.0	38.0	VF	Gas
45	21.5	98	16.9	0.0	34.3	NF	Gas
46	22.5	191	18.1	0.0	34.8	VF	Gas
47	23.5	266	20.3	0.0	31.1	VF	Gas
48	24.5	40	15.3	0.0	22.9	VF	Gas
49	25.5	260	15.1	0.0	11.9	VF	Gas
50	26.5	179	14.0	0.0	21.4	VF	Gas
51	27.5	312	15.6	0.0	28.8	NF	Gas
52	28.5	272	15.5	0.0	34.8	VF	Gas
53	29.5	395	19.4	6.2	25.3	NF	Trans.
54	30.5	405	17.5	13.1	25.7	VF	Oil
55	31.5	275	16.4	17.7	22.5	NF	Oil
56	32.5	852	17.2	19.8	19.2	VF	Oil
57	33.5	610	15.5	21.9	21.3	VF	Oil
58	34.5	406	20.2	16.3	22.3	VF	Oil
59	35.5	535	18.3	19.7	24.6	VF	Oil
60	36.5	663	19.6	19.4	16.3	VF	Oil
61	37.5	597	17.7	17.5	19.8	VF	Oil
62	38.5	434	20.0	14.0	27.5	VF	Oil
63	39.5	339	16.8	20.8	19.7	VF	Oil
64	40.5	216	13.3	18.1	23.3	VF	Oil
65	41.5	332	18.0	15.6	15.6	VF	Oil
66	42.5	295	16.1	19.3	15.5	VF	Oil
67	43.5	882	15.1	19.2	21.2	NF	Oil
68	44.5	600	18.0	20.6	22.2	VF	Oil
69	45.5	407	15.7	15.3	13.4	VF	Oil
70	4847.5	479	17.8	20.8	14.6	VF	Oil
71	48.5	0.0	9.2	14.1	8.7	NF	
72	49.5	139	20.5	0.0	77.1	NF	Water
73	50.5	135	8.4	0.0	57.2	NF	Water
74	51.5	0.0	1.1	0.0	63.6	VF	

FIG. 7-32. Core log illustrating gas-oil and water-oil contacts and permeability data with which to determine net productive sand. (*Courtesy of Core Laboratories, Inc.*)

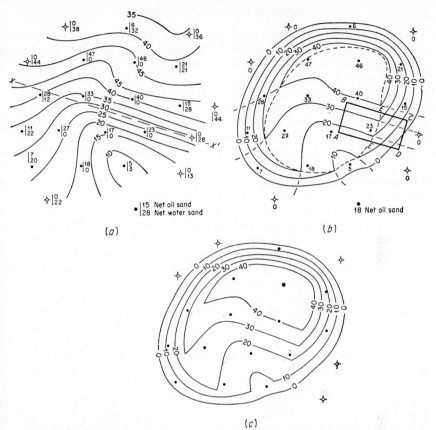

FIG. 7-33. (a) Isopach map drawn on total net sand; (b) isopach map of net oil sand with dashed lines showing location of total net sand contours, inner edge of oil-water wedge shown by dashed curve; (c) completed isopach map of oil reservoir. (*From Wharton.*[15])

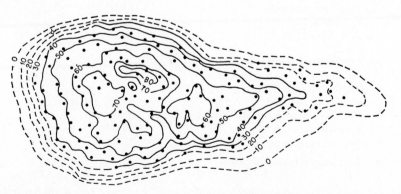

FIG. 7-34. Productive sand thickness, Jones Sand, Shuler Field, Ark. (*From Kaveler.*[10])

531

The porosity is typical of J Sand, ranging from 15 to 20 per cent. Fluid saturations indicate the zone from 4,808.5 to 4,829.0 ft to be dry-gas productive. The absence of any measurable residual oil saturation combined with low total water saturations offers conclusive evidence that the zone is dry-gas productive. The permeability and productive capacity (average permeability × thickness) of the 20.5-ft section would assure high-volume gas production if a completion were made in this zone.

At 4,829.0 to 4,830.0 ft, the first residual oil saturation is observed, marking the point at which the productive characteristics of the J Sand are beginning to change. The residual oil saturation of 6.2 per cent pore space would indicate a transitional interval over which liquid hydrocarbon saturation is developing, and flow of both a free-gas and liquid phase would be expected from this foot. The gas-oil contact is shown by the core analysis to be at 4,829.0 to 4,830.0 ft.

From 4,830.0 to 4,846.0 ft, the residual oil and total water saturations are indicative of an oil-productive zone. High permeability and excellent porosity show the interval to have a large potential and to contain substantial volumes of recoverable oil in place. Both this oil zone and the gas zone above showed vertical fractures in addition to the high permeability of the rock itself. These factors present completion problems discussed later.

From 4,846.0 to 4,847.0 ft, a shale barrier is present, followed by 1 ft of isolated oil sand, which in turn is underlain by 1 ft of sandy shale at 4,848.0 to 4,849.0 ft. The presence of the shale streaks will be helpful in completing the well as a water-free producer.

Two feet of permeable water sand are present from 4,849.0 to 4,851.0 ft. The lack of residual oil saturation combined with high water saturation is conclusive evidence that this 2-ft interval is wet.

Completion should be made near the base of the section, say from 4,840.0 to 4,843.0 ft, in order to diminish the possibility of gas coning from above. The shale streaks underlying the sand should, if laterally continuous, prevent water coning, and completion a few feet above these streaks would be suggested. Here the inter-

INTERPRETATION AND AVERAGE PROPERTIES*

Top of net effective sand, ft	4,807
Base of net effective sand, ft	4,851
Gas-oil contact, ft	4,829
Water-oil contact, ft	Between 4,848 and 4,849
Gross sand thickness, ft	44
Total net effective sand thickness, ft	41
Net effective gas sand thickness, ft	21.5
Net effective oil sand thickness, ft	18
Permeability capacity kh	13,362 md. ft.
Weighted average permeability, millidarcys	326
Porosity capacity ϕh, ft	724.9%
Weighted average porosity, %	17.7

* Additional interpretation by authors.

vals 4,805 to 4,807, 4,808 to 4,808.5, 4,846 to 4,847, 4,848 to 4,849, 4,850.5 to 4,852 are considered to be of such physical character as to be nonproductive even if containing hydrocarbons. The interval 4,850 to 4,851 is considered to consist of one-half permeable effective sand and one-half noneffective sand based on the contrast in the permeability and porosity value.

Gross Rock Volume. The rock volume containing hydrocarbons can be determined for gross or net sections. The rock volume of the gross section can be determined by planimetering the contour maps or the isopach maps. The net rock volume is determined by planimetering the net isopach map.

The areas enclosed by contours of the structure maps based on the top and bottom of the formation are determined by planimetering. These areas (Fig. 7-35) are plotted as a function of depth. The area enclosed by

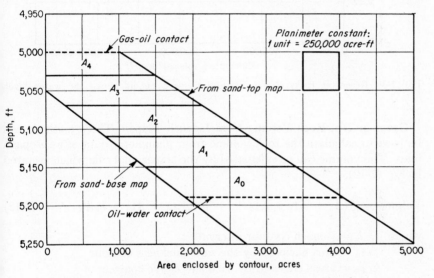

Fig. 7-35. Idealized volume graph for computing gross rock volume.

the two resulting curves represents the gross acre-feet of hydrocarbon-containing rock. The gross rock volume is determined by a graphical integration of the area between the two curves or by planimetering.

In some instances it is desirable to know the rock volume distribution as a function of depth. This can be calculated by dividing the area between the two curves into small segments and calculating the area of each little segment (A_0, A_1, A_2, etc., of Fig. 7-35). The cumulative volume distribution with depth can then be expressed in terms of rock volume above some given depth or rock volume below a given depth. The system chosen depends on whether a gas cap or water drive is to be the predominant source of energy. The gross volume distribution for the data of Fig. 7-35 is shown in Fig. 7-36, cumulating the volumes from the bottom.

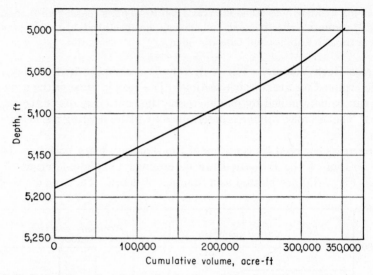

Fig. 7-36. Reservoir rock-volume distribution with height above the water level, field A.

Gross or Net Rock Volume from Isopach Maps. Three methods can be used to calculate the rock volume from planimeter data of an isopach map. The volume can be calculated by the trapezoidal rule which is stated in Eq. (7-12).

$$\text{Volume} = \frac{h}{2}\,(A_0 + 2A_1 + 2A_2 + \cdots + 2A_{n-1} + A_n) + h_n A_n \qquad (7\text{-}12)$$

where h = isopach contour interval
  A_0 = area enclosed by the zero thickness contour
  $A_{1,2,3}$ = areas enclosed by each successive contour
  A_n = area enclosed by the contour line representing the greatest thickness
  h_n = average thickness above the top contour

The gross volume of the isopach map can be calculated from successive calculations of the volume between each contour, as of a frustum of a cone, by use of the pyramidal rule. The volume between any two contour lines is given by Eq. (7-13).

$$V_{0-n} = \frac{h}{3}\,(A_0 + A_n + \sqrt{A_0 A_n}) \qquad (7\text{-}13)$$

where V_{0-n} = volume between 0 and n contours
  h = difference in thickness between two contours
  A_n = area enclosed by n contour
  A_0 = area enclosed by 0 contour

The gross rock volume can also be obtained by plotting a graph of area enclosed by each contour as a function of the thickness represented by that contour (Fig. 7-37). The plotted points are connected by a smooth curve. The rock volume is represented by the area under the curve. This area can be determined by graphical or numerical integration or planimetering. This is probably the most desirable method, as it does not assume any fixed relationship between each contour as is done in the trapezoidal or pyramidal relations.

The rock volume depends on the determination of the thickness of the formation and the areal extent of the reservoir, so that uncertainties in sand thickness are reflected directly in the magnitude of rock volume. An uncertainty of 10 per cent in estimating net effective sand thickness from logs results in an uncertainty of 10 per cent in rock volume. Uncertainties

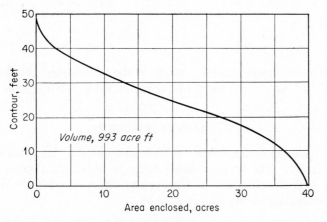

FIG. 7-37. Typical volume graph for determining reservoir bulk volume from isopach maps.

in determining the productive area from interpretation of contact data enter in a more complex fashion and may contribute substantial uncertainties, particularly if the entire oil productive area is underlain by water.

To illustrate further the effect of uncertainties involved in obtaining volumes from an isopach map (considering the map to represent the rock volume accurately), the data from which Fig. 7-37 was prepared can be analyzed with respect to the calculating procedure used:

Contour, ft	Area enclosed, acres
0	39.83
10	36.21
20	27.09
30	12.80
40	2.75
49 (maximum thickness)	

From the volume graph of Fig. 7-37 a rock volume of 993 acre-ft is ob-tained. By application of the trapezoidal rule a volume of 972.5 acre-ft is obtained, and by application of the pyramidal rule a volume of 1,000 acre-ft is obtained. Using the volume graph as a base, the volume from the trapezoidal rule is 2 per cent lower while that from the pyramidal rule is 0.7 per cent high.

Agreement between the trapezoidal rule and the pyramidal rule depends on the ratio of the area of a contour to the next higher contour, for example:

A_n/A_0	Ratio of volumes $\dfrac{pyramidal}{trapezoidal}$
0.8	1.002
0.6	1.011
0.4	1.033
0.2	1.093
0.1	1.169

where A_0 is the lower contour and A_n is the next higher. Differences as high as 5 to 15 per cent exist when the thickness is converging rapidly

HYDROCARBON VOLUME

The calculation of the volume of rock containing hydrocarbons is in itself insufficient. The actual volume and type of hydrocarbons saturating the rock must be determined. The volume of space occupied by hydrocarbons is defined by Eq. (7-14) or (7-15).

$$V_H = V_B\phi(1 - S_w) \qquad (7\text{-}14)$$

where V_H = reservoir hydrocarbon volume
V_B = bulk rock volume containing hydrocarbons
ϕ = mean porosity of hydrocarbon-bearing rock
S_w = mean water saturation of hydrocarbon-bearing rock

$$V_H = \sum_{j=1}^{n} V_j\phi_j(1 - S_{wj}) \qquad (7\text{-}15)$$

where V_j is the rock volume of porosity ϕ_j and water saturation S_{wj} and n is the number of segments of different porosity and water saturation required to define the hydrocarbon volume.

Equation (7-15) can be expressed in terms of gross rock volume or net rock volume. The manner of expressing the volume depends on the data used in calculating the porosity and water-saturation values. To solve either of the volumetric equations it is necessary to determine the values of porosity and water saturation corresponding to some fixed volume of rock. The determination of these values requires a statistical treatment of core-analysis data.

Evaluation of Porosity and Permeability

In evaluation of the permeability and porosity of individual wells, the weighted average permeability and porosity values are computed. That is, each value of porosity or permeability is assumed to represent the interval from which the sample was taken. If all the intervals sampled from a well are of uniform thickness, the weighted average and the arithmetic average are identical. If the intervals differ in both thickness and value of the variable, then the two averages differ. An example calculation of the weighted average porosity and permeability is presented in Example 7-8.

Example 7-8. Calculation of Arithmetic and Weighted Average Porosity and Permeability.

Depth, ft	h	Porosity ϕ, %	ϕh	Permeability, millidarcys	kh
3,690–3,691	1	20	20	10	10
3,691–3,693	2	23	46	100	200
3,693–3,694	1	21	21	50	50
3,694–3,697	3	26	78	200	600
3,697–3,698	1	18	18	70	70
3,698–3,700	2	22	44	120	240
		$\Sigma130$	$\Sigma227$	$\Sigma550$	$\Sigma1{,}170$

Arithmetic average $\phi = \dfrac{130}{6} = 21.7$ per cent

Arithmetic average $k = \dfrac{550}{6} = 91.7$ millidarcys

Weighted average $\phi = \dfrac{227}{10} = 22.7$ per cent

Weighted average $k = \dfrac{1{,}170}{10} = 117$ millidarcys

The weighted average porosity is 4 per cent higher than the arithmetic average, and the weighted average permeability is 22 per cent higher than the arithmetic average. It will be noted that the weighted average permeability is equivalent to considering the intervals to be an array of beds in parallel. The summation of kh is referred to as the permeability capacity, and the summation of ϕh is the porosity or volume capacity of the section.

The weighting of porosity and permeability as described above is satisfactory if the samples are representative of the intervals from which they are obtained. However, in describing the average properties of a reservoir from a large number of samples, statistical methods reduce the work involved and provide additional information with which to describe the physical properties of the system under consideration. Jan Law[16] and A. C. Bulnes[17] have contributed greatly to the application of statistical methods to evaluation of core analysis.

Statistical data are ordinarily classified into classes or ranges of the variable under consideration and the number of occurrences of the variable in each range tabulated. A sample classification of porosity data is shown in Table 7-7. The number of occurrences in a particular range is referred

TABLE 7-7. CLASSIFICATION OF POROSITY DATA INTO RANGES OF
2 PER CENT POROSITY FOR ALL SAMPLES

Porosity range, %	No. of samples	Frequency F, %	Cumulative frequency F_c
Less than 10	161	3.78	3.78
10–12	257	6.04	9.82
12–14	398	9.35	19.17
14–16	493	11.58	30.75
16–18	608	14.28	45.03
18–20	636	14.94	59.97
20–22	623	14.63	74.60
22–24	447	10.50	85.10
24–26	340	7.99	93.09
26–28	176	4.13	97.23
28+	117	2.75	100.00
Totals	4,256	99.92	

to as the frequency. The sum of the frequencies over the number of ranges representing the data is equal to the total number of data points. Frequency, in many applications, is more conveniently expressed as a fraction of the total number of samples permitting comparison of distributions containing different numbers of data points.

Figure 7-38 is a porosity histogram and distribution (cumulative frequency) curve for the data presented in Table 7-7. The histogram is relatively symmetrical as are most porosity distributions. Two statistical measures of central tendency (average values) are shown: (1) the median at a porosity of 17.8 per cent and (2) the arithmetic mean at a porosity of 18.6 per cent. The median is by definition the value of the variable corresponding to the 50 per cent point on the cumulative frequency curve. The median divides the histogram into equal areas. The value of the arithmetic mean depends on the treatment of the data. For unclassified data the arithmetic mean is the sum of the individual values of the variable divided by the total number of such values. For classified data it is convenient to express the arithmetic mean as a summation, Eq. (7-16).

$$\overline{\phi}_a = \sum_{i=1}^{n} \phi_i F_i \qquad (7\text{-}16)$$

where $\overline{\phi}_a$ = arithmetic mean porosity

ϕ_i = class mark (value of porosity at mid-point) of ith-class interval or range

n = number of class intervals

F_i = frequency for ith-class interval, fraction

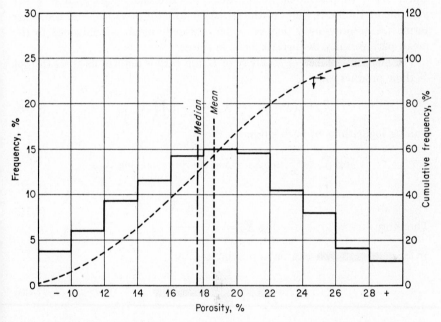

FIG. 7-38. Porosity histogram and distribution for all samples from field A.

This definition is equivalent to the usual statistical definition. For the data presented in Fig. 7-38, the arithmetic mean computed from the classified data using Eq. (7-16) is 18.62 per cent, while that computed from the unclassified data is 18.65 per cent.

Other Statistical Measures. In addition to the measures of central tendency discussed above, three additional measures are encountered at times. These are (1) the mode, (2) the harmonic mean, and (3) the geometric mean. Of these the geometric mean is of greatest importance with respect to statistical evaluation of reservoir rock properties.

The mode is difficult to describe mathematically but can be roughly defined as the value of the variable which occurs most frequently. Therefore, the mode would fall in the class interval having the greatest frequency and is approximated by the class mark of the interval. The mode is of little quantitative significance in evaluation of engineering parameters.

The harmonic mean of an array of numbers is defined as the reciprocal of the arithmetic mean of the reciprocals of the numbers, or

$$\bar{X}_h = \frac{1}{\sum\limits_{i=1}^{n} (F_i/X_i)} \tag{7-17}$$

where $\bar{X}_h$ is the harmonic mean, and X_i is the class mark (value of the variable at the mid-point) of the ith-class interval or range, and the other terms are as previously defined. The harmonic mean is analogous to the mean permeability defined for beds in series.

The geometric mean of an array of L numbers is defined as the Lth root of their product, or

$$\bar{X}_g = (X_1 \cdot X_2 \cdots X_L)^{1/L} = \left[\prod_{i=1}^{L} X_i\right]^{1/L} \tag{7-18}$$

Taking logarithms of both sides,

$$\log \bar{X}_g = \frac{1}{L}(\log X_1 + \log X_2 + \cdots + \log X_L)$$

Therefore
$$\log \bar{X}_g = \frac{\sum\limits_{i=1}^{L} \log X_i}{L} \tag{7-19}$$

or for classified data can be approximated by

$$\log \bar{X}_g = \sum_{i=1}^{n} F_i \log X_i \tag{7-20}$$

where the class interval or range is defined on a logarithmic scale. The geometric mean is required to represent exponential distributions. Its application is illustrated in the evaluation of permeability.

Standard Deviation and Frequency Function. The standard deviation is another important concept in statistics. In effect, it is a measure of the dispersion of the data about the arithmetic mean and for classified data is defined by Eq. (7-21).

$$S = \left[\sum_{i=1}^{n} (X_i - \bar{X}_a)^2 F_i\right]^{1/2} \tag{7-21}$$

where s is the standard deviation and $\bar{X}_a$ is the arithmetic mean. Note that the standard deviation is in the same units as the variable.

For purposes of mathematical analysis it is convenient to replace the histogram by a continuous curve which can be more readily manipulated. The resulting continuous curve is called the normal curve. The normal curve is a symmetrical bell-shaped curve which is completely defined by the arithmetic mean and the standard deviation. The frequency function for a normal distribution is given by

$$f(X) = \frac{1}{S\sqrt{2\pi}}\, e^{-\frac{1}{2}[(X-\overline{X}_a)/S]^2} \qquad (7\text{-}22)$$

where S = standard deviation
e = base of natural logarithm, 2.71828
X = value of the variable
$\overline{X}_a$ = arithmetic mean

A normal curve can be fitted to data conveniently by plotting the variable against the cumulative frequency on arithmetic probability paper. If the data approximate a straight line, then a normal curve is a reasonable fit of the data. The distribution data of Fig. 7-38 plotted on arithmetic probability paper are shown in Fig. 7-39. The dashed line is a smoothed curve through the points. It will be noted that the porosity distribution of field A closely approaches a normal distribution. The equation of the

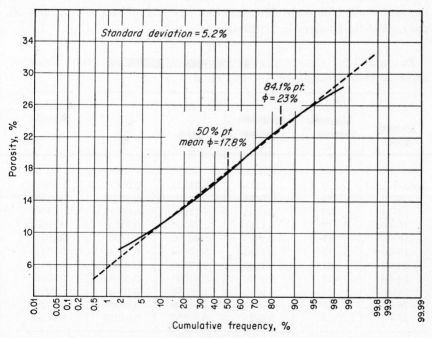

Fig. 7-39. Field A porosity distribution on probability paper.

normal curve has been shown to be defined completely by the arithmetic mean and the standard deviation of the distribution. The standard deviation and mean of a distribution can be estimated from a plot on probability paper in the following manner: The mean is the intercept of the line drawn between the 15.9 and 84.1 per cent values of the curve and the 50 per cent cumulative frequency line. On Fig. 7-39 this value is shown to be 17.8 per cent, which corresponds closely to the median value noted in Fig. 7-38. At the 84.1 per cent cumulative frequency in Fig. 7-39, the porosity ϕ is 23 per cent. The standard deviation is the difference between the porosity at the 84.1 per cent point and the porosity at the 50 per cent point of the cumulative frequency curve. This relation arises from the property of a normal curve that 68.2 per cent of the observations are within one standard deviation of the mean. The standard deviation for this distribution is 5.2 per cent.

In some reservoirs the net productive sand can be determined by a porosity distribution. A "cutoff" value of porosity is selected, so that only samples with porosities greater than the cutoff value are considered "net pay." The cumulative volume capacity for the classified data of field A is calculated in Example 7-9 and plotted in Fig. 7-40. For field A, 98.171 per cent of the storage capacity is represented by samples having porosities of 10 per cent or greater. Thus a cutoff value of 10 per cent porosity could be used to determine net pay and still include at least 98 per cent of the producible hydrocarbons.

Example 7-9. Calculation of Porosity Distribution from Classified Data for Determination of Net Pay Sand.

Porosity range	Mid-value of range, % ϕ_i	Frequency fraction F_i	$\phi_i F_i$	$\dfrac{\phi_i F_i}{\phi_t}$	Cumulative capacity $\displaystyle\sum_{i=1}^{n} \dfrac{\phi_i F_i}{\phi_t}$
Less than 10	9	0.0378	0.34020	0.01827	0.99998
10–12	11	0.0604	0.66440	0.03568	0.98171
12–14	13	0.0935	1.21550	0.06527	0.94603
14–16	15	0.1158	1.73700	0.09328	0.88076
16–18	17	0.1428	2.42760	0.13037	0.78748
18–20	19	0.1494	2.83860	0.15244	0.65711
20–22	21	0.1463	3.07230	0.16500	0.50467
22–24	23	0.1050	2.41500	0.12969	0.33967
24–26	25	0.0799	1.99750	0.10727	0.20998
26–28	27	0.0413	1.11510	0.05988	0.10271
28+	29	0.0275	0.79750	0.04283	0.04283

Average porosity $\phi_t = \Sigma \phi_i F_i = 18.62$ per cent.

Water Saturation

In Chap. 3 the methods for determining the initial water saturation of a small volume of rock were discussed. It was stated that one of the better methods available to the engineer was the use of capillary-pressure data. As capillary-pressure data are measured on extremely small samples, it is possible to have a large number of capillary-pressure curves for the same reservoir. It was suggested in Chap. 3 that the water saturation at several values of capillary pressure be correlated with the permeability of the samples on which the capillary-pressure curves were measured. A straight line for each value of capillary pressure is generally obtained as was shown in Fig. 3-27. Once this family of curves has been constructed, the engineer need only

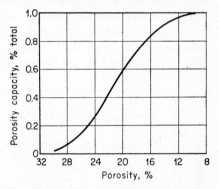

FIG. 7-40. Field *A* distribution of porosity capacity.

calculate the value of permeability and capillary pressure in order to determine the corresponding water saturation.

Classification of Permeability Data. The permeability data obtained from core analysis must be studied and classified before application of statistical methods for its reduction to an average value. In the discussion of porosity it was shown that the samples resulted in a normal-type distribution, so that all the samples could be treated together. The engineer must also analyze the permeability data in the same manner.

The first step in analyzing the porosity data was determining whether or not the samples fitted a normal distribution. This same procedure is followed with the permeability data except in the selection of the permeability ranges. The permeability ranges are selected on equal intervals of the logarithm of permeability in accordance with the recommendation of Law.[16] A histogram of all permeability samples for field *A* is shown in Fig. 7-41, where the range limits are

$$j = \log_2 \frac{k_j}{k_i}$$

or

$$k_j = 2^j k_i$$

where

$$j = 1, 2, 3, 4, \ldots$$

$$k_j = \text{range limits}$$

$$k_i = \text{initial permeability}$$

and the initial value of k_i was selected to be 1.2.

From a study of the permeability histogram and the cumulative frequency curve for field A it is seen that the sample data do not have a normal distribution, as the histogram is unsymmetrical. Statistically the type of data for field A is said to have a skewed distribution. This skewed distribution indicates either that insufficient samples have been taken or that more than one normal distribution exists.

To investigate the data further, it is suggested that the permeability data be plotted on probability paper. Unlike porosity, the logarithm of permeability must be plotted as a function of cumulative percentage samples. The data for field A are presented in Fig. 7-42. The curved line is the

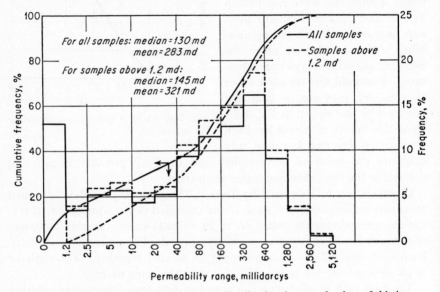

FIG. 7-41. Permeability histogram and distribution for samples from field A.

conventional plotting procedure, whereas the two straight lines result from plotting the logarithm of permeability. These two straight-line segments indicate that two distribution systems are being treated as one. Hence, the permeability data should be further classified.

Two possible permeability variations of interest to the engineer exist: (1) the effect of vertical variations or zoning of different types of material and (2) the effect of areal variations. A closer study of the core, electric, and radioactive logs should indicate if there are two or more distinct permeability systems in the vertical plane of the reservoir. If no vertical variations are indicated, then the core data should be analyzed on an areal basis. In the case of field A it was found that there were two normal permeability systems. The permeability variation was areal and not vertical.

Thus the average permeability of each region must be determined and weighted in accordance with its respective sand volume to determine the field average permeability.

If the classification suggested by Law, $\log_2 (k_j/k_i) = j, j = 1, 2, 3$, etc., yields a normal distribution, then the frequency function [Eq. (7-22)] can be used to express the variation of permeability with frequency or number of samples. Muskat[18] has reported that permeability distributions may not fit a true normal distribution but may be exponential. He suggested that the permeability of each sample or sample range be plotted on semilog paper as a function of the cumulative number of samples having a lesser

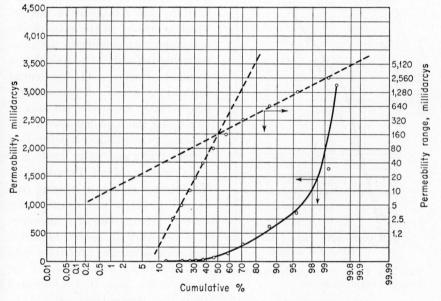

FIG. 7-42. Field A permeability distribution on probability paper.

permeability. The same plot would result if the permeability were plotted as a function of the cumulative frequency. This plot should yield a straight line of the type

$$\log_{10} k = mN + b$$

where k = permeability
N = number of samples having a lesser permeability
m = slope of curve
b = intercept value of log k when N is zero

N could be replaced by the cumulative frequency F, and the only change would be in the magnitude of the slope of the curve.

The cumulative number of samples is plotted against the $\log_{10} k$ for the respective samples of field A in Fig. 7-43. Note that the data can be fitted by three straight-line segments. The dashed curve represents the best single straight line that can be drawn through the data.

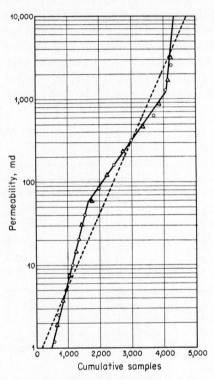

FIG. 7-43. Field A permeability distribution on semilog paper.

The exponential-type variation for a field can, for evaluation purposes, be broken into several straight-line segments such as the three segments shown in Fig. 7-43. The exponential distribution is defined by the stratification ratio r, which is defined as the ratio of the maximum permeability to the minimum permeability of a straight-line segment. Normal and exponential expressions fitted to permeability distributions are useful not only in classifying the data but also in describing permeability stratification in gas cycling and water flooding.

Calculating Mean Permeability. To obtain a statistical average permeability which will describe the over-all performance of a reservoir, it is necessary to determine which statistical averaging procedure is to be used. In the classification of the permeability data, it was indicated that the permeability should be classified on a logarithmic scale. Since the water saturation correlates with the logarithm of permeability, the geometric mean is the average to use with such correlations.

The geometric mean permeability is defined in Eq. (7-19) or (7-20).

$$\log \overline{k}_g = \frac{\sum\limits_{i=1}^{L} \log k_i}{L} \qquad (7\text{-}19)$$

or

$$\log \overline{k}_g = \sum\limits_{j=1}^{n} F_j \log (\overline{k}_a)_j \qquad (7\text{-}20)$$

where $\bar{k}_g$ = geometric mean permeability
        k_i = permeability of sample i
        $(\bar{k}_a)_j$ = arithmetic average permeability of logarithmic class inter-
                val j
        L = total number of samples
        F_j = cumulative frequency of j interval, fraction
        n = total number of classified intervals

For comparison, average permeabilities are calculated for field A by both the geometric and arithmetic mean procedures. These calculations are presented in Examples 7-10 and 7-11. The arithmetic mean procedure is applied to all the samples and to those samples having a permeability greater than 1.2 millidarcys. The geometric procedure is applied only to those samples having a permeability greater than 1.2 millidarcys.

Permeability, like porosity, can be and is used to determine the net sand to be used in volumetric calculations. A cutoff value of permeability can be selected from a permeability capacity curve, so that net sand will be selected on the basis of the samples which have a permeability equal to or greater than the cutoff value. The cumulative permeability capacity for all samples in field A is shown in Fig. 7-44. Eighty per cent of the producing capacity of field A is represented by samples having a permeability greater than 450 millidarcys. Ninety-five per cent of the capacity is represented by samples having permeabilities greater than 100 millidarcys. Thus it would appear that a permeability cutoff value of 1.2 millidarcys would include essentially all the productive sand.

Using a permeability cutoff value of 1.2 millidarcys for field A, the geometric mean permeability calculated in Example 7-11 is 101.29 millidarcys.

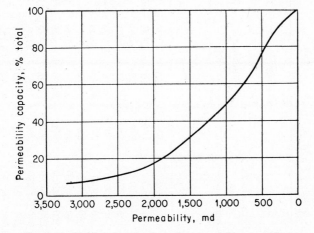

Fig. 7-44. Field A distribution of permeability capacity.

Example 7-10. Calculation of Arithmetic Average Permeability and Permeability Capacity Distribution of All Samples in Field A.

Permeability range, millidarcys	Sample			% sample				Arithmetic			$\dfrac{F_i(\bar{k}_a)_i}{(\bar{k}_a)_i}$ % permeability capacity	
	Number in range	Cumulative number		Range		Cumulative		Average permeability of range $(\bar{k}_a)_i$	All	$F_i(\bar{k}_a)_i$		
		All	$k>1.2$	All	$k>1.2$	All	$k>1.2$				All	$k>1.2$
0-1.2	556	556	0	13.0	0	13.0	0	0.153	0.01989	0	0.0000709	
1.3-2.5	145	701	145	3.4	3.92	16.40	3.92	1.9	0.06460	0.07448	0.00023	0.00023
2.6-5.0	216	917	361	5.07	5.85	21.547	9.77	3.7	0.1876	0.21645	0.00066	0.00067
5.1-10.0	240	1,157	601	5.64	6.5	27.211	16.27	7.5	0.4230	0.4875	0.00199	0.00152
10.1-20.0	197	1,354	798	4.64	5.33	32.215	21.60	14.2	0.6588	0.75686	0.00233	0.00236
20.1-40.0	222	1,576	1,020	5.21	6.0	36.41	27.60	30.2	1.5734	1.8120	0.00556	0.00565
40.1-80.0	396	1,972	1,416	9.3	10.7	45.3626	38.230	58.8	5.4684	6.2916	0.01935	0.01962
80.1-160.0	493	2,465	1,909	11.6	13.3	57.9826	51.560	119	13.804	15.8270	0.04885	0.04935
160.1-320.0	544	3,009	2,455	12.75	14.7	70.561	66.030	234	29.835	34.398	0.10558	0.10225
320.1-640.0	692	3,701	3,145	16.2	18.6	87.0681	87.60.92	462	74.844	85.932	0.26487	0.2679
640.1-1,280.0	386	4,087	3,531	9.06	10.05	96.0587	95.2092	878	79.547	88.239	0.28151	0.2751
1,281.1-2,560.0	142	4,229	3,673	3.34	3.8	99.821	99.18.21	1,663	55.544	63.194	0.19657	0.1970
2,560.1-5,120.0	28	4,257	3,701	0.66	.75	100	100.0	3,120.0	20.592	23.400	0.07287	0.07296

99.87 99.58

548

Example 7-11.   Geometric Mean Permeability for All Samples of Field *A* with a Permeability Greater Than 1.2 Millidarcys.

Permeability range, millidarcys	Average permeability of range $(\bar{k}_a)_j$	$\log_{10} (\bar{k}_a)_j$	Cumulative frequency fraction of range F_j	$F_j \log_{10} (\bar{k}_a)_j$
1.3–2.5	1.9	0.27875	0.0392	0.010927
2.6–5.0	3.7	0.56820	0.0585	0.033239
5.1–10.0	7.5	0.87506	0.0650	0.056878
10.1–20.0	14.2	1.15229	0.0533	0.061417
20.1–40.0	30.2	1.48996	0.0600	0.089397
40.1–80.0	58.8	1.76938	0.1070	0.18932
80.1–160.0	119.C	2.07555	0.1330	0.27604
160.1–320.0	234.0	2.36922	0.1470	0.34827
320.1–640.0	462.0	2.66464	0.1860	0.49562
640.1–1,280.0	878.0	2.94349	0.1005	0.29582
1,280.1–2,560.0	1,663.0	3.22089	0.0380	0.12239
2,560.1–5,120.0	3,120.0	3.4945	0.0075	0.026206
				101.29

The arithmetic mean permeability for samples with a permeability greater than 1.2 is 320.63 millidarcys.

Calculation of Average Water Saturation.   In Chap. 3 the water saturation was shown to be a function of capillary pressure. It was also suggested that capillary pressure could be correlated with permeability and water saturation. The capillary-pressure data reported for a reservoir are obtained on small core samples, each with its own value of porosity and permeability. A series of capillary-pressure curves can be drawn from these data. To reduce the data to one average curve, it was suggested in Chap. 3 that the capillary-pressure data be plotted against water saturation and the logarithm of permeability. A plot of this nature (Fig. 3-27) should yield straight lines for each value of capillary pressure. This series of lines should converge at high permeabilities and low water saturations and diverge at low permeabilities and high water saturations.

There are essentially four approaches to the determination of the water content in a reservoir. In brief the four methods are:

1. Consider the geometric mean permeability to exist throughout the reservoir, and evaluate the water saturation as a function of height above the free water table.

2. Consider the geometric mean permeability to exist throughout, and evaluate the water saturation at a height above the free water table corresponding to the volumetric center of the reservoir.

3. Evaluate the water saturation of each permeability range at the height above the free water table of the volumetric center and weight with respect to the frequency associated with each range.

4. Divide the reservoir into segmental volumes, evaluating the geometric mean permeability of each segment, and determine the water saturation of that segment at a height above the free water table corresponding to the volumetric center of the segment.

The water saturation as a function of height above the free water table is read from the reduced capillary-pressure data of the field or can be obtained from a correlation of oil-base core data. With the use of the geometric

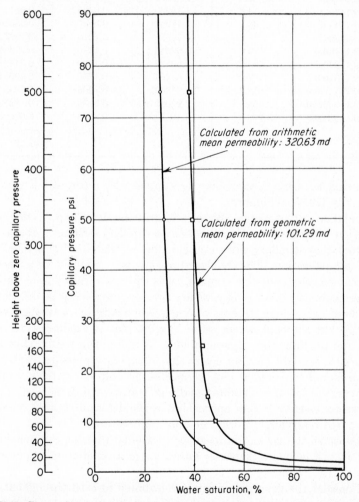

Fig. 7-45. Comparison of calculated water distribution from the geometric and arithmetic mean.

mean permeability, the corresponding values of capillary pressure and water saturation are read from a plot similar to Fig. 3-27. The capillary pressure is converted to height by use of the capillary-pressure equation and the fluid densities. The resulting water-saturation distribution curves for the data of field A are shown in Fig. 7-45. Water-distribution curves are shown for the arithmetic and geometric mean permeabilities of field A. The geometric mean permeability yields water-saturation values approximately 10 per cent greater than the arithmetic mean permeability. The water-distribution curve is used to assign water-saturation values S_{wi} to segmental volumes as a function of the height of these volumes above the free water table.

The second method of calculating the mean water saturation is to calculate the capillary pressure at the volumetric mid-point of the hydrocarbon reservoir. Using this capillary pressure and the geometric mean permeability, the water saturation is read directly from the reduced capillary-pressure data. For field A the mid-point capillary pressure is 25 psia, the geometric mean permeability is 101.29 millidarcys, and the water saturation from Fig. 3-27 is 42.3 per cent.

The third procedure is mathematically identical to the second method provided that the plot of the logarithm of permeability versus water saturation yields a straight line. If this is not the case, the third method should be employed in the solution. The water saturation for the mid-point capillary pressure at the class permeability of each logarithmic range is then read from the field-reduced capillary-pressure data. A water saturation of 42.2 per cent is obtained for field A in Example 7-12 by application of this method.

The fourth procedure is much more detailed. The incremental volumes selected can be as small as desired. The sample data from each incremental volume are analyzed, and an average permeability and mid-point height determined. A corresponding water saturation is then read from the reduced capillary-pressure data. The average water saturation is defined by Eq. (7-23).

$$\overline{S}_{wi} = \frac{\sum\limits_{j=1}^{n} V_j S_{wj}}{\sum\limits_{j=1}^{n} V_j} \tag{7-23}$$

An even further refinement of this procedure is to treat each core sample as being representative of some finite reservoir volume. The volume represented by the samples can be weighted according to the frequency of sampling in a particular area. A water-saturation value is determined for each sample according to its height above the oil-water contact and its permeability. The average water saturation is calculated in Eq. (7-24).

$$\bar{S}_{wi} = \sum_{j=1}^{L} W_j S_{wj} \qquad (7\text{-}24)$$

where L = total number of samples

W_j = weighting factor of jth sample and is defined as ratio of rock volume represented by sample to total rock volume of hydrocarbon reservoir

S_{wj} = water saturation of jth sample

Example 7-12. Weighted Average Water Saturation Based on Average Permeability of Each Logarithmic Range, Field A.

Permeability range, millidarcys	Average permeability $\bar{k}_j$, millidarcys	Frequency of range fraction F_j	Average water saturation $S_{wj}*$	$F_j S_{wi}$
1.2–2.5	1.9	0.0392	0.865	0.03391
2.6–5.0	3.7	0.0585	0.775	0.04534
5.1–10.0	7.5	0.0650	0.712	0.04628
10.1–20.0	14.2	0.0533	0.641	0.03416
20.1–40.0	30.2	0.0600	0.560	0.03360
40.1–80.0	58.8	0.1070	0.485	0.05189
80.1–160.0	119.0	0.1330	0.419	0.05573
160.1–320.0	234.0	0.1470	0.342	0.05027
320.1–640.0	462.0	0.1860	0.255	0.04743
640.1–1,280.0	878.0	0.1005	0.182	0.01829
1,280.1–2,560.0	1,663.0	0.0380	0.115	0.00437
2,560.1–5,120.0	3,120.0	0.0075	0.100	0.00075
				0.422

* From Fig. 3-27, 25-psi curve.

Calculation of Hydrocarbon Volume

Several methods have been discussed by which the water saturation in a hydrocarbon reservoir could be determined. The rock volume of the reservoir, both gross and net, can also be calculated by several different procedures. It is desirable to calculate the hydrocarbon volume on the same basis upon which the rock volume is determined.

If the rock volume is on a net sand (not net productive sand) basis, then the porosity and permeability values used must include all sand samples. The water saturation can be determined by any of the four procedures previously described. The hydrocarbon volume is defined by

$$\text{Surface volume} = \overline{(\text{FVF})} V_R \overline{\phi} (1 - \overline{S}_{wi})$$

or
$$= \overline{(\text{FVF})} \sum_{j=1}^{n} V_{Rj} \overline{\phi} (1 - \overline{S}_{wi})_j \qquad (7\text{-}25)$$

where FVF = formation volume factor (B_g, B_o, etc.) depending on the hydrocarbon present at the initial pressure

V_R = net hydrocarbon sand volume

$\overline{\phi}$ = statistical average porosity of the hydrocarbon sand volume

$\overline{S}_{wi}$ = average water saturation determined from geometric mean permeability of all samples

V_{Rj} = net hydrocarbon sand volume of some segment j

$(1 - \overline{S}_{wi})_j$ = fraction of the pore space not occupied by water in the volume segment j, where $\overline{S}_{wi}$ is determined from height distribution curve constructed using geometric mean permeability of all samples

When the net productive sand volume is used to calculate the hydrocarbon volume, it is necessary to reevaluate either the mean permeability or mean porosity. If, as in the example of field A, a permeability "cut point" is used to determine the net productive sand, then the porosity data for all samples with a permeability greater than the cutoff permeability must be statistically averaged for determination of the hydrocarbon volume.

Two procedures for determining average porosity values are available to the engineer. He can obtain the arithmetic average porosity of all samples with a permeability greater than the cutoff permeability. The same statistical procedure is used as in the calculation of the porosity for all samples. The second approach is to determine the porosity corresponding to each of the logarithmic permeability ranges. This procedure entails more work, as the sample data have to be sorted according to permeability and then the arithmetic average porosity calculated. An example of some of the results of such a procedure is shown in Figs. 7-46 and 7-47. Some of the porosity distribution curves within the permeability ranges are shown in Fig. 7-46. The arithmetic mean porosity of each range is plotted as a function of the mean permeability of the range in Fig. 7-47. The data presented in Fig. 7-47 were prepared from data subdivided into smaller permeability ranges than the permeability ranges shown in Fig. 7-46. Therefore, there are minor discrepancies in average values shown between the two figures. It is interesting to note that the straight-line segments intersect at a permeability of approximately 60 millidarcys. For values of permeability below 60 millidarcys, the change in porosity is small. For permeabilities above 60 millidarcys, porosity increases with increasing per-

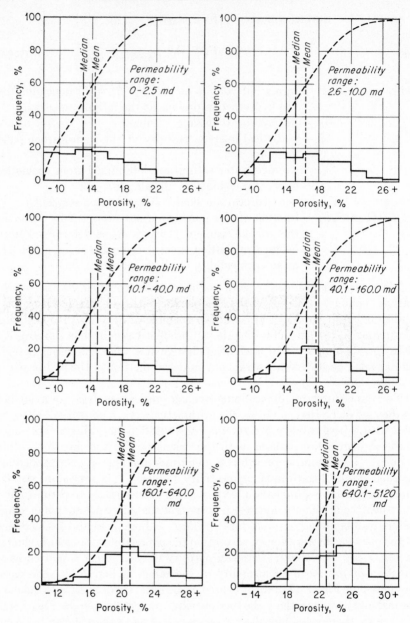

Fig. 7-46. Field *A* porosity distribution by permeability range.

meability at a much faster rate. The change in slope at 60 millidarcys may
be indicative of the difference between data obtained from two different
areas of the field as previously mentioned, or it may reflect the effect of
interstitial clay on porosity and permeability measurements. If the low-
permeability rock has essentially the same grain size and grain structure

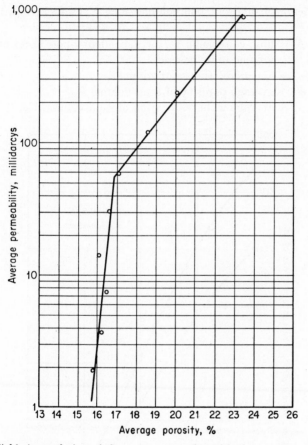

FIG. 7-47. Field A correlation of the average porosity with the average permeability
of a permeability range.

as the high-permeability but differs only in that it contains a great deal of
interstitial clays, then certain laboratory techniques for measuring porosity
would give erroneously high values of porosity for the low-permeability
materials. On the other hand, if the lower permeability resulted simply
from decreased grain size, the small variations in porosity could be attributed to a more uniform grain size in the low-permeability materials.

Once suitable porosity values have been determined, it is only necessary to combine the porosity, water-saturation, and rock-volume terms together to obtain the volume of hydrocarbon in the reservoir. The manner in which the values are combined depends on how each of the values were calculated. All methods of calculating the reservoir hydrocarbon volume can be expressed as

$$\text{Reservoir hydrocarbon volume} = K \sum_{j=1}^{n} V_{Rj}\bar{\phi}_j(1 - \bar{S}_{wi})_j \quad (7\text{-}26)$$

where K = unit conversion factor dependent upon units of V_{Rj} and volume units desired
V_{Rj} = net productive sand volume
$\bar{\phi}_j$ = average porosity of sand volume V_{Rj}
$(\bar{S}_{wi})_j$ = average water saturation of sand volume V_{Rj}

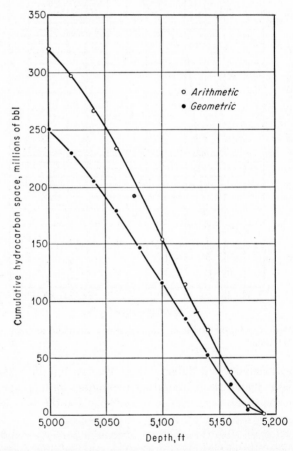

Fig. 7-48. Hydrocarbon space distribution for arithmetic and geometric mean permeabilities of field A.

The manner in which V_{Rj} is defined determines in part the values of porosity and water saturation to be used. A value V_{Rj} can be assigned to every core sample. Some height above the water table can also be assigned to every core sample, in which case each core sample would represent a term in the summation of Eq. (7-26).

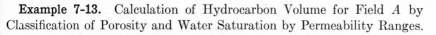

Example 7-13. Calculation of Hydrocarbon Volume for Field A by Classification of Porosity and Water Saturation by Permeability Ranges.

Permeability range, millidarcys	No. of samples	Frequency fraction	Average permeability of range, millidarcys	Average porosity of range fraction	Average water saturation of range *	$F_j \bar{\phi}_j (1 - \overline{S}_{wi})_j$
1.3–2.5	145	0.0392	1.9	0.158	0.865	0.00084
2.6–5.0	216	0.0585	3.7	0.162	0.775	0.00214
5.1–10.0	240	0.0650	7.5	0.165	0.712	0.00308
10.1–20.0	197	0.0533	14.2	0.161	0.641	0.00308
20.1–40.0	222	0.0600	30.2	0.166	0.560	0.00435
40.1–80.0	396	0.1070	58.8	0.171	0.485	0.00942
80.1–160.0	493	0.1330	119.0	0.186	0.419	0.01435
160.1–320.0	544	0.1470	234.0	0.203	0.342	0.01961
320.1–640.0	692	0.1860	462.0	0.218	0.255	0.03017
640.1–1,280.0	386	0.1005	878.0	0.234	0.182	0.01922
1,280.1–2,560.0	142	0.0380	1,663.0	0.243	0.115	0.00814
2,560.1–5,120.0	28	0.0075	3,120.0	0.253	0.100	0.00171
						0.11611

* From Fig. 3-27 for the 25-psi capillary-pressure curve.

$$\text{Reservoir hydrocarbon volume} = \text{rock volume} \dagger \sum_{j=1}^{12} F_j \bar{\phi}_j (1 - \overline{S}_{wi})_j$$

$$= 7{,}758(350{,}000)(0.11611)$$

$$= 315{,}273{,}483.00 \text{ bbl}$$

Mean porosity $\bar{\phi}$ of net productive sand $= 0.1862$

$$\text{Reservoir hydrocarbon volume} = 7{,}758 \sum_{j=1}^{10} V_{Rj} (1 - \overline{S}_{wi})_j \bar{\phi}$$

refers To Ex. 7-14

$$= 7{,}758(32{,}432.28)$$

$$= 251{,}609{,}628.24$$

† Using the cumulative volume data of Fig. 7-36 as net productive sand volume.

Example 7-14. Calculation of Hydrocarbon Volume for Field A by Geometric Mean Permeability and Vertical Rock Volume Distribution.

Height above free water table	Water saturation[a]	Reservoir depth	Cumulative rock volume below, depth in acre-ft	Rock volume between depth readings V_{Ri}	Average water saturation of volume $\bar{S}_{wi}$	$V_{Ri}(1-\bar{S}_{wi})_j\bar{\phi}$	$7,758\Sigma V_R(1-\bar{S}_{wi})_j\bar{\phi}$
0	100	5,200	0[b]				
10	0.68	5,190	0				
20	0.56	5,180	20,000	20,000	0.84	595.84	4,622,526.72
40	0.50	5,160	60,000	40,000	0.62	2,830.24	26,579,528.64
60	0.47	5,140	100,000	40,000	0.53	3,351.60	52,581,241.44
80	0.455	5,120	142,500	42,500	0.485	4,075.40	84,198,194.64
100	0.445	5,100	182,500	40,000	0.4625	4,003.30	115,255,796.04
120	0.435	5,080	222,500	40,000	0.45	4,096.4	147,035,667.24
140	0.430	5,060	262,500	40,000	0.44	4,170.90	179,393,509.44
160	0.425	5,040	296,000	33,500	0.4325	3,528.70	206,769,164.04
180	0.425	5,020	325,000	29,000	0.4275	3,091.50	230,753,021.04
200	0.420	5,000	355,000	25,000	0.4225	2,688.40	251,609,628.24
						32,432.28	

[a] Geometric mean permeability curve, Fig. 7-45.

[b] Using the cumulative volume data of Fig. 7-36 as net productive sand volume.

V_{Rj} can be defined by the permeability ranges and their respective frequencies. A calculation for this definition of V_{Rj} is shown in Example 7-13. The water saturation was determined for each permeability range by the average permeability of the range and the height of the volumetric midpoint of the reservoir above the water table.

The incremental volume V_{Rj} can be determined as a function of height. A single value of porosity is generally used to apply to all incremental volumes. The use of a constant porosity permits its removal from within the summation of Eq. (7-26). The reservoir volume of hydrocarbons for field A is calculated by this procedure in Example 7-14. The reservoir hydrocarbon volume distribution can be calculated as a function of depth by this procedure. The volume distribution for field A is shown in Fig. 7-48.

The reservoir hydrocarbon volume, calculated by any method, can be converted to surface units by means of suitable fluid-volume conversion factors such as B_g, B_o, or B_t. It should be stressed that the values obtained for hydrocarbon volume can be no more accurate than the core-analysis, capillary-pressure, and rock-volume data. If any of these factors is in error by 10 per cent, the hydrocarbon volume will be in error by approximately 10 per cent.

REFERENCES

1. Guthrie, R. K.: Machine Methods for Computing and Data Processing as Applied to Petroleum Engineering and Production Problems, *Drilling and Production Practices*, American Petroleum Institute, 1954.

2. Black, W. Marshall: A Review of Drill Stem Testing Techniques and Analysis, *J. Petrol. Technol.*, June, 1956.

3. Evinger, H. H., and M. Muskat: Calculation of Theoretical Productivity Factor, *Trans. AIME* (reprinted), vols. 146 and 151, 1942 and 1943.

4. Data furnished in private communication from Shell Oil Company.

5. Cupps, C. Q., P. H. Lipstate, and J. Fry: Variance in Characteristics of the Oil in the Weber Sandstone Reservoir, Rangely Field, Colorado, *U.S. Bur. Mines Rept. Invest.* 4761, 1951.

6. Sage, B. H., and W. N. Lacey: Gravitational Concentration Gradients in Static Columns of Hydrocarbon Fluids, *Trans. AIME*, vol. 132, 1939.

7. Espach, Ralph H., and Joseph Fry: Variable Characteristics of the Oil in the Tensleep Sandstone Reservoir, Elk Basin Field, Wyoming and Montana, *U.S. Bur. Mines Rept. Invest.* 4768, 1951.

8. Cook, A. B., G. B. Spencer, F. P. Bobrowski, and Tim Chin: A New Method of Determining Variations in Physical Properties of Oil in a Reservoir, with Application to the Scurry Reef Field, Scurry County, Texas, *U.S. Bur. Mines Rept. Invest.* 5106, 1955.

9. Brochure by Core Labortories, Inc., Dallas, Tex.

10. Kaveler, H. H.: Engineering Features of the Schuler Field and Unit Operation, *Trans. AIME* (reprinted), vols. 155 and 160, 1944 and 1945.

11. Yuster, S. T.: Some Theoretical Considerations of Tilted Water Tables, *Trans. AIME*, 1953.

12. Hubbert, M. King: Entrapment of Petroleum under Hydrodynamic Conditions, *Bull. Am. Assoc. Petrol. Geologists*, August, 1953.

13. Alexander, C. I.: Graphic Representation of Reservoir History, *Bull. Am. Assoc. Petrol. Geologists*, 1950.

14. Vance, Harold: "Petroleum Subsurface Engineering," Education Publishers, St. Louis, Mo., 1950.

15. Wharton, Jay B., Jr.: Isopachous Maps of Sand Reservoirs, *Bull. Am. Assoc. Petrol. Geologists*, vol. 32, no. 7, 1948.

16. Law, Jan: A Statistical Approach to the Interstitial Characteristics of Sand Reservoirs, *Trans. AIME* (reprinted), vols. 155 and 160, 1944 and 1945.

17. Bulnes, A. C., and R. U. Fitting: An Introductory Discussion of the Reservoir Performance of Limestone Formations, *Trans. AIME* (reprinted), vols. 155 and 160, 1944 and 1945.

18. Muskat, M.: "Physical Principles of Oil Production," McGraw-Hill Book Company, Inc., New York, 1949.

CHAPTER 8

THE MATERIAL BALANCE

INTRODUCTION

As has been mentioned previously in this book, the petroleum engineer must be able to make dependable estimates of the initial hydrocarbons in place in a reservoir and predict the future reservoir performance and the ultimate hydrocarbon recovery from the reservoir. This chapter is concerned with the estimation of the initial hydrocarbons in place in the reservoir.

Numerous procedures have been proposed and employed for estimating hydrocarbons in place by volumetric methods (Chap. 7). However, it has become both practical and popular to confirm such estimates by material-balance calculations. The type of material balance used in such estimates is similar to that used in many other fields of engineering for quantity and quality estimate and control. However, custom has established that the material balance be written on a volumetric basis, although this is not necessary. In the simplest form the material-balance equation can be written as *initial volume = volume remaining + volume removed.*

Since oil, gas, and water are present in petroleum reservoirs, it is seen that the material balance can be written for the total fluids or for any one of the fluids present. Furthermore, there are numerous ways of expressing the physical properties of the fluids present and the relationship among these properties. The petroleum literature contains numerous material balances which, to the neophyte engineer, may appear to be different but upon critical examination will be found to be identical.

A concept of material balance for the estimation of hydrocarbons in underground reservoirs was presented by Schilthuis.[1] This work was followed closely by that of Katz[2] and later by Miles.[3] Although much progress has been made in petroleum technology during the past quarter of a century, the material-balance equation of Schilthuis has continued to serve the purpose for which it was intended. The principal improvements in application of the equation to practice have been made possible through refinements in measurements and the continuing efforts of reservoir engineers to expand the equation to encompass the reservoir rock and its contents.

DERIVATION OF MATERIAL-BALANCE EQUATION

The material balance will be developed in this section to illustrate the various relationships existing among variables and the various forms which the balance may take. The derivation presented is based on the total pore volume and its content.

Before deriving the material balance, it has been found expeditious and convenient to denote certain terms by symbols for brevity. The symbols used conform where possible to the standard nomenclature adopted by the Society of Petroleum Engineers in 1956.

B_o = oil formation volume factor = volume at reservoir conditions per volume at stock-tank conditions

B_g = gas formation volume factor = volume at reservoir conditions per volume at standard conditions (used to denote solution-gas volume when more than one type of gas is present)

B_w = water formation volume factor = volume at reservoir conditions per volume at standard conditions

B_{gc} = gas-cap gas formation volume factor = volume at reservoir conditions per volume at standard conditions

B'_g = injected gas formation volume factor = volume at reservoir conditions per volume at standard conditions

$B_t = B_o + (R_{si} - R_s)(B_g/5.61)$ = composite oil or total oil formation volume factor = volume at reservoir conditions per volume at standard conditions

$B_{tw} = B_w + (R_{swi} - R_{sw})(B_g/5.61)$ = composite water or total water formation volume factor = volume at reservoir conditions per volume at standard conditions

c_f = formation (rock) compressibility = pore volume per pore volume per psi

G = initial gas-cap gas volume, scf

G_i = cumulative gas injected, scf

$G_p = G_{ps} + G_{pc}$ = cumulative gas produced, scf

G_{ps} = cumulative solution gas produced, scf

G_{pc} = cumulative gas-cap gas produced, scf

$m = GB_{gi}/5.61NB_{oi}$ = ratio of initial gas-cap–gas-reservoir volume to initial reservoir oil volume

N = initial oil in place, stock-tank bbl

N_p = cumulative oil produced, stock-tank bbl

P = reservoir pressure, psia

P_i = initial reservoir pressure, psia

R_s = solution-gas–oil ratio, scf/stock-tank bbl

R_{si} = initial solution-gas–oil ratio, scf/stock-tank bbl

R_{sw} = solution-gas–water ratio, scf/bbl at standard conditions

R_{swi} = initial solution-gas–water ratio, scf/bbl at standard conditions
R_p = producing gas-oil ratio, scf/stock-tank bbl
R_c = G_p/N_p, scf/stock-tank bbl
R_{cs} = G_{ps}/N_p, scf/stock-tank bbl
S_o = oil saturation, fraction of pore space
S_g = gas saturation, fraction of pore space
S_w = water saturation, fraction of pore space
S_{wi} = initial water saturation, fraction of pore space
S_{wio} = initial water saturation in oil zone, fraction of pore space
S_{wig} = initial water saturation in gas cap, fraction of pore space
W = initial water in place, reservoir bbl
W_e = cumulative water influx, bbl at standard conditions
W_p = cumulative water produced, bbl at standard conditions
W_i = cumulative water injected, bbl at standard conditions

Using the reservoir-engineering terms defined previously and establishing a reservoir volumetric system such as that illustrated in Fig. 8-1,

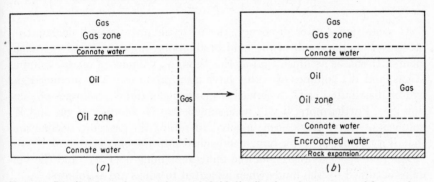

Fig. 8-1. Idealized zonal saturation and fluid-distribution changes caused by production from a hydrocarbon reservoir. (*a*) Initial conditions; (*b*) conditions after producing N_p bbl of oil, G_{ps} and G_{pc} cu ft of gas, and W_p bbl of water.

volumetric balance expressions can be derived to account for all volumetric changes which occur during the natural productive life of the reservoir.

When a reservoir is discovered, it may contain water, gas, and oil, intermingled or segregated into zones. A general material balance should be capable of handling any type of fluid distribution. Therefore, the material or volume balance presented herein will be derived with gas and water in a gas zone; free gas, oil, and water in an oil zone; and a water zone contiguous to the oil zone. The initial reservoir pressure is denoted by P_i, and the total pore space in which hydrocarbons are located is denoted by $V_r\phi$. The volumes of the various other fluids and their location are given below:

Gas-zone pore volume (bbl) = free gas volume (bbl) + water volume (bbl)

$$\frac{GB_{gi}}{5.61}\frac{1}{1 - S_{wig}} = \frac{GB_{gi}}{5.61} + \frac{GB_{gi}}{5.61}\frac{S_{wig}}{1 - S_{wig}} \qquad (8\text{-}1)$$

Oil-zone pore volume (bbl) = oil volume (bbl) + free gas volume (bbl) + water volume (bbl)

$$\frac{NB_{ti}}{1 - S_{gi} - S_{wio}} = NB_{ti} + \frac{NB_{ti}S_{gi}}{1 - S_{gi} - S_{wio}} + \frac{NB_{ti}S_{wio}}{1 - S_{gi} - S_{wio}} \qquad (8\text{-}2)$$

Therefore,

Initial pore volume in which hydrocarbons are contained = gas-zone pore volume + oil-zone pore volume

$$V_r\phi = \frac{GB_{gi}}{5.61}\frac{1}{1 - S_{wig}} + \frac{NB_{ti}}{1 - S_{gi} - S_{wio}} = \left(\frac{GB_{gi}}{5.61} + \frac{GB_{gi}S_{wig}}{5.61(1 - S_{wig})}\right)$$

$$+ \left(NB_{ti} + \frac{NB_{ti}S_{gi}}{1 - S_{gi} - S_{wio}} + \frac{NB_{ti}S_{wio}}{1 - S_{gi} - S_{wio}}\right) \qquad (8\text{-}3)$$

At some time t after discovery, the reservoir pressure will decline to a pressure P as the result of the fluid productions, injections, and encroachments. It will be assumed that at this time, N_p volumes of oil, G_p volumes of gas, and W_p volumes of water have been produced. The produced gas will be distributed as G_{pc} volumes of gas-cap gas and G_{ps} volumes of solution gas. Furthermore, it will be assumed that G_i volumes of gas and W_i volumes of water have been injected to retard the pressure decline and that W_e volumes of water have encroached into the oil zone.

The pore space available for the fluids remaining is $(V_r\phi)'$, and the volumes occupied by the fluids when allocated to zones are as follows:

Pore volume occupied by fluids in initial gas zone

$$= \begin{matrix}\text{remaining initial} \\ \text{free gas volume}\end{matrix} + \begin{matrix}\text{volume of} \\ \text{injected gas}\end{matrix}$$

$$= (G - G_{pc})\frac{B_{gc}}{5.61} + \frac{G_i B_g'}{5.61}$$

$$+ \quad \text{volume change of connate water}$$

$$+ \frac{GB_{gi}}{5.61}\frac{S_{wig}}{1 - S_{wig}}\frac{B_w}{B_{twi}} + \frac{GB_{gi}}{5.61}\frac{S_{wig}}{1 - S_{wig}}\left(\frac{R_{swi} - R_{sw}}{B_{twi}}\right)\frac{B_{gc}}{5.61} \qquad (8\text{-}4)$$

Pore space occupied by fluids in initial oil zone

$$= \begin{matrix}\text{remaining} \\ \text{oil volume}\end{matrix} + \qquad\qquad \text{volume of remaining free}$$

$$= [(N - N_p)B_o] + \left[\frac{S_{gi}NB_{ti}}{1 - S_{wio} - S_{gi}}\frac{5.61}{B_{gi}} + NR_{si} - (N - N_p)R_s - G_{ps}\right.$$

$$\text{gas and gas evolved from solution}$$

$$\left. + \frac{S_{wio}NB_{ti}}{1 - S_{wio} - S_{gi}}\frac{1}{B_{twi}}(R_{swi} - R_{sw})\right]\frac{B_g}{5.61}$$

$$+ \quad \begin{array}{c}\text{volume of}\\\text{initial water}\end{array} \quad + \quad \begin{array}{c}\text{water from}\\\text{contiguous}\\\text{water zone}\end{array} \quad + \quad \begin{array}{c}\text{injected}\\\text{water}\end{array}$$

$$+ \frac{S_{wio}NB_{ti}}{1 - S_{wio} - S_{gi}} \frac{B_w}{B_{twi}} + (W_e - W_p)B_w + W_iB_w \qquad (8\text{-}5)$$

The expression for the pore volume occupied by fluids in the original oil zone assumes that the water produced was the result of water movement from the adjoining water zone. For this reason no allowance is made for gas evolution or free gas production from the encroached or produced water.

The sum of the pore volumes occupied by the fluids remaining at pressure P and those added between pressures P_i and P must equal the volume at pressure P of the initial pore volume.

Remaining pore volume

$$(V_r\phi)' = (G - G_{pc})\overbrace{\left\{\frac{B_{gc}}{5.61} + \frac{G_iB_g'}{5.61} + \frac{GB_{gi}}{5.61}\frac{S_{wig}}{1 - S_{wig}}\frac{B_w}{B_{twi}} + \frac{GB_{gi}}{5.61}\frac{S_{wig}}{1 - S_{wig}}\frac{R_{swi} - R_{sw}}{B_{twi}}\frac{B_{gc}}{5.61}\right)}^{\text{pore space occupied by fluids of initial gas zone}}$$

$$+ \overbrace{\left\{(N - N_p)B_o + \left[\frac{S_{gi}NB_{ti}}{1 - S_{wio} - S_{gi}}\frac{5.61}{B_{gi}} + NR_{si}\right.\right.}^{\substack{\text{pore space occupied by fluids}\\\text{of initial oil zone}}}$$

$$+ \frac{S_{wip}NB_{ti}}{1 - S_{wio} - S_{gi}}\frac{R_{swi} - R_{sw}}{B_{twi}} - (N - N_p)R_s - G_{ps}\bigg]\frac{B_g}{5.61}$$

$$+ \frac{S_{wio}NB_{ti}}{1 - S_{wio} - S_{gi}}\frac{B_w}{B_{twi}} + (W_e - W_p)B_w + W_iB_w\bigg\} \qquad (8\text{-}6)$$

Subtracting Eq. (8-6) from Eq. (8-3) yields the change in initial pore volume. If, by definition,

$$V_r\phi - (V_r\phi)' \overset{\Delta}{=} \Delta(V_r\phi) \qquad (8\text{-}7)$$

and

$$c_f \overset{\Delta}{=} \frac{\Delta(V_r\phi)}{(V_r\phi)(P_i - P)} \qquad (8\text{-}8)$$

then the change in pore volume can be expressed in terms of the original pore volume or in terms of the oil and gas volumes initially in place. Hence

$$V_r\phi - (V_r\phi)' = c_f(V_r\phi)(P_i - P) = c_f(P_i - P)\left(\frac{NB_{ti}}{1 - S_{wio} - S_{gi}} + \frac{GB_{gi}}{5.61}\frac{1}{S_{wig}}\right) \qquad (8\text{-}9)$$

Thus from the difference of Eqs. (8-3) and (8-6)

$$c_f(P_i - P)\left(\frac{NB_{ti}}{1 - S_{wio} - S_{gi}} + \frac{GB_{gi}}{5.61}\frac{1}{S_{wig}}\right) = \left(\frac{GB_{gi}}{5.61} + \frac{GB_{gi}}{5.61}\frac{S_{wig}}{1 - S_{wig}}\right)$$

$$- \left[(G - G_{pc})\frac{B_{gc}}{5.61} + \frac{G_iB_g'}{5.61} + \frac{GB_{gi}}{5.61}\frac{S_{wig}}{1 - S_{wig}}\frac{B_w}{B_{twi}} + \frac{GB_{gi}}{5.61}\frac{S_{wig}}{1 - S_{wig}}\frac{R_{swi} - R_{sw}}{B_{twi}}\frac{B_{gc}}{5.61}\right]$$

$$+ \left(NB_{ti} + \frac{NB_{ti}S_{gi}}{1 - S_{wio} - S_{gi}} + \frac{NB_{ti}S_{wio}}{1 - S_{gi} - S_{wio}}\right) - \left\{(N - N_p)B_o\right.$$

$$+ \left[\frac{S_{gi}NB_{ti}}{1 - S_{wio} - S_{gi}}\frac{5.61}{B_{gi}} + NR_{si} + \frac{S_{wio}NB_{ti}}{1 - S_{wio} - S_{gi}}\frac{R_{swi} - R_{sw}}{B_{twi}} - (N - N_p)R_s - G_{ps}\right]\frac{B_g}{5.61}$$

$$+ \frac{S_{wio}NB_{ti}}{1 - S_{gi} - S_{wio}}\frac{B_w}{B_{twi}} + (W_e - W_p)B_w + W_iB_w\bigg\} \qquad (8\text{-}10)$$

The terms in Eq. (8-10) can be collected to solve for oil in place N or water encroachment W_e. The equation also can be written utilizing certain notations which represent a collection of terms such as

$$R_{cs} = \frac{G_{ps}}{N_p} \quad \text{or} \quad R_c = \frac{G_{ps} + G_{pc}}{N_p}$$

$$B_t = B_o + (R_{si} - R_s)\frac{B_g}{5.61} \quad \text{and} \quad B_{twi} = B_{wi} + (R_{swi} - R_{sw})\frac{B_g}{5.61} \qquad (8\text{-}11)$$

Collecting terms, substituting Eq. (8-11), and assuming that all free gas in the reservoir at the time of discovery is in the gas zone $S_{gi} = 0$, the general material balance reduces to

$$N(B_t - B_{ti}) + \frac{NB_{ti}S_{wio}}{1 - S_{wio}}\left(\frac{B_{tw}}{B_{twi}} - 1\right) + \frac{NB_{ti}}{1 - S_{wio}}c_f(P_i - P) + \frac{G}{5.61}(B_{gc} - B_{gi})$$

$$+ \frac{GB_{gi}}{5.61}\frac{S_{wig}}{1 - S_{wig}}\left(\frac{B_{tw}}{B_{twi}} - 1\right) + \frac{GB_{gi}}{5.61(1 - S_{wig})}c_f(P_i - P)$$

$$= N_p\left[B_o + (R_{cs} - R_s)\frac{B_g}{5.61}\right] + \frac{G_{pc}B_{gc}}{5.61} - \frac{G_iB_g'}{5.61} - (W_e - W_p)B_w - W_iB_w \qquad (8\text{-}12)$$

Equation (8-12) will hereafter be referred to as the general material-balance equation.

Another approach to the derivation of the material-balance equation can be made through consideration of the following relationship expressed at reservoir conditions.

Change in the volume of all reservoir fluids

         + volume of extraneous fluids entering the reservoir = voidages

This approach can be shown expeditiously as follows, using the information shown in Eq. (8-11) and the relationship between gas-zone gas volume and oil-zone oil volume m:

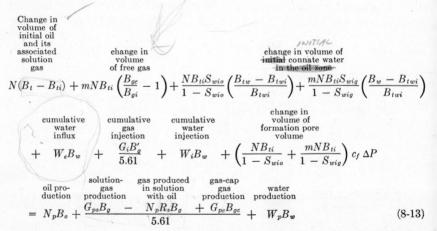

$$N(B_t - B_{ti}) + mNB_{ti}\left(\frac{B_{gc}}{B_{gi}} - 1\right) + \frac{NB_{ti}S_{wio}}{1 - S_{wio}}\left(\frac{B_{tw} - B_{twi}}{B_{twi}}\right) + \frac{mNB_{ti}S_{wig}}{1 - S_{wig}}\left(\frac{B_w - B_{twi}}{B_{twi}}\right)$$

$$+ W_eB_w + \frac{G_iB_g'}{5.61} + W_iB_w + \left(\frac{NB_{ti}}{1 - S_{wio}} + \frac{mNB_{ti}}{1 - S_{wig}}\right)c_f \Delta P$$

$$= N_pB_o + \frac{G_{ps}B_g - N_pR_sB_g + G_{pc}B_{gc}}{5.61} + W_pB_w \qquad (8\text{-}13)$$

The material-balance equation is employed to estimate the volume of oil initially present in a hydrocarbon reservoir when appropriate geologic, production, and laboratory data are available. More specifically, adequate geologic data must be available to permit estimation of the relative size of the gas cap and the oil zone. Field production data including cumulative oil, water, and gas production and reservoir pressure as functions of time are also necessary. Appropriate laboratory data or empirical relationships for the physical properties of oil, water, and gas are also needed to make accurate estimates. The physical-property data needed are primarily volume factors and gas solubilities.

In addition to these data, although it is not necessary, it is highly desirable to know the type of reservoir mechanism which is operative in order to expedite estimation of the volume of the initial hydrocarbons in the reservoir. If the type of reservoir mechanism is not known, lengthy and tedious trial-and-error calculations are necessary. A study of field data, including geologic and well completion information, workover reports, pressure surveys, production data, and decline curve analyses, may give an indication of the type of reservoir mechanism.

The general approach to the estimation of the initial hydrocarbon volume of the reservoir by the material-balance method is to assemble the necessary data and decide what is the probable type of reservoir mechanism.

The geologic, field, and laboratory data are substituted in the equation, and the value of the initial oil volume N is computed for each pressure and time observation. If the data and geologic calculations are accurate and the reservoir mechanism is identified properly, the value of N calculated for each observation will remain approximately constant. In the event that the initial oil volume N is observed to change in one direction with increased cumulative withdrawals from the reservoir, it is probable that the reservoir mechanism has been identified incorrectly or production, pressures, and/or laboratory data are incorrect. Ordinarily, it is a relatively simple matter, although time-consuming, to assume another type of reservoir mechanism and repeat the calculations for N. If the newly assumed mechanism does not give constant values of N, the procedure can be repeated or the data reevaluated.

It should be noted that the material-balance equation does not take into consideration the rate of production of the various fluids. Furthermore, it is obvious that the equation contains three quantities which may not be measured directly: (1) initial oil in place N, (2) initial gas in place G, and (3) the cumulative water influx W_e. As the material-balance equation has a low power of resolution, it is necessary to define at least two of the aforementioned quantities independent of the material-balance equation.

Many times the early pressure history of an oil reservoir is meager or not known. In such a case the material-balance equation will yield very erratic results. By writing the material-balance equation at several time intervals and then obtaining difference equations, it is possible to eliminate as much of the early pressure history as desired. The procedure for deriving a series of difference equations is presented here for a reservoir system with no gas cap, zero initial gas saturation, and no water or gas injection:

$$N(B_{ti} - B_{t1}) = \frac{G}{5.61}(B_{g1} - B_{gi}) - N_{p1}\left[B_{o1} + \frac{B_{g1}}{5.61}(R_{c1} - R_{s1})\right]$$
$$+ (W_{e1} - W_{p1})B_{w1}$$

$$N(B_{ti} - B_{t2}) = \frac{G}{5.61}(B_{g2} - B_{gi}) - N_{p2}\left[B_{o2} + \frac{B_{g2}}{5.61}(R_{c2} - R_{s2})\right]$$
$$+ (W_{e2} - W_{p2})B_{w2}$$

If the equation for *time* 1 is subtracted from the equation for *time* 2, a difference expression is obtained which includes N, W_e, and G but does not include any terms dependent on the initial pressure.

$$N(B_{t1} - B_{t2}) = \frac{G}{5.61}(B_{g2} - B_{g1}) - N_{p2}\left[B_{o2} + \frac{B_{g2}}{5.61}(R_{c2} - R_{s2})\right]$$
$$+ N_{p1}\left[B_{o1} + \frac{B_{g1}}{5.61}(R_{c1} - R_{s1})\right] + (W_{e2} - W_{p2})B_{w2} - (W_{e1} - W_{p1})B_{w1}$$

$$(8\text{-}14)$$

$$N(B_{t2} - B_{t3}) = \frac{G}{5.61}(B_{g3} - B_{g2}) - N_{p3}\left[B_{o3} + \frac{B_{g3}}{5.61}(R_{c3} - R_{s3})\right]$$
$$+ N_{p2}\left[B_{o2} + \frac{B_{g2}}{5.61}(R_{c2} - R_{s2})\right] + (W_{e3} - W_{p3})B_{w3} - (W_{e2} - W_{p2})B_{w2}$$

$$(8\text{-}15)$$

These equations can be continued for as many time intervals as desired to evaluate the problem completely.

Solution-gas Drive

Fortunately, many reservoirs are not nearly so complex as the one to which Eq. (8-12) or (8-13) is applicable in its entirety. The simplest type of reservoir mechanism is the solution (dissolved) gas drive. Equation (8-13) can be reduced to this form assuming that there is no initial gas cap ($G = 0$) or the production from the gas cap G_{pc} equals the gas-cap expansion $G(B_{gc} - B_{gi})$ and there is no water encroachment ($W_e = 0$) or injection ($W_i = 0$). To simplify the equation further let it be assumed that the change in pore and water volumes is so small compared with the oil and solution-gas expansibility that they can be neglected.

Further assume that the following conditions are applicable:

$B_g = B_{gc} = B'_g$ $\qquad\qquad$ $B_{tw} - B_{twi}$ is treated as zero

$S_{wi} = S_{wio} = S_{wig}$ $\qquad\qquad$ $W_e = 0$

$R_c = \dfrac{G_p}{N_p}$ $\qquad\qquad$ $W_p = 0$

$S_{gi} = 0$ $\qquad\qquad$ $W_i = 0$

c_f is treated as zero $\qquad\qquad$ $G_i = 0$

$G = 0 \qquad m = 0$

so $\qquad\qquad N(B_t - B_{ti}) = N_pB_o + \dfrac{N_pR_{cs}B_g}{5.61} - \dfrac{N_pR_sB_g}{5.61}$

hence $\qquad\qquad N = \dfrac{N_p[B_o + (R_{cs} - R_s)(B_g/5.61)]}{B_t - B_{ti}}$ $\qquad\qquad$ (8-16)

since $\qquad\qquad B_t = B_o + (R_{si} - R_s)\dfrac{B_g}{5.61}$

Eq. (8-16) is frequently expressed as follows:

$$N = \dfrac{N_p[B_t + (R_{cs} - R_{si})(B_g/5.61)]}{B_t - B_{ti}} \qquad\qquad (8\text{-}17)$$

Solution-gas–Gas-cap Drive

A somewhat more common case is the one in which there is a combination solution-gas–gas-cap drive mechanism. Equation (8-12) or (8-13) is simplified for a solution-gas–gas-cap drive reservoir by assuming that water encroachment W_e and injection W_i are zero, the change in connate water and pore volume is such a small fraction of the total volume change that it can be treated as zero, and further the fluid properties of the gas cap and solution gas are the same. Symbolically the proceeding assumptions are

$B_g = B_{gc}$ $\qquad\qquad$ $W_e = 0$

$S_{wi} = S_{wio} = S_{wig}$ $\qquad\qquad$ $W_i = 0$

$S_{gi} = 0$ $\qquad\qquad$ $W_p = 0$

c_f is treated as zero $\qquad\qquad$ $G_i = 0$

$B_{tw} - B_{twi}$ is treated as zero

Substituting the above assumptions in Eq. (8-13) yields

$$N(B_t - B_{ti}) + mNB_{ti}\left(\dfrac{B_g}{B_{gi}} - 1\right) = N_pB_o + \dfrac{N_pR_cB_g}{5.61} - \dfrac{N_pR_sB_g}{5.61}$$

then $\qquad\qquad N = \dfrac{N_p[B_o + (R_c - R_s)(B_g/5.61)]}{B_t - B_{ti}[1 - m(B_g - B_{gi})/B_{gi}]}$ $\qquad\qquad$ (8-18)

Simple Solution-gas–Gas-cap–Water-drive Reservoirs

Often petroleum reservoirs are found which have a combination solution-gas–gas-cap–water drive. Equation (8-13) simplifies to Eq. (8-19) whenever the conditions enumerated for the solution-gas–gas-cap drive prevail with the exception that the water influx W_e is finite. The reduced form of Eq. (8-13)

$$N(B_t - B_{ti}) + mNB_{ti}\left(\frac{B_g}{B_{gi}} - 1\right) + W_eB_w = N_pB_o + \frac{N_pR_cB_g}{5.61} - \frac{N_pR_sB_g}{5.61}$$

when solved for initial oil in place gives

$$N = \frac{N_p[B_o + (R_c - R_s)(B_g/5.61)] - W_eB_w}{B_t - B_{ti}[1 - m(B_g - B_{gi})/B_{gi}]} \qquad (8\text{-}19)$$

If there is a good measure of initial oil in place by volumetric methods, the previous equation is frequently rearranged and employed to estimate the water influx into the reservoir as follows:

$$W_eB_w = N_p\left[B_o + (R_c - R_s)\frac{B_g}{5.61}\right]$$
$$- N\left\{B_t - B_{ti}\left[1 - m\left(\frac{B_g - B_{gi}}{B_{gi}}\right)\right]\right\} \qquad (8\text{-}20)$$

and if there has been some water production from the reservoir, the net water influx can be calculated as follows:

$$(W_e - W_p)B_w = N_p\left[B_o + (R_c - R_s)\frac{B_g}{5.61}\right]$$
$$- N\left\{B_t - B_{ti}\left[1 - m\left(\frac{B_g - B_{gi}}{B_{gi}}\right)\right]\right\}$$

Solution-gas–Gas-cap–Water Drive with Fluid Injection

In view of the emphasis on the conservation of reservoir energy, a constantly increasing number of hydrocarbon reservoirs are being pressure-maintained through injection of various fluids including gas and water. Hence, it is quite appropriate to have a material-balance equation for a combination solution-gas–gas-cap–water-drive type of reservoir in which provision is made to account for the injected fluids.

The expansion of the hydrocarbons in the gas and oil zone and the expansion of water in the water zone will be extremely large compared with the expansion of the connate water and shrinkage of the pore space, so that the terms involving changes in volume of the rock and connate water can be treated as zero in the general material-balance equation. If it is further assumed that the connate water saturation is the same in the oil and gas zones and that initially no free gas existed in the oil zone, Eq. (8-12) reduces to

$$N(B_t - B_{ti}) + \frac{G}{5.61}(B_{gc} - B_{gi}) = N_p\left[B_o + (R_{cs} - R_s)\frac{B_g}{5.61}\right]$$

$$+ \frac{G_{pc}B_{gc}}{5.61} - \frac{G_iB_g'}{5.61} - (W_e - W_p)B_w - W_iB_w \quad (8\text{-}21)$$

where $S_{wi} = S_{wio} = S_{wig}$
$\quad S_{gi} = 0$
$\quad c_f$ is treated as zero
$\quad B_{tw} - B_{twi}$ is treated as zero

Equation (8-21) can be rearranged to solve for the oil in place N or the cumulative water influx W_e. The oil in place N is obtained by solving Eq. (8-22), and the water influx is obtained from Eq. (8-23).

$$N = \frac{N_p\left[B_o + (R_{cs} - R_s)\frac{B_g}{5.61}\right] + \frac{G_{pc}B_{gc}}{5.61} - \frac{G}{5.61}(B_{gc} - B_{gi}) - \frac{G_iB_g'}{5.61} - (W_e - W_p)B_w - W_iB_w}{B_t - B_{ti}}$$

$$(8\text{-}22)$$

$$W_e = \left\{N_p\left[B_o + (R_{cs} - R_s)\frac{B_g}{5.61}\right] + \frac{G_{pc}B_{gc}}{5.61} - N(B_t - B_{ti})\right.$$

$$\left. - \frac{G}{5.61}(B_{gc} - B_{gi}) - \frac{G_iB_g'}{5.61}\right\}\frac{1}{B_w} - W_i + W_p \quad (8\text{-}23)$$

Slightly Compressible Hydrocarbon Reservoirs

The general material-balance equation can also be used to estimate the initial oil in place in a reservoir for the conditions in which the reservoir pressure is well above the saturation pressure. When this condition prevails, the oil is undersaturated with gas and it is reasonable to assume that the connate water is also undersaturated. If the oil is undersaturated, there will be no gas cap ($G = 0$), no free gas in the oil zone ($S_{gi} = 0$), and the cumulative produced-gas–oil ratio R_c is equal to the solution-gas–oil ratio R_s. Substituting the above quantities in Eq. (8-13) reduces it to

$$N(B_t - B_{ti}) + \frac{NB_{ti}}{1 - S_{wi}}\left[\frac{S_{wi}}{B_{twi}}(B_{tw} - B_{twi}) + c_f(P_i - P)\right]$$

$$= N_pB_o - (W_e - W_p)B_w \quad (8\text{-}24)$$

All the fluids are above their respective bubble-point pressures ($B_t = B_o$ and $B_{tw} = B_w$), and hence each can be defined in terms of a compressibility factor for a slightly compressible fluid. The volume factors are expressed in terms of the bubble-point volume by Eq. (8-25).

$$B_o = B_{ob}[1 + c_o(P - P_b)]$$

$$B_w = B_{wb}[1 + c_w(P - P_b)] \quad (8\text{-}25)$$

where
$$c_o \overset{\Delta}{=} \frac{B_{oi} - B_{ob}}{B_{ob}(P_i - P_b)}$$

and
$$c_w \overset{\Delta}{=} \frac{B_{wi} - B_{wb}}{B_{wb}(P_i - P_b)}$$

at reservoir temperature.

Substituting the equations for the volume factors into Eq. (8-24), an expression for the oil in place is obtained in terms of oil produced; the rock, oil, and water compressibility factors; the cumulative net water influx; and the total pressure drop.

$$NB_{ob}c_o(P - P_i) + \frac{NB_{oi}}{1 - S_{wi}}\left[\frac{S_{wi}B_{wb}c_w(P - P_i)}{B_{wi}} + c_f(P_i - P)\right]$$
$$= N_pB_{ob}[1 + c_o(P - P_b)] - (W_e - W_p)B_{wb}[1 + c_w(P - P_b)] \quad (8\text{-}26)$$

The only variables in Eq. (8-26) are pressure P, oil production N_p, and net water influx $W_e - W_p$, since N, B_{ob}, B_{wb}, c_o, c_f, c_w, P_b, P_i, and S_{wi} are all constants for a particular reservoir.

Rearranging the terms of Eq. (8-26) to solve for oil in place gives the following:

$$N = \frac{N_pB_{ob}[1 + c_o(P - P_b)] - (W_e - W_p)B_{wb}[1 + c_w(P - P_b)]}{B_{oi}(P_i - P)\left[\frac{1}{1 - S_{wi}}\left(c_f - \frac{S_{wi}}{B_{wi}}B_{wb}c_w\right) - c_o\frac{B_{ob}}{B_{oi}}\right]} \quad (8\text{-}27)$$

By collecting together all constant terms such as

$$B_{oi}\left[\frac{1}{1 - S_{wi}}(c_f - S_{wi}c_w') - c_o\frac{B_{ob}}{B_{oi}}\right] = A$$

where $c_w' = \dfrac{c_wB_{wb}}{B_{wi}}$

$$B_{ob}(1 - c_oP_b) = D$$
$$B_{ob}c_o = E$$
$$B_{wb}[1 - c_wP_b] = F$$

Eq. (8-27) reduces to

$$N = \frac{N_p(D + EP) - (W_e - W_p)(F + B_{wb}c_wP)}{(P_i - P)A} \quad (8\text{-}28)$$

which is an expression in terms of the water influx, cumulative oil and water production, and the reservoir pressure.

If the reservoir has no water influx and no water production, Eq. (8-28) reduces to

$$N = \frac{N_p(D + EP)}{(P_i - P)A} \quad (8\text{-}29)$$

Gas Reservoir

The petroleum engineer must also estimate the hydrocarbon volume in place when only free gas exists in the reservoir. The general material-balance equation can be reduced to a form which will permit the calculation of the initial gas in place. Since there is no liquid petroleum concerned in this evaluation, the oil in place N is zero, the solution-gas production is zero, and the gas-cap production G_{pc} is the total gas production G_p. For a small pressure difference the change in gas volume is so large in comparison with the changes in rock and connate-water volumes that the change in water and rock volume can be treated as zero. For simplification, it is assumed that the water influx, water production, and water and gas injection are all zero. Inserting the above assumptions into Eq. (8-12), it becomes

$$\frac{G}{5.61}(B_{gc} - B_{gi}) = \frac{G_{pc}B_{gc}}{5.61}$$

Rearranging terms gives

$$G = \frac{G_{pc}B_{gc}}{B_{gc} - B_{gi}} = \frac{G_{pc}}{1 - (B_{gi}/B_{gc})} \tag{8-30}$$

From Chap. 4 the gas formation volume factor B_g will be recalled as,

$$B_{gc} = \frac{0.00504TZ}{P}$$

and since it is assumed that the reservoir remains at constant temperature T, Eq. (8-30) becomes

$$G = G_{pc}\left[\frac{1}{1 - (P/Z)(Z_i/P_i)}\right] \tag{8-31}$$

which can be solved with limited production and laboratory data. In the event that there is fluid influx into the reservoir or fluid-condensation phenomena exhibited during the production history, these factors must be taken into consideration and the preceding equation modified appropriately.

Equation (8-31) can be rearranged to express the ratio of reservoir pressure and its compressibility factor as a linear function of the cumulative gas production.

$$\frac{P}{Z} = \left(-\frac{P_i}{GZ_i}\right)G_{pc} + \frac{P_i}{Z_i} \tag{8-32}$$

Equation (8-32) indicates that the gas reserves can be estimated from a graphical plot of P/Z and the cumulative gas production G_{pc}. Extrapolation of the resulting curve to $P = 0$ represents the initial gas in place. whereas extrapolation to the appropriate P/Z ratio corresponding to abandonment indicates the gas reserves.

Comparison of Drives

The general material-balance equation can be arranged to permit the calculation of the fractional part of the total expansion that can be attributed to each type of expansion or energy mechanism. Rearranging Eq. (8-12) so that it equals unity gives

$$\left\{ N(B_t - B_{ti}) + \frac{G}{5.61}(B_{gc} - B_{gi}) + (W_e - W_p)B_w \right.$$

$$+ \left(\frac{NB_{ti}S_{wio}}{1 - S_{wio}} + \frac{GB_{gi}}{5.61}\frac{S_{wig}}{1 - S_{wig}} \right)\left(\frac{B_{tw}}{B_{twi}} - 1 \right)$$

$$+ \left[\frac{GB_{gi}}{(1 - S_{wig})5.61} + \frac{NB_{ti}}{1 - S_{wio}} \right] c_f(P_i - P) + \frac{G_iB_g'}{5.61} + W_iB_w \right\}$$

$$\div \left\{ N_p\left[B_o + (R_{cs} - R_s)\frac{B_g}{5.61} \right] + \frac{G_{pc}B_{gc}}{5.61} \right\} = 1$$

a relation which is nearly identical with that proposed by Pirson[7] to illustrate the fractional energy attributed to each phase. He showed that

$$\frac{N(B_t - B_{ti})}{N_p[B_e + (R_{cs} - R_s)(B_g/5.61)] + (G_{pc}B_{gc}/5.61)}$$

represented the fraction of the total energy derived from the expansion of the oil and its dissolved gas.

$$\frac{(G/5.61)(B_{gc} - B_{gi})}{N_p[B_o + (R_{cs} - R_s)(B_g/5.61)] + (G_{pc}B_{gc}/5.61)}$$

is the fraction of the total energy derived from the expansion of the gas in the gas cap.

$$\frac{(W_e - W_p)B_w}{N_p[B_o + (R_{cs} - R_s)(B_g/5.61)] + (G_{pc}B_{gc}/5.61)}$$

is the fraction of the total energy obtained from the net water influx, and

$$\frac{W_iB_w + (G_iB_g/5.61)}{N_pB_o + (R_{cs} - R_s)(B_g/5.61) + (G_{pc}B_{gc}/5.61)}$$

is the fraction of the total energy derived from any fluids which might be injected into the reservoir.

$$\frac{\left(\frac{NB_{ti}S_{wio}}{1 - S_{wio}} + \frac{GB_{gi}}{5.61}\frac{S_{wig}}{1 - S_{wig}} \right)\left(\frac{B_{tw}}{B_{twi}} - 1 \right) + \left[\frac{NB_{ti}}{1 - S_{wio}} + \frac{GB_{gi}}{5.61(1 - S_{wig})} \right] c_f(P_i - P)}{N_p[B_o + (R_{cs} - R_s)(B_g/5.61)] + (G_{pc}B_{gc}/5.61)}$$

is the fraction of the total energy derived from the expansion of the connate water and the rock in the gas cap and oil zone.

DATA FOR MATERIAL BALANCE

The variables in the previously derived material-balance equation can be grouped according to source or method of measurement. There are essentially four groups: (1) fluid-production data, (2) reservoir pressure and temperature data, (3) fluid-analysis data, (4) core analysis and laboratory rock data. Each group consists of several variables; some are functions of terms in other groups or variables with time.

Fluid-production Data

The methods of measuring and reducing fluid production (oil, water, and gas) to field average values were discussed in Chap. 7. All production data should be recorded with respect to the same time period. If possible, gas-cap and solution-gas production records should be maintained separately.

Gas and oil gravity measurements should be recorded in conjunction with the fluid volume data. Some reservoirs require a more detailed analysis and the material balance is solved for volumetric segments. The produced fluid gravities will aid in the selection of the volumetric segments and also in the averaging of fluid properties.

Reservoir Temperatures

Measurements of temperatures in well bores are made in connection with a number of tests and operations conducted on the well. Some of these temperature observations, such as those made in connection with static pressure tests and well logging, are recorded by maximum-reading thermometers which record the maximum temperature encountered. This maximum temperature is usually that temperature existing at the greatest depth to which the instruments were run. Other temperature observations are made with recording subsurface temperature gauges yielding a continuous record which can be correlated with depth below sea level. Temperature logs or surveys, as discussed earlier, can be used to locate the top of cement behind pipe or to measure the formation temperatures with depth.

The petroleum-reservoir engineer requires an accurate determination of reservoir temperature and frequently the temperature gradient within a formation or group of formations.

Temperatures determined by means of maximum-reading thermometers in general are not of sufficient accuracy for reservoir-engineering purposes. Temperature logs run to detect the top of cement are obviously of little value, as the heat released by the setting reaction of the cement elevates the well-bore temperature. Even if the temperature log is run in open hole prior to setting and cementing casing, errors may occur unless sufficient time is allowed for the mud to attain temperature equilibrium with the formation.

After completion of a well, accurate formation temperature surveys can be made with recording subsurface temperature gauges provided that the

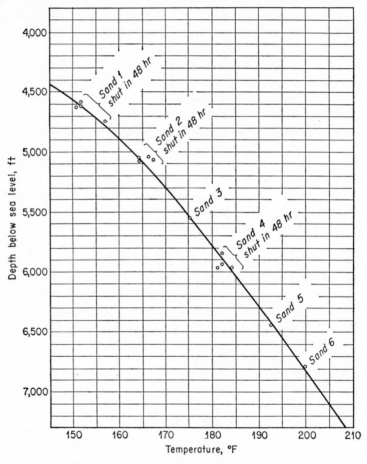

Fig. 8-2. Average temperature gradient, field C.

fluid in the well bore is static and is in thermal equilibrium with the surrounding formations. To satisfy these requirements the well must be closed in for approximately 24 to 48 hr and must be free of casing leaks. Furthermore the recording instrument must be allowed to come to equilibrium with the surrounding well fluid. Equilibrium between the instrument and well fluid is attained more rapidly in a well bore containing liquids than in one containing gas.

Readings made by lowering the instrument to the depth of the completion interval in a well are generally considered to be the most accurate records. The results of a group of such observations are presented in Fig. 8-2. These observations were made in six different sands in a multisand field. It may be noted that the trend of temperature within a sand is similar to that within the total section.

The reservoir temperature is considered to be constant over the life of a reservoir, and all reservoir processes except that of *in situ* combustion

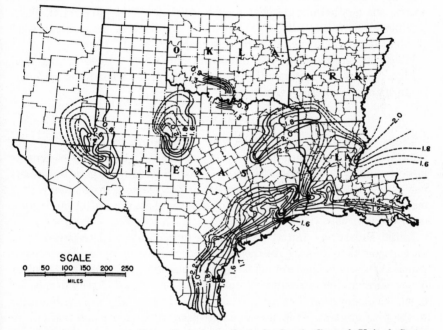

Fig. 8-3. Contour map of geothermal gradients in South Central United States. Contour values are in degrees per 100 ft. Mean surface temperature = 74°F. Estimated subsurface temperature (°F) = (depth in 100 ft) × [(geothermal gradient °F)/100 ft from map] + 74°F. (*After Earl A. Nichols.*[6])

(a recovery stimulation process) are considered to be isothermal processes.

Average reservoir temperatures in low-relief reservoirs are determined by reading from curves similar to Fig. 8-2, the temperature at the volumetric mid-point of the reservoir. In reservoirs of considerable relief the effect of temperature variations can be taken into account by assigning average temperatures to segments of the reservoir defined by depth intervals.

If measured temperature data are not available, regional geothermal temperature gradients can be used to calculate the reservoir temperature. In Fig. 8-3 are presented contours of the geothermal gradients observed in the South Central United States. The temperature can be calculated from the data by the following relation:

$$t = 74°F + g_T(h)$$

where t = formation temperature at the depth h, °F

  h = depth from surface, hundreds of feet

  g_T = geothermal gradient and is read from the contour map of Fig. 8-3, °F per 100 ft

If the reservoir under consideration is near salt intrusions such as salt domes, the geothermal gradient may be altered by the effect of the heat conductivity of the salt. Care must be exercised in using geothermal gradient data where such disturbances may occur.

Reservoir temperature is an important parameter in determining fluid volume factors. The temperature is thus an implicit parameter of material-balance and volumetric calculations.

Reservoir Pressures

Pressure measurements on wells were discussed briefly in Chap. 7. Like the reservoir temperature the reservoir pressure is an important parameter in determining fluid volume factors. Unlike reservoir temperature, reservoir pressure is a variable in most reservoir processes. A distinction is thus made between initial reservoir pressure and those attained after production from the reservoir.

Initial Pressure. Initial reservoir pressure must be determined from measurements made early in the development of a reservoir. The wells in which the observations are made must be closed in to eliminate the effect of pressure gradients around the well bore. The closed-in period must be of sufficient duration to eliminate transient effects of any production prior to the pressure test. It is desirable that accurate pressure observations be made prior to substantial fluid withdrawals from the reservoir.

The earliest pressure observations are usually made in connection with drill-stem tests. Drill-stem test pressure records may include a closed-in pressure observation made prior to the flow test. These observations are an excellent source of data on initial reservoir pressure. The closed-in pressure recorded after the flow test is frequently unreliable unless corrected for the disturbance created by the flow test.

The most reliable initial pressure records are obtained from pressure build-up tests on early wells which are produced until cleaned of completion fluids, then closed in for pressure testing.

Initial reservoir pressures can be verified and, in some instances, determined from correlation of the pressure and production history of the reservoir. For a gas reservoir, it has been shown [Eq. (8-32)] that a plot of P/Z is linear with cumulative gas production. In an oil reservoir the relation between reservoir pressure and cumulative oil production is approximately exponential. The relation between pressure and cumulative oil production for a field is presented in Fig. 8-4. Production-pressure data can be extrapolated to zero production for an estimate of initial reservoir pressure.

Initial subsurface fluid pressures are substantially equal to that of a head of water from the surface to the depth of the oil-water contact of the particular accumulation. Observed pressure gradients range from about 0.43 to nearly 1.0 psi per ft. In the Gulf Coast of Louisiana and Texas the normal gradient is 0.465 psi per ft. Pressure gradients in excess of 0.465 psi per ft are considered abnormal. In many areas, below a limiting

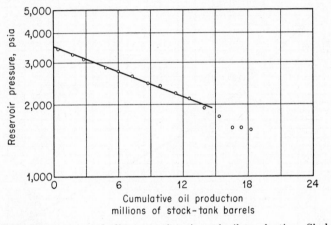

FIG. 8-4. Reservoir pressure decline as a function of oil production, Shuler Field, Ark. (*From Kaveler.*[4])

depth for the area, abnormal pressures become quite common. Two explanations of abnormal reservoir pressures are commonly presented.

If, for example, a reservoir having a closure of 1,000 ft contains gas, the pressure in the gas at the top of the structure will be about 400 psi greater than would be expected for that depth. This is a result of the gas and water being at the same pressure at the gas-water contact but having different densities. The excess pressure in the gas at the top of the structure can be estimated as follows:

$$\Delta P_x = \frac{h_c}{144} (\rho_w - \rho_g) \tag{8-33}$$

where ΔP_x = excess pressure in gas at top of structure
h_c = closure of the structure, ft
ρ_w = density of water, lb/cu ft
ρ_g = density of gas, lb/cu ft

A similar relation holds for oil or oil and gas reservoirs.

If a reservoir exists at great depth in a sequence of sands and shales, the reservoir fluids may support a part of the overburden load. The hydrostatic equivalent of the overburden load is about 1 psi per ft of depth. Thus, fluid pressures may approach 1 psi per ft, particularly if the reservoir is relatively small and is completely enclosed in a thick incompetent shale formation.

The initial reservoir pressures for a multisand field are plotted as a function of depth in Fig. 8-5. The curve represents the "normal" pressure variation for the area. An increase in gradient with depth is apparent.

Average Reservoir Pressures. During the producing life of a hydro-

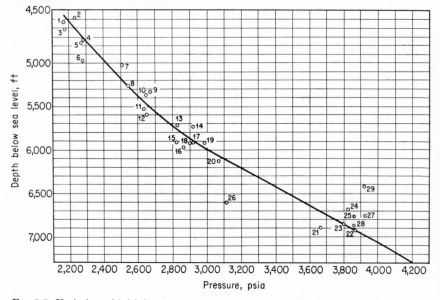

FIG. 8-5. Variation of initial reservoir pressure in field C with depth of measurement.

carbon reservoir the reservoir pressure normally declines. The reservoir pressure must be determined periodically as a function both of time and of production from the reservoir, as most engineering analyses of reservoir performance require a knowledge of the reservoir pressure history.

Fluid volume factors are functions of reservoir pressure. The principal use of reservoir pressure data is to define fluid volume factors for various calculations. Thus the reservoir pressure must be averaged in such a manner as to provide the appropriate fluid volume factors from correlations of fluid properties with pressure.

As discussed in Chap. 2, a producing well can be represented as a radial flow system. Thus, when the well is producing, a large pressure drop occurs in the immediate vicinity of the well bore. The flowing well-bore pressure is not representative of the pressure prevailing in the drainage area of the well. As the desired pressure observation is for determination of the volumetric behavior of the fluids, it is necessary to shut in the well and allow the well-bore pressure to build up to that prevailing in the drainage area. Pressure observations are normally made after the well has been shut in 24 to 48 hr. Such shut-in pressure tests were discussed in Chap. 7.

Pressure surveys in a reservoir are conducted periodically on a group of wells selected to provide adequate areal coverage of the field. The frequency of pressure-survey periods depends on the rate of pressure decline in the field. Semiannual or annual pressure surveys are sufficient if the rate of decline is of the order of a few pounds per square inch per month. The number of wells required depends on the area of the reservoir and the pressure distribution in the reservoir. Large areal pressure differences require a more dense measurement pattern in the pressure survey than more uniform pressure distributions. Pressure observation wells should be selected from the wells with the simplest mechanical equipment, preferably single completions.

In a large field, several days or weeks may be required to conduct a pressure survey. The observed well pressures should be corrected to a common survey date before averaging or for other treatment.

The well pressures are corrected (as in Chap. 7) to various datum depths depending on the use to be made of the pressure data. In reservoirs of low relief, the pressures corrected to the volumetric mid-point of the reservoir are used in determining fluid volume factors to be used in material-balance calculations. This is true of both oil and gas reservoirs. In reservoirs containing oil and an associated gas cap, the pressure data corrected to the gas-oil contact can be used for both the gas-cap area and the oil zone. A more accurate procedure is to correct the oil-zone pressures to the volumetric mid-point of the oil zone and to correct gas-cap pressures to the volumetric mid-point of the gas cap.

An arithmetic average pressure is usually determined from the pressure survey data. If the reservoir is uniform in thickness and the pressure variation in the reservoir is not large, the arithmetic average pressure is satisfactory. Most frequently the above conditions do not prevail, and therefore, other averaging techniques are required. The arithmetic average should always be determined to provide a simple check of the more involved procedures.

Pressure data from a single survey are posted to a map on which the limits of the reservoir have been drawn. For an oil reservoir the limits are defined by the zero contour lines of the oil isopach and any delineating faults. The type of boundary, gas-oil contact on the base of sand, oil-water contact on the top of sand, formation pinchout or fault should be indicated on the map. The pressures are contoured following the general rules of contouring but bearing in mind that only an injection well can be at a higher pressure than all the drainage area about it. Thus a well or group of wells is at a lower pressure than any contour completely enclosing them. Pressure continuity exists among the oil zone, the gas cap, and water zones. A pressure contour does not have to close within the oil zone, but it will close if mapped through a water zone or gas cap. Formation pinchouts, permeability barriers, and sealing faults constitute boundaries across which no flow can occur. For this reason pressure contours may be closed against such a boundary.

A pressure contour or isobaric map of the Shuler Field is shown in Fig. 8-6. The oil accumulation is bounded by an oil-water contact. The arith-

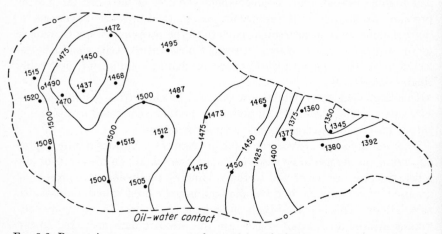

Fig. 8-6. Reservoir pressure survey to be used in calculating average reservoir pressure. (*From Kaveler.*[4])

metic average pressure is 1,461 psi. Two other techniques of averaging the pressure data are (1) areal weighting and (2) volumetric weighting.

In areal weighting, average reservoir pressure is determined by summing the product of pressure and area and dividing the sum by the total area. In mathematical notation,

$$\overline{P}_{AR} = \frac{\Sigma P_i A_i}{A_t} \qquad (8\text{-}34)$$

where $\overline{P}_{AR}$ is the areal weighted pressure, P_i is the pressure prevailing in the area A_i, and A_t is the total area.

In practice, the values of A_i and A_t are determined by planimetering. The area of the reservoir enclosed by the reservoir boundary is planimetered. The areas at a greater pressure than each contour are then successively measured. The average pressure can be determined by a formula similar to the trapezoidal rule for isopachs.

$$A_t \overline{P}_{AR} = A_{c1}P_{c1} + (A_t - A_{c1})\frac{P_{c1} + P_{min}}{2}$$

$$+ \frac{\Delta P}{2}(A_{c1} + 2A_{c2} + 2A_{c3} + \cdots + 2A_{cn-1} + A_{cn})$$

$$+ A_{cn}\frac{P_{max} - P_{cn}}{2} \tag{8-35}$$

where A_t and $\overline{P}_{AR}$ are as defined above; A_{c1} is the area of the reservoir existing at pressures greater than the lowest contour pressure P_{c1}; P_{min} is the minimum well pressure on the map and is less than P_{c1}; A_{c2} is the area existing at pressures greater than P_{c2}, the next higher pressure contour; A_{cn} is the area existing at pressures greater than P_{cn}, the highest contour pressure; P_{max} is the maximum well pressure on the map and is greater than P_{cn}; ΔP is the pressure contour interval. An example calculation of areal weighting is presented in Example 8-1.

Example 8-1. Areal Weighting of Reservoir Pressures (total oil-productive area, 1,890 planimeter units; $P_{min} = 1,345$ psi; $P_{max} = 1,520$ psi).

Area with pressures greater than, psi	Planimeter units
1,350	1,878
1,375	1,808
1,400	1,517
1,425	1,439
1,450	1,320
1,475	977
1,500	297

$$\overline{P}_{AR} = \frac{1}{1,890}\left\{ \begin{array}{l} 1,878(1,350) + (1,890 - 1,878)\frac{1,350 + 1,345}{2} \\ + \frac{25}{2}[1,878 + 2(1,808) + 2(1,517) \\ + 2(1,435) + 2(1,320) + 2(977) + 297] \\ + 297\frac{1,520 - 1,500}{2} \end{array} \right\}$$

$$= \frac{1}{1,890}(2,741,900) = 1,451 \text{ psi}$$

In volumetric weighting, the average reservoir pressure is determined by summing the product of pressure and volume and dividing the sum by the total volume.

In mathematical notation,

$$\overline{P}_{VL} = \frac{\Sigma P_i V_i}{V_t} \tag{8-36}$$

where $\overline{P}_{VL}$ is the volumetrically weighted pressure, P_i is pressure prevailing over the volume V_i, and V_t is the total volume. Volume weighting may be on the basis of sand volume, pore volume, or hydrocarbon volume depending on the accuracy of the data available.

Volumetric weighted pressures can be determined by superimposing the isobaric map on an isopach map and determining the sand volume between successive pressure contours. A somewhat simpler method involves preparing a grid overlay subdivided into squares of convenient size. The boundary of the reservoir is traced to the overlay from an isopach map of the same scale as the isobaric map. The average thickness of each square is estimated from the isopach and recorded on the square. If a square is only partially underlain by the reservoir, the fraction of the square overlying productive area is also estimated and recorded. The volume underlying a square is

$$V_i = f_i h_i A_k \qquad (8\text{-}37)$$

where f_i = fraction of square underlain by productive formation
h_i = average thickness of square
A_k = conversion constant, area of a square, acres

The total volume is the sum of the volume of the squares underlain by productive information:

$$V_t = \Sigma V_i \qquad .$$

(This scheme can be used to determine volumes in lieu of planimetering.) The grid is then superimposed on an isobaric map, and the average pressure P_i for each grid estimated. From these data and Eq. (8-36), the volumetrically weighted pressure is calculated. An example calculation is presented in Example 8-2 and Fig. 8-7. The conversion constant A_k was omitted, as the data were used only for determining volumetrically weighted pressures and A_k divides out.

Although the preceding discussions were in terms of lateral variations in pressure and volume, similar procedures can be used to determine volumetric weighted pressures where vertical variations in pressure and volume must be considered.

If the reservoir contains both a gas cap and an oil zone, the pressure data can be corrected to the gas-oil contact and both the oil-zone pressure data and the gas-cap pressure data contoured on the same map. As pressure continuity prevails, the oil-zone pressure data provides additional control on pressure contours in the gas cap. The pressure in the gas cap is evaluated in a manner similar to that previously described.

For evaluation of the expansion of the aquifer, the pressure at the original oil-water contact is required. The average pressure can be weighted by the length of perimeter at various pressure levels as follows:

$$\bar{P}_{o-w} = \frac{\Sigma P_i L_i}{L_t} + \frac{h_{o-w}}{144}\rho_o \qquad (8\text{-}38)$$

where $\bar{P}_{o-w}$ = weighted pressure along oil-water contact

P_i = pressure of a segment of perimeter of length L_i

L_t = total length of oil-water contact

h_{o-w} = distance oil-zone datum is above original oil-water contact, ft

ρ_o = density of the oil, lb/cu ft

This method of weighting pressures at the oil-water contact is particularly applicable in reservoirs having thin sands where the water is essentially

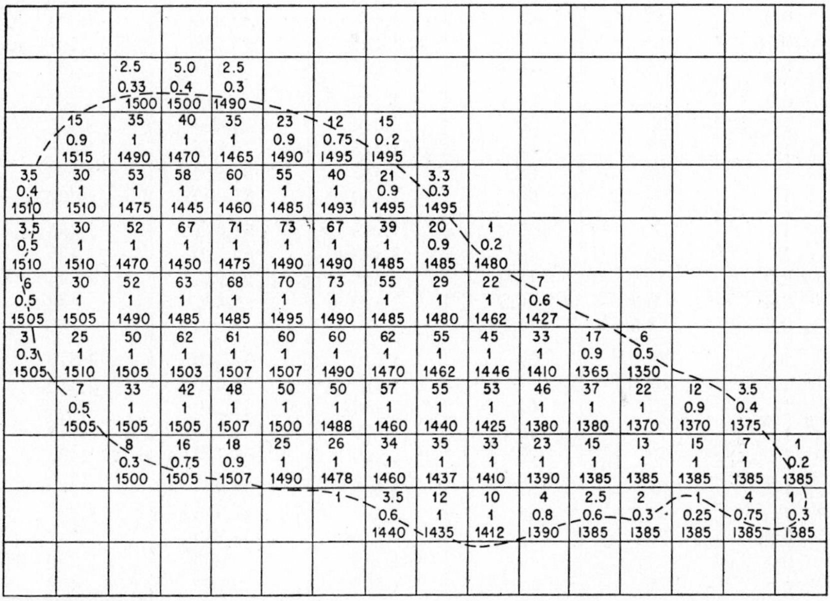

		2.5 / 0.33 / 1500	5.0 / 0.4 / 1500	2.5 / 0.3 / 1490										
	15 / 0.9 / 1515	35 / 1 / 1490	40 / 1 / 1470	35 / 1 / 1465	23 / 0.9 / 1490	12 / 0.75 / 1495	15 / 0.2 / 1495							
3.5 / 0.4 / 1510	30 / 1 / 1510	53 / 1 / 1475	58 / 1 / 1445	60 / 1 / 1460	55 / 1 / 1485	40 / 1 / 1493	21 / 0.9 / 1495	3.3 / 0.3 / 1495						
3.5 / 0.5 / 1510	30 / 1 / 1510	52 / 1 / 1470	67 / 1 / 1450	71 / 1 / 1475	73 / 1 / 1490	67 / 1 / 1490	39 / 1 / 1485	20 / 0.9 / 1485	1 / 0.2 / 1480					
6 / 0.5 / 1505	30 / 1 / 1505	52 / 1 / 1490	63 / 1 / 1485	68 / 1 / 1485	70 / 1 / 1495	73 / 1 / 1490	55 / 1 / 1485	29 / 1 / 1480	22 / 1 / 1462	7 / 0.6 / 1427				
3 / 0.3 / 1505	25 / 1 / 1510	50 / 1 / 1505	62 / 1 / 1503	61 / 1 / 1507	60 / 1 / 1507	60 / 1 / 1490	62 / 1 / 1470	55 / 1 / 1462	45 / 1 / 1446	33 / 1 / 1410	17 / 0.9 / 1365	6 / 0.5 / 1350		
	7 / 0.5 / 1505	33 / 1 / 1505	42 / 1 / 1505	48 / 1 / 1507	50 / 1 / 1500	50 / 1 / 1488	57 / 1 / 1460	55 / 1 / 1440	53 / 1 / 1425	46 / 1 / 1380	37 / 1 / 1380	22 / 1 / 1370	12 / 0.9 / 1370	3.5 / 0.4 / 1375
	8 / 0.3 / 1500	16 / 0.75 / 1505	18 / 0.9 / 1507	25 / 1 / 1490	26 / 1 / 1478	34 / 1 / 1460	35 / 1 / 1437	33 / 1 / 1410	23 / 1 / 1390	15 / 1 / 1385	13 / 1 / 1385	15 / 1 / 1385	7 / 1 / 1385	1 / 0.2 / 1385
					1	3.5 / 0.6 / 1440	12 / 1 / 1435	10 / 1 / 1412	4 / 0.8 / 1390	2.5 / 0.6 / 1385	2 / 0.3 / 1385	1 / 0.25 / 1385	4 / 0.75 / 1385	1 / 0.3 / 1385

FIG. 8-7. Overlay method of averaging pressures using square grid. Top figure in each group represents the sand thickness h in feet; middle figure the fraction f of square underlain by an oil-productive area; bottom figure estimated pressure P of square.

all edge water. If the reservoir is completely underlain by water, the average oil-zone pressure is simply corrected to the oil-water contact. Example 8-3 illustrates peripheral weighting of pressures to determine the average pressure at the oil-water contact.

The determination of appropriate average pressures is always of concern to the reservoir engineer. Every effort should be made to evaluate the data by the procedure or procedures most appropriate to the conditions which exist.

Example 8-2. Determination of Areal and Volumetrically Weighted Pressures by Overlay Method.

(1)	(2)	(3)	(4)	(5) Pfh (3) × (4)	(6) Pf (1) × (4)
f	h	fh	P		
0.33	2.5	0.83	1,500	1,245	495
0.4	5.0	2.0	1,500	3,000	600
0.3	2.5	0.75	1,490	1,118	447
0.9	15	13.5	1,515	20,452	1,354
1	35	35	1,490	52,150	1,490
1	40	40	1,470	58,800	1,470
1	35	35	1,465	51,275	1,465
0.9	23	20.7	1,490	30,843	1,341
0.75	12	9.0	1,495	13,455	1,121
0.2	1.5	0.30	1,495	488	299
0.4	3.5	1.4	1,510	2,114	604
1	30	30	1,510	45,300	1,510
1	53	53	1,475	78,175	1,475
1	58	58	1,445	83,810	1,445
1	60	60	1,460	87,600	1,460
1	55	55	1,485	81,675	1,485
1	40	40	1,493	59,720	1,493
0.9	21	18.9	1,495	28,255	1,345
0.3	3.3	0.99	1,495	1,480	448
0.5	3.5	1.75	1,510	2,643	755
1	30	30	1,510	45,300	1,510
1	52	52	1,470	76,440	1,470
1	67	67	1,450	97,150	1,450
1	71	71	1,475	104,725	1,475
1	73	73	1,490	108,770	1,490
1	67	67	1,490	99,830	1,490
1	39	39	1,485	57,915	1,485
0.9	20	18	1,485	26,730	1,337
0.2	1	0.2	1,480	296	296
0.3	6	1.8	1,505	2,709	452
1	30	30	1,505	45,150	1,505
1	52	52	1,490	77,480	1,490
1	63	63	1,485	93,555	1,485
1	68	68	1,485	100,980	1,485
1	70	70	1,495	104,650	1,495
1	73	73	1,490	108,770	1,490
1	55	55	1,485	81,675	1,485
1	29	29	1,480	42,920	1,480
1	22	22	1,462	32,164	1,462
0.6	7	4.2	1,427	5,993	856
0.3	3	0.9	1,505	1,354	452
1	25	25	1,510	37,750	1,510
1	50	50	1,505	75,250	1,505
1	62	62	1,503	31,186	1,503
1	61	61	1,507	91,927	1,507

Example 8-2. (*Continued*)

(1) f	(2) h	(3) fh	(4) P	(5) Pfh (3) × (4)	(6) Pf (1) × (4)
1	60	60	1,507	9,0420	1,507
1	60	60	1,490	89,400	1,490
1	62	62	1,470	91,140	1,470
1	55	55	1,462	80,410	1,462
1	45	45	1,446	65,070	1,446
1	33	33	1,410	46,530	1,410
0.9	17	15.3	1,365	20,884	1,229
0.5	6	3.0	1,350	4,050	675
0.5	7	3.5	1,505	5,268	753
1	33	33	1,505	99,665	1,505
1	42	42	1,505	63,210	1,505
1	48	48	1,507	72,336	1,507
1	50	50	1,500	75,000	1,500
1	50	50	1,488	74,400	1,488
1	57	57	1,460	83,220	1,460
1	55	55	1,440	79,200	1,440
1	53	53	1,425	75,525	1,425
1	46	46	1,380	63,480	1,380
1	37	37	1,380	51,060	1,380
1	22	22	1,370	30,140	1,370
0.9	12	10.8	1,370	14,796	1,233
0.4	3.5	1.4	1,325	1,855	530
0.3	8	2.4	1,500	3,600	450
0.75	16	12.0	1,505	18,060	1,129
0.9	18	16.2	1,507	24,413	1,356
1	25	25	1,490	37,250	1,490
1	26	26	1,478	38,428	1,478
1	34	34	1,460	49,640	1,460
1	35	35	1,437	50,295	1,437
1	33	33	1,410	46,530	1,410
1	23	23	1,390	31,970	1,390
1	15	15	1,385	20,775	1,385
1	13	13	1,385	18,005	1,385
1	15	15	1,385	20,775	1,385
1	7	7	1,385	9,695	1,385
0.2	1	0.2	1,385	277	277
0.8	3.5	2.8	1,440	4,032	1,152
1	12	12	1,435	17,220	1,435
1	10	10	1,412	14,120	1,435
0.8	4	3.2	1,390	4,448	1,112
0.6	2.5	1.5	1,385	2,078	831
0.3	2	0.60	1,385	831	416
0.25	1	0.25	1,385	347	346
0.75	4	3.0	1,385	4,155	1,039
0.3	4	3	1,385	416	416
75.33		2691.87		3,899,506	109,141

$$\overline{P}_{VL} = \frac{\Sigma P(h)(f)}{\Sigma h(f)} = \frac{3{,}899{,}506}{2{,}691.87} = 1449 \text{ psi}$$

$$\overline{P}_{AR} = \frac{\Sigma P(f)}{\Sigma f} = \frac{109{,}141}{75.33} = 1{,}449 \text{ psi}$$

where P is the estimated pressure of a square determined from the isobaric map, f is the estimated fraction of the square underlain by oil-productive formation determined from the oil isopach, and h is the estimated oil sand thickness of the square determined from the oil isopach.

———•—••—•———

Example 8-3. Peripheral Weighting of Pressures for Pressure at Oil-Water Contact.

Pressure interval, psi	Distance along contact, units
1,345–1,350	1.4
1,350–1,375	2.3
1,375–1,400	10.9
1,400–1,425	2.2
1,425–1,450	2.4
1,450–1,475	2.2
1,475–1,500	13.1
1,500–1,520	13.5
	48.0

$$\overline{P} = \frac{1}{48} \left\{ \frac{1}{2} \left[\begin{array}{l} 1.4(1{,}345 + 1{,}350) + 2.3(1{,}350 + 1{,}375) \\ + 10.9(1{,}375 + 1{,}400) + 2.2(1{,}400 + 1{,}425) \\ + 2.4(1{,}425 + 1{,}450) + 2.2(1{,}450 + 1{,}475) \\ + 13.1(1{,}475 + 1{,}500) + 13.5(1{,}500 + 1{,}520) \end{array} \right] \right\}$$

$$\overline{P} = \frac{1}{96} (139{,}580.5) = 1{,}454 \text{ psi}$$

$$h_{o-w} = 100 \text{ ft} \qquad \rho_o = 43.2 \text{ lb/cu ft} \qquad \rho_w = 62.4 \text{ lb/cu ft}$$

$$\overline{P}_{o-w} = 1{,}454 + \frac{100}{144} (62.4 - 43.2) = 1{,}454 + 15$$

Therefore, $\overline{P}_{o-w} = 1{,}454 + \dfrac{100}{144} (62.4 - 43.2) = 1{,}454 + 15$

$$= 1{,}469 \text{ psi}$$

———•—••—•———

Fluid Analysis

The fluid data are a function of reservoir pressure, temperature, and surface separation conditions. If the surface operating conditions are not altered, then the fluid data need be defined only once. But the fluid-property data should be adjusted for all major changes in surface operating conditions.

The procedure for obtaining and analyzing fluid-property data was discussed in Chap. 5. A method of reducing laboratory fluid data to the basis of a reservoir average was presented in Chap. 7.

Core Analysis and Laboratory Rock Data

Two terms enter the material balance which are dependent on rock properties. The initial water saturation S_{wi} must be defined so as to account for its expansion. Various procedures for obtaining suitable average water-saturation values for use in the material balance are presented in detail in Chap. 7.

The second term in the material-balance equation is the rock compressibility. This value is a function of reservoir pressure and may be defined graphically or by some average compressibility factor. The magnitude of the rock-compressibility factor c_f is measured in the laboratory on core samples from the reservoir. If no laboratory data are available, approximate values can be obtained from Fig. 2-21.

CALCULATION OF OIL IN PLACE
USING THE MATERIAL-BALANCE EQUATION

In the previous sections of this chapter the material-balance equation was derived and methods used in reducing field data for substitution into the material balance were discussed. It was also mentioned that the material balance could be used to evaluate the initial oil in place, the water influx, or the size of the gas cap. The remainder of this chapter is devoted to illustrations of the use of the material balance in determining the initial oil in place for different reservoir mechanisms. The types of reservoirs that are discussed are solution gas, slightly compressible fluid, and a gas reservoir.

Estimation of Oil in Place for a Solution-gas–Drive Reservoir

A reservoir in which fluid productions are obtained solely as the result of the liberation and expansion of gas from solution in the oil is known as a solution-(dissolved) gas type of drive. In this type of drive no fluid influx into the reservoir is assumed, and hence the reservoir is considered to produce under volumetric control.

Since no free gas exists in the oil zone initially and there is no fluid influx or injection into the reservoir, the general material-balance equation reduces to

$$N = \frac{N_p[B_o + (R_{cs} - R_s)(B_g/5.61)] + W_p B_w}{B_t - B_{ti}}$$

which is essentially Eq. (8-16).

A typical example of the estimation of the initial oil in place using the material-balance equation and assuming a solution-gas type drive is presented in Example 8-4. It will be noted that both field and laboratory data are required. The field data include fluid production (oil, water, and gas), fluid-production ratios, and reservoir-pressure data at various times. In the case illustrated in Example 8-4 there was no water production from this reservoir for the time interval under consideration. Laboratory data include the fluid physical properties for oil and gas, primarily formation volume factors and solution ratios. In this particular case, these properties are represented by appropriate equations based on the laboratory investigations.

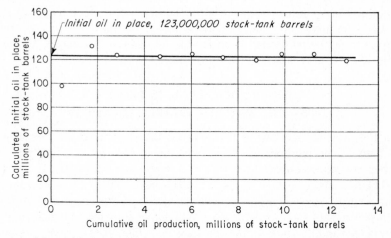

Fig. 8-8. Material-balance calculation for initial oil in place for a solution-gas–drive reservoir. Data points calculated in Example 8-4.

For the sake of expediency these calculations are ordinarily made in tabular form. Once the field and laboratory data are available, it is a simple matter to arrange the data and calculations in tabular form like that shown in Example 8-4. Once the form of the table is established, a calculation for initial oil in place N is made for each time at which the necessary data are available. The results shown in Example 8-4 are typical.

In order to evaluate the results, a plot of the estimates of oil in place N versus the respective cumulative oil productions N_p is prepared. Such a plot is shown in Fig. 8-8. If the resulting calculated values of oil in

Example 8-4. Calculation of Oil in Place by Material Balance for a Solution-gas–Drive Reservoir.

Date	(1) Days after discovery	(2) N_p, millions	(3) P	(4) R_p	(5) R_{ea}	(6) R_s	(7) B_o	(8) B_g	(9) $R_{ea} - R_s$ $\;(5) - (6)$
Sept. 17, 1937	0		3,548	750	750	769	1.452	0.004630	
January, 1938	105	0.4761	3,448	850	820	751	1.444	0.004767	69
April, 1938	195	1.743	3,303	920	880	723	1.432	0.004980	157
July, 1938	286	2.818	3,153	990	917	695	1.420	0.005222	222
October, 1938	378	4.653	2,938	1,020	966	655	1.403	0.005613	311
January, 1939	470	6.030	2,813	1,000	974	632	1.393	0.005868	342
April, 1939	560	7.360	2,678	1,180	999	606	1.382	0.006171	393
July, 1939	651	8.751	2,533	1,420	1,054	579	1.371	0.006534	475
October, 1939	743	9.873	2,453	1,510	1,108	564	1.364	0.006753	544
January, 1940	835	11.250	2,318	1,660	1,170	540	1.353	0.007157	630
April, 1940	925	12.619	2,153	1,920	1,239	508	1.340	0.007723	731

(10) $\dfrac{B_g(R_{ea}-R_s)}{5.61}$ $\;(8)\times(9)\big/5.61$	(11) $R_{si} - R_s$ $\;R_{si}-(6)$	(12) $\dfrac{B_g}{5.61}(R_{si}-R_s)$ $\;\dfrac{(11)\times(8)}{5.61}$	(13) B_t $\;(7)+(12)$	(14) $B_o + \dfrac{B_g}{5.61}(R_{ea}-R_s)$ $\;(7)+(10)$	(15) $B_t - B_{ti}$ $\;(13)-B_{ti}$	(16) $\dfrac{B_o + (B_g/5.61)(R_{ea}-R_s)}{B_t - B_{ti}}$ $\;(14)/(15)$	(17) N, millions $\;(2)\times(16)$
			1.452				
0.0586	18	0.0153	1.4593	1.5026	0.0073	205.836	97.998
0.1390	46	0.0408	1.4728	1.5710	0.0208	75.529	131.647
0.2066	74	0.0689	1.4889	1.6266	0.0369	44.081	124.220
0.3112	114	0.1141	1.5171	1.7142	0.0651	26.332	122.522
0.3577	137	0.1433	1.5363	1.7507	0.0843	20.767	125.225
0.4323	163	0.1793	1.5613	1.8143	0.1093	16.599	122.169
0.5532	190	0.2213	1.5923	1.9242	0.1403	13.715	120.020
0.6548	205	0.2468	1.6108	2.0188	0.1588	12.713	125.515
0.8037	229	0.2921	1.6451	2.1567	0.1931	11.169	125.651
1.0063	261	0.3593	1.6993	2.3463	0.2473	9.488	119.729

$$N = \frac{N_p[B_o + (R_{ea} - R_s)(B_g/5.61)]}{B_t - B_{ti}}$$

$$R_s = 0.18737P + 104.63$$

$$B_o = 8 \times 10^{-5}P + 1.168$$

$$\frac{1}{B_g} = 0.062P - 4$$

place are fairly constant, then an average horizontal line can be drawn through the data indicating the average value for initial oil in place. This also indicates that the type of reservoir mechanism has been identified properly. If the calculated values of oil in place increase with increasing cumulative oil production, this indicates that the reservoir mechanism has not been identified properly and probably energy is being derived from a source other than that assumed. If the calculated values of oil in place decrease with increasing cumulative oil production, this indicates that a greater reservoir voidage is occurring than that which is measured. This result may be due to loss of production to "thief" zones or inaccurate production or laboratory data.

In Figure 8-8 it is seen that the average value of initial oil in place N is 123 million stock-tank barrels. However, it is seen that early in the producing life of the reservoir there was considerable differences in the estimates. This result is not uncommon, since the greatest effects of transient pressure behavior are evident during this period.

Estimation of Oil in Place for Slightly Compressible Fluids

A reservoir which contains only oil and water, both at a pressure such that large pressure drops are possible without evolving gas, is defined as a slightly compressible fluid reservoir. For such a reservoir it is necessary to account for the changes in rock and interstitial water volume as well as the hydrocarbon volume. The expansibility of water may be the order of 1×10^{-6} to 4×10^{-6} vol per vol per psi. The expansibility of the rock or decrease in pore volume may be of the order of 5×10^{-7} to 4×10^{-6} vol per vol per psi. The volume of water available for expansion usually ranges from one-fourth to one times the hydrocarbon volume, while the volume of pore space available for shrinkage ranges from one and one-fourth to two times the hydrocarbon volume.

As no free gas exists in either the oil zone or a contiguous gas cap, the general material-balance equation reduces to Eq. (8-28).

An initial trial set of solutions of Eq. (8-28) can be made assuming no water influx. If the resulting calculated values of oil in place N are consistent and yield a horizontal line when plotted versus cumulative production, then it can be assumed that there is no water influx. When the calculated values of oil in place continually increase with increasing production, then additional energy is being derived from a source not accounted for in the solution.

The procedure for solving the slightly compressible fluid material balance when there is no water influx is demonstrated in Example 8-5. The resulting calculations are plotted in Fig. 8-9 from which the initial oil in place was determined to be 375 million stock-tank barrels. It is noted in the plot of the calculated data of Example 8-5 that the initial values of oil in place are lower than the remaining calculated values. The engineer should expect the initial points in a solution to show the greatest variation because of transient pressure behavior during the early life of the reservoir.

Estimates of Gas in Place from Material Balance

Reservoirs containing only free gas are termed gas reservoirs. Such a reservoir contains a mixture of hydrocarbons which exists wholly in the gaseous state. The mixture may be a "dry," "wet," or "condensate" gas, depending on the mixture and the pressure and temperature at which the accumulation exists.

Gas reservoirs may have water influx from a contiguous water-bearing portion of the formation or may be volumetric (i.e., have no water influx). The general material balance applied to a volumetric gas reservoir has been shown [Eq. (8-30)] to reduce to

$$G_{pc} = G \left(1 - \frac{B_{gi}}{B_{gc}} \right) \tag{8-30}$$

Furthermore for a dry or wet gas reservoir, as no liquid condensation occurs in the reservoir, Eq. (8-30) simplifies to Eq. (8-32):

$$\frac{P}{Z} = \frac{P_i}{Z_i} + \left(-\frac{P_i}{GZ_i} \right) G_{pc} \tag{8-32}$$

The form of Eq. (8-32) is linear; thus a plot of P/Z versus G_{pc} is linear on rectangular coordinate paper.

For condensate gases, liquid condensation occurs in the reservoir so that B_g cannot be replaced by a simple relation of pressure and compressibility factor. However, Eq. (8-30) is linear in G_{pc} and the reciprocal of B_{gc}. A plot of G_{pc} as a function of $1/B_{gc}$ is linear and when extrapolated to $1/B_{gc} = 0$ will intersect the gas-production axis at the value of G.

In the event that liquid production is obtained at the surface, the liquid volumes must be converted to equivalent gas volumes and added to the measured gas production.

An example calculation of the gas initially in place for a volumetric dry gas reservoir is shown in Example 8-6. As shown in the example the gas in place can be calculated from the data using the slopes or from a graphical analysis as shown in Fig. 8-10. The gas in place can also be calculated from a least-squares fit of the data when substituted in Eq. (8-30).

———— •◆• ————

Example 8-5. Calculation of Oil in Place by Material Balance for a Slightly Compressible Fluid Reservoir with No Water Influx. The reser-

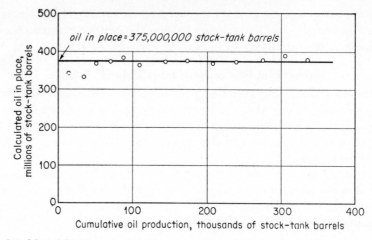

Fig. 8-9. Material-balance calculation of oil in place for a slightly compressible hydrocarbon reservoir with no water influx. Data points calculated in Example 8-5.

voir contains slightly compressible hydrocarbon fluids with no water influx or water production; hence its oil in place can be calculated by Eq. (8-29).

$$N = \frac{N_p(EP + D)}{(P_i - P)A}$$

From laboratory data

$c'_w = 2.9 \times 10^{-6} \text{ psi}^{-1}$

$c_f = 2.0 \times 10^{-6} \text{ psi}^{-1}$

$S_{wi} = 0.25$

$B_{oi} = 1.480$ at 4,934 psia

$B_{ob} = 1.492$ at 4,330 psia

$$c_o = \frac{B_{ob} - B_{oi}}{B_{ob}(P_b - P_i)} = \frac{0.012}{-1.492(604)} = -13.31 \times 10^{-6} \text{ psi}^{-1}$$

From the laboratory data

$$A = 1.480 \left[13.31 \times 10^{-6} (1.00^{81}) - \frac{0.25}{0.75} 2.9 \times 10^{-6} + \frac{2.0 \times 10^{-6}}{0.75} \right]$$

$$= 1.480 \times 5.127 \times 10^{-6}$$

$$= 7.588 \times 10^{-6} \text{ psi}^{-1}$$

$$D = {}'1.492 [1 - (-13.31 \times 10^{-6}) 4,330]$$

$$= 1.492 (1 + 0.0596)$$

$$= 1.578$$

$$E = 1.492 (-13.31) 10^{-6}$$

$$= -1.986 \times 10^{-5}$$

Example 8-5 (*Continued*). Calculation of Oil in Place.

Time since discovery, days t	Cumulative oil production, stock-tank bbl N_p	Average reservoir pressure, psia P	Cumulative pressure drop, psi $P_i - P$	Cumulative expansion per unit volume, $A(P_i - P) \times 10^{-5}$	EP	Oil volume factor B_o $EP + D$	Cumulative voidage, bbls $N_p(EP + D)$	Oil in place, M stock-tank bbl N
0	0	4,934	0					
91.2	14,045	4,926	8	6.070	0.0978	1.4802	20,789	342,490
182.4	35,731	4,913	21	15.935	0.0976	1.4804	52,896	331,950
273.6	51,127	4,907	27	20.488	0.0974	1.4806	75,699	369,480
364.8	70,958	4,897	37	28.076	0.0972	1.4808	105,075	374,250
456.0	88,656	4,889	45	34.146	0.0971	1.4809	131,291	384,500
547.2	108,032	4,846	58	44.010	0.0968	1.4812	160,017	363,590
638.4	143,938	4,859	75	56.910	0.0965	1.4815	213,244	374,700
729.6	173,630	4,844	90	68.292	0.0962	1.4818	257,285	376,740
820.8	207,985	4,824	110	83.468	0.0958	1.4822	308,275	369,330
912.0	239,178	4,809	125	94.850	0.0955	1.4825	354,581	373,830
1003.2	275,556	4,792	142	107.750	0.0952	1.4828	408,594	379,200
1094.4	306,533	4,780	154	116.855	0.0949	1.4831	454,619	389,040
1185.6	336,996	4,760	174	132.031	0.0945	1.4835	499,934	378,650

595

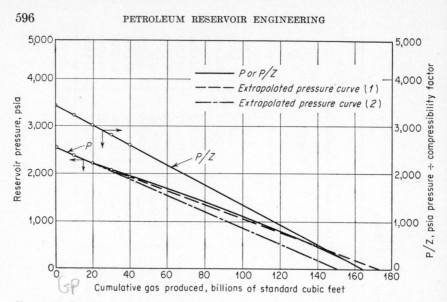

FIG. 8-10. Graphical solution of material balance for a volumetric gas reservoir. (*Adapted from Gruy.*[5])

Fitting a horizontal line to a plot of cumulative oil production (N_p) and calculated oil in place (Fig. 8-9) yields a value of oil in place of 375 million stock-tank barrels.

Example 8-6. Determination of Gas Initially in Place for a Volumetric Gas Reservoir.

Data:

Gas gravity = 0.80
Reservoir temperature = 149°F or 609°R
Initial reservoir pressure = ~~2,778~~ 2538 psia

Cumulative gas production G_{pc}, billion scf	Reservoir pressure, psia
0	~~2751~~ 2538
10	2381
20	2223
30	2085
40	1940
50	1801

Calculations:

$$_pT_c = 420°R \text{ (from Fig. 4-30)}$$
$$_pP_c = 663°R \text{ (from Fig. 4-30)}$$
$$T_r = 609 \div 420 = 1.45$$
$$P_r = P \div 663$$

G_{pc}	P	P_r	Z	P/Z	$\Delta(P/Z)$	$\Delta(P/Z)/\Delta G_{pc}$
0	2,538	3.83	0.742	3,420		
10	2,381	3.59	0.739	3,223	-197	-19.7
20	2,223	3.37	0.739	3,021	-202	-20.2
30	2,085	3.14	0.742	2,811	-210	-21.0
40	1,940	2.93	0.747	2,597	-214	-21.4
50	1,801	2.72	0.755	2,385	-212	-21.2
						-103.5

From Eq. (8-32) $\text{Av} = -20.7$

$$\frac{P}{Z} = \frac{P_i}{Z_i} + \left(-\frac{P_i}{GZ_i}\right) G_{pc}$$

differentiating with respect to G_{pc}

$$\frac{d(P/Z)}{dG_{pc}} = -\frac{P_i}{GZ_i}$$

Therefore, $\dfrac{\Delta(P/Z)}{\Delta G_{pc}} = -\dfrac{P_i}{GZ_i}$

thus

$$G = \left(-\frac{P_i}{Z_i}\right) \frac{\Delta G_{pc}}{\Delta P/Z}$$

G_{pc}	G	Average slope	G
0		-20.7	165.2
10	173.6		
20	169.3		
30	162.9		
40	159.8		
50	161.3		

Since Eq. (8-32) is linear, a plot of P/Z as a function of G_{pc} is a straight line. This is illustrated in Fig. 8-10. When

$$P = 0 \qquad P/Z = 0$$

and from Eq. (8-32)

$$G = G_{pc}$$

Therefore the extrapolation of the line through the data points to $P/Z = 0$ intersects the G_{pc} axis at the value of G. From Fig. 8-10, $G = 165$ billion standard cubic feet.

———•••———

Figure 8-10 is a plot of reservoir pressure as a function of cumulative gas production. The solid pressure line represents the actual variation of reservoir pressure with production throughout the life of the reservoir. It will be noted that the line is curved, lying below the P/Z line at all points except at zero pressure, where the two lines must intercept. If the gas were ideal (i.e., $Z = 1$), the pressure would be linear with production. An extrapolation of the pressure-production curve as a straight line may lead to overestimates (extrapolated pressure curve 1) or underestimates (extrapolated pressure curve 2) of the gas initially in place.

Water influx into a gas reservoir will cause the P/Z curve to deviate from linearity with production. Inaccuracies in pressure or production data will also cause irregularities.

REFERENCES

1. Schilthuis, R. J.: Active Oil and Reservoir Energy, *Trans. AIME*, vol. 118, p. 33, 1936.

2. Katz, D. L.: Methods of Estimating Oil and Gas Reserves, *Trans. AIME*, vol. 118, p. 18, 1936.

3. Miles, A. J.: Private report, 1936.

4. Kaveler, H. H.: Engineering Features of the Shuler Field and Unit Operation, *Trans. AIME*, vol. 155, p. 58, 1944.

5. Gruy, H. J.: "A Critical Review of Methods Used in the Estimation of Natural Gas Reserves," Thesis, A and M College of Texas, 1956.

6. Nichols, E. A.: Geothermal Gradients in Mid-Continent and Gulf Coast Oil Fields, *Trans. AIME*, vol. 170, p. 44, 1947.

7. Pirson, S. J.: "Oil Reservoir Engineering," 2d ed., McGraw-Hill Book Company, Inc., New York, 1958.

NAME INDEX

SUBJECT INDEX

NOT VALID WITH FREE FLOWING WATER.

Material Balance

$$B_g = C\frac{z}{P} = \frac{RTz}{V_m P}$$

$$= .028\frac{Tz}{P}$$

Gas

Source Point

G = initial Amt. of gas in Res SCF

$G B_{g i}$ = Res gas in place

$B_{g i}$ = initial Gas Formation Vol. Factor $\frac{RCF}{SCF}$ 5.61 $\frac{RCF}{RBBL}$

G_p = Gas Produced SCF

EXPANSION = PRODUCTION

$G(B_g - B_{g i}) = G_p B_g$ ③ $G = \dfrac{G_p}{1 - \frac{z_{i/p}}{z/p}}$ ⑤ $-G + \dfrac{P}{z}\left(\dfrac{G z_i}{P_i}\right) = -G_p$

$G = \dfrac{G_p}{1 - \frac{B_{g i}}{B_g}}$ ④ $G = \dfrac{G_p}{1 - \frac{P z_i}{z P_i}}$ ⑥ $\dfrac{P_i}{z_i} - \dfrac{G_p}{G}\dfrac{P_i}{z_i} = \dfrac{P}{z}$

(b) $+ m x = y$

slope $m = -\dfrac{P_i}{G z_i}$

WATER FLOW

VOLUMETRIC EXPANSION _ABOVE_ BUBBLE POINT   POINT SOURCE

INITIAL RES VOL = FINAL RES VOLUME

$N(B_{o i}) = (N - N_p) B_o$

$N(B_{o i} - B_o) = -N_p B_o$

$$\boxed{N = \dfrac{-N_p B_o}{B_{o i} - B_o}}$$

N = INITIAL AMT OF OIL (STB)

$B_{o i}$ = INITIAL FVF $\left(\frac{RBBL}{STB}\right)$

N_p = oil Produced (STB)

$N_p R_{s i}$ = Gas Produced $\frac{STB (SCF)}{STB}$

slope of line most import. parameter

have water drive

(volumetric expansion)

VOLUMETRIC EXPANSION RESERVOIR _AT_ BUBBLE POINT   NO WATER INFLUX

$$V_{o i} = V_o + V_g$$

(RBBL)   (RBBL)

$$(N) B_{o i} = (N - N_p) B_o + G_f \dfrac{B_g}{5.61}$$

moles of FREE GAS in RES = moles of gas INITIALLY IN RES. − moles of gas PRODUCED − moles of gas IN SOL.

$$G_f (SCF) = N R_{s i} - N_p R_p - (N - N_p) R_s$$

$$\boxed{N = \dfrac{N_p \left[B_o + B_g (R_p - R_s) \right]}{B_o - B_{o i} + B_g (R_{s i} - R_s)}}$$

N(STB) + N(R_s) (SCF)

↓Produce

$G_f (SCF)$ FREE GAS

(N-N_p) STB   (N-N_p)R_s GAS IN OIL SCF

↑N_p ↑$G_p = N_p R_p$ ↑P_i

$R_p = \dfrac{G_p}{N_p}$

FROM PROD. DATA ONLY

R_p = NET CUMULATIVE PROD. G.O.R.

$R_p = \dfrac{CUM\ GAS\ PRODUCED\ (SCF)}{CUM\ OIL\ PRODUCED\ (STB)}$

WANT R_p AS SMALL AS POSSIBLE TO GET OUT MOST OIL

CAN ① Reinject (EXPENSIVE) $R_p = 0$
② WORK OVER HIGH GOR WELL
③ Reduce FLOW RATE

$$r = \dfrac{N_p}{N} = \text{RECOVERY } \%$$

OIL & GAS SATURATION

S_w = WATER VOLUME

$$S_0 = \frac{\text{OIL VOL. IN RES.}}{\text{PORE VOLUME}} = \frac{(N-N_p)B_0}{\frac{N(B_{0i})}{(1-S_w)}}$$

$$S_0 = \frac{(N-N_p)B_0(1-S_w)}{N(B_{0i})}$$

$$S_g = 1 - S_0 - S_w$$

/

MATERIAL BALANCE

NO ORIGINAL GAS CAP
NO WATER DRIVE

① OIL PORE VOLUME IN RES. = 75 MM cuft. (RCF)

R_s ② Solubility of GAS IN CRUDE = .42 $\frac{SCF}{STB-PSI}$

P_i ③ INITIAL BOTTOM HOLE PRESSURE = 3500 PSIA

T ④ BOTTOM HOLE TEMP. = 140°F

P_B ⑤ Saturation Pressure of the reservoir = 2400 PSIA (P_B)

B_{oi} ⑥ FORMATION VOLUME FACTOR AT 3500 PSIA = 1.333 $\frac{RBBL}{STB}$

Z ⑦ Compressibility Factor of Gas AT 1500PSIA & 140°F = .95

N_p ⑧ OIL PRODUCED WHEN PRESSURE IS 1500PSIA = 1.0 MM STB

R_p ⑨ Net Cumulative produced G.O.R. = 2800 $\frac{SCF}{STB}$

a) Calculate initial STB OF OIL IN RES.

$$N = \frac{OIL\ PORE\ VOL\ RCF}{FVF\ AT\ 3500PSIA\ \frac{RBBL}{STB}\ 5.61\frac{RBBL}{RCF}} = \frac{75\ MM}{1.33(5.61)} = 10\ MM\ STB$$

$$N = 10\ MM\ STB$$

b) CALCULATE INITIAL SCF OF GAS IN RES.

$$G.I.P. = N(R_{si}) = 10\ MM\ STB\ \frac{.42\ SCF}{STB}\ \frac{2400PSIA}{PSIA} = 10\ MMM\ SCF$$

$$G.I.P. = 10\ MMM\ SCF$$

c) Calculate The INITIAL DISSOLVED G.O.R.

$$R_{si} = G.O.R. = \frac{10\ MMM\ SCF}{10\ MM\ STB} = \frac{G.I.P.}{N} = 1000\ \frac{SCF}{STB} = R_{si} = (.42)(2400)$$

d) Calculate The SCF OF GAS REMAINING IN THE RES AT 1500 PSIA

GAS remaining = TOTAL GAS - GAS PRODUCED

$$G_{1500} = G.I.P. - N_p\ R_p$$
$$= 10\ MMM\ SCF - 1.0\ MMSTB\ 2800\ \frac{SCF}{STB}$$
$$G = 7.2\ MMM\ SCF$$

e) Calculate The SCF OF FREE GAS IN the RES. AT 1500 PSIA

G_F = Total GAS - GAS PROD. - GAS IN SOL.

$$G.I.P. - N_p R_p - R_s(N - N_p)$$

$$G_F = 10\ MMM\ SCF - 2.8\ MMM\ SCF - (.42)(1500)(10 - 1)\ MMMS\ CF$$

$$G_F = 1.53 \times 10^9\ SCF$$

f) Calculate The Gas Volume Factor oF The Escaped Gas at 1500 PSIA AT STANDARD CON 14.7 PSIA & 60°F

$$B_g = \frac{R}{V_m} \cdot \frac{ZT}{P} = \frac{10.74(.95)(140+460)}{(380.69)(1500)}$$

$$\underline{B_g = .01064 \frac{RCF}{SCF}}$$

g) Calculate The Res. Vol. of Free gas AT 1500 PSIA

$$V = G_F \, B_g$$

$$(1.53 \, MMMSCF)(.01064 \frac{RCF}{SCF})$$

$$\underline{V = 16.28 \, MM \, RCF}$$

h) Calculate The Total Reservoir GOR AT 1500 PSIA

$$R_T = \text{Total GOR} = \frac{\text{TOTAL GAS IN RES}}{\text{OIL IN RES}} = \frac{7.2 \, MMSCF}{9.0 \, STB} = \underline{800 \frac{SCF}{STB}}$$

i) Calculate The dissolved GOR AT 1500 PSIA

$$R_S = (.42 \, SCF/STB \, PSI)(1500 \, PSIA)$$

$$\underline{R_S = 630 \, SCF/STB}$$

j) Calculate The liquid Volume Factor of The oil at 1500 PSI

$$B_O = \frac{\text{Total Res Vol} - G_F}{\text{S.T. Vol of OIL AT 1500}} \overset{(RBBL)}{\underset{S.T. Vol of OIL AT 1500}{}} = \frac{25 \, MM \, RCF(RBBL) - 1.53 \, MMMSCF}{5.61 \, RCF} \overset{RBBL}{}$$
$$= \frac{25MM \, RCF(RBBL)/5.61 \, RCF - 1.53 MMMSCF/5.61}{9.0 \, MM \, STB}$$

$$\underline{B_{O_{1500}} = 1.153 \frac{RBBL}{STB}}$$

k) Calculate The Total oil Volume Factor oF The oil AND iTS INITIAL COMPLIMENT OF DISSOLVED GAS AT 1500 PSIA.

$$B_T = B_O + (R_{S_i} - R_S) B_g$$

$$B_T = 1.153 \frac{RBBL}{STB} + (800-630) \frac{SCF}{STB} . 01064 \frac{RCF}{SCF} . \frac{BBBL}{5.61 \, RCF}$$

$$\underline{B_T = 1.865 \, \frac{RBBL}{STB}}$$

EXAMPLE PROBLEM

$$T_{res} = 74 + 1200\left(\frac{2.2}{100}\right) = 100.4\,°F$$

$$P_{res} = .43(1200) = 516\,PSIA$$

Given:
reservoir 1200 fT
surface Temp = 74°F
Geothermal Gradient = $2.2\,°F/100FT$
Hydraulic Gradient = .43 PSI/FT
GAS GRAVITY = .7
OIL GRAVITY = 20°API

1) Solution G.O.R. AT Bubble Point
   use Beals Correlation P.432

   Solution GOR. ≈ 95 $\frac{RCF}{Rbbl}$

2) Bubble point Pressure

   $$P_{res} = P_b = 516 + 14.7 = 530\,PSIA$$

3) OIL FORMATION VOLUME FACTOR
   USE Chart 5-31   P434

   $$B_o = 1.046\ \frac{RBBL}{STB}$$

4) Solution GOR. AT 400 PSIA
   Beals Correlation

   $$Sol\ G.O.R._{400} = 75\ CuFT/bbl$$

5) OIL FORMATION VOLUME FACTOR AT 400 PSIA
   USE FIG 5-31   P434

   $$B_{o_{400}} = 1.038\ RBBL/STB$$

6) Oil Viscosity AT Bubble Point

   M = Viscosity 34 cp

7) OIL VISCOSITY AT 400 PSIA

   $$M = 37\ cp$$